Extreme Multivalued Waves as a Model of Everything

Shamil Usmanovich Galiev

Extreme Multivalued Waves as a Model of Everything

From Euler Figures to Faraday Waves, Particle-Waves and Universes

Springer

Shamil Usmanovich Galiev
Department of Mechanical Engineering
University of Auckland
Auckland, New Zealand

ISBN 978-3-031-67026-8 ISBN 978-3-031-67027-5 (eBook)
https://doi.org/10.1007/978-3-031-67027-5

This Springer imprint is published by the registered company Springer Nature Switzerland AG
The registered company address is: Gewerbestrasse 11, 6330 Cham, Switzerland

I dedicate this book to the memory of my parents, Valitova Ajar Khayrutdinovna and Galiev Usman Zakirovich

Preface

Nonlinearity, Sounds, and Waves

The title of this book was not formed immediately. At first, the title "Extreme Waves in Scalar Fields" was planned. The idea of extending the results obtained to the widest possible range of phenomena did not immediately arise, although the idea of the wave nature of the entire world around us has existed for many millennia. In particular, the Pythagoreans (classical Greece) believed that there was a musical harmony of all things reflecting the deep structure of the real world. They identified the sound of objects to some extent with the object itself. According to this, the planets, when moving, create "music of the spheres." The movement of planets can be studied by studying this music. If you look at these ideas, which are about three thousand years old, from a modern perspective, you can see the analogy of these views with modern ideas that all bodies consist of clots of energy described by certain oscillating energy fields.

When looking at the Universe from these positions, it appears as an infinite variety of waves. With this consideration, the Universe is the sum of these waves. On the other hand, it seems to be filled with the sounds of these waves. It sounds like an unimaginably complex, unimaginably polyphonic orchestra. The instruments in it do not play independent parts. In it, everything and everyone is connected to everything, including everything that happened in the past. Turning on, turning off, and turning on various tools again depends on unimaginably many circumstances. The sound of instruments changes greatly over time [1].

To some approximation, depending on the distance of the instrument and the time it was played, the sounds produced by the instruments can be considered harmonic and continuous. However, at different stages of the development of the Universe and in different areas of it, non-harmonic waves can also sound. In particular, shock waves, solitons, and so-called ossilons can sound. All these waves can be called unambiguous. They form, perhaps, terrible or beautiful, but single-layer music. Perhaps, in order to understand the sound of the orchestra called the Universe, it is necessary to take into consideration the waves that are called multivalued. These are waves that are described by essentially nonlinear equations, that is, equations that have several simultaneously existing, interdependent solutions.

Nature is inherently nonlinear. Everything in it is connected by nonlinear dependencies, although they do not always manifest themselves clearly. In the simplest case of describing the Universe as some kind of continuous medium, we must take into account the fact that pressure, density, and temperature are quantities that do not exist independently. In other cases, the quantities characterizing nature are not so clearly related. For example, we can speak with almost complete confidence about the connection between quantum and gravitational phenomena, but so far this connection has not been clearly established.

Apparently, some such connections manifest themselves under the most extreme conditions, when the equations governing the processes are extremely highly nonlinear. In weakly nonlinear processes, these equations break down into approximately independent equations that describe changes in only one or several quantities. For example, the equations describe only gravitational or quantum processes.

The author does not have an equation for studying the combined, highly nonlinear sound of gravitational and quantum effects. At the same time, he is confident that due to high nonlinearity, these equations can sound in a completely unfamiliar way, namely as multi-valued solutions.

In this book, the author decided to explore the sound of one instrument in the orchestra of the Universe. The sound of this instrument is determined by the nonlinear Klein-Gordon equation (NKGE). In the general case, this equation has several independent solutions, but under certain, resonant conditions, these solutions merge into a single stream of sounds, in which each previously independent note sounds with full rights. The author strives to show that this stream of sounds can sound at all levels: quantum, in the macroworld, and at the birth of the Universe. During the explore, it faces difficult problems, since there is little experimental and observational data associated with multi-valued wave processes. The fact is that multivalued waves arise only during nonlinear, high-energy processes. As an example, consider electromagnetic waves. If conditions existed on Earth for the emergence of their multivalued variants, then, apparently, the very existence of life on Earth would be impossible. As another option, we indicate gravitational waves. They can be very diverse in space, close to their sources, but they reach the Earth in an extremely weakened state, that is, in a linear version. Therefore, their deep, cosmic nonlinearity is not perceived by us.

However, in terrestrial conditions, there is a different type of gravitational waves. These are gravitational surface waves, with which for thousands of years humanity has associated the very concept of waves (only about 500 years ago, Leonardo da Vinci connected the propagation of sounds in the atmosphere with the propagation of waves in it). It was through the example of ocean and water waves that humanity realized the possibility of the emergence of harmonic and shock waves, as well as solitons.

The author began his research many years ago precisely with the study of nonlinear, gravitational surface waves [2]. This allowed him in further research to stand, at least to some extent, on the solid ground of experiments and not fall into myth-making.

However, the experimental study of high-energy surface waves also poses enormous difficulties. The fact is that in the conditions of earth's gravity, they are unstable, although they exist everywhere. Indeed, almost everyone observed their appearance and disappearance on the banks of rivers, lakes, seas, and oceans.

The situation has begun to change over the past 20 years. In particular, 2 years ago an article was published quite definitely describing the origin, development, and disappearance of multivalued surface waves [3].

The above determined the difficulties of preparing for the publication of this book, which contains the results of almost 25 years of research work. The author published the previously obtained results in the form of articles, preprints, and even books devoted to nonlinear, extreme waves, where the results obtained for multivalued waves were presented more as a theory interesting only for mathematicians [4].

A Very Brief History of the Development of Research into the Acoustics of the Early Universe

Thus, the author of the book tries to cover a very wide range of problems with his theory, in particular, to extend it to the wave dynamics of the early Universe. The first work that opened up the possibility of acoustic sounding of the expanding Universe was published in 1946 [5]. The problem seemed so far from reality and at the same time so grandiose that it did not receive strong development for almost 20 years. Indeed, the second fundamental work devoted in this case to acoustic waves in the early Universe was published in 1965 [6]. That is, in the same year when the cosmic microwave background radiation was experimentally discovered. Perhaps, as a response to this grandiose discovery, fundamental research appeared, presented in [7, 8]. However, this discovery was not enough for many publications devoted to the acoustics of the early Universe to appear. The situation changed dramatically at the beginning of this century, when several publications appeared [9, 10], which laid the foundation for a comprehensive study of the extremely difficult problem of studying the acoustics of the very early Universe.

Of course, the waves that arose in the initial spherical Universe were highly nonlinear, and possibly multivalued. These waves are modeled in Chaps. 5 and 7 of the book. However, in the process of expanding the sphere, they are smoothed out and, apparently, are well approximated by harmonics. But perhaps some manifestations of nonlinearities exist in the cosmic microwave background radiation and in photographs of the very early Universe. In particular, for many years, many telescopes have observed some circular structures of the early Universe. We have presented one such structure below.

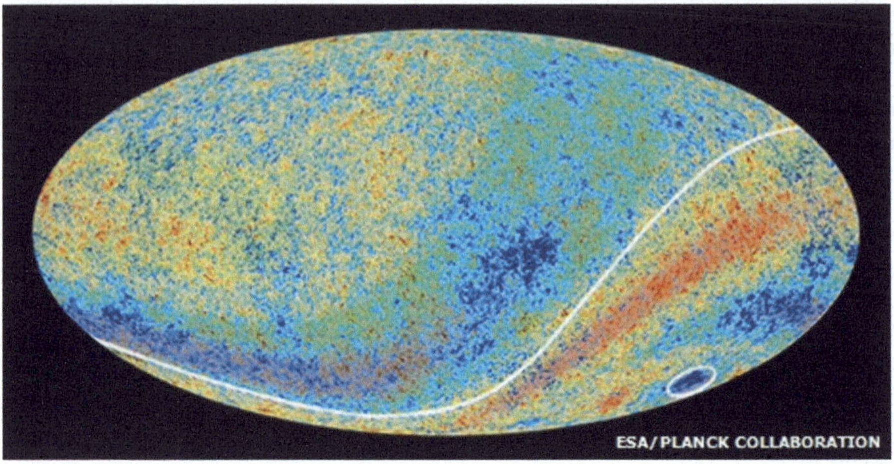

Planck's map and the north/south differences, and a "cold spot" in the cosmic microwave background (CMB) radiation [3]

Perhaps similar circular structures correspond in some nonlinear way to the elliptical and circular structures shown in Figs. 5.6, 5.7, 7.23, and 7.25. On the other hand, such structures could also be a signal of Conformal Cyclic Cosmology. This theory was developed by physicists Roger Penrose and Vahe Gurzadyan and proposes that the Universe goes through infinite cycles [10].

The acoustics of the early Universe are very important for understanding its evolution, since they describe the network waves, in accordance with which the web of the galaxies of our modern Universe may have been formed. It is important to emphasize that this web is not uniform. For example, it includes the Hercules-Corona Borealis Great Wall (HCBGW), which spans about 10 billion light years. This size categorically does not correspond to the so-called cosmological principle, which states that the matter of the Universe is, on average, distributed throughout it uniformly. To fulfill this principle, it is necessary that the dimension of the existing cosmic structure does not exceed 1.2 billion light years. Thus, HCBGW increases the size allowed by the cosmological principle by almost 10 times. Its visible area is about 10% of the visible area of the Universe. We emphasize that the structure like HCBGWs (similar size) has been found in many places. In particular, cosmic (space) voids located between similar structures can have dimensions that are larger than requires of the cosmological principle. The largest of this void has a diameter of 2 billion light years.

It's also important to note that HCBGW is approximately 10 billion light years away from us. Thus, this structure already existed at the time of the final formation of our planet (Earth). The incredibly rapid structuring of the HCBGW, consisting of an unimaginably large number of galaxies, stars, and planets, poses intractable problems for the current cosmological model.

We believe that the creation of a new cosmological model may involve taking into account highly nonlinear, possibly multivalued wave effects [10].

Structure of the Book

The above determined the structure of the book. It consists of two parts.

In the first, the author focuses on the description of surface waves based on NKGE, namely, multivalued waves experimentally discovered over the past 20–25 years. After, in his opinion, he managed to convince the reader of their existence, he moves on to the second part of the book.

In it, he extends the results obtained in the first part to the problems of the microworld, as well as the emergence of the Universe and very early space.

In this way, he begins to determine the second purpose of the book, reflected in the next words of its title "as a Model of Everything."

The fact is that from the beginning of writing the book, she focused on the possibility of describing the Universe on the basis of a certain single equation, which has a relatively simple form. This idea may have belonged to Hermann Weyl, who wrote "I am bold enough to believe that the whole of physical phenomena may be derived from one single universal world-law of the greatest mathematical simplicity" [11].

Albert Einstein apparently began to think about this idea at the same time as Weyl [12]. Einstein famously spent the last half of his physics working life on his program of unified field theories. His conviction was that a unification program not only could combine his GR gravitational field theory with Maxwell's equations but also shed light on the quantum mystery.

Since then, a large number of prominent authors have considered the problem of the "Theory of Everything" from different points of view and understanding [13–19]. In particular, Paul Dirac, in the late 1920s, explored the relativistic generalization of the Schrödinger equation (in the terminology of this book, he used an analogue of the NKGE). He hoped to derive a "Theory of Everything" based on it—that is, to extend this equation to all elementary particles. These calculations by Dirac date back to 1928, when he discovered that this wave equation allows for many solutions. He only had to take a small step to be the first to describe multivalued waves. But he used his solutions to predict the existence of new elementary particles, for which he received the Nobel Prize in Physics.

The possibility of describing Everything using a limited set of equations and ideas has been discussed by many works [11–19]. When you think about it, you ask the question: Is this possible? There is apparently only one answer—impossible!

The question can be reformulated: Is it possible to describe the Everything, albeit approximate, but comprehensive? And in this case, the author comes to the conclusion—impossible, at least at the present time!

Truly modern science arose approximately 500 years ago. While the first large human settlements arose about 10,000 years ago. Thus, modern science, as an element of universal human culture, exists for a relatively short period of time during the existence of a collective, highly developed human society.

An even more striking picture emerges if we look into the future. It apparently extends for humanity over thousands and perhaps millions of years. Have we really already solved all the fundamental problems of the existence of the Universe and left no fundamental problems for the next generations? It is hard to believe!

It seems to the author that future generations will look at the highest modern achievements with the same sense of admiration with which we look at the achievements of the classical Greeks in predicting the existence of atoms or in describing the movements of the planets! They will probably be admired not by the practical results of modern scientific achievements, but by the courage of scientific thought and the beauty of scientific constructions based on such a small amount of experimental data.

Indeed, the atomic theory of the ancient Greeks implied a description of Everything and was no less bold than Rene Descartes' Theory of Everything or the field theory on which Albert Einstein worked. Here we must understand that many physicists usually understand the Theory of Everything as the idea of unification.

Theory of Everything or Unification

In general, one can see the division of researchers into two mutually complementary and overlapping groups. The first strives to describe nature. At the same time, it often ignores the fundamental explanation of the connections and causes of the processes and phenomena they describe. This approach is determined by the mathematization of science, especially its physical disciplines. It became possible to simulate processes without entering into their physical essence. Indeed, almost no one thinks in terms of atoms, electrons, or molecules when studying the flow of liquids or the propagation of waves in solids. Moreover, we move even further away from fundamental reality when we turn to solutions to model equations. Approximate methods are often used to solve them. When applied problems are solved, we can judge the accuracy of the solutions by comparing their consequences with observations and experiments. The situation becomes more complicated when we plunge more and more into the depths (mysticism) of the microworld and space. Apparently, there is no doubt that both of these areas are filled with certain sounds. But are they completely described by solutions, for example, in the form of Fourier series? For example, we have the Schrödinger equation, which is undoubtedly an approximate one, then we wrote down its approximate solution, but how far have we moved away from the fundamental reality with such steps? It may be unknown.

Perhaps the highest achievement of this group of researchers is associated with the name of Charles Darwin [20]. Often this is the path of pioneers who formulate problems that inspire and stimulate other generations of researchers. Is it possible to find more inspiring researchers than Aristotle or the modern creators of the theories of supersymmetry or string theory?

On the other hand, the development of science is determined by an increasingly deeper understanding, and not just a description, of processes in the Universe. Many people believe that the main task of science is to create fundamental knowledge in which there is no doubt. It was on this basis that Rene Descartes envisioned the further development of knowledge. Elements of knowledge are not created instantly. Science is approaching them, testing and polishing them, literally for decades, under the fire of criticism. An example of the difficult entry of a new concept into science was given by Galileo Galilei. In his book *Discourses and Mathematical Demonstrations Concerning Two New Sciences*, Galileo said about the acceleration of natural motion: "...The present does not seem to be the proper time to investigate the cause of the acceleration of natural motion concerning which various opinions have been expressed by various philosophers,.... At present it is the purpose of our Author merely to investigate and to demonstrate some of the properties of accelerated motion (whatever the cause of this acceleration may be)-meaning thereby a motion, ..." Initial reactions to this idea were negative. Even Descartes protested at Galileo's decision to seek descriptive formulas: "...Everything that Galileo says about bodies falling in empty space is built without foundation: he ought first to have determined the nature of weight ..." [20].

This approach is basic for fundamental science. Thanks to him, such concepts as waves, inertia, energy, evolution, atom, and many others arose.

These two approaches complement each other. It is difficult to say how they will interact in the future. Indeed, experimental research is becoming so expensive that human society may not be able to finance it any time soon.

Summary

The author considers himself to be in the first group of researchers. Following the thoughts of Einstein and Weyl, he builds some very approximate mathematical model of multi-valued wave processes, which is confirmed by experiments performed in terrestrial, macro-sized experiments. He then extends the theory, which may be true under terrestrial conditions, to areas very distant from those in which the theory was confirmed [21, 22].

He believes that the Universe is full of multivalued, highly nonlinear processes, which are not taken into account by theoretical physics or taken into account on the basis of non-physical hypotheses [23]. Of course, the author understands the crudeness and approximation of his constructions describing Everything, but he hopes that the very idea of the existence of multivalued waves and the possibility of looking at the Universe from this position may interest the reader [24, 25].

Finally, with this book the author celebrates the 100th anniversary of the beginning of the development of quantum mechanics [26] and modern cosmology [27].

Auckland, New Zealand — Shamil Usmanovich Galiev

References

1. Bartusiak M (2000) Einstein's unfinished symphony: listening to the sounds of space-time. National Academies Press, Washington DC
2. Galiev ShU, Galiev ShT (1998) Resonant travelling surface waves. Phys Lett A 246: 299–305; Galiev ShU (1999) Topographic effect in a Faraday experiment. J Phys A Math Gen 32:6963–7000; Galiev ShU (2003) The theory of non-linear transresonant wave phenomena and an examination of Charles Darwin's earthquake reports. Geophys J Int 154:300–354; Galiev ShU (2005) Strongly nonlinear two-speed wave equations for coastal waves and their application. Phys D 208:147–171; Galiev ShU (2008) Strongly-nonlinear wave phenomena in restricted liquids and their mathematical description. In: Perlidze T (ed) New nonlinear phenomena research. Nova Science Publishers, New York, pp 109–300
3. Liu D, Lin P (2022) Two-layer liquids Interface instabilities in Faraday waves of two-layer liquids with free surface. J Fluid Mech 941:A33. https://doi.org/10.1017/jfm.2022.259
4. Galiev ShU (2015) Darwin, geodynamics and extreme waves. Springer; Galiev ShU (2020) Evolution of extreme waves and resonances. CRC Press, 508 pp.
5. Lifshitz E (1946) On the gravitational stability of the expanding universe. J Phys USSR 10: 116–129; Golda ZA, Woszczyna A (2001) Acoustics of early universe-Lifshitz versus gauge-invariant theories. J Math Phys 42(2):856–862
6. Sakharov AD (1965) The Initial stage of the expansion of the Universe and the occurrence of inhomogeneity of the distribution of matter. Zh Exp Theor Phys 49(1): 345–357 (Сахаров А (1965) Д. Начальная стадия расширения Вселенной и возникновение неоднородности распределения вещества. Журн. эксперим. и теорет. Физики 49(1): 345–357
7. Zel'dovich YaB, Sunyaev RA (1969) The interaction of matter and radiation in a hot-model Universe. Astrophys Space Sci 4:301–316; Sunyaev RA, Zel'dovich YaB (1970) Small-scale fluctuations of relic radiation. Astrophys Space Sci 7(1):3–19; Sunyaev RA, Zel'dovich YaB (1980) Microwave background radiation as a probe of the contemporary structure and history of the universe. Ann Rev Astron Astrophys 18: 537–560
8. Peebles PJE, Yu JT (1970) Primeval adiabatic perturbation in an expanding universe. Astrophys J 162:815–836
9. Galiev ShU, Galiyev TSh (2001) Nonlinear transresonant waves, vortices and patterns: from microresonators to the early Universe. Chaos 11:686–704; Miller CJ, Nichol RC, Butuski DJ (2001) Acoustic oscillations in the early Universe and today. Science 292:2302–2303; Seife C (2001) Echoes of the Big Bang put theories in tune. Science 292:823; Golda ZA, Woszczyna A (2001) Acoustics of early universe-Lifshitz versus gauge-invariant theories. J Math Phys 42 (2):856–862; Wright EL (2007) Acoustic waves in the early Universe. J Phys Conf Ser 118:012007; Eisenstein DJ, Bennett C (2008) Cosmic sound waves rule. Physics Today 61 (4):44–50
10. Penrose R (2010) Cycles of time: an extraodinary new view of the universe. Bodley Hea; Horváth I, Hakkila J, Bagoly Z (2014) Possible structure in the GRB sky distribution at redshift two. A&A 561, L12; Vernstrom T et al (2023) Polarized accretion shocks from the cosmic web. Sci Adv 9(7). https://doi.org/10.1126/sciadv.ade7233
11. Goenner HFV. On the history of unified field theories (1914–1933)
12. Einstein A (1949, 1951) Philosopher-scientist/ (ed. Schilpp PA). Harper and Brothers, New York
13. Weinberg S (1992) Dreams of a final theory: the search for the fundamental laws of nature. Random House
14. Hawking S The theory of everything Phoenix Books 9465 Wilshire Boulevard, Suite 315 Beverly Hills, CA 90212
15. Fock V (1959). The theory of space, time and gravitation. Pergamon Press, London
16. Barrow JD (2008) New theories of everything. Oxford University Press, Oxford
17. Søvik AO (2022) A basic theory of –everything. De Gruyter

18. McGrath AE (2019) A theory of everything (that matters). Hodder & Stoughton
19. Davies PCW, Brown J (eds) (1988) Uperstrings: a theory of everything? Cambridge University Press
20. Brake ML (2009) Revolution in science: how Galileo and Darwin changed our world. Palgrave Macmillan, New York
21. Galiev ShU, Galiyev TSh (2014) Nonlinear scalar field as a model describing the birth of the Universe. Известия Уфимского Научного Центра Российской Академии Наук (Herald of Ufa Scientific Center, Russian Academy of Sciences (RAS)) 2: 7; Galiev ShU, Galiyev TSh (2016) Simple model of the origin of the Universe. Известия Уфимского Научного Центра Российской Академии Наук (Herald of Ufa Scientific Center, Russian Academy of Sciences (RAS)) 4: 27; Galiev ShU, Galiyev TSh (2021) Modeling of an appearance of particle-waves and the Universe as the extreme wave resonant events. Известия Уфимского Научного Центра РАН (Herald of Ufa Scientific Center, Russian Academy of Sciences (RAS)), N2, С. 5-23. http://journal.ufaras.ru/wp-content/uploads/2021/10/5-23.pdf
22. Galiev ShU, Galiyev TSh (2013) Scalar field as a model describing the birth of the Universe. InterNet. Galiev – Galiyev 18–12–2013 (2013) https://researchspace.auckland.ac.nz/docs/uoa-docs/rights.htm; Galiev ShU, Galiyev TSh (2014) Coherent model of the births of the UNIVERSE, the dark matter and the dark energy. http://hdl.handle.net/2292/23783; Galiev ShU, Galiyev TSh (2016) Eruption of the Universe Out of a Pre-universe. https://researchspace.auckland.ac.nz/handle/2292/30899; Galiev ShU, Galiyev TSh (2017) Eruption of Matter, Dark Energy and the Universe from a Pre-universe. https://researchspace.auckland.ac.nz/handle/2292/31792; Galiev SU, Galiyev TS (2017) Nonlinear quantum waves in the light of recent slit experiments. https://researchspace.auckland.ac.nz/handle/2292/36751; Galiev SU, Galiyev TS (2017) 'Elastica'-like waves and particles: from Bernoulli, Euler, Laplace and Faraday to the eruption of the Universe. http://hdl.handle.net/2292/31785
23. Van Vechten LND, Graham N (1973). DeWitt, Bryce; Graham, R. Neill (eds) The many-worlds interpretation of quantum mechanics. Princeton series in physics. -99. Princeton University Press, Princeton. p. v. ISBN 0-691-08131-X
24. Galiev ShU (2024) Multivalued waves and the Richtmyer-Meshkov instability as the causes of the formation of galaxies. Herald of Ufa Scientific Center Russian Academy of Sciences 1: 5–17. https://doi.org/10.31040/2222-8349-2024-0-1-5-17
25. Ilgamov MA (2023) Annotation book Galiev S.U. "Extreme Multivalued Waves in Scalar Fields" planned by Springer for publication in early 2024. Herald of Ufa Scientific Center, Russian Academy of Sciences. 2023. № 3. С. 115–120. https://doi.org/10.31040/2222-8349-2021-0-2-115-120
26. de Broglie L (1925) Recherches sur la théorie des quanta (Researches on the quantum theory), Thesis, Paris, 1924, Ann. de Physique 10(3):22
27. Friedmann A (1924) Über die Möglichkeit einer Welt mit konstanter negativer Krümmung des Raumes. Zeitschrift für Physik 21:326–332

Acknowledgments

I thank the University of Auckland and the Department of Mechanical Engineering that provided me with conditions for independent scientific work on this book during almost 30 years of work. I thank Vice-Chancellor, Professor Dawn Freshwater for her interest in my research. I would also like to thank Professors D. Bhattacharyya, B. Mace, K. Jayaraman, and S. Bickerton (University of Auckland), who have always supported me, which was especially important for my trips abroad to scientific conferences.

In general, work on the book began many years ago due to the initiatives of my scientific teacher Professor Ilgamov Marat Aksanovich, as well as Professors Ganiev R.F. and Nigmatullin R.I. (the Academy of Science of Russia). They have largely determined the research directions presented in this book. I want to remember with deep gratitude Professors Novozhilov V.V. (the former Academy of Science of USSR), Pisarenko G.S. (the former Academy of Science of USSR), and of my friend Sakhabutdinov Zh.M. (the former Academy of Science of USSR). They supported all my scientific beginnings. It is my great pleasure to acknowledge my friend Professor Akhmediev N. (the Australian National University). His support and attention were very important for both beginning and completing of this project. Professor E. Pelinovsky's favorable opinion about the book was also very important for the Springer Publishing House (Federal Research Center Institute of Applied Physics of RAScience, Russia).

I thank also my son, Timur Galiyev, who helped me in writing Part 2. During a few years we worked together on the new theory of the origin of the Universe. We called it as the model Galiev-Galiyev (MGG). It is presented in this book. I asked him to be a co-author of this book, but he did not agree.

Finally, I am grateful for the contribution of all scientists whose results and ideas were used in this book. This project could hardly be realized without the generous permission of the following publishers for using material from their publications:

Springer (Heidelberg, New York, Dordrecht, London);
CRC Press (Taylor & Francis Group—Mechanical, Aerospace, Nuclear & Energy Engineering);
Ufa Center of the Russian Academy of Sciences (UFIC RAS) (Ufa, Russia);
Center of Modern Education (Moscow, Russia).
To all of them—my sincere thanks! It is hard for a writer to even dream of a better publishing team than the Springer Nature Press team. Many thanks Margaret Deignan (Senior Publishing Editor) and Fairle T. Thattil (Project Coordinator) for their professional service and turning my manuscript into this book. Finally, I would like also to thank Yamini Balabaskar, Production Supervisor, whose keen eye, good taste, and sense of structure improved this book in many ways.

Contents

About the Author

Shamil Usmanovich Galiev was born on April 18, 1942 in Kazan, Tatar Republic (Russia), and is the son of Usman Zakirovich Galiev and Ajar Khayrutdinovna Valitova. He was a researcher in the Academy Sciences of The Union of Soviet Socialist Republics (USSR), Kazan, in 1965–1972 and in Academy Sciences Ukraine, Kiev, in 1972–1981, a head of the department in Academy Sciences Ukraine, Kiev, in 1981–1995, a professor in Kiev Politech. University in 1986–1989, and a visiting professor at Brown University, Providence, Rhode Island, United States, in 1992. Since 1996, he has been a professor and honorary academician at the University of Auckland, New Zealand.

Part I
From Euler Figures to Faraday Waves and Particle-Waves

All extraordinary, unusual phenomena in the world are associated with one or another strongly nonlinear properties of Nature. Let me clarify this idea. A huge number of processes and phenomena occurring around us can be approximately interpreted as linear and described by linear equations. The solution to every linear equation is unique, smooth, and continuous so that the solution describes a well-defined process. Figuratively speaking, the solution can only describe one cat of a certain color. The cat can't change color. The cat can change its size, but not very much. For example, a linear equation cannot describe the transformation of the cat into a tiger or describe the evolution of a smooth, imperceptible wave into a relentless, terrifying tsunami.

If we want to describe reality more accurately, we must use nonlinear equations. Unlike linear equations, they can have many solutions. As a result, the researcher is faced with the task of choosing a solution that corresponds to the real process. In general, the researcher has to use observational or experimental data or rely on intuition and a sense of beauty when choosing from a variety of solutions a solution that corresponds to the process under study. In some cases, these solutions may overlap each other and form complex structures. According to our research, this usually occurs in the vicinity of some singular points of the corresponding differential equation. In the case of dynamic and wave problems, singular points are often associated with resonances. It is in their vicinity that multivalued waves can arise.

Thus, to describe multivalued waves, it is necessary to learn how to study wave processes in zones of singularities (resonances). This task is, to some extent, similar to the one that faced the creators of the "renormalization method."

This method, created by R. Feynman, J. Schwinger, and F. Dyson in 1944–1949, made it possible to eliminate from the theory the divergences that arise when solving equations by the perturbation method. Note that this method formed the basis of the theory of quantum electrodynamics, which has fantastic accuracy and predictive power.

So, the concept of resonance is very important to the book. In it, this concept is identified with the case when the most important linear terms of the nonlinear wave equation reduce to zero. For example, this occurs in a finite resonator when an integer number of waves are excited along its length. Then the linear terms in the corresponding wave equation can disappear near the boundaries. In other cases, multivalued waves arise if some disturbance begins to move at a certain critical speed.

This part of the book contains a large number of equations, their solutions, calculations, and, finally, experimental data describing the emergence, propagation, and evolution of multivalued waves. The connection between their occurrence and resonance processes and, in particular, with the existence of singular points in the corresponding nonlinear wave equations is demonstrated.

Chapter 1
Basic Equations of Scalar Fields and Examples of Nonlinear Waves Arising in Them

The concept of a field was introduced by Faraday to explain electromagnetic connections at distances that cannot be explained from the standpoint of mechanics. Now the concept he introduced defines a huge area of scientific knowledge that is intensively developing. It is used to describe both physical and mechanical processes.

The fields differ in their levels of difficulty. Based on their level of difficulty, the scalar, vector, tensor, and other fields are divided. The simplest processes are described by scalar field equations.

Its peculiarity is that all field characteristics are determined through one sought-after function (e.g., velocity potential, temperature, or scalar potential).

In this chapter, we present a number of wave equations of mathematical physics and show that in some cases, they are reduced to the same expressions and can have the same solutions.

1.1 Governing Equations of Scalar Fields

Perhaps, Da Vinci was the first to understand that surface water waves, sound waves and light waves are all propagated according to the same laws. It is known now that similar waves may be generated in different physical, chemical, and biological systems.

It is known that physical processes of generation and transformation of waves can differ dramatically; nevertheless, equations and analytical solutions describing these processes are often similar. For example, shock-like, soliton-like, and cnoidal-like solutions are well known in nonlinear wave dynamics.

S. U. Galiev, *Extreme Multivalued Waves as a Model of Everything*,
https://doi.org/10.1007/978-3-031-67027-5_1

1.1.1 Examples of Equations Describing Waves in Scalar Fields

It is shown here how the introduction of such concepts as velocity potential, scalar potential, electromagnetic field, and the like makes it possible to reduce the study of wave scalar fields to the solution of wave equations having, qualitatively, similar structures.

Wave Equations for Ideal Liquids and Gases These equations follow from the classical equations of motion, continuity, and state, which relate the speed, pressure, and density of a continuous medium [1, 2]. Thus, there are five equations and five unknowns. If the motion of the medium is irrotational, then the velocity potential $\overline{v} = \nabla\psi$ can be entered (see also, the Sect. 5.8.2).

Substituting the last expression into the equations of motion and eliminating the operator ∇, we derive

$$\frac{\partial\psi}{\partial t} + \frac{1}{2}(\nabla\psi)^2 + \int_{p_0}^{p} \rho^{-1} dp = f(t). \tag{1.1}$$

Here $f(t)$ is an arbitrary function. It is found that $f(t) = 0$. The equation of state can have different forms. As an example, we give the equation of state for liquids:

$$\frac{p + B}{p_0 + B} = \left(\frac{\rho}{\rho_0}\right)^{\gamma}. \tag{1.2}$$

Here p_0 and ρ_0 are the pressure and density in a fluid at rest, respectively; γ is the adiabatic exponent and B is constant. The equation is valid at normal temperature and not very high pressure. In the latter case, $\gamma = 7.15$. We emphasize that an equation of the form (1.2) describes the properties of many media. Of course, the values of γ and B can vary greatly. For example, if $B = 0$ and $\gamma = 0$, then (1.2) defines a medium without pressure, the properties of which are close to the properties of foam or fog. For a monatomic gas, $\gamma \approx 1.4$. For photon gas, $\gamma \approx 3$.

When using $\overline{v} = \nabla\psi$, (1.1), and (1.2), we can find $p = p(\psi)$ and $\rho = \rho(\psi)$. Thus, the assumption of an irrotational motion has allowed the determination of the velocity, the pressure, and the density using one arbitrary function ψ. We can obtain a strongly nonlinear equation for ψ by satisfying the continuity equation. Finally, we find the equation

$$c^2\nabla^2\psi = \psi_{tt} + \left[(\nabla\psi)^2\right]_t + (\nabla\psi)^2\nabla^2\psi, \tag{1.3}$$

where ∇^2 is the Laplacian operator and

$$c^2 = c_0^2 - (X-1)\left[\psi_t + \tfrac{1}{2}(\nabla\psi)^2\right], \tag{1.4}$$

where c_0^2 is the undisturbed sound velocity.

1. *Spherical waves.* We have in the case of pure spherical quadratic nonlinear waves

$$\left[c_0^2 - (X-1)\psi_t\right]\left(\psi_{rr} + 2r^{-1}\psi_r\right) = \psi_{tt} + 2\psi_r\psi_{rt}, \tag{1.5}$$

   where r is the radial Euler coordinate.
2. *Longitudinal waves.* We have in the case of pure longitudinal quadratic nonlinear waves

$$\psi_{tt} - \left[c_0^2 - (\gamma-1)\psi_t\right]\psi_{xx} + 2\psi_x\psi_{xt} = 0, \tag{1.6}$$

   where x is the Euler coordinate.

Longitudinal waves (the Lagrangian coordinate). We have in this case for longitudinal displacement u

$$u_{tt} - c_0^2(1+u_a)^\gamma u_{aa} = 0. \tag{1.7}$$

Different environments correspond to different γ. For example, Eq. (1.7) takes the form (1.8) for photon gas

$$u_{tt} - c_0^2(1+u_a)^3 u_{aa} = 0. \tag{1.8}$$

Here c_0 is the speed of light. The last equation resembles that describing nonlinear one-dimensional (1D) electromagnetic waves.

The Perturbed Maxwell Wave Equation One-dimensional propagation of the electromagnetic waves E is described by the following equation [3, 4]:

$$E_{tt} - c_0^2 E_{xx} = \left(a_1 E + a_2 E^2 + a_3 E^3 + \mu^* E_t + k^* E_{xx}\right)_{tt}, \tag{1.9}$$

where linear phenomenological dissipative $\mu^* E_{t\,tt}$ and dispersive $k^* E_{xxtt}$ terms were added in (1.9). Coefficients c_0, a_1, a_2, a_3 μ, and k may depend on t and x. Eq. (1.9) gives opportunities to study the light propagation in spatially and temporally inhomogeneous media and microresonators. If c_0 is a function of t, then Eq. (1.9) describes parametrically excited waves. Nonlinear Schrödinger-type equations follow from the perturbed Maxwell equation (1.9).

Linear Klein-Gordon Field Equation This equation has the form

$$\varphi_{tt} - c_0^2 \nabla^2 \varphi + m^2 \varphi = 0. \tag{1.10}$$

Physicists are well aware that the fundamental properties of the world around us are determined by certain physical fields.

Equation (1.10) was the first equation of relativistic quantum mechanics for the wave function of a particle. It was proposed in 1926 by the Austrian physicist Schrödinger as a relativistic generalization of the Schrödinger equation. It was also independently proposed by the Swedish physicist Klein, the German physicist Gordon, and the Russian physicist Fock [3, 5].

Klein and Gordon started the derivation of (1.10) from the well-known expression $E^2 = p^2 c_0^2 + m^2 c_0^4$. It is natural to use the identity from special relativity describing the energy. After quantization, they had

$$\hbar^2 c_0^2 \nabla^2 \varphi - m^2 c_0^4 \varphi = \hbar^2 \varphi_{tt}. \tag{1.11}$$

The speed of light c_0 and Planck constant $\hbar$ are often seen to clutter the equations, so they are therefore often expressed in natural units where $c_0 = \hbar = 1$. We will further use Eq. (1.10) and its generalizations to cases of nonlinear problems.

Equation (1.10) can be considered as describing some scalar fields, similar to how Maxwell's equations describe electromagnetic fields. This analogy to a certain extent determines the very widespread use of (1.10). Equation (1.10), as well as the well-known equation of Paul Dirac, became examples of the unification of special relativity and quantum mechanics.

Nonliner Klein-Gordon Equation (NKGE) In particle physics, the Klein-Gordon equation represents the wave equation of a free relativistic particle. In this case, m is the mass of the particle. We shall consider the following form of this equation:

$$\varphi_{tt} - c_0^2 \nabla^2 \varphi = -\partial V(\varphi)/\partial \varphi + \mu \nabla^2 \varphi_t + k \nabla^4 \varphi. \tag{1.12}$$

The phenomenological dissipative $\mu \nabla^2 \phi_t$ and dispersive $k \nabla^4 \phi$ terms were added in (1.12). Many of these are stable or metastable due to a dynamic balance between energy injection and dissipation.

The function V is the scalar field potential. Different expressions for the function V can be found in [6–8]. The potential may have higher-order terms

$$V(\varphi) = V_0 + \frac{1}{2} m^2 \varphi^2 - \frac{1}{4} \lambda \varphi^4 + \lambda_6 \Lambda^{-2} \varphi^6 + \lambda_8 \Lambda^{-4} \varphi^8 + \ldots . C^* \varphi. \tag{1.13}$$

Here V_0, m^2, λ, λ_6, λ_8, Λ, and C^* are constants. If the scalar field and the potential are known, one can calculate the pressure p and the energy density ρ according to expressions in [6–8]:

$$p=\frac{1}{2}\varphi_t^2-\frac{1}{6}(\nabla\varphi)^2-V(\varphi), \tag{1.14}$$

$$\rho=\frac{1}{2}\varphi_t^2+\frac{1}{6}(\nabla\varphi)^2+V(\varphi). \tag{1.15}$$

It is interesting that the Klein-Gordon equation may be reduced to the nonlinear Schrödinger equation in the small-momentum limit. On the other hand, the Klein-Gordon equation is also known in astrophysics. Perhaps the most fundamental properties of the Universe are determined by scalar wave fields. As an example, let us cite the recently discovered Higgs particles described by a scalar field. Another example is field theory in curved space-time (the Robert-Walker metric).

$$\varphi_{tt}-a^{-2}\nabla^2\varphi+3H\varphi_t+\partial V(\varphi)/\partial\varphi=0. \tag{1.16}$$

Here $a = a(t)$ is the scalar factor which determines the variation of the Universe with time. The term $3H\phi_t$ corresponds in the mechanical analogy to friction, and its effect is called Hubble drag or damping [6–8]. Let us imagine that the equation describes a huge number of extreme wave processes. In particular, this equation describes oscillon clumps of waves that are localized in space but oscillate in time and do not disperse. These oscillons existed in space in the very first moments of the creation of the Universe [9]. Naturally, the thought arises that in those moments, other extreme waves also arose, in particular extreme multivalued waves, the study of which this book is devoted to.

Nonlinear Schrödinger Equation (NSE) This equation may be derived from the nonlinear Maxwell wave equation, the Klein-Gordon equation, or equations of gas dynamics. We will demonstrate this below.

1.1.2 Nonlinear Versions of Schrödinger and Klein-Gordon Equations

Let us write NKGE in the so-called φ^4 field form

$$\varphi_{tt}-c_0^2\varphi_{xx}+m^2\varphi-\lambda\varphi^3=0 \tag{1.17}$$

We will seek the solution in the form of the quasi-harmonic wave with slowly varying complex amplitude $A(x, t)$ (an envelope function) and high-wave number modulated function $e^{i\theta}$ (so-called carrier wave) [3]:

$$\varphi(x,t)=A(x,t)e^{i\theta}+A^*(x,t)e^{-i\theta}. \tag{1.18}$$

Here

$$\theta = \omega t - kx, \tag{1.19}$$

where ω and k are constants. We assumed that the wave φ may spontaneously self-modulate. Generally speaking, the modulation, which arises due to the overtones induced by nonlinearity, can split the wave into "wave packets" which behave like solitons. These solitons are made of a carrier wave modulated by an envelope signal, and this is why they are called envelope solitons. In (1.18), the function $A(x, t)$ corresponds to the envelope of the wave. We think that similar soliton-like waves can describe the forms of "quantum particles" and its motion.

A modulated wave includes space and time scales: (1) a fast time and space variation of the carrier wave and (2) a much slower variation of the envelope.

We found that

$$\begin{aligned} &\varphi_{tt} = A_{tt}e^{i\theta} + 2A_t i\omega e^{i\theta} - A\omega^2 e^{i\theta}, \quad \varphi_{xx} = A_{xx}e^{i\theta} - 2A_x ike^{i\theta} - Ak^2 e^{i\theta}, \\ &\varphi^3(x,t) = A^3 e^{3i\theta} + 3A^2A^* e^{i\theta} + 3AA^{*2}e^{-i\theta} + A^{*3}e^{-3i\theta}. \end{aligned} \tag{1.20}$$

These expressions are substituted into (1.17), and the coefficients of the exponentials $e^{i\theta}$ are equated. This yields an equation for the complex amplitude $A(x, t)$:

$$\begin{aligned} &A_{tt} - c_0^2 A_{xx} + 2i\omega\left(A_t + c_0^2 k\omega^{-1}A_x\right) + \\ &\left(c_0^2k^2 + m^2 - \omega^2\right)A - 3\lambda A|A|^2 = 0. \end{aligned} \tag{1.21}$$

Let

$$c_0^2k^2 + m^2 - \omega^2 = 0. \tag{1.22}$$

Then

$$A_{tt} - c_0^2 A_{xx} + 2i\omega\left(A_t + c_0^2 k\omega^{-1}A_x\right) - 3\lambda A|A|^2 = 0. \tag{1.23}$$

Let $X = x - v_{gr}t$ and $T = t$, then we can get nonlinear Schrödinger equation (NSE):

$$iA_T + 0.5\omega^{-1}\left(v_{gr}^2 - c_0^2\right)A_{XX} - 1.5\lambda\omega^{-1}A|A|^2 = 0. \tag{1.24}$$

Thus, we showed that the nonlinear Klein-Gordon equation is the generalization of the NSE. If $\beta = 0.5c_0^2\omega_0^{-3}\omega_p^2$, the last equation yields

$$iA_T - \beta A_{xx} - 1.5\lambda\omega^{-1}A|A|^2 = 0. \tag{1.24.1}$$

Thus, instead of Eq. (1.17) for the function φ written in variables x and t, we got Eq. (1.24.1). The latter is written in variables X and T for coefficient A (1.18). Now let us transform (1.24.1) into hydrodynamic form.

Hydrodynamic Equations Let

$$A = b\exp(i\phi). \tag{1.24.2}$$

Here b and ϕ are the unknown ones. In this case, Eq. (1.24.1) is divided into system of two equations

$$1.\ \phi_T - \beta\phi_X^2 + \beta\left(b^{-1}b_{XX}\right) + 1.5\lambda\omega^{-1}b^2 = C, \tag{1.25}$$

$$2.\ \left(b^2\right)_T - 2\beta\left(b^2\phi_X\right)_X = 0. \tag{1.26}$$

Here the first equation corresponds to the equation of fluid motion and the second to the continuity equation.

The Strongly Nonlinear Acoustics Equation or Equation Airy Then we assume that $q = -3\lambda\omega^{-1}\beta$ and

$$X = a + u(a,t),\ U = -2\beta\phi_X,\ U_a = U_X(1+u_a),\ \text{and}\quad U = u_t \tag{1.27}$$

$$U_X = U_a(1+u_a)^{-1} = u_{ta}(1+u_a)^{-1}. \tag{1.28}$$

Therefore, from (1.26), we have $\ln\left[b^2(1+u_a) = \ln\left(b_0^2\right)\right.$ and

$$b = b_0(1+u_a)^{-0.5},\ u_{TT} = -q\left(b^2\right)_X + 2\beta^2\left(b^{-1}b_{XX}\right)_X. \tag{1.29}$$

Finally, after some additional calculations, we find

$$u_{tT} = qa_0^2u_{aa}(1+u_a)^{-3} + 2\beta^2\left\{0.25(1+u_a)^{-5}u_{aa}\left[-u_{aaa} + 0.6125(1+u_a)^{-1}{u_{aa}}^2\right]\right.$$
$$\left. + 0.5\left[-0.5u_{aaaa} + 3u_{aaa}(1+u_a)^{-1}u_{aa} - 6.625 * \ 4.5(1+u_a)^{-2}u_{aa}{u_{aa}}^2\right](1+u_a)^{-4}\right\}. \tag{1.30}$$

Let $\beta^2 \approx 0$. Then

$$u_{tT} = qa_0^2u_{aa}(1+u_a)^{-3}. \tag{1.31}$$

We obtained the equation similar to (1.7). Eq. (1.7) was first obtained by Airy, who used it to study the propagation of waves on the surface of liquids [10–13].

Let us list the main stages of our derivation of Eq. (1.31). We started with Eq. (1.17) written for φ. This equation was then rewritten for A ((1.24) and (1.24.1)). This is how NSE is obtained. The last equation is transformed using (1.24.2) into two equations of hydrodynamic typewritten for the unknowns a and ϕ. Next, we enter the transformation $X = a + u(a, t)$, $U = -2\beta\phi_X$ (1.27). As a result, the hydrodynamic equations are reduced to (1.31).

1.2 Surface Water Waves

In this section, we will present a very condensed diagram for deriving the surface wave equations. These equations are very important for the book, since most of the experiments discussed below were performed with surface waves, and these equations will be used in describing the experiments. At the same time, a detailed derivation of these equations is given in other books by the author, so we will give here only a very condensed scheme for their derivation.

The traditional way of the mathematical description of nonlinear surface waves is based on the use of Euler method [1, 2]. Another approach to the description of these waves is based on the method of Lagrange [1, 10–15]. Generally speaking, the Lagrangian approach is less convenient than Euler method if hydrodynamical problems are considered. However, it is completely true only for three-dimensional problems which are not considered here. In studying one-dimensional and two-dimensional (2D) problems, the Lagrangian approach simplifies the analysis and the solution of difficult nonlinear wave problems.

1.2.1 Lagrangian Description and Initial Equations

We will consider the two-dimensional (2D) motion. The Cartesian orthogonal coordinate system is used. It is assumed that the horizontal axis is located on the free surface of the medium, and the vertical axis is drawn vertically downward. The initial horizontal coordinate of a particle of the medium is a and c is its initial vertical coordinate. Coordinates a and c are the Lagrangian coordinates. We also introduce the Euler coordinates x and z, which indicate the particle coordinates at time t.

The current coordinates x and z are determined by the Lagrangian coordinates, according to the following formulas [11–15]:

$$x = a + u\ , z = c + w. \tag{1.32}$$

Here $u = u(a, c, t)$ is the horizontal displacement and $w = w(a, c, t)$ is the vertical displacement of the particles. Let us consider the surface layer of the medium. The undisturbed (initial) state of the upper surface of the layer is determined by the equation $c = 0$, and the bottom surface geometry is determined by the equation $c = h(a)$, where $h = h(a)$ is the thickness of the layer. This thickness depends on the vertical displacement of the upper surface. In terms of the Lagrangian coordinates and the initial depth $h_0(x)$, we have $h(a) = h_0(x) - w$.

Equations of Motion Using the noted above coordinates, we write the following 2D equations of the motion of the mass:

$$\rho(x_{tt} - X) = p_{11,x} + p_{31,z}\,, \quad \rho(z_{tt} - g) = p_{13,x} + p_{33,z}. \tag{1.33}$$

The subscripts t, x, and z indicate the derivatives with respect to time and the Euler coordinates. In [1], these equations were written for liquid. In the presented form, they are valid for many models of continuum mechanics. Here X and g are the components of external forces per unit mass. We assume that $g = g_0 + g_d$, where g_0 is the gravity acceleration and $g_d = g_d(t)$ is the forced acceleration (e.g., g_d may be generated by the vertical motion, as in Faraday's experiments). p_{11} and p_{33} are the direct stress components and p_{31} and p_{13} are the shear stress components.

Equation of Continuity The equation of continuity [1] is written in the form

$$1 = (1 + w_c)(1 + u_a). \tag{1.34}$$

State Equation The wave nonlinearity is often determined by the physical properties of the medium, described by the equation of state which connects the pressure p, the density ρ, and the temperature. For water surface waves, the incompressible model for fluid may be assumed. The state equation has the form $\rho = \text{const}$.

Boundary Conditions The displacements of the liquid particles have to be equal to the displacements of the water-free surface ($c = 0$). Therefore, we postulate that

$$w = -\eta \text{ and } u = U \text{ at } c = 0, \tag{1.35}$$

where $\eta = \eta(a, t)$ is the surface elevation and $U = u(c = 0, a, t)$.

Now we can write the dynamic boundary condition for the stress at $c = 0$:

$$\begin{aligned} p(t, a, c = 0) &= P_{33}(t, a, c = 0) + p_0 \\ &- \sigma[(1 + U_a)\eta_{aa} - U_{aa}\eta_a]\left[(1 + U_a)^2 + \eta_a^2\right]^{-3/2}, \end{aligned} \tag{1.36}$$

where σ is the coefficient of the surface tension.

1.2.2 Long Waves

A basic expression for the pressure and a basic strongly nonlinear wave equation.

The liquid pressure may be determined from the equation of the vertical motion (1.33). Integrating this equation between 0 and c and using the boundary conditions (1.35) and (1.36), one can obtain after some simple calculations following the works [11–15].

$$p = \int_c^0 \rho(u_{tt} - X)u_c dc + \int_c^0 \rho w_{tt}(1 + w_c)dc + \rho gc + \rho g\,\eta + \rho gw + p_0$$
$$- \sigma[(1 + U_a)\eta_{aa} - U_{aa}\eta_a]\left[(1 + U_a)^2 + \eta_a^2\right]^{-3/2} - \int_c^0 p_{13,a}dc + P_{33}(c, a, t) + \overline{C}. \tag{1.37}$$

Here $\overline{C} = \overline{C}(a, t)$ is the integration function. This equation takes into account g_d, nonlinear effects, vertical acceleration, surface tension, stresses, and X. Nevertheless, if $u_c = X = g_d = w_{tt} = \sigma = p_{13,\ a} = P_{33} = 0$, then (4.23) yields the classical expression for the pressure. Indeed, using the Euler coordinate z (1.32) and assuming $\overline{C} = 0$, we get from (1.37) that $p - p_0 = \rho g_0(z + \eta)$.

Substituting (1.37) into (1.33), we derive the basic governing equation:

$$(u_{tt} - X)(1 + u_a) + g\eta_a + w_{tt}w_a =$$
$$-\left\{\int_c^0 (u_{tt} - X)u_c dc + \int_c^0 w_{tt}(1 + w_c)dc + \rho^{-1}p_0 - \rho^{-1}\sigma[(1 + U_a)\eta_{aa} - U_{aa}\eta_a]\right.$$
$$\left.[(1 + U_a)^2 + \eta_a^2]^{-3/2}\right\}_a + \rho^{-1}\left(\int_c^0 p_{13,aa}dc - P_{33,a} + P_{11,a} + p_{31,c}\right). \tag{1.38}$$

Now we have Eqs. (1.34) and (1.38) for the three unknown quantities η, u, and w.

We have written very complex equations. Below, we will consider certain particular cases of them. Integrating (4.24) between 0 and h, we simplify the problem to the following 1D formulation:

$$\int_h^0 [(u_{tt} - X)(1 + u_a) + g\eta_a + w_{tt}\ w_a]dc =$$
$$- \int_h^0 \left\{ \int_c^0 (u_{tt} - X)u_c dc + \int_c^0 w_{tt}(1 + w_c)dc + \rho^{-1} p_0 \right.$$
$$- \rho^{-1}\sigma\eta_{aa}\left[(1 + U_a)^2 + \eta_a^2\right]^{-3/2} - \rho^{-1}\sigma(U_a\eta_{aa} - U_{aa}\eta_a)[(1$$
$$\left. + U_a)^2 + \eta_a^2]^{-3/2} \right\}_a dc + \rho^{-1} \int_h^0 \left(\int_c^0 p_{13,aa} dc - P_{33,a} + P_{11,a} + p_{31,c} \right) dc. \tag{1.39}$$

It is assumed here that

$$w = (c^n h^{-n} - 1)\eta + h_a c^n h^{-n} \widehat{U}. \tag{1.40}$$

η **is an** elevation above the undisturbed surface of the liquid.

Bottom Friction, Governing Equations, and Remarks First, we focus on the bottom friction effect. Above, we derived the equations describing the viscous effects through the components of the stress tensor. Now we should determine expressions for these components. Following [14, 15], we will use the stress tensor in the form of Navie. According to this model, the deviatoric components P_{11}, P_{33} and the shear stress components p_{13}, p_{31} determine the viscous effects as

$$P_{11} = -\frac{2}{3}\ \mu_*(u_{ta} + w_{tc}) + 2\mu_* u_{ta}, P_{33} = -\frac{2}{3}\ \mu_*(u_{ta} + w_{tc}) + 2\ \mu_* w_{tc},$$
$$p_{13} = p_{31} = \mu_*(u_{tc} + w_{ta}), \tag{1.41}$$

where μ_* is the coefficient of the kinematic viscosity [1]. Using (1.41) and the boundary conditions, we find

$$\rho^{-1} \int_h^0 (-P_{33,a} + P_{11,a})dc = -2\mu_*\rho^{-1}(2hu_{taa} + 2h_a u_{at} + h_{aa}u_t) \tag{1.42}$$

and

$$\rho^{-1}h^{-1}\int\limits_{h}^{0}\int\limits_{c}^{0}p_{13,aa}dcdc=\mu_{*}\rho^{-1}h^{-1}\int\limits_{h}^{0}\int\limits_{c}^{0}w_{taaa}dcdc$$
$$=\mu_{*}\rho^{-1}h\left[\frac{1}{6}\left(h^{-1}\eta_{t}\right)_{aaa}h-0.5\eta_{taaa}+\frac{1}{6}\left(h_{a}h^{-1}u_{t}\right)_{aaa}h\right]$$
$$=-\mu_{*}h\rho^{-1}\left(\frac{1}{6}hu_{taaaa}+0.5\eta_{taaa}\right). \tag{1.43}$$

Here $\eta\approx-hu_{a}-h_{a}u$. Obviously, the value of (1.43) is much less than (1.42). Therefore, the expression (1.43) will be ignored going forward. Considering the abovementioned cases, we can rewrite Eq. (1.39) in the form

$$(1+u_{a})u_{tt}-gh(1+u_{a})^{-2}u_{aa}+gh_{a}\left[(1+u_{a})^{-1}-1\right]-gh(h_{aa}u+h_{a}u_{a})$$
$$+\frac{1}{3}h^{2}[(1\ \ +u_{a})^{-1}]_{tta}(1+u_{a})^{-1}+\frac{1}{2}hh_{a}\left[(1+u_{a})^{-1}\right]_{tt}(1+u_{a})^{-1}$$
$$-\ \frac{1}{2}h\left[h_{a}u_{tt}(1+u_{a})^{-1}\right]_{a}+\frac{1}{2}hh_{a}u_{tt}\left[(1+u_{a})^{-1}\right]_{a}$$
$$-\ \frac{1}{2}h\left[(1+u_{a})^{-1}\right]_{tt}\left\{h_{a}\left[(1+u_{a})^{-1}-1\right]-h_{aa}u-h_{a}u_{a}\right\}$$
$$+h_{a}u_{tt}\left\{h_{a}\left[(1+u_{a})^{-1}-1\right]-h_{aa}u-h_{a}u_{a}\right\}+\rho^{-1}p_{0,a}-\rho^{-1}\sigma\eta_{aaaa}$$
$$-\ 2\mu_{*}h^{-1}\rho^{-1}(2hu_{taa}+2h_{a}u_{at}+h_{aa}u_{t})+\rho^{-1}h^{-1}p_{31}(a,t,c=0)$$
$$-\ \rho^{-1}h^{-1}p_{31}(a,t,c=h)$$
$$=0. \tag{1.44}$$

Here the term $2\mu_{*}\rho^{-1}h^{-1}(2hu_{taa}+2h_{a}u_{at}+h_{aa}u_{t})$ is determined by the internal viscosity of liquid. The seabed friction is determined by the bottom layer model. The water motion may be either laminar or turbulent in the bottom layer. The latter case may be described by the eddy viscosity model $p_{31}(a,t,c=h)=\overline{s}u_{t}|u_{t}|$, where $\overline{s}$ is the friction coefficient. The more simple (linearized) form of the bottom friction model (laminar model) is $p_{31}(a,t,c=h)=\overline{s}u_{t}$ [14, 15].

We have presented a concise scheme for deriving one of the equations describing the propagation of long waves along the water surface. A more detailed description of the main ideas of the developed theory and the derivation of the basic equations is in [14, 15]. In general, the equations obtained there are quite complex. Eq. (1.44) is not the most complex of them. As an example of the simplest equation of the theory, we have the following:

$$\begin{aligned}&(1+u_a)u_{tt}-ghu_{aa}(1+u_a)^{-2}-\frac{1}{3}h^2(1+u_a)^{-3}u_{aatt}+\rho^{-1}p_{0,a}\\&+\rho^{-1}h^{-1}[\,p_{31}(a,t,c=0)-p_{31}(a,t,c=h)]=0.\end{aligned} \tag{1.45}$$

Here ρ is the density, p_0 is the atmospheric pressure, the term $p_{31}(a,t,c=0)$ determines the bottom friction, and $p_{31}(a,t,c=h)$ takes into account the wind friction. This equation takes into account highly nonlinear effects, the influence of the atmosphere, and the most important effects of friction.

The cases discussed above are valid when the wavelength is significantly more than the depth of the ocean. There are various estimates for which equations of the form (1.45) are valid. According to Lamb [1], they can be used if the ratio η/h is smaller than 0.5. Our calculations are consistent with Lamb's estimate [1].

1.2.3 Short Waves

When deriving the equation for short waves, we assumed [14, 15] that the vertical displacement changes not according to a linear law, but according to the law following from the linear solution for surface waves. Thus, when deriving the equation for short surface waves, instead of (1.40) and $\eta \approx -hu_a - h_a u$ we used the following expression:

$$w=\eta\sinh K(h-c)/\sinh Kh). \tag{1.46}$$

Here $K=2\pi\lambda^{-1}$ is the wavelength. Based on this approximation, we obtained the following equation for short ocean waves:

$$\begin{aligned}&\Phi_{tt}-gh\Phi_{aa}-hK^{-1}\Phi_{ttaa}=\frac{1}{2}hK^2(\Phi_{tt}\Phi)_a\\&-\frac{1}{3}hK^3\left[\Phi\int\left(\Phi_a^2\right)_{tt}da+2\Phi_{tt}\int\Phi_a^2da\right]_a\\&+\rho^{-1}p_{31}(a,t,c=0)+\rho^{-1}hp_{0,a}.\end{aligned} \tag{1.47}$$

We put here $p_{31}(a,t,c=h)=0$ since th effect of the bottom friction is extremely low for the deep water. The quadratic nonlinear version of this equation is

$$\Phi_{tt}-gh\Phi_{aa}-hK^{-1}\Phi_{ttaa}=\frac{1}{2}hK^2(\Phi_{tt}\Phi)_a+\ \rho^{-1}p_{31}(a,t,c=0). \tag{1.48}$$

General Comment It is easy to see that if we discard the right-hand sides in Eqs. (1.47) and (1.48), we obtain linear wave equations taking into account dispersion. Thus, the influence of depth is determined by the fact that the dispersion effect can no longer be neglected. If the wave is long, then the dispersion term is small. If, in addition to this, the influence of surface effects is small, then the Airy equation follows from (1.45):

$$u_{tt} = ghu_{aa}(1 + u_a)^{-3}. \tag{1.49}$$

Comment Based on the analysis of his equation, Airy denied the possibility of the existence of solitons on the surface of water, although they were first observed in 1834 on the surface of a canal by Scott Russell [16]. It was only in 1871 that Boussinesq, when he introduced dispersion, described solitons mathematically [17].

1.2.4 Mathematical Models of Waves at the Interface of Liquids

Next, we consider waves propagating along a flat interface between two liquids and also write down the equation of nonlinear oscillations of an absolutely flexible string. It is suggested that these equations describe identical, strongly nonlinear waves.

The classical result is known [1] that the speed of wave propagation along the interface depends on the density of the contacting liquids in accordance with the formula

$$C^2 = \frac{g(\rho_1 - \rho_2)}{k(\rho_1 + \rho_2)}. \tag{1.50}$$

Here ρ_1 is the density of the lower liquid, ρ_2 is the density of the upper liquid, $k = 2\pi\lambda^{-1}$, and λ is the wavelength. The presence of the upper liquid causes a decrease in wave speed. It is also necessary to note that there may be a discontinuity in the movement of liquids at the interface. The vertical component of the movement, of course, remains continuous, but the horizontal component can even change its sign. This can provoke the appearance of breaks, shifts, and the appearance of vortices on the surface wave profiles (Fig. 1.1).

The initial wave is in the upper layer. Particles in this layer move relative to the low layer particles. The formation of the wave turbulence is defined by a distinction of the speeds of these particles. If this distinction is very small, it does not perturb the interface and no vortex motions arise. If the distinction is big enough, the vortices can be formed in the interface.

We will consider all these processes in Chaps. 2 and 3 of this book.

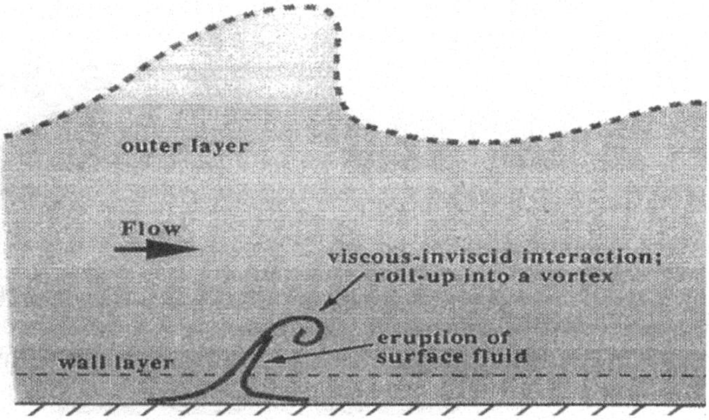

Bad Image

Fig. 1.1 Schematic of turbulent bursting process (not to scale) [14, 18]

Models of Wave Processes The equation describing the waves on the intersurface of two liquids is derived in [19]:

$$N_{tt} - a_0^2 N_{xx} = \frac{1}{3} a_0^2 h_1 h_2 N_{xxx} + (h_1 - h_2)(h_1 h_2)^{-1} \left(2N_t N_{xt} + a_0^2 N_x N_{xx}\right). \qquad (1.51)$$

Here

$$a_0 = \left(\frac{g h_1 h_2}{h_1 + h_2} (\rho_2 - \rho_1)/(\rho_2 + \rho_1)\right)^{0.5}, \quad N_x = \eta \qquad (1.52)$$

We use notation from [19]. According to (1.51), on the interface can be some weakly nonlinear waves determined by the quadratic nonlinearity and a weak dispersion. However, (1.51) does not take into account the cubic nonlinearity. Therefore, we should introduce the cubic nonlinear term there to model the highly nonlinear evolution of the interface.

Now consider unidirectional waves on the surface in question. For them, we write the following equation taking into account the cubic nonlinear term [20]:

$$\eta_t + \left(c_0 + \alpha_1 \eta + \alpha_2 \eta^2\right)\eta_x + \frac{1}{6} c_0 h_1 h_2 \eta_{xxx} = 0. \qquad (1.53)$$

Here

$$c_0 = \left(\frac{g\, \sigma h_1 h_2}{h_1 + h_2}\right)^{0.5}, \quad \alpha_1 = \frac{3}{2} c_0 \frac{h_1 - h_2}{h_1 h_2}, \quad \alpha_2 = \frac{3c_0}{(h_1 h_2)^2}\left[\frac{7}{8}(h_1 - h_2)^2 - \frac{h_1^3 + h_2^3}{h_1 + h_2}\right],$$
$$\sigma = 2(\rho_2 - \rho_1)/(\rho_2 + \rho_1) \ll 1. \qquad (1.54)$$

Bearing in mind (1.51), (1.52), and (1.53), we rewrite Eq. (1.51) by introducing into it the phenomenological cubic nonlinear term $\beta_1 N_x^2 N_{xx}$:

$$N_{tt} - a_0^2 N_{xx} = \frac{1}{3} a_0^2 h_1 h_2 N_{xxxx} + (h_1 - h_2)(h_1 h_2)^{-1}\left(2N_t N_{xt} + a_0^2 N_x N_{xx}\right) + \beta_1 N_x^2 N_{xx}. \tag{1.55}$$

The coefficient β_1 is unknown. It is assumed that it can be derived from experiments when solving each specific problem. However, some idea of its magnitude can be obtained as follows: Let us consider the case when $h_1 = h_2$ and the quadratic terms in (1.55) disappear. In addition, we will not take into account dispersion effects. We enter the coordinate

$$r = ct - x. \tag{1.56}$$

We further take into account that $\eta = -N_r$ (1.52). When the above conditions are met, Eq. (1.55) gives

$$c\eta_t + a_0^2 \eta_x = \beta_1 \eta^2 \eta_x. \tag{1.57}$$

On the other hand, Eq. (1.53) gives

$$\eta_t + c_0 \eta_x + \alpha_2 \eta^2 \eta_x + = 0. \tag{1.58}$$

From a comparison of these equations, we find

$$c = a_0^2 c_0^{-1} \text{ and } \beta_1 = -\alpha_2 a_0^2 c_0^{-1}. \tag{1.59}$$

In this way, the unknown coefficient of Eq. (1.55) is determined. It will take the form as a result of

$$N_{tt} - a_0^2 N_{xx} = \frac{1}{3} a_0^2 h_1 h_2 N_{xxxx} + (h_1 - h_2)(h_1 h_2)^{-1}\left(2N_t N_{xt} + a_0^2 N_x N_{xx}\right) - \alpha_2 a_0^2 c_0^{-1} N_x^2 N_{xx}. \tag{1.60}$$

It is easy to see that we have reduced the study of waves at the interface of liquids to an equation similar to the equations for surface waves (1.45), (1.47), and (1.49). Therefore, they can have similar solutions. However, there is a fundamental difference between these cases. Thus, waves on the contact surface can be practically independent of the gravitational acceleration of the Earth. That is why, under terrestrial conditions, the appearance of extreme multivalued waves is much easier to observe at the interfaces of liquids.

We emphasize that when deriving Eq. (1.60), Eqs. (1.51) and (1.53) obtained by other authors were used. They obtained different expressions for wave velocities, namely a_0 (1.52) and c_0 (1.54). The difference is not very large, about 1.4 times. The wave velocities at the contacts under consideration are usually not large, and the indicated discrepancy is not very important for modeling experiments.

Let us now consider the most important case for us, when the surface of the upper liquid becomes free. It is this case that is usually studied in experiments. In this case, the system has, as it were, two degrees of freedom, and in some cases, the layers can vibrate almost independently of each other.

Two cases can be distinguished in experiments. The first is when mainly the upper layer vibrates, and the lower layer seems to have solidified. The second case is when vibrations of the lower liquid occur with a weak influence of the upper layer. Let us note again that the peculiarity of the contact surface is not only in reducing the influence of gravitational forces but also in the emergence of the possibility of sliding of layers relative to each other. This is precisely what is associated with the appearance of folds, breaks, and vortices on the wave profiles of waves observed on the contact surface of liquid media. There extreme multivalued waves were most often observed experimentally, which draws our attention to the corresponding experiments. We also note that the problem of the evolution of waves into vortices is of great importance for understanding many processes, starting with the problem of the emergence of the Universe and galaxies and ending with the problem of the emergence of turbulence (see the Chaps. 2 and 7).

Of course, for waves at the interface of the liquids, the description of the influence of gravity remains the most important. Interestingly, its effect becomes very weak if the thickness of the top layer of liquid is much less than the thickness of the other layer (see note at the end of this section)! Another very interesting case is when the thicknesses of the layers are equal to each other. Then

$$N_{tt} - a_0^2 N_{xx} = \frac{1}{3} a_0^2 h_1 h_2 N_{xxxx} + \beta_1 N_x^2 N_{xx}. \tag{1.61}$$

That is, the influence of cubic nonlinearity becomes much greater than quadratic nonlinearity, and a very strong amplification of waves occurs at the interface of liquids.

String model It is known that the equation of the strong bend of a rod was received by Euler and presented in the appendix to his book *De Curvis Elasticis* [21, 22]. Adding to this equation the inertial term, we write

$$\eta_{tt} = a_0^2 \eta_{xx} \left(1 + \eta_x^2\right)^{-1.5}. \tag{1.62}$$

Here a_0^2 is a speed of waves in the rod, η is a deviation from the undisturbed direct form of the rod. If rod deviations are not too great, we can rewrite (1.62) in the form:

$$\eta_{tt} = a_0^2 \eta_{xx} \left(1 - \frac{3}{2}\eta_x^2 + \frac{15}{8}\eta_x^4\right). \tag{1.63}$$

In Eqs. (1.62) and (1.63), we denote the Lagrangian coordinate as x. We took into account nonlinearities of Eq. (1.63) to the fifth degree. If we limit ourselves to taking into account only cubic nonlinearity, then (1.63) gives the simplest equation for a thin layer on the surface of a deep liquid. High nonlinearities are taken into account in the work [23]. There the following equation was presented.

$$\eta_{tt} - \beta_2\eta_{xx} = \beta_1\eta_{xx}\left(1 + \eta_x^2\right)^{-0.5}\left[1 + \left(1 + \eta_x^2\right)^{-1}\right]. \tag{1.64}$$

We get this if we discard quantities of the sixth or more orders

$$\eta_{tt} - \beta_2\eta_{xx}\left(1 - \frac{3}{2}\eta_x^2 + \frac{15}{8}\eta_x^4\right) = 0. \tag{1.65}$$

This equation coincides with (1.63). Thus, we can assume that when choosing the appropriate coefficients, the equations for flexible strings can describe waves in two-layer media [24].

Nonlinear Klein-Gordon Equations for the Motion of the String and Some Surface Waves Nonlinear Klein-Gordon equations were used to describe the motion of a string in the work [24], when studying the theory of surfaces in differential geometry. The connection between the equations of curvature of surfaces and solitons is also shown [25]. Indeed, at the moment of the "Big Bang," the fabric of space-time was deformed by gigantic, extremely compact gravitational waves of the soliton type. Layering them on top of each other may have created extremely long and extremely thin threads of energy called "cosmic strings." That is, unexpectedly for many, the familiar concept of a string has grown to be used on a cosmic scale. Just as unexpectedly, the concept of a string previously formed the basis of "string theory," which was very popular among physicists [25–28].

Resonant "Dead Water" The effect of wave amplification in two-layer ocean systems is well known to sailors. In particular, a layer of fresh or warm water may be on the ocean surface, and as a result, a local two-layer system is formed. This system has some specific properties and its own wave speed (resonant speed). The upper fluid usually reduces the wave speed. Resonant effects can take place if some vessel (or some wave) crosses such a system with the resonant speed. In particular, the resistance to movement of the vessel is amplified. The famous researcher Horace Lamb wrote in his fundamental book *Hydrodynamics* [1, p. 371] "near the mouths of some of the Norwegian fjords there is a layer of fresh over salt water. Waves of considerable height in this boundary are easily produced. To this cause is ascribed the abnormal resistance occasionally experienced by ships in those waters". The effect of the upper fresh water, so-called 'dead water', was described by Pliny the Elder at the beginning of our era, and it was well-known to Vikings. One of their outstanding descendants Fritjof Nansen wrote "we did not move almost due to dead water; the vessel was carrying away all surface layer of water. Dead water forms a swell or even waves of big or smaller sizes". The upper layer slides on the salty water, as on a solid surface, and moves together with the vessel. We think that waves arising in this manner, even in the deep ocean, can be described by the theory of long waves developed in [14, 15].

Nansen described this phenomenon to Professor Vilhelm Bjerknes. Special experiments showed that if the vessel speed is far from the resonant speed, the specified effect does not arise. Of course, the effect takes place while a two-layer system exists. Such bands of surface fresh water arise near mouths of the big rivers, near ice fields, and in fjords, notably at Fiordland in New Zealand.

It is possible to assume that similar effects can occur in places of the fast rising of highly salty and cold ocean waters to the surface. It can take place in zones of interaction of ocean currents (e.g., the Agulhas leakage around the southern tip of Africa), in places of uneven seabed (e.g., the Bermuda region), in areas of meetings of warm and cold waters (e.g., the Cape Horn or the North Sea). Such two-layer systems are unstable, but the period of their existence can be sufficient to form a extreme wave. In the case of considerable amplification and overturning, such a wave $c^2 = c_0^2$ can increase the instability of the system.

1.3 Unidirectional Scalar Waves and Representation of the Basic Wave Equations in a Generalized Form

We have looked at a wide range of wave equations. The purpose of this chapter was to show that these equations have a fairly standard form and contain well-defined components. These are the d'Alembert operator and nonlinear terms. The equations may also contain dispersion and viscous terms.

This structure of the equations makes it possible to significantly simplify their representation in the case of unidirectional waves. Then the indicated wave equations can be reduced to one, unified representation. Let us take a closer look at this idea.

Klein-Gordon Equation As the most popular version of (1.12), we will consider the equation describing the so-called φ^4 scalar field. This equation has the form

$$\varphi_{tt} - c_0^2\varphi_{xx} + m^2\varphi - \lambda\varphi^3 = C_*. \tag{1.66}$$

New coordinates r and s are introduced:

$$r = ct - x, \qquad s = ct + x. \tag{1.67}$$

In this case, Eq. (1.65) is rewritten in the form

$$\left(c^2 - c_0^2\right)(\varphi_{rr} + \varphi_{ss}) + 2\left(c^2 + c_0^2\right)\varphi_{rs} + m^2\varphi - \lambda\varphi^3 = C_*. \tag{1.68}$$

For one-direction traveling waves, the last equation yields

$$\left(c^2 - c_0^2\right)\varphi_{rr} + m^2\varphi - \lambda\varphi^3 = C_*. \tag{1.69}$$

Equation (1.69) is multiplied on φ_r and then is integrated. As a result, we have

$$\varphi_r^2 - \frac{1}{2}\lambda\left(c^2 - c_0^2\right)^{-1}\left(\varphi^4 - 2\lambda^{-1}m^2\varphi^2 + 4\lambda^{-1}C\right) = C_* r. \tag{1.70}$$

Here C is a constant of integration. Solutions of (1.70) may be expressed using elliptic integrals (see Sect. 1.4.1). We emphasize Eqs. (1.69) and (1.70) have a singularity at $c^2 = c_0^2$.

Equations (1.69) and (1.70) will be considered carefully in Chaps. 2 and 3.

The Airy-type equations

1. Let us consider the equation for long surface waves

$$u_{tt} - c_0^2 u_{aa} = \beta u_a u_{aa} + \beta_1 u_a^2 u_{aa} + k u_{aatt}. \tag{1.71}$$

The common solution of equations similar to (1.71) is not known. However, the particular solutions of Eq. (1.71) can be received for unidirectional waves, when $r = a - ct$. In this case, Eq. (1.71) is transformed into the ordinary differential equation which has the first integral

$$B(u_r)_{rr} = B_3\left(B_1 B_3^{-1} u_r + B_2 B_3^{-1} u_r^2 + u_r^3 + C_1 B_3^{-1}\right). \tag{1.72}$$

Here B, B_1, B_2, and B_3 are the certain constants and C_1 is a constant of integration. Now a new unknown value Φ is introduced:

$$\Phi = u_r + \frac{1}{3}B_2 B_3{}^{-1}. \tag{1.73}$$

In this case, (1.72) yields

$$BB_3^{-1}\Phi_{rr} = \Phi^3 + p\Phi + q. \tag{1.74}$$

Here $p = B_1 B_3^{-1} - \frac{1}{3}\left(B_2 B_3^{-1}\right)^2$, $q = \frac{2}{27}\left(B_2 B_3^{-1}\right)^3 - \frac{1}{3}B_1 B_2 B_3^{-2} + C_1 B_3^{-1}$. Finally, we can transform (1.72) to the basic form (1.69):

$$B\Phi_{rr} + m^2\Phi - \lambda\Phi^3 = C. \tag{1.75}$$

Here $m^2 = -pB_3$, $\lambda = B_3$ and $C = qB_3$.

2. Let us consider another example of the long wave equation (see also (1.63))

$$u_{tt} - c_0^2 u_{aa} = \beta u_{aa} u_a^2 + \beta_1 u_{aa} u_a^4 + k u_{aatt}. \tag{1.76}$$

Remarkably Eq. (1.76) contains only the even nonlinear terms. For unidirectional waves, Eq. (1.76) yields

$$B_1 u_r = \frac{1}{3}\beta u_r^3 + \frac{1}{5}\beta_1 u_r^5 + B u_{rrr} + C_1. \tag{1.77}$$

Let

$$\Phi = u_r, \tag{1.78}$$

$\beta_1 \approx 0$ and a constant of integration $C_1 = 0$. In this case, Eq. (1.77) may be rewritten as (1.69) and (1.74).

The Nonlinear Maxwell Equation For unidirectional waves, we can write (1.9) in the form

$$k^* E_{rr} + \left(1 - c^{-2} c_0^2 + a_1\right) E + a_2 E^2 + a_3 E^3 + Cr + C_1 = 0. \tag{1.79}$$

Here $\mu^* = k^* = 0$, C and C_1 are constants of integration. Let $C = C_1 = a_2 = 0$. In this case, Eq. (1.79) may be transformed to the basic form (1.69):

$$B E_{rr} + m^2 E - \lambda E^3 = 0. \tag{1.80}$$

Here $B = k^*$, $m^2 = 1 - c^{-2} c_0^{-2} + a_1$, $\lambda = a_3$.

Another Example Many of the wave equations may be simplified to (1.69) form. Let us consider the Gardnet equation [14]. This cubic nonlinear equation is written here in the usual dimensionless form

$$u_t + 6 u u_x + 6 u^2 u_x + u_{xxx} = 0. \tag{1.81}$$

It is known that the linear transformation

$$\Phi = u\left(x - \frac{3}{2} t, t\right) + \frac{1}{2} \tag{1.82}$$

simplifies a solution of (1.81) to the solution of the modified Korteweg-de-Vries equation

$$\Phi_t + 6\Phi^2 \Phi_x + \Phi_{xxx} = 0. \tag{1.83}$$

For unidirectional waves the Eq. (1.83) yields

$$\Phi_{rr} - c\Phi + 2\Phi^3 = C. \tag{1.84}$$

This equation practically coincides with (1.74).

Conclusions Fundamental equations describing nonlinear one-direction traveling waves in scalar and electromagnetic fields and equations describing long waves in different continuum may be transformed to Eqs. (1.69) or (1.74). The solutions to this equation are well known. In simplified form, they can be written as

$$\varphi = \pm A \sec h\Delta r \quad \text{or} \quad \varphi = \pm B \tanh \nabla r. \tag{1.85}$$

Here A, B, Δ, and ∇ are constants. Thus, nonlinear equations may have ambiguous solutions. In the general case, which is the focus of this book, nonlinear wave equations can have multivalued solutions that describe waves that have a multivalued shape. The shape may have jumps, folds, and swirls. Thus, we will consider cases of complex waves. Their profiles may resemble the curves studied in the theory of catastrophes.

The above applies to many nonlinear wave equations of mathematical physics. The above does not depend on whether waves are considered in structures of quantum dimensions or whether they are waves in the Universe.

1.4 Exact Continuous and Discontinuous Solutions of Klein-Gordon Equations

Nonlinear wave equations are found in all areas of science, from mechanics and physics to biochemistry and fundamental questions of the origin of the Universe and living matter. Despite their diversity, an applied researcher usually divides them into two classes—integrable (or close to them) and non-integrable. There are not many integrable equations. That is why it is desirable to have as many of their solutions as possible, since the vast majority are non-integrable equations or close to integrable ones. The value of exact solutions is that by using them you can construct approximate solutions for equations that are close in form to almost integrable. Their solutions often describe periodic or quasi-periodic wave phenomena. Thus, there is a class of almost integrable equations when one can use solutions of linearized equations or exact solutions of other nonlinear equations.

However, the vast majority of nonlinear wave equations encountered in science are not integrable. This presents so far insurmountable difficulties for the development of general mathematical methods for solving them. The complexity of constructing solutions is aggravated by the fact that non-integrable wave systems become stochastic at a sufficiently large initial energy. In particular, after a certain time has elapsed, solutions to equations can become multivalued. For example, this occurs during the evolution of extreme waves into vortices and turbulization of the wave process. For example, this occurs in zones of singularities, bifurcations and various resonances. There, in particular, extreme multivalued waves can evolve into vortices and turbulize the wave processes.

The most important task associated with the study of extremal multivalued waves is to construct, albeit approximate, solutions to highly nonlinear wave nonintegrable equations that describe observational and experimental data. The book is dedicated to this problem.

Below, as an introduction to the problem, are given examples of exact or approximately exact solutions of the equations discussed in the previous section. The main attention is paid to the Klein-Gordon equation.

1.4.1 Examples of Exact Solutions of the Nonlinear Klein-Gordon Equation

1. The equation having the following form will be solved:

$$\varphi_{tt} - c_0^2\nabla^2\varphi + m^2\varphi - \lambda\varphi^3 = 0. \tag{1.86}$$

Next, we will consider plane waves. A solution of (1.86) is sought in the form $\varphi = \varphi(r)$, where

$$r = ct - \sum_{n}^{N} k_n x_n. \tag{1.87}$$

Here x_n are coordinates, $N = 1, 2, 3, 4, \ldots$ and

$$c = c_0 + \overline{c}, \tag{1.88}$$

where $\overline{c}$ is the perturbation of the speed c. In this case, Eq. (1.86) yields

$$\left(c^2 - c_0^2 \sum_{n}^{N} k_n^2\right)\varphi_{rr} + m^2\varphi - \lambda\varphi^3 = 0. \tag{1.89}$$

For simplicity of the analysis, we assume $\sum_{n}^{N} k_n = 1$. Then, if $c = c_0$, we have three solutions: $\varphi = 0$ and $\varphi = \pm m\lambda^{-0.5}$ (see (1.97)). On the other hand, solutions of (1.89) are different for slow ($c < c_0$) and fast ($c > c_0$) waves (see (1.95) and (1.97)).

One can find analytical expressions describing these types of waves. We can integrate (1.89) once and transform it into the first-order equation

$$(\varphi_r)^2 + \Pi(\varphi) = C, \tag{1.90}$$

where $\Pi(\Phi) = \frac{1}{2}\left(2m^2\varphi^2 - \lambda\varphi^4\right)\left(c^2 - c_0^2\sum_n^N k_n^2\right)^{-1}$ and C is the integration constant. This equation may be considered as the conservation law for the total energy of the "particle" of the field moving in the potential well $\Pi(\varphi)$. Now we can write

$$\varphi - \varphi_0 = \pm\int\frac{d\varphi}{\sqrt{C - \Pi(\varphi)}}. \tag{1.91}$$

Here φ_0 is another arbitrary constant. In our case, the expression $C - \Pi(\varphi)$ is the fourth-degree polynomial, and therefore the right-hand side of (1.91) transforms into the elliptic integral.

In Figs. 1.2 and 1.3, we presented a few curves which are determined by some solutions of Eq. (1.89). The curves are valid for fast ($c > c_0$) waves

If $c < c_0$, then slow stationary waves are generated in the system. Four forms of them are presented in Fig. 1.3.

Thus, we have shown that the Klein-Gordon equation has exact solutions. These solutions and the curves of Fig. 1.3 are very important for us. With these curves in mind, we can construct solutions without using elliptic integrals. Let us consider a few of these solutions.

2. Let

$$\varphi = A\,\mathrm{sec}\,hKr. \tag{1.92}$$

Let us substitute (1.92) into Eq. (1.89). Then we equate the terms containing sechKr and sec hKr $\tanh^2 Kr$. As a result, we obtain two equations for determining A and K. From them, we find three solutions for amplitude A:

$$A_\pm = \pm\sqrt{2\lambda^{-1}m^2} \quad \text{or} \quad A_0 = 0. \tag{1.93}$$

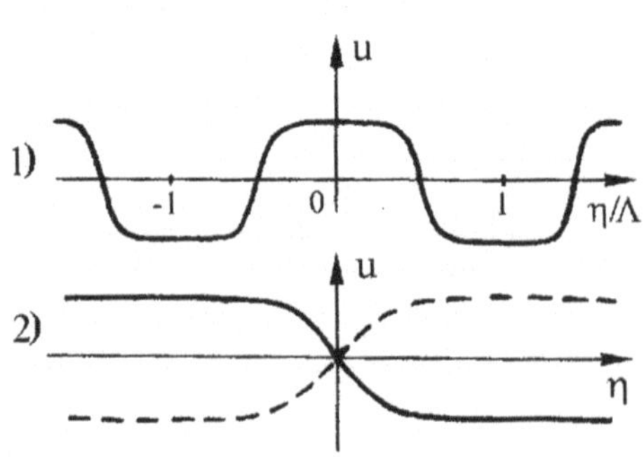

Fig. 1.2 The forms of fast stationary waves ($c > c_0$, $-\lambda < 0$); (1) is the step-like wave: (2) is the shock-like wave (kick) [3, 15]

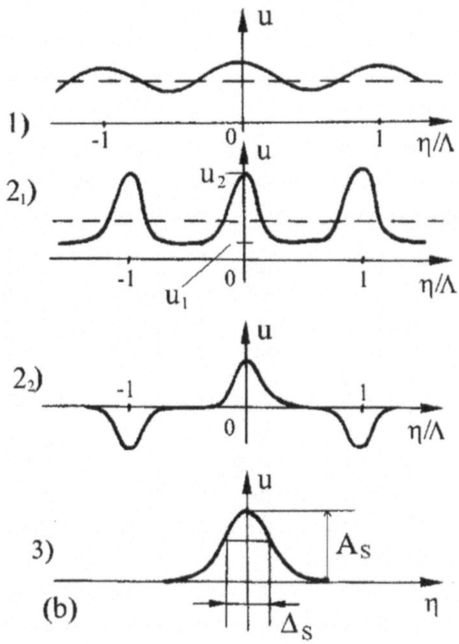

Fig. 1.3 The forms of slow stationary waves ($c < c_0$, $-\lambda < 0$); (1) are nearly sinusoidal waves; (2_1) is cnoidal wave; (2_2) is oscillon-like cnoidal wave [9, 29–37]; (3) is solitary wave [1, 16, 17]

The expression for K has the form

$$K = \pm\sqrt{m^2\left(c^2 - c_0^2\sum_n^N k_n^2\right)^{-1}}. \tag{1.94}$$

Finally,

$$\varphi_\pm = \pm\sqrt{2\lambda^{-1}m^2}\,\mathrm{sec}\,h\left\{\pm\sqrt{m^2\left(c^2 - c_0^2\sum_n^N k_n^2\right)^{-1}}\right\}r \text{ or } \varphi = 0. \tag{1.95}$$

3. Let

$$\varphi = A \quad \tanh \quad Kr. \tag{1.96}$$

In this case, the complete solution of Eq. (1.89) is written as $\varphi = 0$ and

$$\varphi_\pm = \pm\sqrt{m^2\lambda^{-1}} \quad \tanh\left\{\pm\sqrt{-\frac{1}{2}m^2\left(c^2 - c_0^2\sum_n^N k_n^2\right)^{-1}}\right\}r. \tag{1.97}$$

It can be assumed that the last solution describes the solitary wave in Fig. 1.2, and solution (1.95) describes the solitary wave in Fig. 1.3.

As a final example, consider the following model equation:

$$Gy_{xx} + ay + by^3 + cy^k = 0. \tag{1.98}$$

The coefficients of the equation are constant. It can be shown that different versions of this equation have solutions of the same type. This is shown in the table below:

$$\begin{aligned}
&1.\ Gy_{xx} + ay + by^3 = 0.\ \text{If}\ y \rightarrow \sec\mathrm{h}x = \cosh^{-1}x;\\
&2.\ Gy_{xx} + ay + cy^5 = 0.\ \text{If}\ y \rightarrow \sec\mathrm{h}^{1/2}x = \cosh^{-1/2}x;\\
&3.\ Gy_{xx} + ay + cy^{-3} = 0.\ \text{If}\ y \rightarrow \sec\mathrm{h}^{-1/2}x = \cosh^{1/2}x;\\
&4.\ Gy_{xx} + ay + cy^{-5} = 0.\ \text{If}\ y \rightarrow \sec\mathrm{h}^{-1/3}x = \cosh^{1/3}x.
\end{aligned} \tag{1.99}$$

The sign $\rightarrow$ denotes "corresponds" to the solution. It means that the constants of the solution are determined from the corresponding equation.

The considered examples can help in constructing solutions in the form of isolated waves.

1.4.2 Examples of Exact Discontinuous and Multivalued Solutions in Klein-Gordon Equations

Let us look at a very particular and at the same time very important and general for subsequent, the case of wave equations. As we have emphasized, many of them may be written as

$$\left(c^2 - c_0^2\right)\varphi_{rr} + m^2\varphi - \lambda\varphi^3 = C_*. \tag{1.100}$$

In the case $c^2 = c_0^2,\ C_* = 0$, we have that

$$\varphi_{\pm} = \pm m\lambda^{-0.5}, \quad \varphi_0 = 0. \tag{1.101}$$

Using these solutions and taking into account that they must describe traveling waves, we can construct various wave profiles. Let's consider three of them and the waves they describe.

Based on our initial assumptions, three values (1.101) correspond to some unidirectional waves. Let us take this into account the fact that these solutions cannot transform into each other without corresponding continuity jumps. There is nothing surprising in the coexistence of such jumps and waves. We assume that these waves are some variants of discontinuous (shock) waves, well known in hydrodynamics.

Using (1.101), we constructed three variants of these types of waves. They are shown in Fig. 1.4. It is assumed that $m\lambda^{-0.5} = 1$ in calculations.

Curve a corresponds to a traveling stepped wave. When constructing it, we neglected the contribution to the wave of the zero solution (1.101). The wave amplitude takes only two values corresponding to $\varphi_{\pm} = \pm m\lambda^{-0.5}$. Wave b is an example of a multivalued wave that will receive a lot of attention in the book. The presented multivalued curve can be interpreted as an infinite chain of rectangular scalar field concentrators moving in one-dimensional space. The last chain of concentrators c can be interpreted in different ways. We interpret it here as rectangular isolated concentrators of a scalar field moving in space.

We have outlined a certain interpretation of the decisions under consideration when discussing Fig. 1.4. You may not agree with it. Indeed, although the existence of fairly stable shock waves is a well-established fact, this cannot be said about rarefaction waves following the compression zone. Sustained rarefaction waves are practically not observed in nature [38, 39]. Apparently, they arise only if the phase state of the medium in which the wave propagates changes. For example, the instantaneous appearance of a cavitation zone (rarefaction zone) in a liquid is possible when a shock wave is reflected from a free or slightly deformable surface of the liquid. In general, we believe that the appearance of discontinuous rarefaction shock waves is possible in media whose properties are described by specially selected, complex, highly nonlinear equations of state.

The situation with waves b and c in Fig. 1.4 is even more incomprehensible. According to it, the possibility of the appearance of a chain of particle-waves arises. Of course, in our time, when almost 100 years have passed since the creation of the field of quantum mechanics, this fact will not surprise everyone. We believe that, according to the solutions under consideration, two waves can arise simultaneously in the scalar field. There could even be three of them. Indeed, if $c^2 = c_0^2$ and $C_* \neq 0$, then we can write the following algebraic equation:

$$\lambda\varphi^3 - m^2\varphi + C_* = 0. \tag{1.102}$$

Let $C_* \neq 0$ and all the roots of this equation are real. Using them, you can construct other interesting discontinuous wave structures corresponding to Eq. (1.100).

Let us note that if nonlinear terms containing φ^5, φ^7 and φ^9 are in Eq. (1.100), we could have 5, 7, or 9 values for wave amplitudes, respectively. We could be able to construct very complex wave structures for the scalar field using this variety of waves.

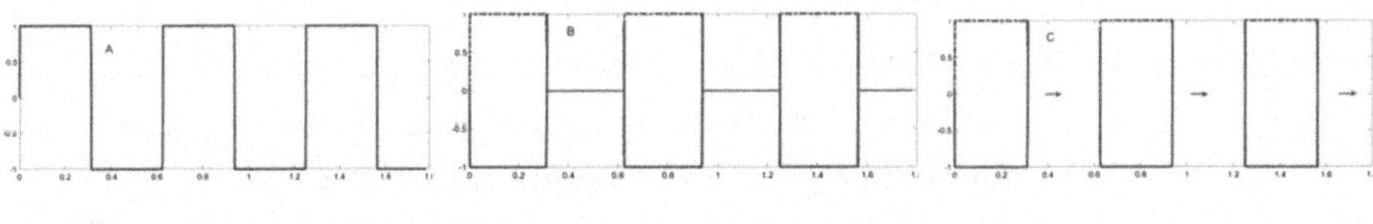

Fig. 1.4 Three variants of waves traveling in one direction (**A**—a stepped wave, **B**—a chain of coupled multivalued waves (bubbles) moving in a scalar field, **C**—a chain of uncoupled multivalued waves (bubbles) formed by a scalar field)

This concludes our consideration of examples of exact solutions of the wave equation (1.100). Let us emphasize that the presented solutions are qualitatively valid for all wave equations considered in this book. We will often refer back to the questions posed above and to their solutions in this book.

Next, we move on to considering approximate solutions of wave equations. We emphasize that the same equation can have both an exact and an approximate solution, depending on the problem under consideration. For example, studying waves in limited volumes is usually more difficult than studying them in an unlimited environment. Therefore, when studying waves in limited media, one often has to use assumptions without which it is sometimes impossible to study waves in infinite media.

1.5 Approximate Solutions and Highly Localized Wave Structures

Let us build an approximate solution to Eq. (1.86), which at the same time describes the curves in Fig. 1.2 and 1.3 and is consistent with experimental data.

1.5.1 Oscillons and Granular Surface Waves

Let

$$J(r) = A \sec h(e \sin Kr) + C. \tag{1.103}$$

From (1.103), it follows

$$J_{rr}(r) \approx A \sec h^3(e \sin Kr) \sinh^2(e \sin Kr) e^2 K^2 - \frac{1}{2} A \sec h(e \sin Kr) e^2 K^2. \tag{1.104}$$

The expressions (1.103) and (1.104) are substituted into (1.86). As a result, we have

$$\begin{aligned} &Ae^2K^2\left(c^2 - c_0^2 \sum_n^N k_n^2\right)\left[\sec h^2(e \sin Kr) \sinh^2(e \sin Kr) - \frac{1}{2}\right] \sec h(e \sin Kr) \\ &= -m^2[A \sec h(e \sin Kr) + C] + \lambda\left\{A^3 \sec h(e \sin Kr)\left[1 - \tanh^2(e \sin Kr)\right]\right. \\ &\quad \left. + 3CA^2\left[1 - \tanh^2(e \sin Kr)\right] + 3AC^2 \sec h(e \sin Kr) + C^3\right\}. \end{aligned} \tag{1.105}$$

Here we equate similar quantities and discard the small terms. As a result, we get these equations:

$$-\frac{1}{2}Ae^2K^2\left(c^2-c_0^2\sum_n^N k_n^2\right)+m^2-3\lambda C^2=\lambda A^2;\quad m^2-3\lambda A^2=\lambda C^2\ ;\quad A_0=0. \tag{1.106}$$

Having solved them together, we find that

$$\begin{gathered}A_\pm=\pm\sqrt{\frac{1}{3}\left(m^2\lambda^{-1}-C^2\right)}\ ,\quad A_0=0.\\ e^2K^2=\frac{4}{3}\left(m^2-4\lambda C^2\right)\left(c^2-c_0^2\sum_n^N k_n^2\right)^{-1}.\end{gathered} \tag{1.107}$$

Here e or K remain uncertain. Let us consider special cases of the obtained solution.

1. First, we will consider the case of unidirectional waves. Curves determined by expressions (1.103), (1.106), and (1.107) are presented in Fig. 1.5. We assume that $A = 1$ or $A = -1$ for simplicity. The solid sections of the curves are determined by $A = 1$, and the interrupted sections are determined by $A = -1$. Let us compare the waveforms presented in Fig. 1.5 with curves following from the exact solution (1.91). These curves are shown in Fig. 1.3. It is easy to see that if the value e in (1.103) is small, then the resulting solution describes curve (1) in Fig. 1.3. If the value e increases and $A = 1$, then curve D describes the curve (2_1) in Fig. 1.3. We emphasize that we can build solutions in which the value of A can change abruptly. This case is represented by curve C in Fig. 1.5. This curve corresponds to the curve (2_1) in Fig. 1.3 corresponding to the exact solution.

Finally, using the obtained approximate solution, we can construct a chain of localized clumps (bubbles) of a scalar field propagating in one direction (curve A). This chain is a some analogue of the chain presented in Fig. 1.4.

However, in the present case, we have the case of a multidimensional completely closed structure A in Fig. 1.5. It can be called a kind of pulsating bubble, periodically compressing to zero volume and periodically reaching maximum volume [32].

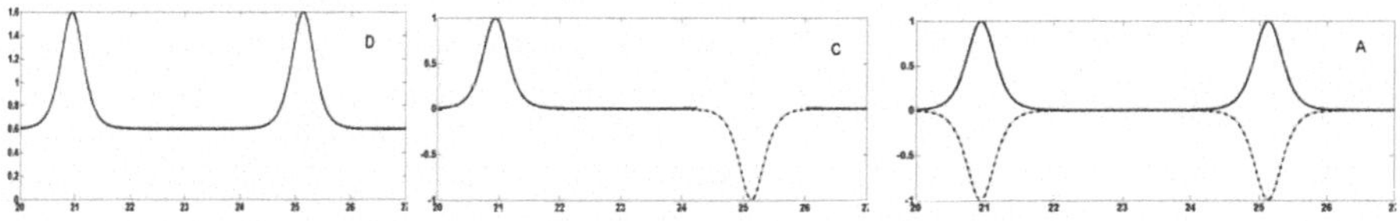

Fig. 1.5 Examples of unidirectional oscillating waves

2. Now we consider the resonant multidirectional waves formed by the sum

$$\varphi = J(r) + J(s) \tag{1.108a}$$

Here $r = ct - \sum_{n}^{N} k_n x_n$, $s = ct + \sum_{n}^{N} k_n x_n$. During the calculations, we assume that $A = 1$, $e = 1.25$, K = 0.5 (left) and $A = 1$, $e = 5$, K = 0.5 (right). Results of the calculations are shown in Fig. 1.6. The curves describe the full period of oscillations.

Curve 1 of Fig. 1.6 shows the crater formation within the physical field. Curves 2 and 3 illustrate the plume (peak) formation because of converging two waves.

Here we must emphasize that the wave structures presented in Fig. 1.6 are for $A = 1$. Similar structures, but of the opposite sign, are built for $A = -1$. Using these structures, it is possible to construct a third one, which corresponds to some oscillating bubble in some multidimensional space.

We have repeatedly emphasized that wave processes in different media should not differ greatly if similar conditions occur. Therefore, it is not surprising for us that the approximate solution obtained for scalar field waves (1.103) qualitatively describes waves on the surface of granular media. They are demonstrated in Fig. 1.7.

We obtained a solution (1.103) which in the one-dimensional case is in qualitative agreement with the exact solution. This also applies to discontinuous and multivalued wave patterns, which can be constructed using three possible values for amplitude A (1.107). These patterns can be especially interesting in the multidimensional case. However, it is difficult to construct and comprehend the patterns in the case of many variables. In order to somewhat simplify the problem and to some extent understand it, we will consider spatial radial oscillations of a wave structure in the next section, which can be called a bubble.

1.5.2 Oscillons and Bose-Einstein Condensate Waves

For us it is important that the solution (1.103) is describing qualitatively the curves of the exact solution. On the other hand, the curve C of Fig. 1.5 describes qualitatively oscillons on a surface of the granular materials presented in Fig. 1.8.

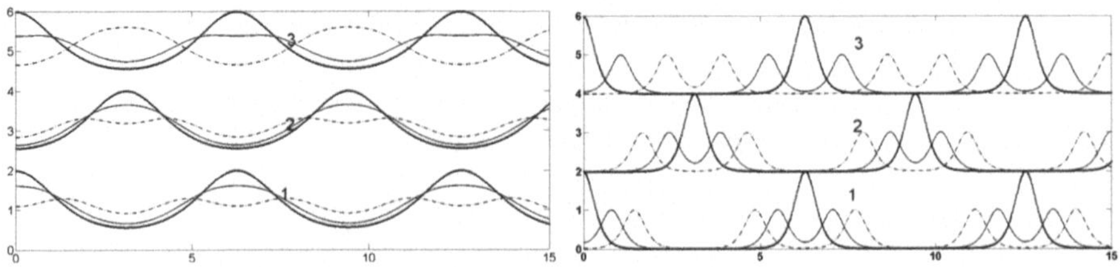

Fig. 1.6 Convergence and divergence of scalar waves. Curves 1, 2, and 3 describe the wave dynamics over the oscillation period. These curves are placed at different distances from the horizontal axis for ease of analysis and comparison

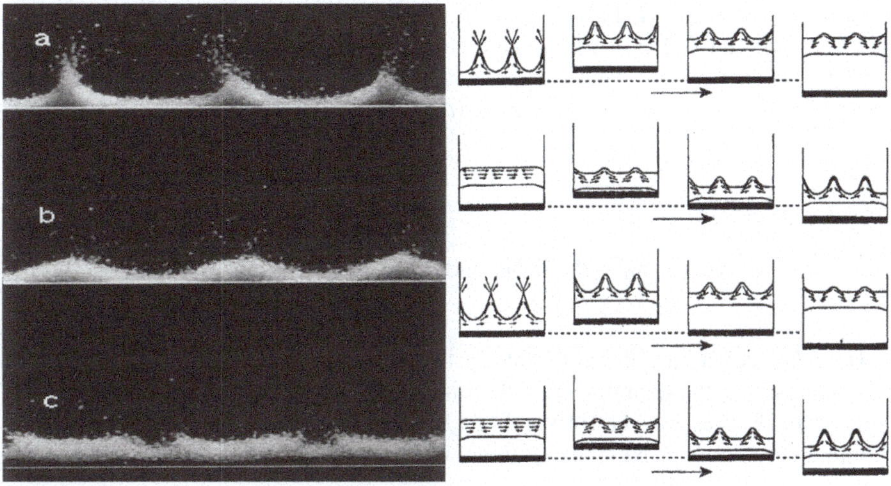

Fig. 1.7 Granular waves: experiment (the left) and theoretical modeling (the right). The letters **a**, **b**, **c** define different moments in time. The corresponding drawings show the dynamics of the waves [14, 15, 33, 34, 40]

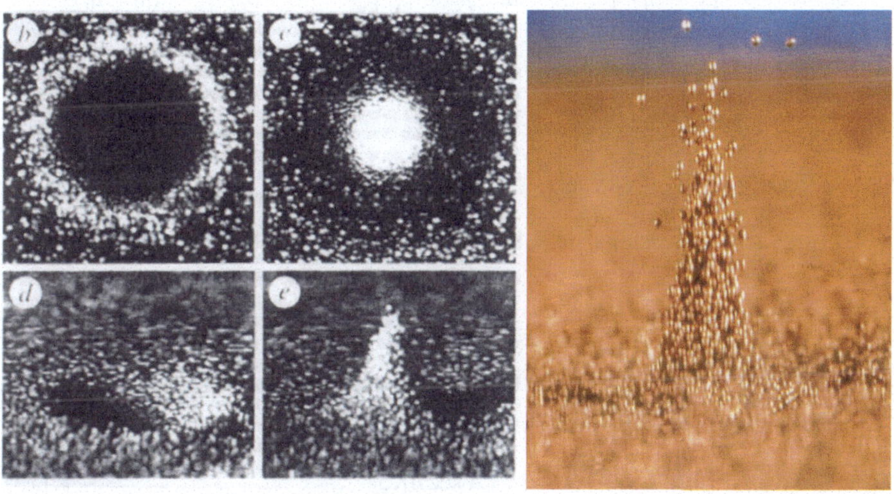

Fig. 1.8 The periodical granular column and the crater on the surface of the vertically excited layer: *b* and *c* are a bird's-eye view, and *d* and *e* are a side view (left). The column and two craters (right) [14, 15, 41]

Many physical objects in nature can be modeled as a thin-walled bubble or a spherical clot of energy. Let it produce radial vibrations. Let us describe these oscillations based on the Klein-Gordon equation:

$$\varphi_{tt} - c_0^2 \sum_{i=1}^{I} \varphi_{ii} + m^2\varphi - \lambda\varphi^3 = 0. \tag{1.108b}$$

Here $i = 1; 1; 3;\ldots I$. Index i corresponds to the derivative with respect to the spatial coordinate x_i. Let us introduce a new variable:

$$\xi = a(t) - \alpha, \tag{1.109}$$

Here

$$a(t) = \overline{R}_0 + R_0 \sin^n \omega t \text{ and } \alpha = \left[\sum_i^I (x_i - a_i)^2\right]^{0.5}. \tag{1.110}$$

Here $\overline{R}_0$ and R_0 are the constants, n is an integer, x_i determines the Cartesian coordinates, and the quantities a_i determine the point relative to which the oscillations occur. Expressions (1.109) and (1.110) connect the time and space familiar to us into a certain unity. Next, using the new variable, we find

$$\begin{aligned}
\varphi_{tt} &= \frac{1}{4} R_0^2 n^2 \omega^2 \sin^{2n-4} \omega t \sin^2 2\omega t \Phi_{\xi\xi} + R_0 n \omega^2 \left[(n-1)\sin^{n-2}\omega t - n \sin^n \omega t\right]\Phi_\xi, \\
\varphi_{ii} &= \left[\sum_i^I (x_i - a_i)^2\right]^{-1} (x_i - a_i)^2 \varphi_{\xi\xi} + \left\{\left[\sum_i^I (x_i - a_i)^2\right]^{-1.5} (x_i - a_i)^2 \right. \\
&\quad \left. - \left[\sum_i^I (x_i - a_i)^2\right]^{-0.5}\right\}\varphi_\xi.
\end{aligned} \tag{1.111}$$

Then the last expressions are substituted into (1.108b). We get

$$\begin{aligned}
&\left(R_0^2 n^2 \omega^2 \sin^{2n-2} \omega t \cos^2 \omega t - c_0^2\right)\varphi_{\xi\xi} \\
&+ R_0 n \omega^2 \left[(n-1)\sin^{n-2}\omega t - n \sin^n \omega t\right]\varphi_\xi = -m^2 \varphi + \lambda \varphi \Phi^3.
\end{aligned} \tag{1.112}$$

In what follows we will assume $n = 2$. Let

$$\varphi = A \sec hK\xi. \tag{1.113}$$

Equation (1.112) yields

$$\begin{aligned}
&\left(4R_0^2 \omega^2 \sin^2 \omega t \cos^2 \omega t - c_*^2\right)\left(2K^2 A \sec hK\xi \tanh^2 K\xi - K^2 A \sec hK\xi\right) \\
&- 2R_0^2 \omega^2 \left[1 - 2\sin^2 \omega t\right] \tanh K\xi \sec hK\xi = -m^2 A \sec hK\xi + \lambda A^3 \sec h^3 K\xi.
\end{aligned} \tag{1.114}$$

This leads to two algebraic equations that give

$$\begin{aligned}\left(c_*^2 - 4R_0^2\omega^2 \sin^2\omega t \cos^2\omega t\right)K^2 &= -m^2 + \lambda A^2,\\ \left(4R_0^2\omega^2 \sin^2\omega t \cos^2\omega t - c_*^2\right)K^2 &= -\frac{1}{2}\lambda A^2.\end{aligned} \tag{1.115}$$

Now let us find expressions for K and A. We neglect the dynamic components in them compared to the static ones and get

$$K_\pm = \pm\sqrt{\lambda A^2 - m^2\left(c_0^2 - 2R_0^2\omega^2\right)^{-1}} \quad \text{and} \quad A_\pm = \pm m\sqrt{2\lambda^{-1}}, \quad A_0 = 0. \tag{1.116}$$

We have obtained expressions corresponding to those found in the previous section. Therefore, the results of calculations using (1.116) will resemble, qualitatively, those presented in Fig. 1.7. However, there are significant differences, since here the structure is motionless in space and can be multidimensional. We have obtained 3 solutions (1.116). Using them, we can construct various pictures of the evolution of the scalar field. Some of them are similar to those that will be considered in Chap. 7. There is also a possibility of the emergence of multivalued multidimensional wave systems corresponding to the one-dimensional chains and bubbles shown in Fig. 1.4. Generally speaking, solution (1.113) opens up the possibility of describing many wave structures. Here we will restrict ourselves to considering oscillons.

The calculation results are presented in Fig. 1.9. There are two-dimensional paintings constructed. Calculations were carried out for $A_+ > 0$ and coordinate ξ (1.109). The paintings $A_+ < 0$ are completely symmetrical to those presented.

The entire pulsation cycle is shown as the localization transforms into a ring structure, which then collapses back into the localization. Of course, this figure gives only the main thing that characterizes the oscillations of the volume of the scalar field relative to the center.

There is a certain similarity between the oscillations of the scalar field and the data from the experiments studying the wave processes in the Bose-Einstein condensate (BEC) [14, 15, 41, 41] and on the granular layers shown in Figs. 1.7 and 1.8.

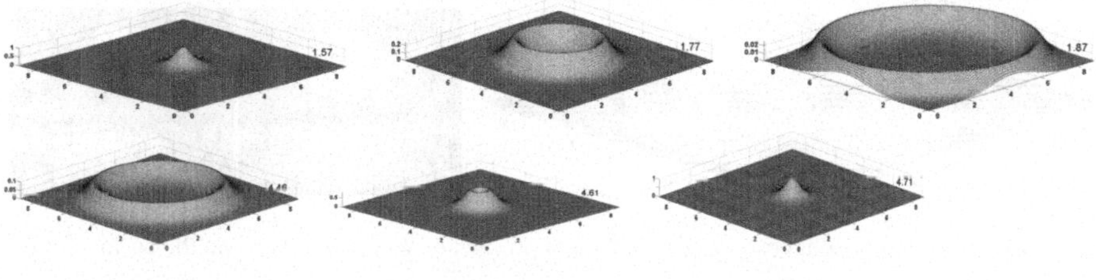

Fig. 1.9 Results of calculations performed according to (1.113) at $A = 1$, $n = 2$, and $K = 2$. The numbers define certain moments of time within the pulsation period

Dynamic wave processes observed for BEC are shown in Figs. 1.10 and 1.11. It is possible to see periodic formation of peaks, they transform into craters which generate new peaks.

At large K solution (1.113) describes the periodic oscillations of a thin ring and its collapse into a very small volume (Fig. 1.11, left).

Traditionally, eigenmodes of the Bose-Einstein condensate are described by the nonlinear Schrödinger (Gross-Pitaevskii) equation. However, we underlined (the Sect. 1.1.2) that the nonlinear Schrödinger may be considered as a particular case of the nonlinear Klein-Gordon equation. Thus, the last can also describe the evolution of condensate wave functions; this allows us to consider Figs. 1.10 and 1.11 as tests for solution (1.113).

1.6 Conclusion

We have considered a wide range of nonlinear wave equations and presented them in the case of unidirectional waves in some unifying form. Using the last simplification, the author showed qualitatively that the unifying equation can have many single-

Fig. 1.10 Oscillations of a rotating Bose-Einstein condensate [14, 15, 41, 41] (see http://www.lkb.ens.fr/recherche/atfroids/anglais/vortex_an.html)

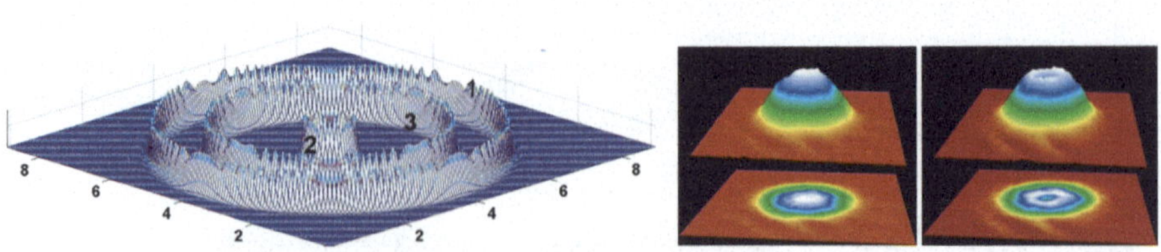

Fig. 1.11 Two-dimensional calculation (left) and profiles of waves observed on the surface of the BEC
The side view (upper) and the bird's-eye view (bottom) (right) [14, 15, 41]
A hole is clearly visible at the center of the condensate

valued and multivalued solutions. Of these, the chapter focuses on single-valued continuous resonant solutions. The results obtained create the opportunity to move on to the description of extreme multivalued waves. The study of the occurrence and propagation of which in various physical processes and circumstances is the main goal of the book.

References

1. Lamb H (1932) Hydrodynamics, 6th edn. Dover Publications, New York. Landau LD, Lifshitz EM. The Classical Theory of Fields. Pergamon, New York; Nigmatulin RI (2014) Mechanics of continuous media. Moscow (in Russian) (Нигматуллин Р.И. 2014. Механика Сплошной Среды. Москва, ГЭОТАР-Медиа)
2. Landau LD, Lifshitz EM (1987) Fluid mechanics. Pergamon, New York
3. Ostrovsky LA, Potapov AI (1999) Modulated waves. Theory and applications. Johns Hopkins University Press, Baltimore and London
4. Bullough RK, Jack PM, Kitchenside PW, Saunders R (1979) Soliton in laser physics. Phys Scr 20:364
5. Kneubhl FK (1997) Oscillations and waves. Springer
6. Mo H, van den Bosch F, White S (2011) Galaxy formation and evolution. Cambridge University Press
7. Linde A (1990) Particle physics and inflationary cosmology. CRC Press, Boca Raton
8. Lyth DH, Liddle AR (2009) The primordial density perturbation. Cosmology, inflation and origin of structure. Cambridge University Press, Cambridge
9. Galiev SU, Galiyev TS (2001) Nonlinear trans-resonant waves, vortices and patterns: from microresonators to the early universe. Chaos: an interdisciplinary. Journal of Nonlinear Science 11(3):686–704
10. Airy GB (1845) Tides and waves. In: Encyclopaedia metropolitana, London
11. Galiev SU (2005) Strongly nonlinear two-speed wave equations for coastal waves and their application. Phys D 208:147–171
12. Galiev SU (2008) Strongly-nonlinear wave phenomena in restricted liquids and their mathematical description. In: Perlidze T (ed) New nonlinear phenomena research. Nova Science Publishers, New York, pp 109–300
13. Galiev SU (2009) Modelling of Charles Darwin's earthquake reports as catastrophic wave phenomena. The University of Auckland
14. Galiev SU (2010) Charles Darwin's geophysical reports as models of the theory of catastrophic waves. The University of Auckland, p 655
15. Galiev SH (2020) Evolution of extreme waves and resonances. CRC Press, p 508
16. Russell JS (1844) Report on waves. Report: British Association for the Advancement of Science, 14th Meeting, Murray, London, p 311–390
17. Boussinesq J (1872) Théorie des ondes et remous qui se propagent le long d'un canal rectangulaire horizontal, en communiquant au liquide contenu dans ce canal des vitesses sensiblement pareilles de la surface au fond. J Math Pures Appl 17:55–108
18. Smith CR, Walker JDA (1995) Turbulent wall-layer vortices. In: Green SL (ed) Fluid vortices. Kluwer Academic Publishers
19. Horn DA, Imberger J, Ivey GN, Redekopp LG (2002) A weakly nonlinear model of long internal waves in closed basins. J Fluid Mech 467:269–287
20. Pelinovsky E, Talipova T, Didenkulova E (2024) The Hopf equation with certain modular nonlinearities. Phys Lett A 507:129489. Slunyaev AV, Pelinovsky DE, Pelinovsky EN (2023) Rogue waves in the sea: observations, physics, and mathematics. Phys Usp 66(2):148–172; Abrashkin AA, Pelinovsky EN (2022) Gerstner waves and their generalizations in

hydrodynamics and geophysics. Phys Usp 65(5):453–467; Choi W, Camassa R (1999) Fully nonlinear internal waves in a two-fluid system. J Fluid Mech 396:1–36
21. Euler L (1952) Methodus inveniendi lineas curvas maximi minimive proprietate gaudentes sive solutio problematis isoperimetrici latissimo sensu accepti, vol 1. Springer Science & Business Media
22. Galiev SU, Galiyev TS (2017) 'Elastica'-like waves and particles: from Bernoulli, Euler, Laplace and Faraday to the eruption of the Universe. http://hdl.handle.net/2292/31785
23. Devaud M, Hocquet T (2017) An essay on the vibrating string and its nonlinearities. There the following equation was presented. https://hal.archives-ouvertes.fr/hal-01532358v2
24. Zhao D-M et al (2018) Nonlinear vibration of an elastic soft string: large amplitude and large curvature. Math Probl Eng 2018. https://doi.org/10.1155/2018/7909876
25. Kobayashi K (1984) Nonlinear Klein-Gordon equations for the motion of the string. Prog Theor Phys 71(2):424–426. https://doi.org/10.1143/PTP.71.424
26. Serjeant S (2010) Observational cosmology. Cambridge University Press, Cambridge
27. Green MB et al (2012) Superstring theory. Cambridge University Press
28. Saplakoglu Y (2020) "Cosmic string" gravitational waves could solve antimatter mystery. Scientific American
29. Laposky BF (1969) Oscillons: electronic abstractions. Leonardo 2(4):345–354
30. Gleiser M (1994) Pseudostable bubbles. Phys Rev D 49:2978
31. Umbanhowar PB, Melo F, Swinney HL (1996) Localized excitations in a vertically vibrated granular layer. Nature 382:793–796
32. Belova TI, Kudryavtsev AE (1997) Solitons and their interactions in classical field theory. Phys Usp 40:359
33. Cerda E, Melo F, Rica S (1997) Model for subharmonic waves in granular materials. Phys Rev Lett 79(23):4570. https://doi.org/10.1103/PhysRevLett.79.4570
34. Clement E, Labous L (2000) Pattern formation in a vibrated granular layer: the pattern selection issue. Phys Rev E 62(6):8314–8323. Clement E, Vanel L, Rajchenbach J, Duran J (1996) Pattern formation in a vibrated granular layer. Phys Rev E 53:2972–2975
35. Galiev SU (1999) Topographic effect in a Faraday experiment. J Phys A Math Gen 32: 6963–7000
36. Graham N, Stamatopoulos N (2006) Unnatural oscillons lifetimes in an expanding background. Phys Lett B639:541–545. https://doi.org/10.1016/j.physletb.2006.06.070
37. Alexeeva NV, Barashenkov IV, Bogolubskaya AA, Zemlyanaya EV (2023) Understanding oscillons: standing waves in a ball. Phys Rev D 107(7):076023
38. Zel'dovich YB (1946) On the possibility of rarefaction shock waves. Zh Eksp Teor Fiz 4: 363–364
39. Borisov AA, Borisov AA, Kutateladze SS, Nakoriakov VE (1983) Rarefaction shock wave near the critical liquid-vapour point. J Fluid Mech 126:59–73
40. Galiev SU (2015) Darwin, geodynamics and extreme waves. Springer
41. Stock S, Bretin V, Chevy F, Dalibard J (2004) Shape oscillation of a rotating Bose-Einstein condensate. Europhys Lett 65:594–600

Chapter 2
Euler Figures, Unidirectional Multivalued Waves, Quanta, and Vortices

In this chapter, we will consider mainly unidirectional waves. We are interested in questions of how, in the process of one or another evolution, they can be transformed into multivalued waves. In particular, how the unidirectional waves transform into the Euler figures, particle-waves, or a train of particles.

Euler's Elastica Forms The elastica forms of strips attracted the attention of James Bernoulli, Leonhard Euler, Pierre Laplace, Gustav Kirchhoff, Max Born, and others [1–18]. In particular, James Bernoulli posed the precise problem of the elastica in 1691 [1]. Euler characterized the family of curves known as the "elastica" (Fig. 2.1) [2].

Laplace's Capillary Waves Apparently, the elastica forms can exist in many highly nonlinear wave systems. We suppose that the elastica forms can describe qualitatively periodic appearance of splashes, drops, and bubbles on a liquid surface as the result of highly nonlinear effects and the surface tension. In particular, the forms which intended for the description of different forms of rods may be used for the analysis of highly nonlinear capillary waves (Fig. 2.2).

Pierre Simon Laplace investigated the shape of the capillary in his 1807 work [3]. We think that the elastica-like waves and their dynamics describe processes which have been observed in many experiments. In particular, parametrically (vertically) forced surface waves, which may be known as Faraday waves [4], can describe as elastica forms [19, 20].

Experiments of Sir Taylor and an "Elastica" Let us consider the results of the interesting experiments performed by Sir G. I. Taylor (Fig. 2.3) [9].

Taylor wrote in 1969, "The analogy between the equations of motion of a viscous fluid and that of an incompressible elastic solid is well-known. It was pointed out to me by Brooke Benjamin that when a thin rod is subjected to end compression it forms an elastica and that Love's *Theory of Elasticity* contains a picture of the set of elasticas. I, therefore, found a very viscous fluid, known as SAIB (Sucrose Acetate

S. U. Galiev, *Extreme Multivalued Waves as a Model of Everything*,
https://doi.org/10.1007/978-3-031-67027-5_2

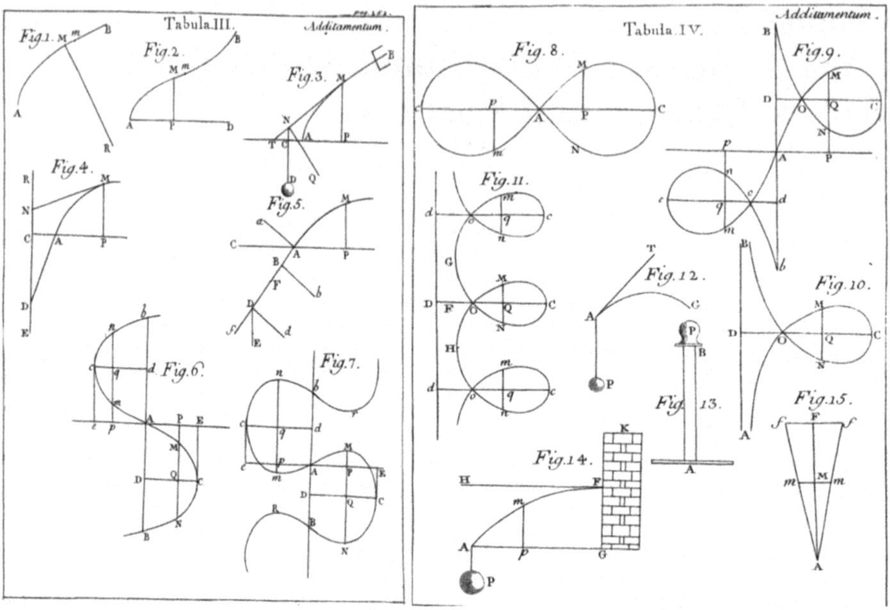

Fig. 2.1 Six of Euler's elastica figures: Figs. 6–11 (left and right) [2]

Fig. 2.2 Some elastica forms can describe the water surface forms which can appear because of the capillarity [3]

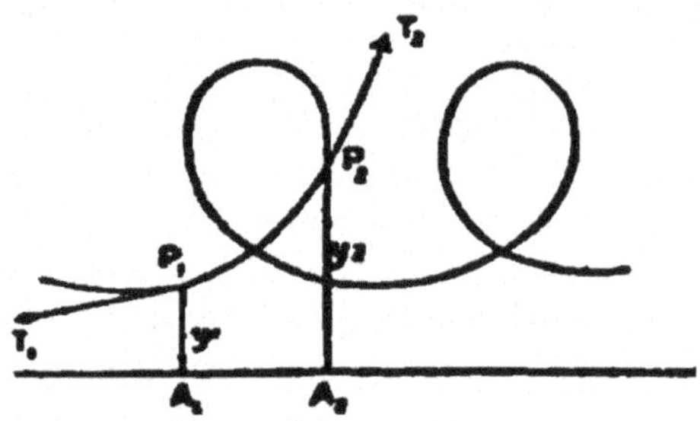

Isobutirate) and stretched a thread of it between two sticks. I then laid this thread on the surface of a dish of mercury and either pushed the sticks toward one another or laid two matches on the thread and pushed them toward one another, and photographed the thread" (see Fig. 2.4, $\lambda = 1$, 0.5 and 0.4). Sir G. I. Taylor also observed a very viscous stream of the fluid falling onto a plate [9, 19, 20]. A short distance above the plate the stream begins to wave or rotate into a spiral form. More complex folding was observed when the stream falls into a fluid of less density than itself. In the last case, the stream can fold and form knots. The folding of the stream resembles the buckling of the elastic bodies which have been showed in Fig. 2.4.

The folding and formation of the elastica (Euler's elastica figures) are well-known processes of the instability of many long flexible systems (Fig. 2.4). In particular, folding and knotting of thin structural elements can take place under an axial impact [11, 12].

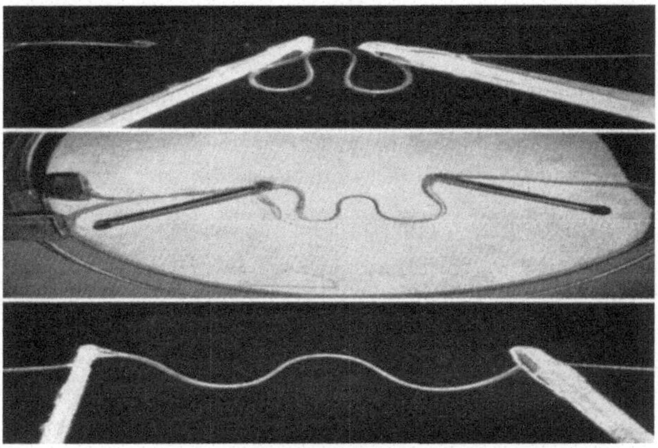

Fig. 2.3 Threads of SAIB, floating on mercury [9, 19, 20]

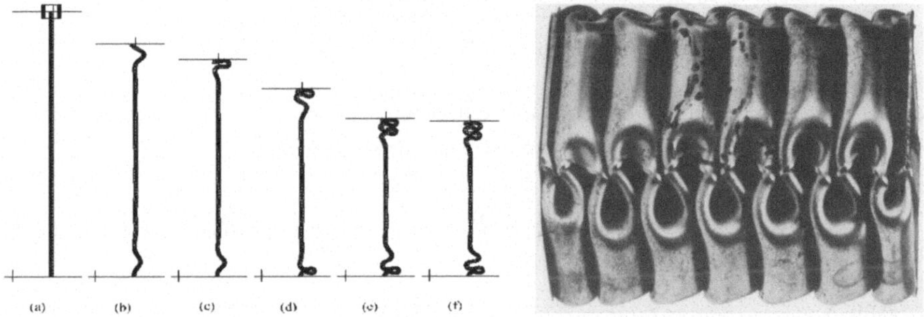

Fig. 2.4 Development of the dynamic buckling form of a long structure under an impact loading (*a*) initial, (*b*) $t = 0.322$ ms, (*c*) $t = 0.504$ ms, (*d*) $t = 0.918$ ms, (*e*) $t = 1.598$ ms, (*f*) $t = 2.612$ ms [12] (left). The folding can take place during the crushing of square tubes (right) [11, 12, 19, 20]

The shapes presented in Fig. 2.4 (right) form by the compression of very flexible rods. The forms in Fig. 2.4 (left) correspond to the tucks arising at the axial crushing of the square tube.

Multivalued Waves The emergence of multivalued stable waves began to be studied by Sibgatullin [21, 22]. He associated the appearances of them with resonant effects. However, he did not develop his research. In particular, he did not notice the connection between his results and Euler's research. This was done by Filipov, who introduced such a concept as the Euler soliton [23]. Subsequently, multivalued waves were studied in many publications [24–27]. The connection between the theory of these waves and Euler's results was not emphasized in all publications.

2.1 Euler Figures as an Exact Solution to the Nonlinear Klein-Gordon Equation (NKGE)

The basic equations used here were obtained in the previous chapter. As an example, we give the Klein-Gordon equation (1.12) written in the form

$$\varphi_{tt} - c_0^2 \nabla^2 \varphi + m^2 \varphi - \lambda \varphi^3 = 0. \tag{2.1}$$

In a generalized form for the case of one-dimensional unidirectional waves, they can be represented as follows:

$$y_{rr} = F(y_r, y, r). \tag{2.2}$$

or in a slightly different form $y_r^2 = F\left(\int y_r^2 dr, y, r\right)$. For example, for this case, the Klein-Gordon equation is written in the first chapter as follows (1.70):

$$\varphi_r^2 = \frac{1}{2}\lambda\left(c^2 - c_0^2\right)^{-1}\left(\varphi^4 - 2\lambda^{-1}m^2\varphi^2 + 4\lambda^{-1}C\right) + C_* r. \tag{2.3}$$

A wide class of Eq. (2.1) can be presented in the form

$$y_{xx} = B + B_1 y + B_2 y^2 + B_3 y^3 + \ldots + B_n y^n + F(x). \tag{2.4}$$

Let us consider special cases of this equation.

Let $F(x) = 0$ in (2.4). In this case, we can easily find that

$$\frac{1}{2} y_x^2 = By + \frac{1}{2} B_1 y^2 + \frac{1}{3} B_2 y^3 + \frac{1}{4} B_3 y^4 + \ldots + \frac{1}{(n+1)} B_n y^{n+1} + C. \tag{2.5}$$

And

$$\frac{dx}{dy} = \frac{1}{\sqrt{2\left(By + \frac{1}{2}B_1 y + \frac{1}{3}B_2 y^2 + \frac{1}{4}B_3 y^3 + \ldots + \frac{1}{(n+1)}B_n y^{n+1} + C\right)}}. \tag{2.6}$$

Here C is a constant of integration. Then we can find that

$$\int dx = \int \frac{dy}{\left[2\left(By + \frac{1}{2}B_1 y^2 + \frac{1}{2}B_2 y^2 + \ldots + \frac{1}{(n+1)}B_n y^{n+1} + C\right]^{0.5}}. \tag{2.7}$$

The problem lies in evaluating the integral on the right-hand of (2.7). For the functions with a highest nonlinearity of y^3, the solution can be expressed in terms of what are known as elliptic functions.

1. Elastica forms. Euler used the following expressions to describe the elastic forms [13, 16, 21, 22]

$$\frac{dx}{dy} = \frac{a^2}{\sqrt{(c^2 - y^2)(2a^2 - c^2 + y^2)}} - \frac{c^2 - y^2}{\sqrt{(c^2 - y^2)(2a^2 - c^2 + y^2)}} \tag{2.8}$$

and

$$\frac{ds}{dy} = \frac{a^2}{\sqrt{(c^2 - y^2)(2a^2 - c^2 + y^2)}}, \tag{2.9}$$

where a and c are constants and $s = \sqrt{dx^2 + dy^2}$. Solutions of (2.8) and (2.9) may be expressed using elliptic functions. Euler classifies the solutions to (2.8) and (2.9) using the parameters a and c.

To simplify the analysis of (2.8), Levien [16] proposes the use of a single λ to replace both a and c:

$$\lambda = \frac{1}{2} a^2 c^{-2}. \tag{2.10}$$

Using only λ, Levien constructed the forms presented in Fig. 2.5 (right). We emphasize that all elastica forms obtained by Euler and many strongly nonlinear traveling waves may be described by Eq. (2.4) [28].

Thus, it is shown that Eq. (2.9) describes a wide variety of Euler figures. Therefore, let us consider it more carefully; thus, we write (2.9) in the form

$$y_s^2 = a^{-4}\left[c^2\left(2a^2 - c^2\right) + y^2\left(2c^2 - 2a^2\right) + y^4\right]. \tag{2.11}$$

Let

$$z^4 = y^4 c^{-2}\left(2a^2 - c^2\right)^{-1}. \tag{2.12}$$

Then Eq. (2.11) gives

$$z_s^2 = a^{-4} c^{-4}\left[1 - z^2 c^{-1}\left(2c^2 - 2a^2\right)\left(2a^2 - c^2\right)^{-0.5} + z^4\right]. \tag{2.13}$$

It is important that Eq. (2.13) has a real representation only if

$$a^2 \geq \frac{1}{2} c^2. \tag{2.14}$$

Taking this into account, we find that Eq. (2.13) by virtue of (2.10) can only describe the curves $\lambda \geq 0.25$ in Fig. 2.5. Consequently, Eq. (2.13) we obtained does not fully correspond to Eq. (2.9) and does not describe the curves $\lambda = 0.249$, $\lambda = 0.2$, and $\lambda = 0.1$.

We do not know the reason for this discrepancy.

In Chap. 1 (Sect. 1.3), it is shown that many equations considered in this book, in the case of considering unidirectional waves, are reduced to the form (2.11) or (2.13).

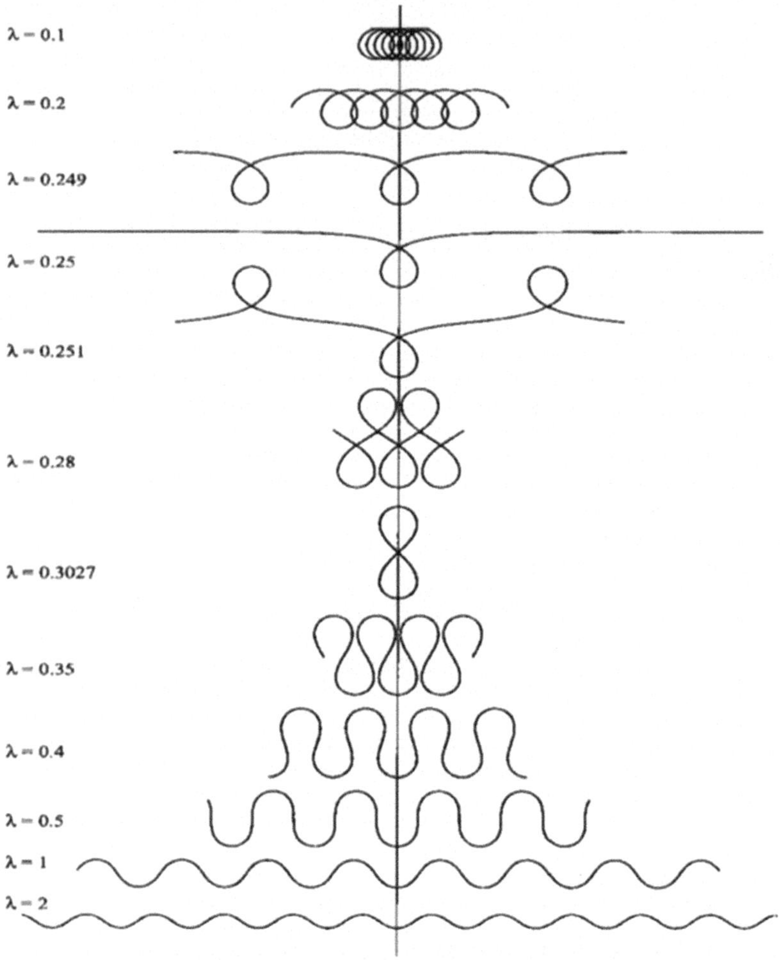

Fig. 2.5 The evolution of the elastica forms as function of certain parameters a, c (left), and λ: $\lambda = 0.1$; 0.2; 0.249; 0.25; 0.251; 0.28; 0.3027; 0.35; 0.4; 0.5; 1; and 2 (from top to bottom of the picture) (right) [16, 19, 20]

2. Let us demonstrate this using the example of nondimensional scalar waves. We write the corresponding equation (2.1) here in Cartesian coordinates

$$\varphi_{tt} - c_0^2 \sum_n^N \varphi_{nn} + m^2 \varphi - \lambda \varphi^3 = 0. \tag{2.15}$$

In this way, plane multidimensional waves are considered. Let us introduce a new variable

$$r = ct - \sum_{n}^{N} k_n x_n. \tag{2.16}$$

As a result, we obtain the equation

$$\left(c^2 - c_0^2 \sum_{n}^{N} k_n^2\right) \varphi_{rr} + m^2 \varphi - \lambda \varphi^3 = 0. \tag{2.17}$$

It can be brought to the form

$$\varphi_r^2 = \frac{1}{2} \lambda \left(c^2 - c_0^2 \sum_{n}^{N} k_n^2\right)^{-1} \left(\varphi^4 - 2\lambda^{-1} m^2 \varphi^2 + 4\lambda^{-1} C^*\right). \tag{2.18}$$

Here C^* is the constant of integration. It is necessary to emphasize that in (2.18), we use notations for constants λ and c, which, of course, do not correspond in physical meaning to the constants λ and c included in relation (2.8)–(2.14). We allowed the indicated overlap of notations so that, on the one hand, we do not deviate from the notations of other authors used in (2.8)–(2.14), and on the other hand, we do not deviate from the notations introduced in Sect. 1.4.1. Let

$$\frac{1}{2}\lambda \left(c^2 - c_0^2 \sum_{n}^{N} k_n^2\right)^{-1} = a^{-4} c^{-4},$$

$$\lambda^{-1} m^2 = c^{-1} \left(2c^2 - 2a^2\right) \left(2a^2 - c^2\right)^{0.5} \quad \text{and} \quad 4\lambda^{-1} C^* = 1. \tag{2.19}$$

Then Eq. (2.18) coincides with (2.13).

Thus, using the example of the Klein-Gordon equation, it is shown that typical wave equations considered in this book can be presented in the form (2.9) or (2.11). Consequently, a wide range of wave equations can have solutions describing the Euler figures in Fig. 2.5.

Let us consider special cases.

3. It is easy to see that in the one-dimensional case, Eq. (2.1) is reduced to the form

$$\begin{aligned} &\varphi_r^2 + \frac{1}{2} \lambda \left(-c^2 + c_0^2\right)^{-1} \left(\varphi^2 - \lambda^{-1} m^2\right)^2 \\ &+ \frac{1}{2} \lambda \left(-c^2 + c_0^2\right)^{-1} \left(4\lambda^{-1} C - \lambda^{-2} m^4\right) = 0. \end{aligned} \tag{2.20}$$

Here $r = ct - x$. Let

$$\xi = \sqrt{\frac{1}{2}\lambda\left(-c^2 + c_0^2\right)^{-1}}r. \tag{2.21}$$

Then

$$\varphi_{\xi}^2 = -\left(\varphi^2 - \lambda^{-1}m^2\right)^2 - \left(4\lambda^{-1}C - \lambda^{-2}m^4\right). \tag{2.22}$$

Let $-\lambda^{-1}m^2 = 2c$ и $4\lambda^{-1}C - \lambda^{-2}m^4 = B^2$.
In this case, we can rewrite Eq. (2.22) in the form

$$\varphi_{\xi}^2 = B^2 - \left(\varphi^2 + 2c\right)^2. \tag{2.23}$$

Solutions are built using elliptic functions. The cases may be classified as follows:

3.1. $c = -1$ and $B = 2$ (Fig. 2.6)

It is easy to see that this figure corresponds well to Euler figure 10 in Fig. 2.1.

3.2. $c = -1$ and $0 < B < 2$ (Fig. 2.7)

3.3. $c = -1$ and $B > 2$ (Fig. 2.8)

It is easy to see that Fig. 2.8 corresponds well to Euler figures 7, 8, and 9 in Fig. 2.1. We have shown that the Klein-Gordon equation describes the elastica-like forms. Thus, the different elastica-like waves can be in the scalar fields.

In certain cases, the surface tension is responsible for the appearance of such waves. Thus, a behavior of the systems consisting of two or more adjoining environments (phases) can be determined by a thin layer of the contact between these environments (phases). For example, in the case of very short surface waves, mechanical properties of such surface are characterized, first of all, by the surface

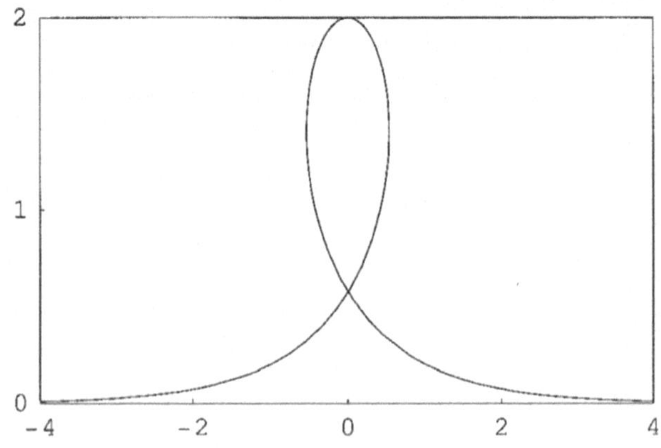

Fig. 2.6 Euler soliton [19, 23]

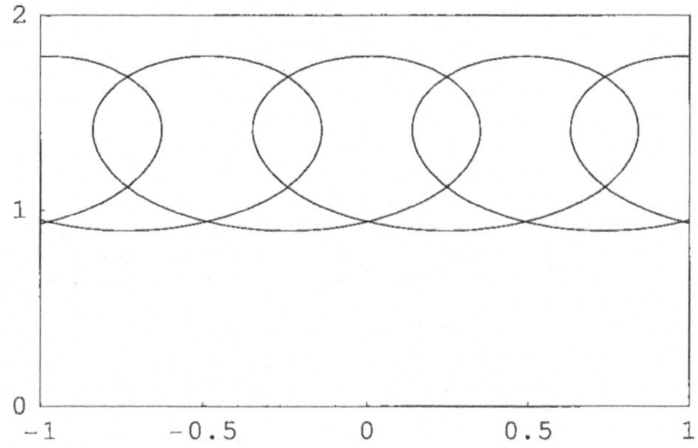

Fig. 2.7 The elastica form modeling the case $\lambda = 0.2$ in Fig. 2.5 (right) [19, 20, 26]

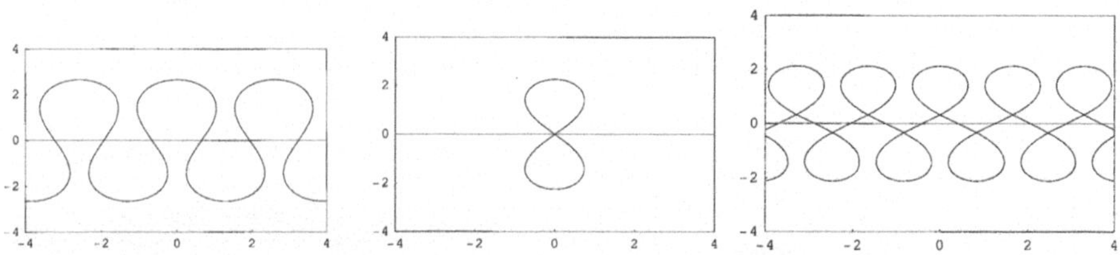

Fig. 2.8 The elastica forms modeling the cases $\lambda = 0.4$, 0.3027 and 0.28 in Fig. 2.5 (right) [19, 20, 26]

tension. This tension defines properties of a surface similar to properties of a thread (string).

2.2 Euler Figures and Extreme Multivalued Waves: Examples, Equations, and Unified Solutions

We presented above the set of equations. It is possible to develop further the theory of extreme multivalued волны (catastrophic) waves being based on any of them. A well-known example of similar waves are the overturning waves. However, they are unstable. Internal catastrophic waves and catastrophic waves on an interface of two different phases of liquid may be more stable; for example, internal oceanic waves can exist for a long time. Their stability is explained by the influence of the environment surrounding a wave. Figure 2.9 shows the appearance of similar waves in the liquid flowing in the channel. The originally smooth profile is transformed into a mushroom-like form. Then the last is supplcmented by the curls. Further, the flow loses stability and transforms into turbulence. These processes are not shown in Fig. 2.9.

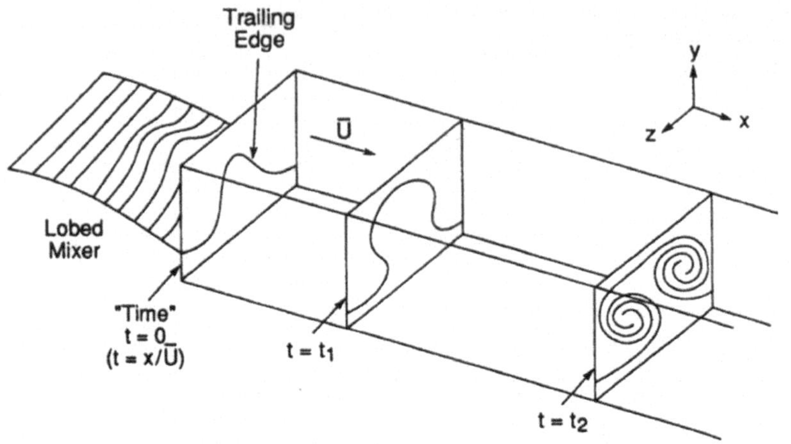

Fig. 2.9 Development of internal waves in a liquid flowing through a channel [29]

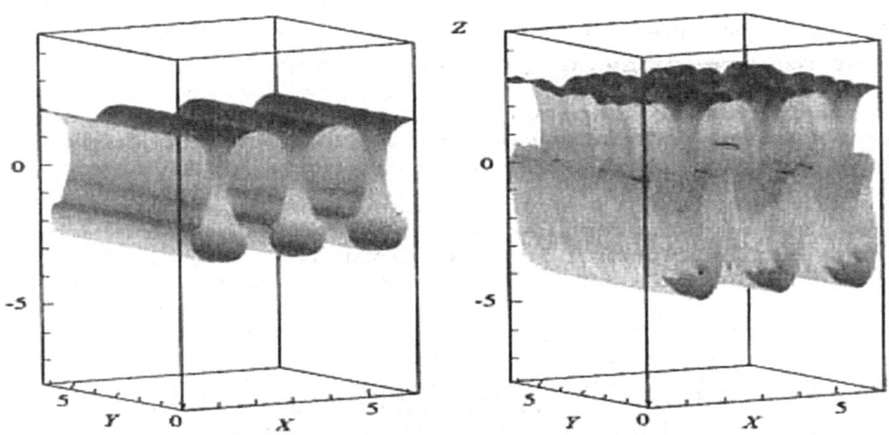

Fig. 2.10 Two moments of an evolution of an interface of two media. The evolution of the initially harmonic perturbation (the amplitude 0.013) generates the "elastica" form with the amplitude 2.5 (at the left), which then is transformed into the three-dimensional wave pattern (on the right) [30] (see also Figs. 2.5 and 2.8)

Figure 2.10 demonstrates the instabilities of an interface of two liquids which leads to the turbulence of the interface surface. In particular, the "elastica" form and the mushroom-like waves can serve as some elements of this process.

Initial perturbations evolve into vortices in wave processes presented in Figs. 2.11 and 2.12.

Many studies are devoted to similar processes. In particular, the evolution of the "elastic" forms into vortices is possible during the impulsive and periodic processes [20].

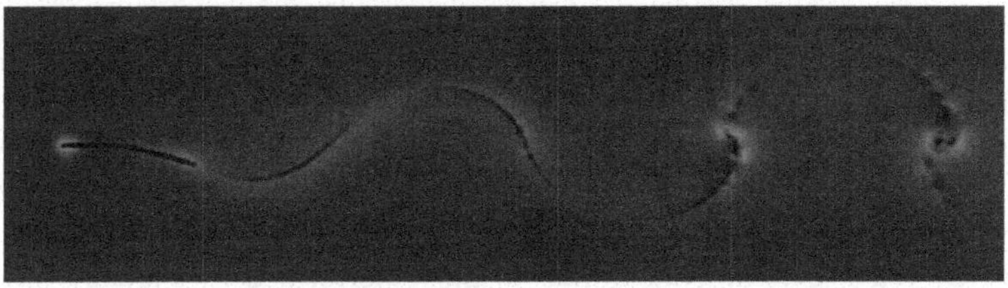

Fig. 2.11 Photo of Mark Murray and Laurens Howl (Duke University, CNCS) [20]

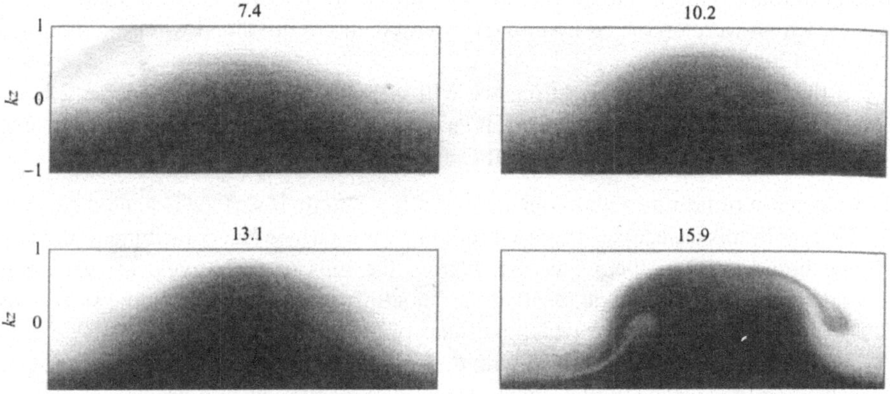

Fig. 2.12 Evolution of a two-dimensional breaking interfacial wave and the generation of vortices. The wave propagates from left to right [19, 20]

The wave dynamics of Figs. 2.9, 2.10, 2.11, and 2.12 resembles the evolution of figure −2 to figure −0.35 as shown in Fig. 2.5.

Multivalued waves and their dynamics describe processes which were observed in many experiments. In particular, this dynamics describes a transformation of waves into vortices. More complex processes will be shown below (see, for example, Figs. 2.39, 3.30, 3.34).

2.2.1 Strongly Nonlinear Effects and Unification of Their Description and Analysis

During the last two decades, some surprising nonlinear wave phenomena have been observed on the free surfaces of some vertically excited media. We have earlier considered some of those phenomena. Now we continue our consideration by studying surface forced waves in resonators.

Our task in this section is to give an idea of the possible mechanisms of the evolution of surface waves into jets and particles, in other words, the mechanisms of evolution of smooth waves into discontinuous and multivalued ones.

These mechanisms are demonstrated by the example of vertically excited surface waves. It is important to note that these waves were first explored by Michael Faraday in 1831 [4].

In recent years, experiments similar to those performed by Faraday have been carried out for cases of strong and very strong vertical oscillations. Both waves on the free surfaces of liquids and on the interfaces of different liquids were studied. It was found that with an increase in the amplitude of the vertical excitation, smooth waves, which were observed by Faraday, evolve into multivalued waves in the form of Euler figures.

At the beginning of this section, we will very briefly describe the schemes and results of experiments known to us. It is easy to imagine that their multivalued results could take place in cases where significant capillary or surface forces exist.

1. In the case of liquids, situations are possible when their surface is heated close to boiling. In this case, even slight vibrations of the surface can contribute to the fact that liquid particles begin to detach from the tops of the waves. In particular, under the action of vertical oscillations, standing waves can form on the surface of the liquid, from the top of which particles of the liquid can separate. This process can be associated and explained with the simultaneous existence of short capillary waves on the surface, which, when focused, generate jets and drops.

Some surprising nonlinear waves may be observed on free surfaces of some vertically excited liquids (Fig. 2.13; see also Fig. 3.29). The curvature of the spherical meniscus (crater) is reversed violently as a result of the impact. The multivalued wave is formed, and atomization of the surface can have a place. In particular, research showed the possibility of pyramidal waves forming on a liquid surface. Crests of the pyramidal waves can sometimes erupt drops and jets.

Figure 2.14 shows a sequence of video frames of a drop which was vibrated in the vertical direction in its fundamental axisymmetric mode [31]. The excited amplitude is slowly increased from image (a) to image (f). The amplitude held fixed for a period long enough so that transient motions, associated with the increasing amplitude, can disappear. The frequency is fixed at 987 Hz. The forms of the surface waves and the

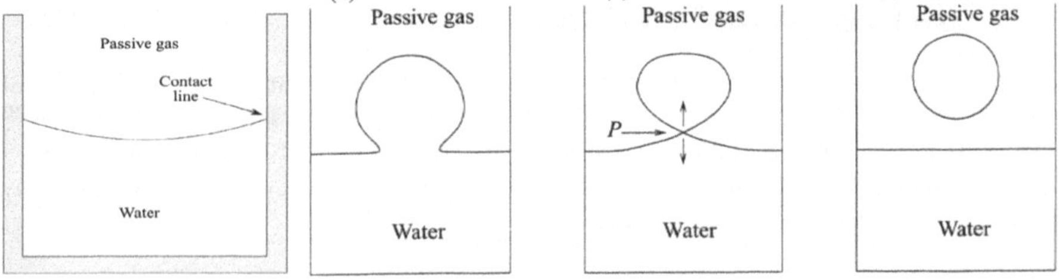

Fig. 2.13 Scheme of axial impact of a cylindrical tube filled with a liquid wetting the tube wall

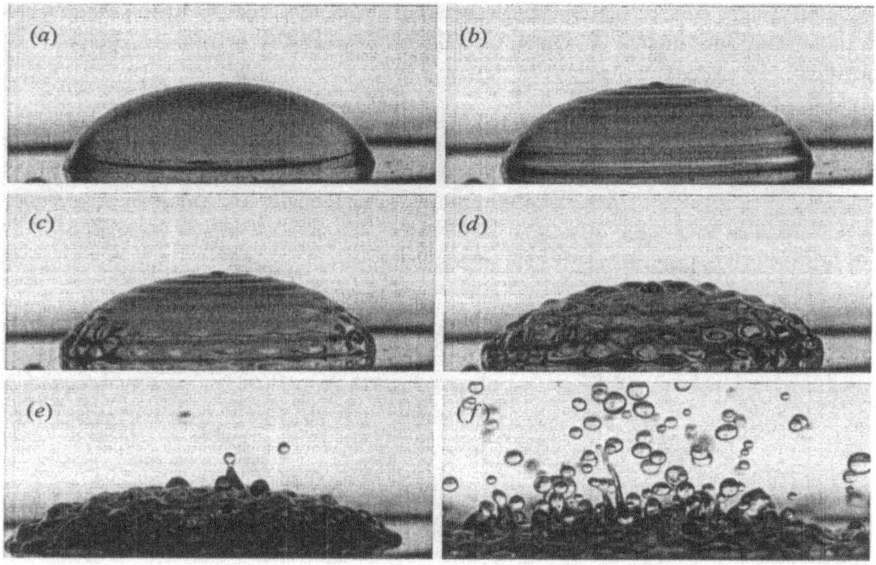

a

b

Fig. 2.14 Examples of atomization and surface highly nonlinear waves. Vertically excited waves and drops on the surface of a 100 microliter liquid drop (diameter about 8 mm). (*a*) The undisturbed drop, (*b*) axisymmetric standing waves on the free surface at small excitation amplitude, (*c*) the appearance of stationary and then slowly rotating azimuthal waves on the free surface, (*d*) intense time-dependent wave motion over the entire drop free surface, (*e*) the appearance of time-dependent craters, liquid spikes, and pyramid-like waves, (*f*) the rapid ejection process of small secondary droplets (bursting) and the formation of mushroom-like waves [31] (**a**). A qualitative pattern of an appearance of the virtual particle in vacuum or a separation of new closed Universe from a big volume of space [20, 32] (**b**)

small drops, shown in Fig. 2.14, are determined by surface tension and some resonant conditions. The process can serve as some model for many natural wave phenomena. In particular, we can image the appearance of similar waves, craters, and spikes on the sea surface during seaquakes.

Perhaps a similar, unexpected appearance of particles can occur not only on the surfaces of various media but also inside their depths, for example, during some unknown resonant processes occurring in a vacuum. It is known that vacuum is no static environment. Vacuum permanently fluctuates in subatomic scales because of unpredictable quantum pulses. There are virtual particles which spontaneously appear and immediately disappear. There is the concept that the origin of Universe can remind a generation of virtual particles. Can we, using the given analogy, put in compliance with speculative theories of the origin of the Universe of the visible resonant process of the atomization of a water drop? The theory of this analogy is presented in monographs [19, 20, 32]. Its development is given in the last chapter of this book.

The process of atomization of the liquid surface as a result of the transresonant process was modeled in [20, 30]. A parameter R which was introduced there determined the width of the resonant band as R varied from 1 to -1, with $R = 0$ at resonance. The results of the calculations for four values R located very near $R = 0$ are presented in Fig. 2.15.

It is emphasized that Fig. 2.15 describes qualitatively some peculiarities of wave processes shown in Fig. 2.14. In particular, the eruption of water jets from the wave tops is determined by some resonant condition. However, the reader can also see

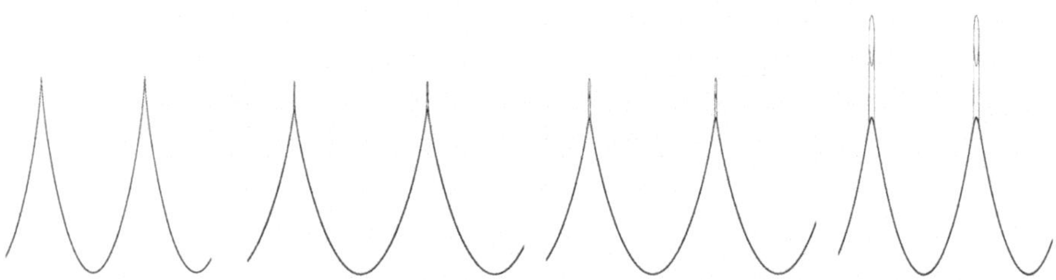

Fig. 2.15 The eruption of particles and jets (streams) from the top of waves when the transresonant parameter R varies from positive to negative values very near $R = 0$ [20, 30]

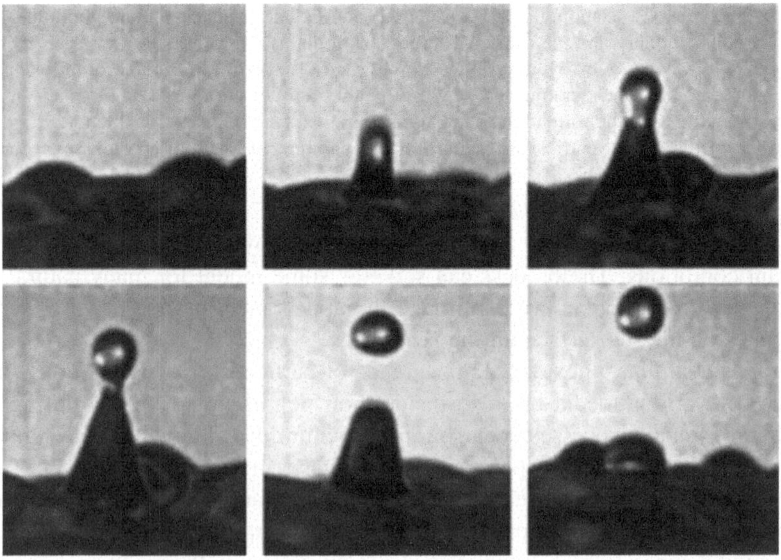

Fig. 2.16 A time sequence of images of a small part of the surface (2 × 2 mm) [20, 30, 31]

from Figs. 2.14 and 2.16 that the jets are formed by collapses of water craters (see also figures in Chap. 3).

When droplet ejections begin from the free drop surface (Fig. 2.14), the wave motion on it appears to be mostly chaotic. During this motion, the free surface depressions, or craters, are formed. They collapse and ultimately eject the secondary droplets (Fig. 2.16).

Thus, we have shown that structures resembling Euler figures arise not only during the deformation of elongated structural elements but also in liquid media. This is very important for this study. Of course, no less important is the question of ways to mathematically describe such wave structures.

Below we will give an analytical technique for the mathematical study of extremal multivalued waves, which will be widely used in cases of one-dimensional waves, and in some cases, this technique and its results will be used in relation to non-one-dimensional problems.

It is known that the analytical solution of nonlinear equations is associated in the general case with almost insurmountable difficulties, which are often associated with the interaction of linear and nonlinear effects. Another difficulty is related to the presence of areas where the solution gives infinity for certain parameters of the problem. These are the so-called singular or resonant regions. However, in these singular regions, the influence of nonlinearities increases so strongly that the influence of the most important linear terms can be neglected (the influence of the d'Alembert operator practically disappears). It greatly simplifies the mathematical solution of the equations. Additionally, in this chapter, we limited ourselves to considering unidirectional waves. In this case, as shown in the previous chapter, the study of a wide range of strongly nonlinear wave problems is reduced to solving an equation of the form

$$\varepsilon J_{rr} + B_1 J + B_2 J^2 + B_3 J^3 + \ldots. + B_n J^n + F(r) = 0. \tag{2.24}$$

Here $F(r)$ is an arbitrary function. In the vicinity of resonances and singularities, the linear effects become much weaker than the nonlinear ones. This allowed us to limit ourselves to solving Eq. (2.24) for cases when $\varepsilon J_{rr} = 0$ and the value $B_1 J$ is small. The smallness of this value, in fact, determines the range of resonances of the cases considered in this chapter.

Now, we rewrite Eq. (2.24) in the form:

$$r^* J^5 + J^3 + r_* J^2 + RJ + F(r) = 0. \tag{2.25}$$

We have received the algebraic equation which can be solved by simple numerical methods. If the term $r^* J^5$ is zero, this equation has analytical solutions.

The Wave Solution of the Cubic Algebraic Equation Let the term r^*J^5 be ignorable in (2.25). In this case, (2.25) takes the form

$$J_*^3 + r_*J_*^2 + RJ_* + F(r) = 0. \tag{2.26}$$

Further, periodical resonant wave processes will be considered. We assume that $F(r) = -\sin \omega r$, $r = t - c_0^{-1}x$ and c_0 is the sound speed. If $R = 0$, we have exact linear resonance. Let us enter the designations

$$J_* = J - r_*/3 \tag{2.27}$$

and

$$p_* = R - r_*/3\ , q_* = 2r_*^3/27 - r_*R/3 - F(r). \tag{2.28}$$

As a result, Eq. (2.26) yields

$$J^3 + p_*J + q_* = 0. \tag{2.29}$$

It is necessary to distinguish four cases, when Eq. (2.29) has real solutions:

1. Let $p_* = 0$, then Eq. (2.29) is satisfied if

$$J = (-q_*)^{1/3}. \tag{2.30}$$

2. Let $p_* > 0$, then the function J is unique, single-valued, and continuous

$$J = -2D \sinh\left[1/3 \operatorname{arcsin} h\left(0.5q_* D^{-3}\right)\right], \tag{2.31}$$

   where $D = [sign(q_*)](|p_*|/3)^{0.5}$.
3. Let $p_* < 0$ and $q_*^2/4 + p_*^2/27 \leq 0$. In this case, there is no continuous single-valued solution, and a solution with discontinuities may be constructed. However, nature often manifests multivalued solutions (e.g., breaking waves and turbulence). Indeed, nonlinear systems often exhibit two or more dynamic equilibrium states for the same values of parameters. Some states may be chaotic, whereas others are periodic. We will construct here a regular multivalued solution with the help of the following smooth single-valued solutions:

$$J_i = -2D \cos\left[1/3 \arccos\left(0.5q_* D^{-3}\right) + 2N\pi/3\right], \tag{2.32}$$

   where $N = 0$, 1, and 2. Using these continuous solutions, we can construct multivalued solution.

4. Let $p_* < 0$ and $q_*^2/4 + p_*^2/27 > 0$, then we have

$$J = -2D\cosh\left[1/3\operatorname{arccos}h\left(0.5q_*D^{-3}\right)\right]. \quad (2.33)$$

We presented only the real solutions of (2.29).

Remark The presented solutions define a wave traveling on one side. The wave traveling on other side is being determined in a similar way.

We will be mainly interested in cases where $p_* < 0$ and $q_*^2/4 + p_*^2/27 \leq 0$, and we have harmonic solutions (2.32). It is very important that they are solutions not only to the Eq. (2.29), but they also approximately satisfy the linear terms of the wave equations of the form $\varphi_{tt} - c_0^2\varphi_{xx} = 0$.

2.2.2 Euler Figures and Transresonant Development of Multivalued Waves into Mushroom-Shaped Structures, Plumes (Jets), and Bubbles

This section is devoted to the calculation of multivalued resonance waves and the qualitative description of the strongly nonlinear recently observed surface waves. Thus, we begin to use the solutions (2.30), (2.31), (2.32), and (2.33).

The curves of Figs. 2.17, 2.18, and 2.19 are calculated according to $r_* = 0$ in (2.26). In Figs. 2.17a, b and 2.18a, b, the curves separated by the intervals are calculated according to the solutions (2.32) and (2.3). These curves allow to con struct the continuous multivalued periodic waves as shown in Figs. 2.17c and 2.18c. The drawing technique is similarly used by us during drawing of the cubic nonlinear waves presented in Fig. 2.15. Harmonic curves (b) in Figs. 2.17 and 2.18 are determined by the linearized equation (2.26).

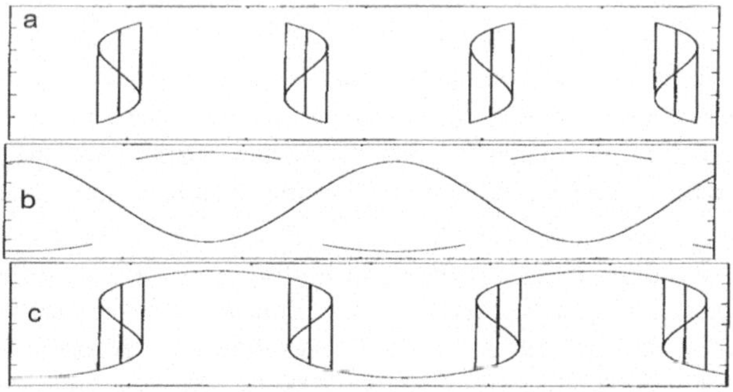

Fig. 2.17 Multivalued continuous mushroom-like waves (**c**) determined by solutions of Eq. (2.26) and parameters $R = -0.5$, $t = 0$. The waves are constructed by the multivalued solution (2.32) (**a**) and the discontinuous solution (2.33) (**b**). The harmonic wave (**b**) is determined by the linearized equation (2.26) and parameters $R = -0.5$, $t = 0$ [19, 30, 33]

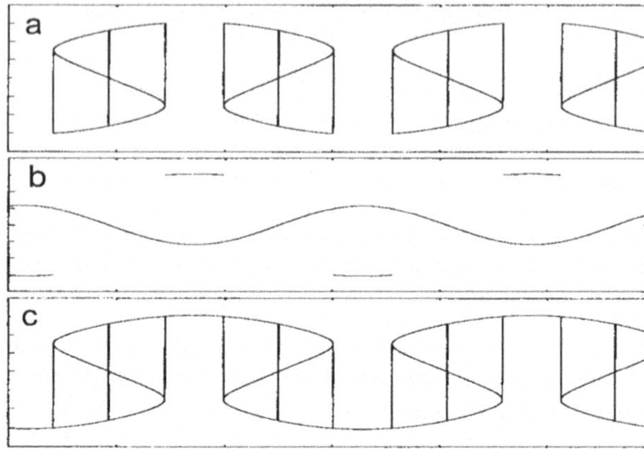

Fig. 2.18 Multivalued continuous mushroom-like waves (**c**) determined by solutions of Eq. (2.26) and parameters R = −0.9, t = 0. The waves are constructed by the multivalued solution (2.32) (**a**) and the discontinuous solution (2.33) (**b**). The harmonic wave (**b**) is determined by the linearized equation (2.26) and parameters R = −0.9, t = 0 [19, 30, 33]

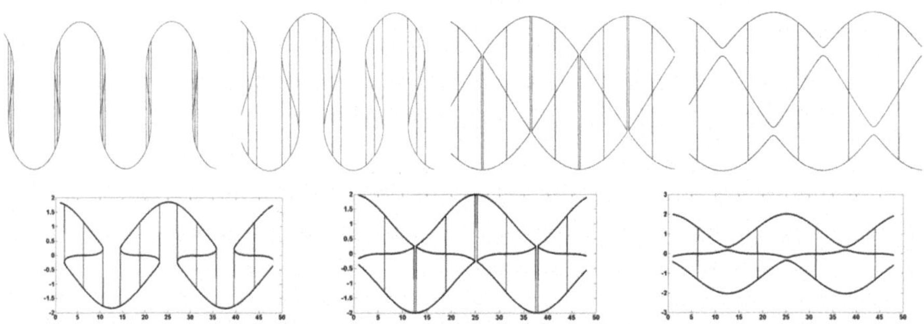

Fig. 2.19 Multivalued wave calculated according to the analytical solutions of Eq. (2.26). The upper line demonstrates the generation ($R = -0.2$, the left side), the transresonant evolution ($R = -0.7; -0.999$), and the bifurcation ($R = -1.01$. the right side) of the waves. It is assumed that $r_* = 0$ and $t = 0$ in (2.26). The waves having jets on the tops may appear near the point of bifurcation ($R = -0.999$). The cnoidal-type and pyramid-type waves may be generated as a result of the bifurcation ($R = -1.01$). The low line of the curves are calculated for another wave parameter and $R = -0.7$; -0.89 and -0.92 [30, 34]

Let us underline that the continuous curves having the folds in pictures (a) of Figs. 2.17 and 2.18 are formed by three solutions (2.32). The vertical lines determine the points where the instantaneous jump from one solution to another solution is possible. As a whole, the specified curves and the lines form the closed circuits. We can assume that small particles of the wave surface move along these curves and lines. In particular, they may be particles of liquids (gas). It is possible that some of them make the jump while the others form the continuous wave profile with the fold. The jumps can be upward and downward. Therefore, it is possible to consider the closed, insulated circuits a in Figs. 2.17 and 2.18 as some rough models of curls (see also Figs. 2.9 and 2.24). We will often use this analogy further.

It is possible to consider curves (c) in Figs. 2.17 and 2.18, as models of waves having vortices (curls) in the both sides of the waves [19, 20, 30] (see, also, Figs. 3.34, 4.23 and 4.25).

Very roughly, these models describe the waves shown in Figs. 2.9, 2.10, 2.11, and 2.12. It is important for us that according to the developed theory, an appearance of such waves is determined by the transresonant process. We will continue to study this process in different chapters of this book. We also note that the curves in Figs. 2.17c and 2.18c remind the elastica forms as shown in Figs. 2.1 and 2.5. Thus, the presented mushroom-like waves can be generated also during the impulse compression of highly deformable flexible bodies (Fig. 2.4). More possibly, such waves can arise on a liquid surface or on an interface of two media (Figs. 2.9, 2.10, 2.11, and 2.12). The curvature of the waves depends on proximity to an exact resonance as it follows from a comparison of Figs. 2.17c and 2.18c. We assumed that the exact linear resonance corresponds to $R = 0$.

The transresonant evolution of the mushroom-like waves is shown in Fig. 2.19. This form of the waves exists in an interval $0 < R < -1$. Cnoidal and pyramid-like waves arise near the resonant band ($R = -1.01$). Our calculations have also shown that near $R \approx -1$, at the moment of formation of the pyramidal waves, there is a possibility of eruption of the jet from the wave top. Thus, Fig. 2.19 demonstrates the transresonant evolution of harmonic waves into elastica forms and the appearance of pyramidal and cnoidal waves near the boundary of resonant band, where $R \approx -1$.

In Fig. 2.20, we take into consideration quadratic nonlinear properties of the interface. Thus, the quadratic nonlinear term is taken into account in (2.26). Those data show the dependence of waveforms from r^* and R. In Fig. 2.20 (left) the value r^* changes from -0.3 (Fig. 2.20f) to 0.75 (Fig. 2.20c). This change transforms a mushroom-like wave into particles and bubble-like structures (Fig. 2.20b, c, f), and to cnoidal (Fig. 2.20d) and harmonic waves (Fig. 2.20e). Forms of the particles also depend on R.

The dependence from r_* and R is shown in Fig. 2.20 (right). Parameter R changes from -0.3 to -1.2. The evolution also depends on the quadratic nonlinearity. In particular, for $r^* = 0.1$, there is one structure similar to a drop, but for $r^* = 0.4$, three structures having a drop are formed during the transresonant development of the mushroom-like waves. If $-1 > R < -1.2$, then the pyramidal wave is transforming into the harmonious wave.

Thus, the transresonant development of the mushroom-like waves into particles, drops, bubbles, and waves varied in the form as shown Fig. 2.20. Perhaps, this dynamics is connected with transform waves into vortices. Really, we have given the most simple interpretation of the solutions of Eq. (2.26). More complex interpretation of them includes the possibility of jumps in zones of multivalues of the wave profiles. Similar jumps may provide the effect of dry friction. At last, because of the catastrophic nature of these waves, the transresonant evolution of them in a direction from $R = -1$ to $R = 1$ differs from the evolution in a direction from $R = 1$ to $R = -1$. It very much complicates the analysis of extreme waves. At the same time, this complexity inspires a hope that "elastica" forms have much more big area of applicability than the description of a strong bend of rods and strings.

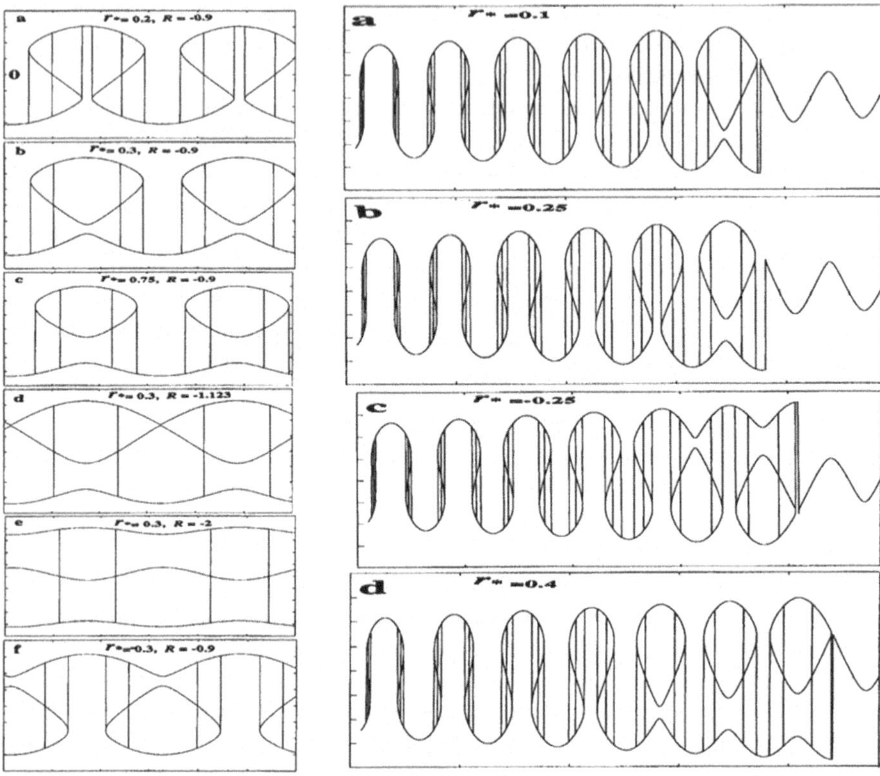

Fig. 2.20 The appearance of the mushroom-like waves and a transformation of them into particles (drops) above the wave crests (**b**, **c**) or into the pyramidal waves and the harmonic waves (**d**, **e**), or bubbles below the wave troughs (**f**) (left). The transresonant evolution of the catastrophic multivalued waves and effect of the quadratic nonlinearity. The transresonant evolution of the mushroom-like waves into particles (drops) (**a**, **b**, **d**) or bubble-like structures (**c**), if $R > -1$, and the pyramidal and harmonic waves, if $R < -1$ (right) [19, 30, 33]

Conclusion According to the cubic nonlinear equation (2.26), particles leave the interface if $r^* > 0$. The bubbles appear if $r^* < 0$. Occurrence of the streams over tops of waves is possible. These processes can be defined by the capillary effect (cubic nonlinearity) and by the elastic properties of the surface. Change of the exciting periodic force can strongly change the result of the calculations. However, according all our calculations, the atomization of the surface may be determined by its resonant excitation.

Let us notice that sizes of the particles can be rather small. The shorter a wave, the stronger the capillary effect and the smaller the arising particles. It is emphasized that the occurrence of particles, drops, bubbles, and streams on the interface during free or forced vibrations is a problem of big practical, scientific, and technological interest. The capillary effect has often crucial importance for these processes.

2.2.3 Atomization of Liquid Surfaces, the Emergence of Particle-Waves and Vortices as a Manifestation of Multivalued of Nature's Decisions

The indicated features of extreme multivalued waves (EMW) were studied on the basis of the cubic algebraic equation. That is, based on the cubic algebraic equation (2.26), which qualitatively, in our opinion, describes the resonant properties of the cubic nonlinear Klein-Gordon equation. Let the last equation also contain a nonlinear term of the fifth degree. It will correspond to an algebraic equation of the fifth degree. Below we will consider and compare solutions for the two indicated cases, following the first work devoted to EMW and published in 2001 [28].

Thus, we consider the equations

$$(J')^3 + \left(3R/2^{2/3}\right)J' + \cos \omega r = 0 \tag{2.34}$$

и

$$A_2(J')^5 + (J')^3 + \left(3R/2^{2/3}\right)J' + \cos \omega r = 0\ . \tag{2.35}$$

First, we consider transresonant evolution of waves. It is assumed that $R = 0.03\omega r$, where ωr varies from -2 to -45 (Fig. 2.21a), or $R = -0.8 + 0.2 \cos (\omega r/4)$ (Fig. 2.21b, c).

The cubic nonlinearity generates loop-like (mushroom-like) structures which bifurcate into three harmonic curves when $R \approx -1$ (see Fig. 2.21a). These harmonic curves are tied by vertical lines. The fifth-order nonlinearity defines the more

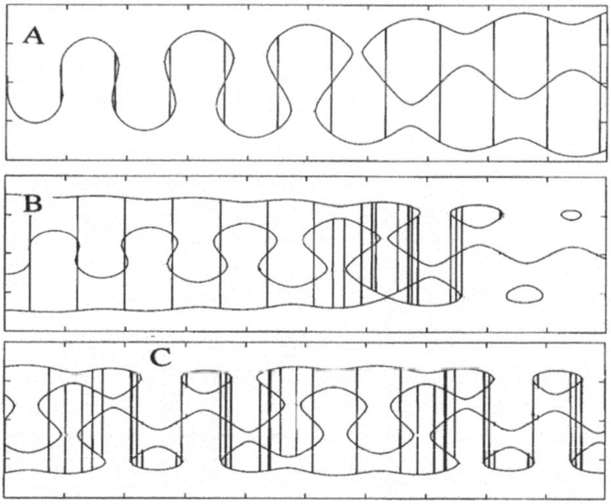

Fig. 2.21 The transition of ripples into mushroom-like waves and cloud-like structures. It was calculated according to cubic equation (2.34) (**a**) or fifth-order equation (2.35), where A2 = −0.11 (**b** and **c**) [28]

interesting picture. The mushroom-like formations generate the ellipsoid-like clouds when $R \approx -1$. The fifth-order nonlinearity periodically generates these clouds (see Fig. 2.21c), if R oscillates near $R \approx -1$. More detailed results of the numerical calculations can be seen in Fig. 2.22, where assumed $R = const.$

One can see the transformation of the complex structure ($R = -0.81$, $A_2 = -0.1095$) into a chain of particle-waves similar to the Karman vortex street ($R = -0.81, A_2 = -0.15$).

Then we are showing in Fig. 2.23 that the complex structure from Fig. 2.22 ($R = -0.81, A_2 = -0.1095$) contains mushroom- and breaker-like waves, and spiral structures.

We showed that under the simple harmonic excitation, the systems can manifest very complex behavior [35–37].

Thus, it was shown that in highly nonlinear media, the ripples can be smoothly transformed and amplified into mushroom-like waves (Euler Fig. 7 (Fig. 2.1)), spiral- and ellipsoidal-like structures, vortices, and jets. Of course, these are only preliminary results. They will be clarified in the future. We will focus on it in

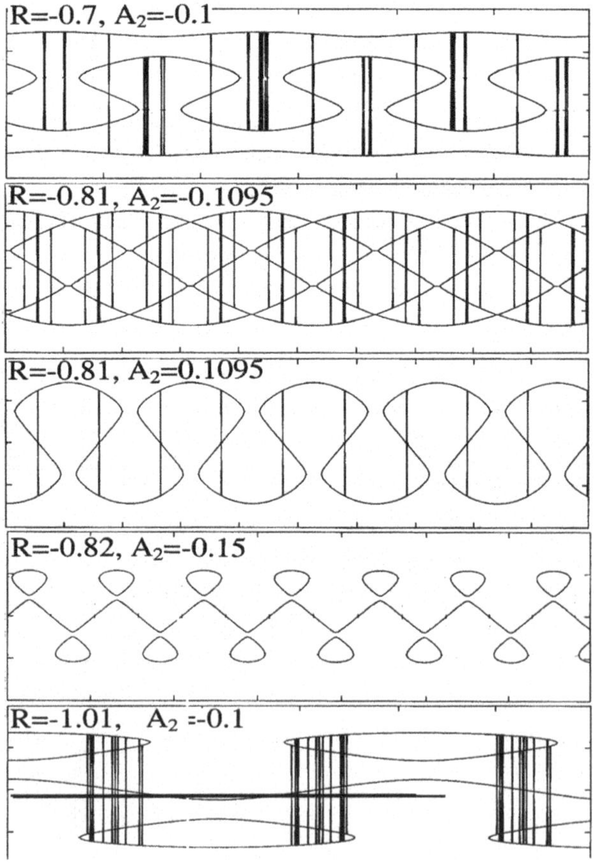

Fig. 2.22 Effect of the fifth-order nonlinearity on mushroom- and cloud-like structures [28]

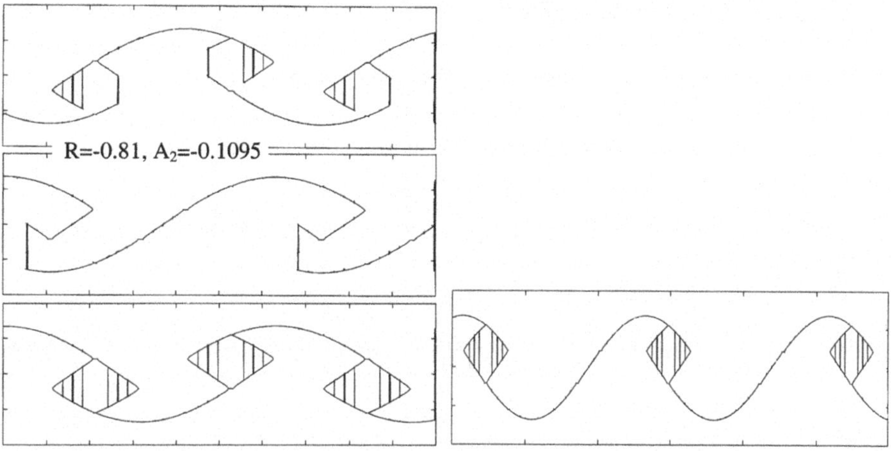

Fig. 2.23 Resonant mushroom- and breaking-like waves, and spiral-like structures generated by the fifth-order nonlinearity if $R \approx -1$ [28]

particular in the next section using published theoretical and experimental data concerning the evolution of waves into vortices in the gas filling the pipe.

2.3 Effects of Viscosity on the Evolution of Multivalued Waves into Vortices

The problem of the evolution of harmonic perturbations into eddies is very difficult for analytical research. It is enough to say that it is very close to the problem of the origin and evolution of turbulence from a laminar stationary flow.

The considered examples show that taking into account the cubic nonlinearity of wave processes makes it possible to simulate the evolution of waves into eddies. This requires certain resonance conditions. Additionally, the influence of friction on the development of the process may be important. If these conditions are satisfied, many wave equations can describe the process of arising wave turbulence.

2.3.1 Simulation of Chester and Moser Results [38]

Let us consider the influence of viscosity on the wave evolution.

1. First, consider a two-layer liquid having approximately equal layer thicknesses (see Eq. (1.61)). We will not take into account the dispersion, but we will introduce the eddy viscosity into the equation following [38–40]. Thus, we take into account the friction between liquids on the contact surface.

As a result, we get the equation

$$N_{tt} - a_0^2 N_{xx} = \beta_1 N_x^2 N_{xx} + s_* N_t |N_t|. \tag{2.36}$$

Here $\beta_1 = -\alpha_2 a_0^2 c_0^{-1}$. We will consider unidirectional waves. Let

$$r = at - x. \tag{2.37}$$

Then, (2.36) gives

$$\left(a^2 - a_0^2\right) N_r = \frac{1}{3}\,\beta_1 N_r^3 + s_* a^2 \int N_r |N_r| dr. \tag{2.38}$$

Let

$$N = J(r) + J_1(r). \tag{2.39}$$

We will consider $J(r)$ as some seed term or, in other words, some initial perturbation of the system unknown to us. Subsequent consideration will allow us to judge the deviations of the system from the initial state. We accept that $J(r) \ll J_1(r)$. In addition, we assume that $a^2 \approx a_0^2$. Thus, we consider the case when the resonance condition $a^2 \approx a_0^2$ is satisfied. Since the initial perturbation is very small, (2.38) and (2.39) give

$$\left(a^2 - a_0^2\right) J_r + \left(a^2 - a_0^2\right) J_{1r} = \frac{1}{3}\,\beta_1 J_{1r}^3 + s_* a^2 \int J_{1r} |J_{1r}| dr. \tag{2.40}$$

If $J_r = A \sin kr$, the equation is similar to the Chester equation (2.41). The last is the cubic nonlinear model of the viscous gas oscillated in the tube:

$$-\sin\tau = g^3 + 3R(2)^{-2/3} g + s_2 \int g|g| d\tau. \tag{2.41}$$

Here the integral corresponds to the eddy viscosity. Equation (2.41) describes the occurrence of vortices in the gas (Fig. 2.24) if some resonant condition is satisfied.

2. Now let us consider the Klein-Gordon equation. First, we introduce in it a term taking into account the eddy viscosity and assume $r = ct - x$.

$$\left(c^2 - c_0^2\right) \varphi_{rr} + m^2 \varphi - \lambda \varphi^3 + s_* \int \varphi |\varphi| dr = 0. \tag{2.42}$$

We further accept that $\varphi = J(r) + J_1(r)$ and $J(r) \ll J_1(r)$. Thus, $J(r)$ is considered as the seed perturbation. As a result, we obtain the required equation which has the form (2.40). If $J(r) = A \sin kr$, then (2.42) yields

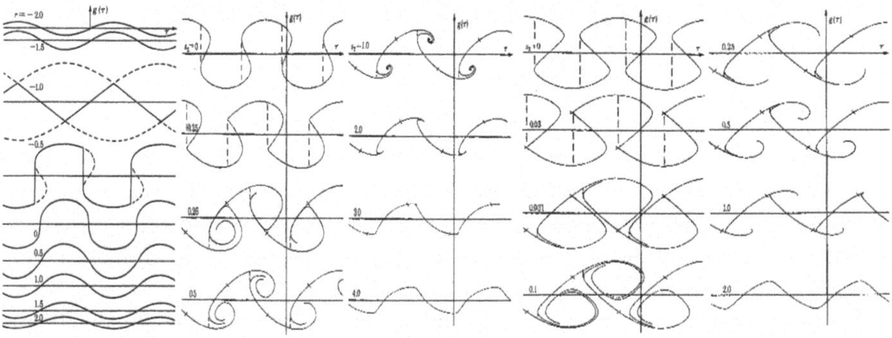

Fig. 2.24 The appearance of the mushroom-like waves within the resonant band according to Chester's model (left). Effects of s_2 on the wave evolution into vortices: $R = -0.5$ (center) and $R = -0.9$ (right)

$$-\left(c^2 - c_0^2\right)Ak^2 \sin kr + m^2 J_1 - \lambda\varphi J_1^3 + s_* \int J_1|J_1|dr = 0. \tag{2.43}$$

Now following the results of Chester and Moser [38], we can illustrate the effect of viscosity on extreme multivalued waves. Of course, this influence is greatly increasing in areas of resonance.

3. Probably, many strong nonlinear wave phenomena may be explained by various wave resonances. A simple example is presented in [41]. The resonant oscillations of gas in pipes were researched. The waves were excited by the piston. It was found that periodical vortices can appear in the gas near the wall of the tube [41]. The wave turbulence arose at some frequencies lying within the resonant band.

Although many similar phenomena are looked as extremely complex, they can be described sometimes by simple mathematical expressions. For example, Chester and Moser [19, 38] have used the simple equation (2.41) to describe the evolution of waves into vortices in the tube (Fig. 2.24).

Examples of the transresonant evolution of waves are shown in Fig. 2.24. Within the resonance band, the harmonic wave transforms into the elastica-like (mushroom-like) waves. These waves may be transformed into vortices. Results of the transformation depend on coefficients s_2 and R (2.41). Thus, Fig. 2.24 illustrates some scenario of the evolution of waves into vortices. This scenario explained qualitatively the appearance of the vortices which may be important elements of the wave turbulence. In our case, this evolution corresponds to the succession: the harmonic perturbation, the weakly nonlinear waves, the overturning elastica-like waves, and the vortices and the turbulence.

The authors have special interest in the results of Chester and Moser [38]. We decided to generalize these results to research the transformation of the elastica-like waves into vortex structures.

2.3.2 *Simulation of the Richtmyer-Meshkov Instability Results (One-Frequency Seed)*

Above are the results of many amazing experiments that can only be described by developing the theory of extreme multivalued waves. But perhaps most of the results of experimental studies of the Richtmyer-Meshkov instability [42] are surprising. We will return to these results more than once in this book (see Sect. 3.4). Here we present only some results of the instability directly related to the problem of the emergence of vortices (Figs. 2.9, 2.11, 2.24, and 2.25).

For many years, I have been trying to understand and describe mathematically the results of an experimental study of the Richtmyer-Meshkov instability. Some results of mathematical modeling are shown in Fig. 2.25. This section is devoted, first of all, to analytical modeling of the evolution of waves into vortices.

Sinusoidal Seed Disturbance At first, we notice that, modeling vortices, Chester [38, 40] used the nonlinear (eddy) viscosity. However, it complicates both the subsequent calculation and the analysis of the origin of vortices. Therefore, we here use the linear viscosity. We begin our study from the following equation:

$$J_*^3/3 + 2^{-2/3}RJ_* + \mu \int J_* dr \; + 2^{-2/3}J = 0. \tag{2.44}$$

In this section, we assume that $r = at - x$. The function J_* is an unknown traveling wave, J is the forcing wave (initial perturbation), the term $\mu \int J_* dr$ takes into account the linear viscosity, and $R = R(x, t)$ is the transresonance parameter. It is important to underline that R depends on the coordinate and time. For simplicity, we have assumed that the forcing function can be defined by one harmonic. It was assumed that $J = AR \sin \omega r$, $A = 0.3$, $R = A_1 - \alpha\omega r$, and $0 < \alpha\omega r < 0.7$. Thus, $J = AR \sin \omega r$

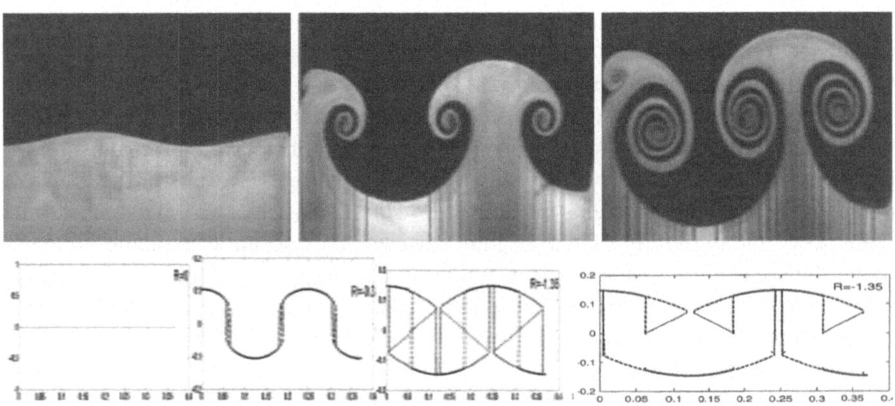

Fig. 2.25 The comparison of experimental data [19, 20, 42] with calculations. Here R is some transresonant parameter (see also Figs. 2.23 and 2.24)

is considered as seed disturbance and R is changed in the process of wave motion. A case is considered when the wave runs in some resonant band. At $r = 0$, the transresonant parameter R is positive, but further, in the process of wave motion, it becomes negative. It is possible to expect that this variability and the weak viscosity define the possibility of an occurrence of vortices.

First of all, we underline that the viscous effect may be important often only near some singular points, for example, near the shock fronts (see [38]). Therefore, the basic profile of a wave is being described by the equation:

$$J_*^3/3 + 2^{-2/3}RJ_* + 2^{-2/3}RJ = 0, \tag{2.45}$$

far enough from these points.

Now we return to Eq. 2.44. Let us begin with its differentiation. As a result, we have

$$\begin{aligned} &\left(J_*^2 + R/2^{2/3}\right)J'_* + 2^{-2/3}R'J_* + \mu J_* \\ &+ 2^{-2/3}\omega[-A\alpha \sin \omega r + A(A_1 - \alpha\omega r)\cos \omega r] = 0\ . \end{aligned} \tag{2.46}$$

Hereinafter, the stroke designates a derivative, $J'_* = \partial J_*/\partial r$. Eq. (2.46) has the singularities on the lines defined by R, where

$$J_* = J^+ = \sqrt{-R/2^{2/3}}\ , \tag{2.47}$$

$$J_* = J^- = -\sqrt{-R/2^{2/3}}\ . \tag{2.48}$$

The lines defined by Eqs. (2.47) and (2.48) are qualitatively shown in Fig. 2.26 as the straight lines. Generally speaking, they can cross the thick lines defined by Eq. (2.45).

In Fig. 2.26, these crossings are showed by the black spots (points). Let us remind that we investigate an occurrence of the vortices as a result of crossing of the resonant band. Therefore, the transresonant parameter R determines the wave evolution as shown in Fig. 2.26. In particular, the elastica-like waves begin to form only if $R < 0$. The points of crossings and the black spots in Fig. 2.26 define those parameters of the process at which the origin of the vortices is possible.

Further, we will investigate Eq. (2.46) closely near the singular lines there, where it is possible to write,

$$r = r_0 + z, \tag{2.49}$$

Here r_0 is some crossing point near which Eq. (2.46) is studied. For simplicity, we will write below r instead of r_0. Then, we use following representations J^+ or J^-:

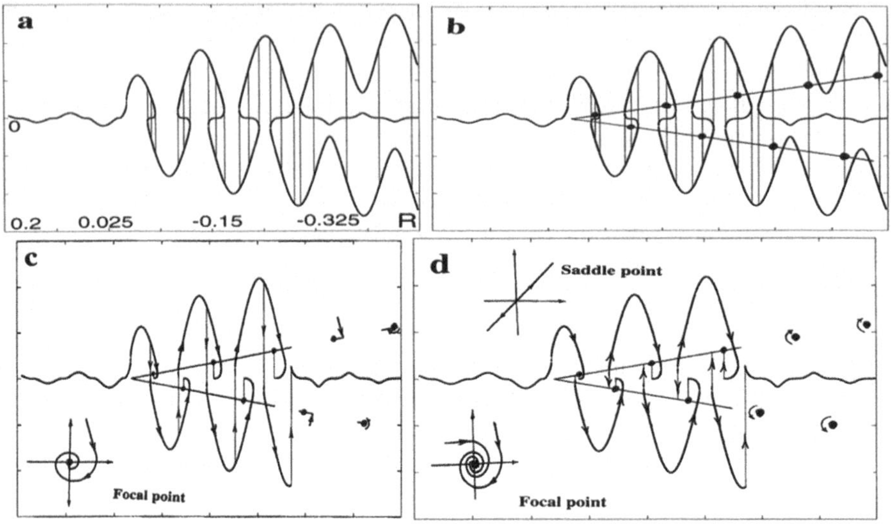

Fig. 2.26 Transresonant evolution of a harmonic disturbance into elastica-like waves and vortices is described by Eq. (2.45). Thick lines are the sum of $J_* + J$. The straight thin lines on the pictures **b**, **c**, and **d** are the singular lines. The singular points are shown by the black spots. Nonlinearity and friction can begin to form the vortices, when $R \approx 0$ [43]. The pictures **a**, **b**, **c**, **d** define different models of calculations

$$J_* = J^+ = g + \sqrt{-R/2^{2/3}}, \tag{2.50}$$

$$J_* = J^- = g - \sqrt{-R/2^{2/3}}. \tag{2.51}$$

Here g is a very small perturbation of $\sqrt{-R/2^{2/3}}$. We substitute the last expressions in (2.46). It is important that g is a quick variable function at the vicinity of a considered point and R is almost a constant. Therefore, in the considered vicinity, $R' \approx 0$. As a result, Eq. (2.46) gives

$$\begin{gathered}\left(g^2 \mp 2g\sqrt{-2^{-2/3}R}\right)g' + \mu\left(g \mp \sqrt{-2^{-2/3}R}\right) \\ +2^{-2/3}\omega(-A\alpha \sin \omega r + AA_1 \cos \omega r - A\alpha\omega r \cos \omega r) = 0.\end{gathered} \tag{2.52}$$

We will ignore here the small cubic nonlinear term $g^2 g'$. Then

$$\begin{gathered}\mp 2\sqrt{-2^{-2/3}R}gg' + \mu\left(g \mp \sqrt{-2^{-2/3}R}\right) \\ + 2^{-2/3}\omega(-A\alpha \sin \omega r + AA_1 \cos \omega r - A\alpha\omega r \cos \omega r) = 0.\end{gathered} \tag{2.53}$$

Let us consider the following expression in (2.53):

$$\mp 2^{1/3}\,\mu\sqrt{-\;R} + \omega A(-\,\alpha \sin \omega r + A_1 \cos \omega r). \tag{2.54}$$

It connects the initial disturbance with R and viscosity. In the process of the wave motion, the value (2.54) is changing. In particular, (2.54) may be a zero. To illustrate these cases, we assume that the term $\mp 2^{1/3}\mu\sqrt{-R}$ and the value A_1 in (2.54) are small. Then, (2.54) gives

$$\omega r \approx \pi N\ . \tag{2.55}$$

Here $N = 0, \pm 1, \pm 2, \pm 3, \ldots$ Thus, if the viscous effect is very small, Eq. (2.53) is simplified as

$$g' = \pm 2^{-2/3}(-\,R)^{-1/2}\left(\mu g - 2^{-2/3}\omega^2 A\alpha\omega r \cos \omega r\right)g^{-1}. \tag{2.56}$$

Let

$$E \approx 2^{-2/3}(-\,R)^{-0.5}\mu \;\text{ and }\; a = 2^{-4/3}\omega^3 A(-\,R)^{-0.5}\cos \omega r. \tag{2.57}$$

We study a small vicinity of a point lying on singular lines. Therefore, $a = 2^{-4/3}\omega^3 A(-R)^{-0.5}$ cos ωr in (2.57) may be considered as a constant. Eq. (2.56) becomes a bilinear equation

$$g' = (\mp ar \pm Eg)/g. \tag{2.58}$$

Equation (2.58) has the singular points. They are defined by eigenvalues λ of the Eq. (2.58):

$$\lambda^-_{1,2} = 0.5\left(-\,E \pm \sqrt{E^2 + 4a}\right), \tag{2.59}$$

$$\lambda^+_{1,2} = 0.5\left(E \pm \sqrt{E^2 - 4a}\right). \tag{2.60}$$

We underline that the eigenvalues $\lambda^-_{1,2}$ (2.59) correspond to the singular line (2.50), and $\lambda^+_{1,2}$ (2.60) correspond to the singular line (2.51). The behavior of the solutions of Eq. (2.58) depends on $\lambda^{\mp}_{1,2}$. According to (2.59) and (2.60), the singular points correspond to nodes or saddles, if $\sqrt{E^2 \pm 4a}$ is real. It is the case of large viscosity or nonresonant waves. When the root becomes imaginary (E reduces), the unstable nodes can be transformed into the spirals (focuses). This evolution is determined by R, A, A_1 and ω (2.54).

For example, if $E = 0$ (exact resonance), we have $\lambda^-_{1,2} = \pm 2i\sqrt{|a|}$. Then the singular point is a focus. If we have $\lambda^+_{1,2} = \pm 2\sqrt{|a|}$, a singular point is a saddle. If

E (or R) is decreased from zero and $a < 0$, then the focus can be transformed into a spiral (vortex). If $N = 1$, the spiral is located on the singular lines (2.50), and the saddle is located on (2.51). The opposite situation takes place if $N = 2$ and $a > 0$. Generally speaking, if N is odd, then the spirals (vortices) are located on (2.50) and the saddles on (2.51). If N is even, then the spirals (vortices) are located on (2.51) and the saddles on (2.50).

The formation and the development of the vortices are qualitatively shown in Fig. 2.26c, d. It is found that Fig. 2.26c, d describes the results of experiments (Figs. 2.9, 2.11, 2.24, 2.25, and 2.27) better than Fig. 2.26a, b. In particular, the profiles in Fig. 2.26c, d qualitatively describe the formation and the evolution of the waves and the vortices in Figs. 2.9, 2.10, 2.11, and 2.12. In Fig. 2.26c, d, the vortices form the structure similar to the Karman vortex street.

Thus, waves and vortex waves having various amplitudes and forms can be generated within a resonant band. However, the vortex waves can arise only if the transresonant parameter R is a negative value. In addition to many figures of this part of the book, Fig. 2.26 shows the hysteresis effect. Thus, because of the vertical jumps in multivalued zones (folds), the different dynamics arises during increasing or decreasing of R. Therefore, the results of calculations can be interpreted differently because they depend on the direction of crossing of the resonant band.

We showed the essential influence of nonlinearity on the appearance of vortices. The effect of viscosity is also very important. However, according to the presented theory, the most important effect is determined by a resonance. Negligible small

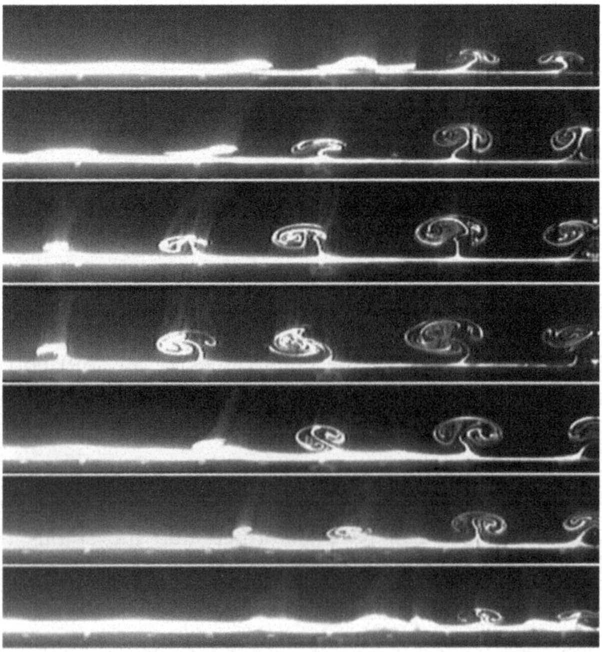

Fig. 2.27 Photographs of some sequences of developments of small disturbances (not for equal time intervals) into mushroom-like structures. They grow in size and number when the flow speed increases [19, 30, 44]

disturbances and viscous effects can lead to the appearance of "explosive" turbulence in a resonant band.

2.3.3 Simulation of the Richtmyer-Meshkov Instability Results (Two-Frequency Seed)

Below, we will attempt to model the wave evolution demonstrated in Fig. 2.25 additionally. We take into account that Fig. 2.26 describes the Richtmyer-Meshkov instability excited due to one harmonic on the interface of media. The occurrence of more complex vortex structures (Figs. 2.25, 2.52 and 3.30) may be explained by two seed harmonics [19, 30].

First, following [19, 28], we will consider the following equation:

$$-0.11J_*^5 + J_*^3 + 3R/2^{2/3}J_* + 3R/2^{2/3}J = 0. \tag{2.61}$$

The results of the numerical solution of this equation are presented in Fig. 2.28. It is assumed that $R = -0.2 + 0.002Y(r) + \tanh[Y(r)/50]$, where Y changes according to the linear law from 30 to -120. Waves in Fig. 2.28a, c are calculated for the initial disturbance $J = A\cos 0.5r$, where $A = 0.3$ (Fig. 2.28a) or $A = 0.7$ (Fig. 2.28c). Waves in Fig. 2.28b are found for $J = 0.3(1 + 0.75\cos 0.5r)\cos 0.5r$.

It is visible that within a resonant band, the continuous initial disturbances can bifurcate into the closed loops which can be interpreted as vortices. They form a semblance reminding Karman's vortex street. The growth of amplitude A from 0.3 to 0.7 (Fig. 2.28c) strongly complicates the transresonant wave evolution. As a result, within the resonant band, shock-like waves, breakers, and mushroom-like waves can be generated. However, behind a resonant band, the picture becomes simpler and may be interpreted as Karman's vortex street, pyramidal waves, and saw-like waves.

Different interpretations of the results presented in Fig. 2.28 are possible. In the context of our research, we connect them with the appearance of vortices. It is easy to see some analogy of the wave evolution presented in Fig. 2.28c, and an instability of the interface of two liquids (Figs. 2.25, 2.52 and 3.30) which generates the beautiful vortex structures. The impressing result of Richtmyer-Meshkov's instability

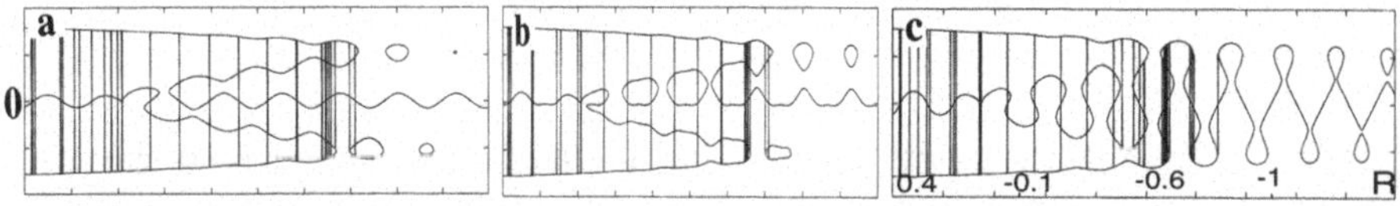

Fig. 2.28 Transresonant development of the initial disturbances and generation of the mushroom-like waves, breakers, and a cluster of the vortices described by the fifth-order algebraic equation (2.61) [28]. The pictures **a**, **b**, **c** define different models of calculations

considered together with Fig. 2.28c illustrates the strong influence of high nonlinearity on the occurrence of vortices and turbulence. One can see certain correlations between experimental and theoretical data.

We have decided that two forcing harmonics in Eq. (2.46) allow us to describe qualitatively the evolution shown in Fig. 2.56 and Chap. 3. Let us check this assumption. For this purpose, we used the algorithm of calculation which generalizes a technique of Sect. 2.3.2 on a case of two seed harmonics. Let $J = A(A_1 - \alpha\omega r)(\sin\omega r + A_3 \sin 3\omega r)$ in (2.46):

$$\left(J_*^2 + 2^{-2/3}R\right)J_{*'} + 2^{-2/3}R'J_* + \mu J_* - 2^{-2/3}A\alpha\omega(\sin\omega r + A_3 \sin 3\omega r)$$
$$+2^{-2/3}\omega AA_1(\cos\omega r + 3A_3\cos 3\omega r) - 2^{-2/3}\alpha\omega^2 rA(\cos\omega r + 3A_3\cos 3\omega r) = 0. \tag{2.62}$$

According to Sect. 2.3.2, we found very near the singular points

$$\mp 2\sqrt{-2^{-2/3}R}\,gg' + \mu\left(g \mp \sqrt{-2^{-2/3}R}\right) - 2^{-2/3}A\alpha\omega(\sin\omega r + A_3\sin 3\omega r)$$
$$+2^{-2/3}\omega AA_1(\cos\omega r + 3A_3\cos 3\omega r) - 2^{-2/3}\alpha\omega^2 rA(\cos\omega r + 3A_3\cos 3\omega r) = 0. \tag{2.63}$$

Further, the expression is written

$$\mp\sqrt{-2^{-2/3}R\mu} - 2^{-2/3}A\alpha\omega(\sin\omega r + A_3\sin 3\omega r)$$
$$+\,2^{-2/3}\omega AA_1(\cos\omega r + 3A_3\cos 3\omega r). \tag{2.64}$$

In the vicinity of the singular points, the value $(\cos\omega r + 3A_3 \cos 3\omega r)$ changes slowly and may be considered as a constant. In this case, the expression (2.64) may be given zero for some values A_1, A_3, α, μ, ωr, and R .

As a result, for this case, Eq. (2.63) gives

$$\mp 2\sqrt{-2^{-2/3}R}\,gg\ + \mu g - 2^{-2/3}\alpha\omega^2 rA(\cos\omega r + 3A_3\cos 3\omega r) = 0. \tag{2.65}$$

Then (2.65) is rewritten in the form (2.58):

$$g' = (\mp ar \pm Eg)/g, \tag{2.66}$$

$$E \approx 2^{-2/3}(-R)^{-0.5}\mu \quad \text{and} \quad a = 2^{-4/3}\omega^3 A(-R)^{-0.5}(\cos\omega r + 3A_3\cos 3\omega r). \tag{2.67}$$

Thus, modeling of the evolution of two initial harmonics into vortices was simplified to Eq. (2.66). The details of the analysis are given in Sect. 2.3.2. The results of calculations are presented in [19, 30, 43].

These results are also presented in Figs. 2.29, 2.30, and 2.31 (thick lines). The thin lines determine the singular lines $J^{\pm}$.

Figure 2.29 demonstrates the appearance on the wave profile ($R = -0.094$) of the closed loops which can be interpreted as some prototype of vortices. The complex mushroom-like waves appear further when ($R < -0.1$).

The transresonant evolution of the initial disturbances is shown in Fig. 2.30. This figure illustrates additionally the dynamics of Fig. 2.29.

The vortex formation is connected with viscosity. If the viscosity is big enough, then the expression (2.64) is always much more zero. In this case, the expression (2.64) defines the curve which does not cross the singular lines (2.50) and (2.51) [19, 30, 43]. The corresponding curve is presented Fig. 2.30b (the wavy line). Therefore, the research is not reduced to Eq. (2.66), and the formation of vortices is impossible. However, if the friction is insignificant, the amplitude of the wavy line can be large. In this case, the wave profiles of Figs. 2.29 and 2.30 can be unfair at the points of intersection of thick and thin lines (at the singular points). There, an appearance of vortices is possible.

Results of further research are presented in Fig. 2.31. Thus, the wavy line can begin to cross the singular lines (2.50) and (2.51) near the black spots (Figs. 2.29,

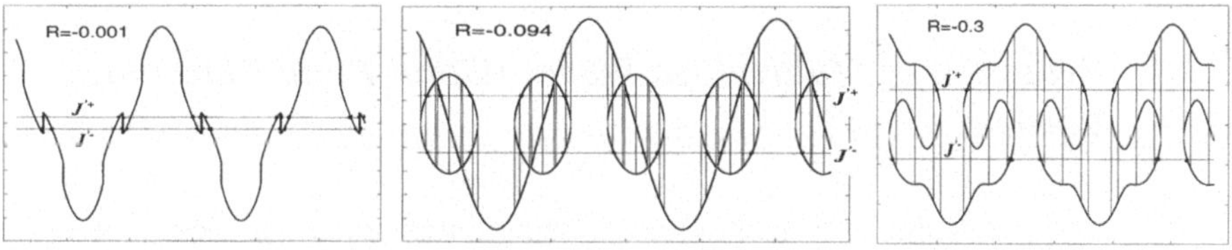

Fig. 2.29 The wave resonant mushroom-like structures corresponding to various values of the transresonant parameter R. Singular (thin) lines are parallel to the axis [43]

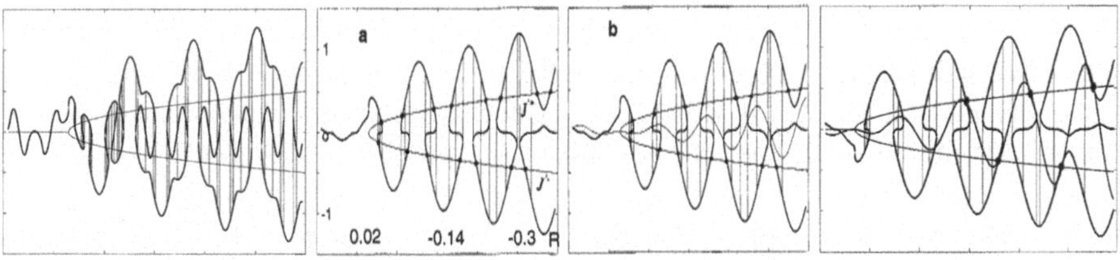

Fig. 2.30 An example of the transresonant evolution of two harmonic disturbances into mushroom-like structure within the range $-0.4 < R < 0.1$. The point of the bifurcation of the straight line into "the two-tooth plug" formed by singular lines is well distinguishable (left). An example of the transresonant evolution of one harmonic disturbance into mushroom-like structure within the range $-0.4 < R < 0.1$ (**a**, **b**). The point of the bifurcation of the straight line into "the two-tooth plug" formed by singular lines is well distinguishable (right) [19, 30, 43]

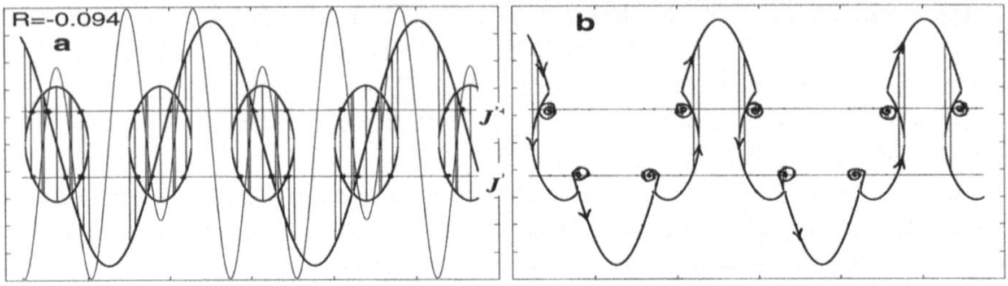

Fig. 2.31 Localization of vortices (**a**) and qualitative simulation of Richtmyer-Mechkov instability (**b**) [19, 43]

2.30, and 2.31). Near these spots, Eq. (2.63) is transformed into (2.66), which can have the spiral solution. Physically, it corresponds to the formation of vortices.

Of course, we have examined here qualitatively the connection between the appearance of extreme multivalued waves, the possibility of their evolution in the turbulence, and the connection of these processes with turbulence. We believe that all these processes are well adjusted by the Faraday waves and the Richtmyer-Meshkov instability. We will return to these very interesting connections many times. In particular, the Richtmyer-Meshkov instability has a dedicated section in Chap. 3.

2.3.4 Evolution of Multivalued Waves Within Fluid Flow into Vortices

Above, using the example of the nonlinear Klein-Gordon equation, we showed that if certain resonance conditions are met, almost any nonlinear wave equation can be reduced to a nonlinear ordinary differential equation. This equation may have a solution that describes the multivalued wave (MW). In cases where this solution contains a singularly growing term, the MW can evolve into a vortex structure. All of the above can be extended to waves propagating in flows. This is a very interesting and important problem. However, we do not have the opportunity to study it in any detail. We will present here only some of the results following [19, 30, 34].

Let us look at Eq. (1.3) which approximately describes waves and flows of various continuous mediums. In one-dimensional case, Eq. (1.6) gives

$$\varphi_{tt} - c_0^2\varphi_{xx} + 2\varphi_x\varphi_{xt} + (X-1)\varphi_t\varphi_{xx} + \tfrac{1}{2}(X+1)\varphi_x^2\varphi_{xx} = 0. \qquad (2.67.1)$$

We wrote down a very complex equation. To obtain even a very approximate solution, it must be greatly simplified. Moreover, we need to obtain solutions that describe processes resembling turbulent ones. The simplest way to find such solutions is to reduce Eq. (2.67.1) to an equation of the form (2.45). To realize this idea we will accept

$$\varphi = \varphi^{(1)} + \varphi^{(2)} + U - u_0 x. \tag{2.67.2}$$

Here $\varphi^{(1)}$, $\varphi^{(2)}$, and U are unknown functions. We assume that $\varphi^{(1)} \gg \varphi^{(2)}$ and $\varphi^{(2)} \gg U$. Value u_0 defines the average wave speed. Equation (2.67.2) is substituted into Eq. (2.67.1). As a result, we simplify our research to the analysis of the following system of equations:

$$\varphi^{(1)}{}_{tt} - c_0^2 \varphi^{(1)}{}_{xx} - 2u_0 \varphi^{(1)}{}_{xt} = 0, \tag{2.67.3}$$

$$\varphi^{(2)}{}_{tt} - c_0^2 \varphi^{(2)}{}_{xx} - 2u_0 \varphi^{(2)}{}_{xt} + 2\varphi^{(1)}{}_{xt}\varphi^{(1)}{}_{xt} + (X-1)\varphi^{(1)}{}_{t}\varphi^{(1)}{}_{xx} = 0, \tag{2.67.4}$$

$$\begin{aligned}
&U_{tt} - c_0^2 U_{xx} - 2u_0 U_{xt} + 2\left(\varphi^{(2)} + U\right)_x \varphi^{(1)}{}_{xt} \\
&\quad + 2\left(\varphi^{(1)} + \varphi^{(2)} + U\right)_x \varphi^{(2)}{}_{xt} + 2\left(\varphi^{(1)} + \varphi^{(2)} + U\right)_x U_{xt} + (X-1)\Big[\left(\varphi^{(2)}\right. \\
&\quad \left. + U\right)_t \varphi^{(1)}{}_{xx} + \left(\varphi^{(1)} + \varphi^{(2)} + U\right)_t \varphi^{(2)}{}_{xx} + \left(\varphi^{(1)} + \varphi^{(2)} + U\right)_t U_{xx}\Big] \\
&\quad + ½(X+1)\Big[\left(\varphi^{(1)} + \varphi^{(2)} + U\right)_x^2 \varphi^{(1)}{}_{xx} + \left(\varphi^{(1)} + \varphi^{(2)} + U\right)_x^2 \varphi^{(2)}{}_{xx} \\
&\quad + \left(\varphi^{(1)} + \varphi^{(2)} + U\right)_x^2 U_{xx}\Big] = 0
\end{aligned} \tag{2.67.5}$$

Now our goal is to write Eq. (2.67.5) in the simplest possible form. Let's look at the first two equations first. They are solved by the method proposed in [30, 45]. It is presented in an expanded form in Sect. 3.1. We enter the variables

$$r = (C_* - u_0)t - x\,, \quad s = (C_* + u_0)t + x. \tag{2.67.6}$$

Here

$$C_*^2 = c_0^2 + u_0^2. \tag{2.67.7}$$

Using (2.67.6) and following [30, 45], we receive the following solutions of Eqs. (2.67.3) and (2.67.4):

$$\varphi^{(1)} = J(r) + j(s), \tag{2.67.8}$$

$$\begin{aligned}
\varphi^{(2)} &= J_2(r) + j_2(s) - \frac{1}{4} C_*^{-1}(X+1) J_r^2 x + \frac{1}{4} C_*^{-1}(X+1) j_s^2 x \\
&\quad + \frac{1}{48} C_*^{-2}[2(C_* - u_0) - (X-1)(C_* + u_0)] jJ_r \\
&\quad + \frac{1}{4} C_*^{-2}[2(C_* + u_0) - (X-1)(C_* - u_0)] Jj_s.
\end{aligned} \tag{2.67.9}$$

Expressions (2.67.8) and (2.67.9) determine the coefficients of Eq. (2.67.5). If we take into account only the resonance terms in (2.67.9), then (2.67.9) yields

$$\varphi^{(2)} = J^*(r) + j^*(s) - \frac{1}{4}C_*^{-1}(X+1)J_r^2x + \frac{1}{4}C_*^{-1}(X+1)j_s^2x. \tag{2.67.10}$$

Next, we will consider unidirectional waves. For an unidirectional wave

$$\varphi^{(1)} + \varphi^{(2)} = J(r) - \frac{1}{4}C_*^{-1}(X+1)J_r^2x. \tag{2.67.11}$$

We introduce (2.67.11) into (2.67.5). Then approximately

$$\begin{aligned} &U_x^3 + 3(c_* + u_0A)U_x^2 + 6\Big\{(X+1)^{-1}\left(\overline{C}_*^2 - c_0^2 + 2u_0\overline{C}_*\right) \\ &+ (c_* + u_0A)\left[J - \frac{1}{4}C_*^{-1}(X+1)J_r^2x\right]_x\Big\}U_x \\ &= \frac{3}{2}(c_* + u_0A)C_*^{-1}(X+1)\left(J_r^2x\right)_x J_x + \overline{C} \end{aligned} \tag{2.67.12}$$

The next task is to solve it. To do this, consider unidirectional waves, the speed of which $\overline{C}_*$ is very close to the difference $C_* - u_0$. Thus, the resonance is studied. The influence of some nonlinearities, dissipation, and dispersion was not considered. We are interested in the influence of the initial perturbation on the wave evolution, first of all. Discarding small terms, we write (2.67.12) in the following three forms:

$$U_x{}^3 + 3\beta_1^{-1}\left(-\widehat{C}^2 + C^2\right)U_x + J_x{}^3 = 0, \tag{2.67.13}$$

$$U_x^3 + 3\left[\beta_1^{-1}\left(-\widehat{C}^2 + C^2\right) + \beta\beta_1^{-1}J_x\right]U_x + J_x^3 = 0, \tag{2.67.14}$$

$$U_x^3 + 3\left[\beta_1^{-1}\left(-\widehat{C}^2 + C^2\right) + \beta\beta_1^{-1}J_x + J_x^2\right]U_x + J_x^3 = 0, \tag{2.67.15}$$

A value U_x was calculated. The value $u_a^{(2)}$ was not considered. Thus, the following formula was used:

$$\varphi_x = -\varepsilon\cos Kr + U_x. \tag{2.67.16}$$

We will not dwell further on the description of constants and calculation techniques. Their details can be found in [19, 30].

Thus, a complex of the simplifications outlined will be used in the analysis. We also remind that the governing equation (2.67.12) was introduced as the model equation for the rough description of many experiments. Therefore (2.67.12) pretends to describe many experiments. Thus, it can describe each of them very roughly. Certainly, there are small chances to describe the complex experiments successfully if to use such a simplified approach. At last, during the calculations, we will meet a problem of uncertainty of the equation coefficients. All these restrictions of used mathematical apparatus aggravate difficulties in the description of experiments and observations by the developed theory. However, we have no other way of estimation of justice of the theory, besides comparing its results with experiments. Thus, during the calculations, we will be orientated on the data of experiments.

Let us now present some calculation results.

The results of the calculation according to (2.67.13) are presented in Fig. 2.32. Curves 1 and 2 were calculated for the initial (seed) perturbations ($\varepsilon = 0.235$ or 0.35) and $K = 0.6$. They demonstrate the evolution of the initial small perturbation into mushroom-like waves. These waves break up in the end of a resonance band into three curves. The interesting fork well viewed in curves 1 is being formed. The fork middle (the mushroom-like waves) is defined by the fast wave evolution. Below curves 1 and 2, the experiment data [19, 30, 44] are allocated. As a whole, the experimental curves coordinate with the calculation.

Results of an attempt at more detailed research of the vortex wave evolution are presented in Fig. 2.33. The calculations model the experiments presented in this figure, and Eq. (2.67.14) was used. Curves A are defined by the calculations, and curves A1 are defined by our interpretation of the calculations.

According to Fig. 2.33, curves A1, qualitatively, describe some stages of the wave evolution into vortices observed in the experiment (Fig. 2.33, the top).

Let us now consider the calculations and the experiments presented in Fig. 2.33 jointly. We remind that various interpretations of the results are possible. As a matter of fact, it is the situation which we repeatedly met above. On the whole, the calculations, roughly, correspond to the experiments with waves spreading in a flow near solid boundary. This idea is supported by the calculations presented in Fig. 2.34.

The upper pattern corresponds to the experiment. We will stop on the results of calculations and their analysis. At the beginning, we consider curves A and A1. They are calculated according to (2.67.15). Curves A1 are some interpretations of A.

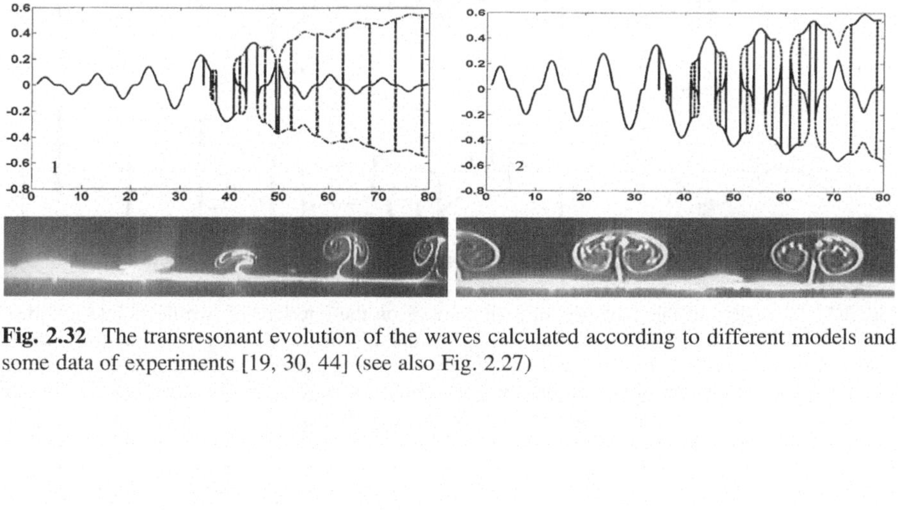

Fig. 2.32 The transresonant evolution of the waves calculated according to different models and some data of experiments [19, 30, 44] (see also Fig. 2.27)

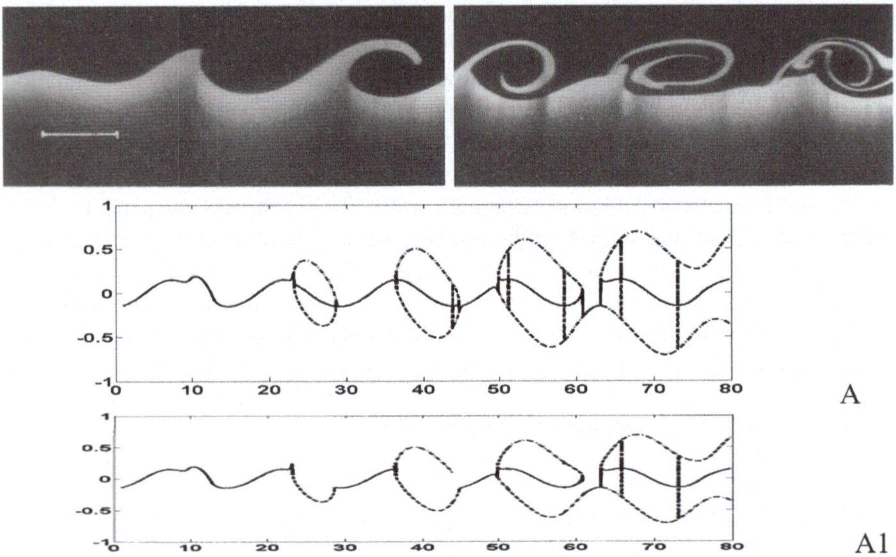

Fig. 2.33 The transresonant evolution of waves into vortices. The experimental data (left) [30, 46] and their modeling (right). Figures **A** and **A1** are versions of calculations

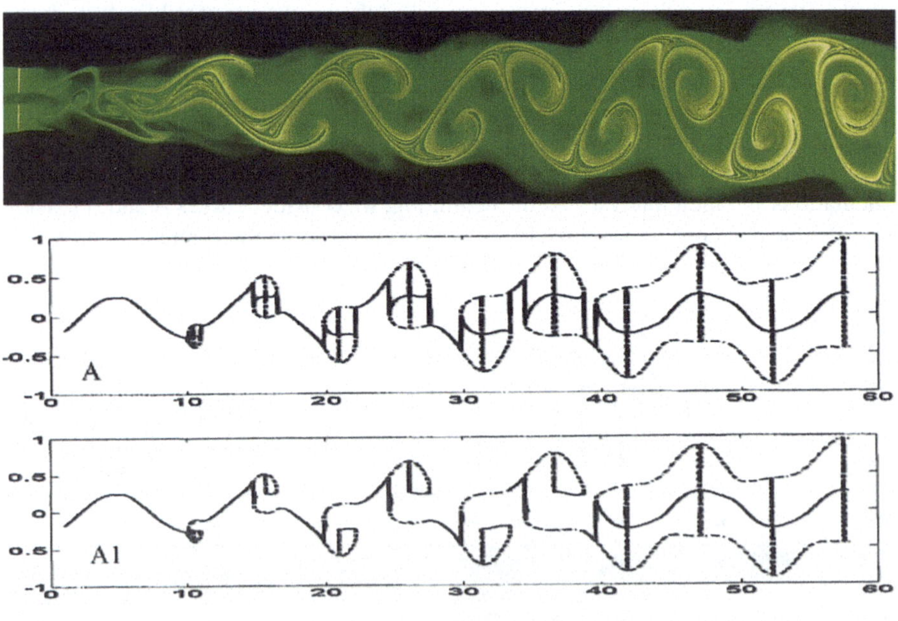

Fig. 2.34 A pattern of the evolution of perturbations on the interface of two flows into a vortex motion (upper). A result of analytical modeling (*A*) and a scheme of mainstream line (*A1*) which models the vortex motion. Though the initial perturbation is not known and, strictly speaking, a final configuration of flow too, it is possible to speak about the wave process and about the certain consent of the analytical analysis and the experiments [19, 30, 47]

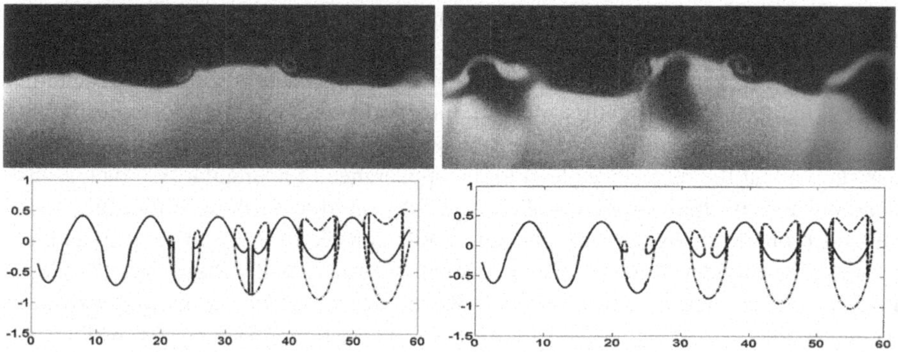

Fig. 2.35 The experimental data [19, 30, 46] and evolution of vortex wave structures

These curves are defined by the algebraic equation (2.67.15) which has three decisions. The curves have the folds and the vertical lines. They define the zones in which jumps and bifurcations can take place. Thus, jumps from one solution to another are possible. Above, we connected the appearance of these zones with the formation of vortex waves. This idea is illustrated by the picture A1. There we gave the interpretation of curves A to describe roughly the experimental curves. Parts of the curves which do not correspond to the experiment are eliminated. We suppose that there is a certain similarity of the theoretical and experimental curves, although our approaches were very rough and the experimental curves are very complex.

The following calculations model the waves framed by vortices. Such waves sometimes appear on the interface of two liquids. We have devoted to these waves two previous sections. There we tried to model some results of Richtmyer-Mechkov instability [32]. Here we will continue our modeling of the vortex waves.

Experimental and calculated waves are given in Fig. 2.35. The experimental waves are located on the top line. The curves on the following line model the experiment. On the left side, the calculated curves are placed. On the right, their interpretation is placed. During the calculations, Eq. (2.67.15) was used. Within the selected resonant zone, the forced waves practically were not amplified but can be framed by the loops ("ears") which explicitly correspond to vortices observed in experiments (the top of Fig. 2.35).

2.4 Transresonant Phenomena

For us, it is very important that the generation of extreme multivalued waves (EMW) is associated with the passage of a wave through a resonant region. This forces us to pay special attention to the features inherent in resonant zones.

Apparently, the phenomenon of resonance was known to Galileo and Huygens nearly four centuries ago. The resonance was well known to eighteenth-century astronomers. Nevertheless, the modern story of the resonances actually began at the end of the nineteenth century due to the works of Poincaré. Considering the problems of the celestial mechanics (the so-called multi-body problem), Poincaré has shown that the resonances lead to the appearance of a singularity ("dangerous" denominator) in these problems. Poincaré [48] referred to these difficulties as "a general problem of dynamics." Indeed, similar problems exist for any dynamical system. The founder of the theory of dissipative structures Ilja Prigogine (the Nobel Prize winner in chemistry of 1977) [49, 50], analyzing the origin of complexities in *Nature*, writes "... Whether we can 'exclude' interactions? Systems for which such exception is possible, Poincaré named 'integrated' ... the majority of dynamic systems are not integrated and interactions cannot be excluded: the reason of that and another is covered in occurrence of resonances ...". It also relates to waves as Prigogine emphasized. In any complex system, a large quantity of resonances is possible. Certainly, the conditions for the occurrence of the resonances are not yet equivalent to their realizations in actual situations. All of it takes place, both in the Universe and in such a complex system, as the infinite quantity of interacting oceanic waves.

We are all more or less familiar with the concepts of the resonance. If we stretch a spring (thus, we will deduce it from the stable state), and then we release, the spring will start to make oscillations with some defined, characteristic frequency. We then apply to the spring the external force changing with a frequency which can vary. If the relation of two frequencies (the frequency of free oscillations of the spring and the frequency of changes of the external force) is equal to any simple number (if one frequency is equal to another or more it in two, three, four, ... times), then the amplitude of oscillations of the spring sharply increases. The same takes place with a string when we play on a musical instrument. We hear harmonics which are generated by the resonant oscillations of the string. Still, ancient Greeks have suggested that harmonious accords are uttered by the strings, whose lengths are related among themselves, as some integers. It is possible to put these numbers in conformity with the modern resonant frequencies.

Greeks also tried to compare the movements of the planets with numerical relations. According to their representations, each relation conforms to some sound. The sounds of the planets are defined by the speed of their movement on the corresponding spheres and form some harmony. All tones of the heavenly spheres form "harmony of spheres" in the sense of some musical scale. Copernicus and Kepler also began their reflections with searches of some mathematical relations defining "harmony of spheres." Despite the speculativeness of all these great ideas, the thought appears involuntarily about the connection of the reflections of ancient Greeks, Copernicus and Kepler, about some general harmony with modern concepts of the resonance, the matter waves of de Broglie, and the theory of everything [51].

2.4.1 Linear Resonance Phenomena

Let us consider the resonance zone using the linear equation

$$J_{rr} + \Omega^2 J = \varepsilon \cos \omega r. \tag{2.68}$$

The general solution of (2.68) has the form

$$J = A\sin\left(\Omega r + \delta\right) + \frac{\varepsilon}{\Omega^2 - \omega^2}\cos \omega r, \tag{2.69}$$

where the amplitude A and the phase δ are defined by initial conditions. Thus, the solution represents the superposition of the waves, which have the frequency Ω and the frequency ω. The solution contains the small divider. It is clear that the solution is invalid, when the following resonant condition is satisfied.

$$\Omega^2 = \omega^2. \tag{2.70}$$

It is possible to determine the behavior of the solution in the limit $\Omega \to \omega$, having written down (2.69) in the form

$$J = \overline{A}\sin\left(\Omega r + \overline{\delta}\right) + \frac{\varepsilon}{\Omega^2 - \omega^2}\left(\cos \omega r - \cos \Omega r\right). \tag{2.71}$$

We redefined the amplitude $\overline{A}$ and phase $\overline{\delta}$ in (2.71). The limit $\omega \to \Omega$ is found by L'Hôpital's rule

$$J = \overline{A}\sin\left(\Omega r + \overline{\delta}\right) + \frac{\varepsilon}{2\Omega} r\sin \Omega r. \tag{2.72}$$

The solution contains the singular term. We emphasize here only that this term demonstrates the unlimited linear growth of the amplitude of the wave J at the resonance. This term describes the absorption of energy from the initial wave.

"Dangerous" Dividers For example, following Prigogine [49, 51], we will consider the system characterized by two frequencies. By definition, we have the resonance, if $n_1\omega_1 + n_2\omega_2 = 0$, where ω_1 and ω_2 are frequencies and n_1 and n_2 are integers distinct from a zero. Resonances lead to an occurrence in the solution of the terms with "dangerous" dividers. In the case of harmonious excitation, these dividers look like $1/(n_1\omega_1 + n_2\omega_2)$. The corresponding terms increase up to the infinity in the case of the exact resonance. A similar problem can arise in any dynamic system.

We consider Lamb's result [17, pp. 263–264] as the illustration of the problem of the dangerous divider for pure forced waves. Lamb used the following equation to describe surface waves on water:

$$u_{tt} + g\eta_a = -\rho^{-1} p_{0,a}. \tag{2.73}$$

Here $\eta = -hu_a$. In the case of a traveling pressure $p_0 = \rho f(Ut - a)$ moving with the speed U, Lamb found

$$\frac{\eta}{h} = \frac{p_0}{\rho\left(U^2 - gh\right)}. \tag{2.74}$$

When $U^2 \rightarrow gh$, we have $\eta \rightarrow \infty$. For the real situation, the wave elevation grows up only during finite time. After some time, the wave breaks as a result of nonlinear effects.

Usually, a condition of the resonance may be derived from linear analysis of the process being studied. For example, Lamb [17] used Eq. (2.73). These resonances may be called linear. However, nonlinear mathematical models must be often used for the analysis of observed anomalous phenomena. The nonlinear theory can bring to light the resonant nature of them. In this case the revealed resonances may be called as nonlinear.

2.4.2 Nonlinear Resonance Phenomena

We looked at various forms of extreme multivalued waves above. It is shown that the appearance of these waves is associated with various resonant phenomena. In particular, within the resonant band, a wave of initially small amplitude can be strongly amplified and transformed into shock waves, then into multivalued waves, particle–waves, and even vortex structures. It is remarkable that in some cases, the shape of multivalued waves coincides with the famous Euler figures which are known in the theory of stability of structures.

Thus, inside the resonant bands can be sources of extreme phenomena. In particular, we are interested in jumps inside the waveforms. Let us consider the jumps that occur during wave processes using the example of the experimental data of Natanzon [52].

Figure 2.36 (left) shows the sketch of the experimental apparatus which Natanzon used. During these experiments, the water column was excited by the oscillating piston. The length of the column and the piston frequency were changed. Figure 2.36 summarizes some results of experimental and analytic investigations. The amplitude-frequency curves were measured for the tube length $L = 3.7$ m. The forced amplitude a was 0.0011 or 0.002 m.

The dependence of the amplitude of forced oscillations on the frequency ω is shown. The points define the experimental data. The solid lines are the theoretical curves. The arrows directed along the curves show the direction of the frequency change. Vertical arrows determine the points at which the solution may be unstable (amplitudes and frequencies at which a jump from one branch of the solution to another is possible).

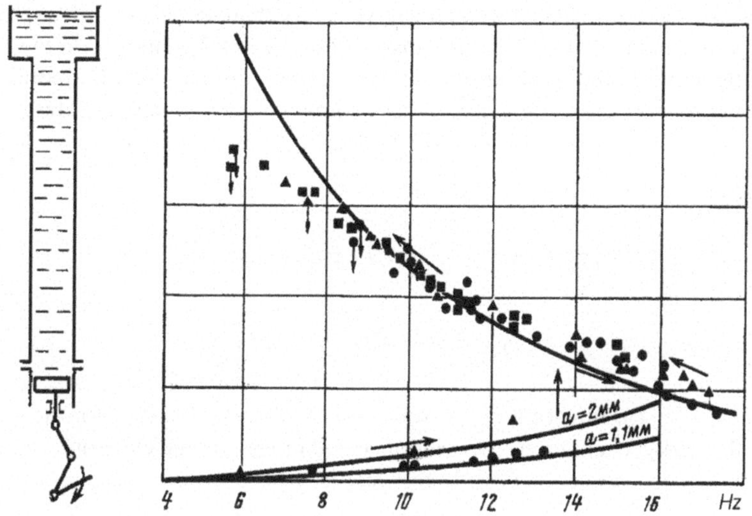

Fig. 2.36 Data of Natanzon's experiments. Amplitude-frequency curves of the water column measured for open tube [19, 30, 52]

Figure 2.36 agrees well with amplitude-frequency curves obtained for many cubically nonlinear systems. The hysteresis nature of the oscillations is clearly visible. Such oscillations are described well by ordinary differential equations:

$$u_{tt} + \mu u_t + \omega_*^2 u + \beta_* u^3 = l \sin \omega t, \tag{2.75}$$

which describes the forced oscillations of a nonlinear system with viscous resistance and cubic quasi-elastic characteristics. The dependence of the amplitude of oscillations excited in the system on the frequency ω is determined by the following expression:

$$(\omega^2 - \omega_*^2)A - \beta_* A^3 \pm l\sqrt{1 - \mu^2\omega^2 A^2 a^{-2}} = 0. \tag{2.76}$$

Equation (2.76) can have one real root for certain parameters. With increasing frequency, it can fall into the resonance band, where there are three real roots. Each of them corresponds to oscillations with a specific amplitude. Thus, in the resonance band, there can be several real solutions. The multiplicity of the solution means that at a given frequency of the disturbing force, oscillations with different amplitudes are possible, but not all possible types of oscillations are stable.

Singularities as Fields of Euler's Elastic Figures Generation It is that many Euler figures have folds where jumps are possible, well known in the catastrophe theory. In particular, due to a strong nonlinearity, the oscillations become relaxation, consisting of sections of fast and slow movements. In the case of waves, the wave profile is divided into parts of smooth and very fast changes. In a rough approximation, we can assume that in this case, discontinuities or folds appear on the profile. Outside these areas, the wave profile remains smooth, in particular harmonic.

Let us consider periodical waves described by a model equation of the form

$$\varphi_{tt} - c_0^2 \varphi_{aa} + \varphi = -\alpha\left(\varphi_a - \frac{1}{3}\varphi_a^3\right). \tag{2.77}$$

Let $r = ct - a$ (2.77). In this case, Eq. (2.77) yields

$$\varepsilon\varphi_{rr} + \varphi = \alpha\left(\varphi_r - \frac{1}{3}\varphi_r^3\right). \tag{2.78}$$

Here $\varepsilon = c^2 - c_0^2$. Equation (2.78) was used by Van der Pol [53] for the explanation of the jerky oscillations (relaxation oscillations) of an electronic circuit. If $\nu = \varphi_r$, $x = \varphi\alpha^{-1}$, and $\varepsilon = \alpha^{-2}$, (2.78) becomes

$$\varepsilon\nu_x = \left(\nu - \frac{1}{3}\nu^3 - x\right)\nu^{-1}. \tag{2.79}$$

Now we consider (2.79) together with Fig. 2.37, where $x = \nu - \frac{1}{3}\nu^3$ [54]. We assume that ε is very small, but different from zero, and consider a solution curve that starts at P. Since P is away from Γ, ν_x is approximately $-\infty$ down to P_1 where it reaches Γ. At this point, $\nu_x = 0$. Since ν_x is approximately $+\infty$ away from Γ, the solution curve tends to follow Γ, staying below it, until it reaches the vicinity of P_2.

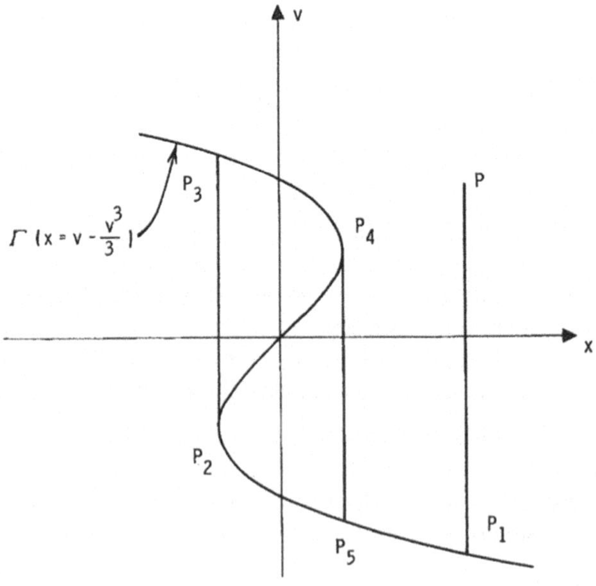

Fig. 2.37 Illustration to an analysis of Eq. (2.79) [19, 54]

At this point, the solution curve turns almost vertically upward until it intersects Γ at P_3.

Since $v_x \approx \pm\infty$ away from Γ, the solution curve tends to follow Γ from P_3 clockwise, staying above Γ, until it reaches the vicinity of P_4, where it turns almost vertically downward to intersect Γ at P_5. Then it tends to follow the path from P_5 to P_2. Therefore, the limit of the periodic solution $\varepsilon \to 0$ consists of two segments P_5P_2 and P_3P_4 of Γ, and the two vertical lines P_4P_5 and P_2P_3. Thus, the limit solution as $\varepsilon \to 0$ satisfies $\alpha\left(\varphi_r - \frac{1}{3}\varphi_r^3\right) - \varphi = 0$ except at certain points where $v = \varphi_r$ has jump discontinuities.

However, the analysis does not take into account the solutions of the equation of the corresponding P_4P_2 line. In this case, you can build the solution without jump discontinuities.

Thus, we have shown that Eq. (2.78) can have multivalued solutions. The existence of these solutions, apparently, is limited to an area where x varies from P_3P_2 to P_5P_4. Of course, if the zone of multivalued solutions is very narrow, then it can be approximated as the jump in the sought value.

For us, it is important that we begin the analysis with wave equation (2.77). Let us look at a more complex wave equation and the case of forced oscillations.

Transresonant Origin of Chaotic Waves Let the wave equation written for a unidirectional wave have the form

$$\varphi_{rr} + \beta\varphi_r + \varphi - \varphi^3 = F \sin \omega r. \tag{2.80}$$

If $\beta = 0$ and φ^3 are negligible, then there is a wave resonance at a frequency $\omega = 1$. Figure 2.38 illustrates the dynamics of the phenomena, which may take place in the resonant band if there is the influence φ^3 and viscosity.

Calculations were made for different F. The value of damping losses was not changed ($\beta = 0.1$). In Fig. 2.38, the curves are presented and calculated for three cases F. In the case $F = 0.0562$, within the resonant band, the oscillations (waves) may be of two different amplitudes. Both amplitudes are obviously different from each other. The frequency band where there are upward-downward jumps is accurately defined. If F is larger, then the amplitude of the forced oscillations grows and the frequency band, where the jumps occur, is reduced. In the case $F = 0.08$, the jump band is very small. The amplitude may become uncertain (Fig. 2.38b). There, according to the calculations, a very small change of F can give strongly different amplitudes of the oscillations (waves). If F was increased further, then the frequencies appear, when the chaotic forced oscillations and the infinite amplitudes may be generated. When ω gets to this very narrow band, any small change of the parameters can lead to big changes of the oscillations (waves) (Fig. 2.38c). We emphasize that the calculations presented in Fig. 2.38 were received for the mechanical system. However, according to the theory developed in the book, they are qualitatively fair for the cubic nonlinear wave systems.

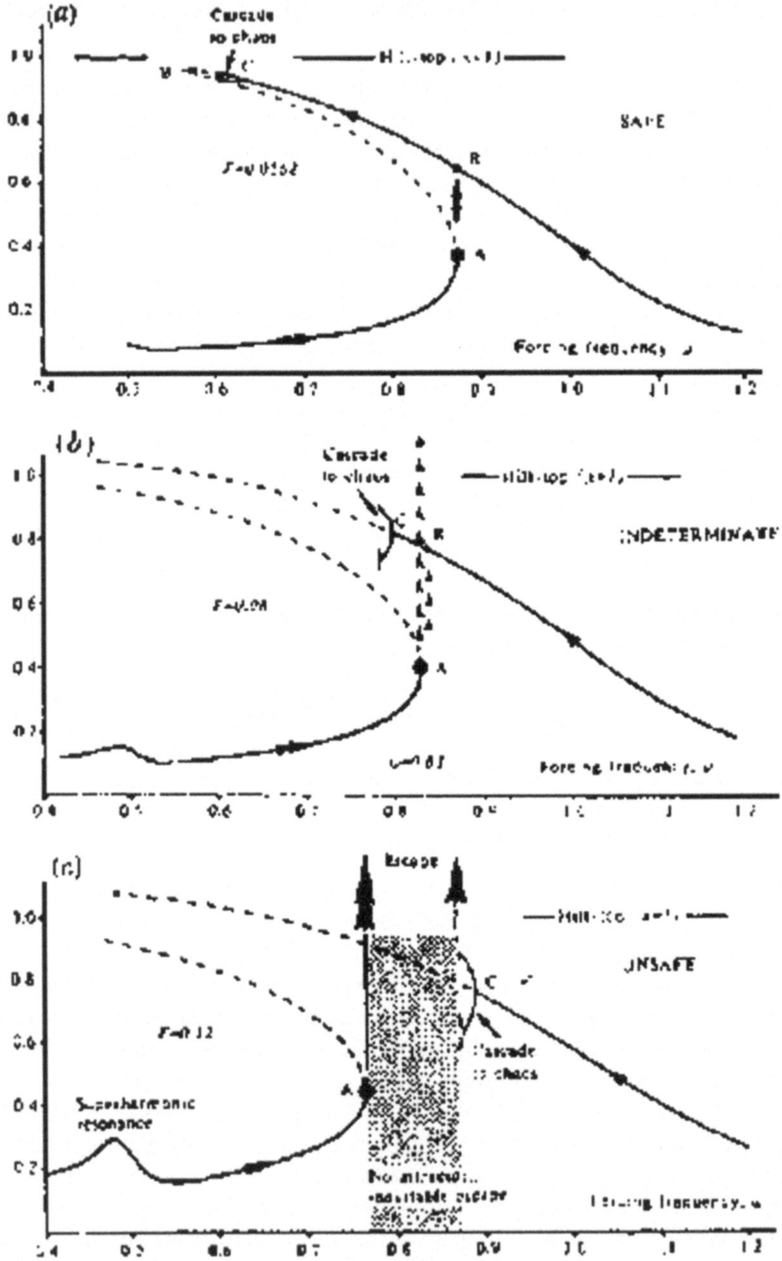

Fig. 2.38 Resonant curves corresponding to Eq. (2.80), which show complex behavior of the wave system crossing the resonant band [19, 30, 55]. Figures **a–c** are different presentations of the resonant curves

2.4.3 *Method of "Resonant" Perturbations*

Exact analytical solutions of nonlinear problems are usually impossible. Of course, analytical solutions have great advantages. They are compact and beautiful. However, they exist only for so-called integrable equations. As an example of such a situation, let us return to the consideration of Eq. (1.66).

$$\varphi_{tt} - c_0^2\varphi_{xx} + m^2\varphi - \lambda\varphi^3 + C = 0. \tag{2.81}$$

Here C is an arbitrary constant. The solution of the equation, obtained in Sects. 1.4 and 2.1, used elliptic integrals. It showed us the connection between wave processes and Euler figures. However, this solution is not valid for all values of the equation parameters. The situation is even worse in the case of non-integrable equations. When solving them, we are forced to use approximate solution methods. When assessing the accuracy of approximate solutions, we usually rely on experimental results or known exact solutions.

Perturbation Methods Approximate methods of the solution of the nonlinear wave equations are widely used. In particular, the so-called perturbation methods or the methods of the small parameter are well known. According to these methods, the unknown value is presented in the form of the expansion

$$u = u^{(1)} + u^{(2)} + u^{(3)} + \ldots., \tag{2.82}$$

here $u^{(1)} \gg u^{(2)} \gg u^{(3)}$. Usually, some approximate solution of the problem may be described by the first terms of the expansion. The first term of the expansion is usually determined by the solution of the linearized problem. The subsequent terms are determined by the corresponding nonhomogeneous linear equations. Though the expansion (2.82) can be divergent, it may be useful, as both for qualitative and quantitative descriptions of nonlinear phenomena.

The perturbation methods often give a good description of reality if the first term $u^{(1)}$ already describes it well enough. Other members of the procedure then only correct it. However, how to find a successful first approximation if we know that the wave is strongly nonlinear, for example, its surface contains folds, cavities, and particles of matter break away from it?

Physically, expansion (2.82) is equivalent to a description of a nonlinear wave by a sum of linear waves. Therefore, the specified method based on the representation of a nonlinear wave in the form of the sum of linear waves does not allow to describe the formation of elastica-like (extreme multivalued) waves.

The generally accepted position is that the study of strongly nonlinear waves is a very complex problem. However, maybe the complexity is determined also by the fact that during the study, mathematical methods are used that do not correspond to the nature of the strongly nonlinear phenomenon.

It may be necessary to change the methods themselves. Maybe instead of the method (2.82), which describes well weakly nonlinear waves, we should use another algorithm?

Indeed, the use of Arabic numerals strongly simplified the solution of complex arithmetic problems in comparison with the use of Roman numerals which were very successfully used in the simplest arithmetic operations. In the case of strongly nonlinear waves, the same situation can occur. Usually, we try to use methods that are well proven during the analysis of linear and weakly nonlinear problems, even for the study of strongly nonlinear problems. In particular, the perturbation method is used which is qualitatively determined by the expression (2.82).

Thus, the study of the occurrence of highly nonlinear extreme waves requires the development of unconventional algorithms for the integration of the corresponding wave equations.

Reduction of the Problem to Self-Excitation of Catastrophic (Extreme Multivalued) Waves We have offered such an algorithm and have developed it in several articles [28, 39, 56] and books [19, 20, 30]. First, the algorithm was intended for the research of shock waves excited in the vicinity of the nonlinear resonance [28, 56, 57]. According to the algorithm in the expansion (2.82), the second term can reach the value of the first term and even surpass it. For example, it may be during the catastrophic amplification of the extreme ocean waves. Certainly, in this case, the condition $u^{(1)} \gg u^{(2)} \gg u^{(3)}$, which is a basis of the traditional application of the expansion (2.82) is not satisfied.

An essence of the algorithm is described very shortly below. We search for the solution of the governing equation, as the sum of functions

$$u = u^{(1)} + u^{(2)} + \ldots. \tag{2.83}$$

The sum is substituted in the governing equation from which the equations are deduced for each term of the sum.

The peculiarity of the algorithm is that for the last term $u^{(2)}$, a nonlinear equation is used. We assume that it is important when physical systems excited in the vicinity of resonance frequencies are considered. At the same time, we consider u_1 as a small seed disturbance. According to this algorithm, the problem of the extreme waves may be presented as the research of self-excitation of the nonlinear waves in the vicinity of some resonant conditions [19, 20, 30]. Subsequently, situations are taken as resonance when the linear terms in the wave equations become much smaller than the nonlinear terms.

We will use the algorithm (2.83) and its variants several times in this book when studying various nonlinear waves produced in resonators. In particular, in this section, the version of this algorithm will be used to study the elastica-like waves described by the Eq. (2.81).

2.4.4 Application of the Method of the "Resonant" Perturbations

Let us illustrate the main variants of the approximate method (2.83) used in the book. Let

$$\varphi = \varphi_1 + \varphi_2. \tag{2.84}$$

No assumptions are made about the order of magnitudes included in the sum. Expression (2.84) is substituted into (2.81). As a result, it takes the form

$$(\varphi_1 + \varphi_2)_{tt} - c_0^2(\varphi_1 + \varphi_2)_{xx} + m^2(\varphi_1 + \varphi_2) - \lambda(\varphi_1 + \varphi_2)^3 + C = 0. \tag{2.85}$$

Our goal is to obtain equations for the defining functions φ_1 and φ_2. Next, new variables are introduced:

$$s = ct + x \quad r = ct - x. \tag{2.86}$$

Here $c = c_0 + \Delta$ and $c_0 \gg \Delta$. We point out that Δ is a very small value. In general, it can be equal to zero. We will assume that the case $\Delta = 0$ corresponds to exact linear resonance. As a result of using (2.86), we have

$$\begin{aligned}&\left(2\Delta c_0 + \Delta^2\right)\left[(\varphi_1 + \varphi_2)_{rr} + (\varphi_1 + \varphi_2)_{ss}\right] + 2\left(c^2 + c_0^2\right)(\varphi_1 + \varphi_2)_{rs}\\&+m^2(\varphi_1 + \varphi_2) - \lambda(\varphi_1 + \varphi_2)^3 + C = 0.\end{aligned} \tag{2.87}$$

We will proceed from the fact that we are interested in solutions in the form of traveling waves, which interact very weakly despite their significant amplitude and have a multivalued profile. In addition, we will consider only unidirectional waves. Finally, let us assume for the bare perturbation that

$$\varphi_1 = A\cos kr. \tag{2.88}$$

We write the following equation taking into account all that has been said

$$\begin{aligned}&\left(2\Delta c_0 + \Delta^2\right)(A\cos kr + \varphi_2)_{rr} + m^2(A\cos kr + \varphi_2) - \lambda(A\cos kr + \varphi_2)^3\\&+ C = 0.\end{aligned} \tag{2.89}$$

It is not possible to solve it analytically, so we will consider options for its approximate solution.

1. Equation (2.89) is divided into the following two equations in the first version,

$$-\left(2\Delta c_0+\Delta^2\right)Ak^2\cos kr=0,$$

$$\left(2\Delta c_0+\Delta^2\right)\varphi_{2rr}+m^2(A\cos kr+\varphi_2)-\lambda(A\cos kr+\varphi_2)^3+C=0. \tag{2.90}$$

The amplitude of the seed disturbance is not determined with this option. It is usually assumed to be very small in calculations.

2. In the second variant, the term $m^2\varphi$ is taken into account from the very beginning.

Then the analysis is reduced to the consideration of equations slightly different from (2.90):

$$-\left(2\Delta c_0+\Delta^2\right)Ak^2\cos kr+m^2A\cos kr=0,$$

$$\left(2\Delta c_0+\Delta^2\right)\varphi_{2rr}+m^2\varphi_2-\lambda(A\cos kr+\varphi_2)^3+C=0. \tag{2.91}$$

From the first equation, $k^2=m^2(2\Delta c_0+\Delta^2)^{-1}$. In this option, the amplitude of the seed perturbation is not determined, as in the first case. Considering the second equation, we accept that $(2\Delta c_0+\Delta^2)\varphi_{2rr}\ll m^2\varphi_2$.Then

$$\varphi_2^3+3A\cos kr\varphi_2^2+\left(3A^2\cos^2kr-\lambda^{-1}m^2\right)\varphi_2+A^3\cos^3kr+C=0. \tag{2.92}$$

Examples of waves described by this equation are shown in several subsequent sections.

We emphasize that according to (2.92), the exact resonance occurs when $3A^2\cos^2kr=\lambda^{-1}m^2$.

3. In the third option, the influence of nonlinearity is taken into account in the first equation. As a result, the study is reduced to solving the following equations:

$$-\left(2\Delta c_0+\Delta^2\right)k^2+m^2-\frac{3}{4}\lambda A^2=0,$$

$$\left(2\Delta c_0+\Delta^2\right)\varphi_{2rr}+m^2\varphi_2-\lambda\left(\frac{1}{4}A^3\cos 3kr+3\varphi_1^2\varphi_2+3\varphi_1\varphi_2^3+\varphi_2^3\right)=0. \tag{2.93}$$

We completely determine the seed perturbation from the first equation:

$$A=\pm\sqrt{\frac{4}{3}\lambda^{-1}\left(m^2-\left(2\Delta c_0+\Delta^2\right)k^2\right)}. \tag{2.94}$$

after which it is determined φ_2.

Comment According to the above in the proposed theory, classical infinity at resonance (see Sect. 2.4.1), from an inevitable disaster, turns into a saving opportunity for solving the nonlinear wave equation and describing reality.

2.5 Study of Cubic Nonlinear Unidirectional Multivalued Waves

Extreme multivalued waves are a new, practically unexplored area of nonlinear wave science associated with the transresonant evolution of waves. Let us explore some more the assumption about the resonant nature of their emergence, development, and sustainability. In the region of resonances, the influence of nonlinearities on the shape and amplitude of the waves greatly increases. Indeed, ultimately, our latest research came down to solving Eq. (2.92) reminiscent of the Duffing equation (2.80). This equation in resonance regions can describe very complex, even chaotic, behavior of various mechanical systems (Fig. 2.38). Therefore, we can expect that Eq. (2.92), despite its apparent simplicity, can describe a wide variety of wave processes.

Let me imagine the resonance region as a kind of "black box." In materialized form, it corresponds to the black area in Fig. 2.38. We usually know the wave processes that enter in this "box," that is, the resonance region. But we have little idea what is happening there and what may come out from it.

2.5.1 Influence of Seed Perturbation and Wave Phase

1. We have already shown that very strong nonlinear, multivalued effects can take place in resonance zones. In particular, wave profiles can become ambiguous, discontinuous, and contain vortices. A wave can disintegrate into individual elements, in particular, and evolve into particle-waves. In this section, we decided to consider the case of transresonant evolution, when the amplitude of external excitation grows according to a linear law. First we want to consider the dynamics of wave formation when the amplitudes of the seed and nonlinear waves differ greatly. In this case the "seed" disturbance φ_1 has practically no effect on the sum of these two waves. Further, the contribution of the "seed" disturbance increases and even begins to influence the profile of the nonlinear wave φ_2. Finally, we reach the limiting case when the "seed" disturbance approaches the amplitude of strongly nonlinear waves.
2. We emphasize that the coefficients $3A \cos kr$ and $3A^2\cos^2kr - \lambda^{-1}m^2$ in (2.92) periodically pass through 0. Moreover, they do not pass through 0 simultaneously if $\lambda^{-1}m^2 \neq 0$. At the same time, calculations and theoretical considerations [28] show that the passage through zero of the linear term in (2.92) is especially

important for the evolution of waves. Thus, the emergence of wave profiles similar to Euler figures occurs after the linear term passes zero. In this case, as is known from the theory of linear resonant oscillations, the phase of oscillations changes. In the resonant region, it changes relative to the "seed" oscillations. The phase changes from zero to π.

Experimental and theoretical studies of resonant gas oscillations in a pipe have shown that this effect remains valid in the case of quadratic (i.e., nonlinear) resonance [57, 58]. In the case of quadratic resonance, the amplitude of the excited nonlinear wave is of the order of the amplitude of the forcing oscillations. Therefore, the amplitude of the waves arising in the resonator depends both on the amplitude of the forcing oscillations at the boundary of the resonator and on the amplitude of the waves arising only due to nonlinear effects. Therefore, in the case of quadratic resonance, the above phase shift is extremely important and is taken into account when modeling the experimental results [57, 58]. In the case of linear resonances, the above phase shift does not play a significant role, since the amplitude of the "seed" oscillations is usually much smaller than the amplitude of the waves excited in the resonator, which is determined almost exclusively by nonlinear relations.

A completely different situation arises in the case of excitation of extreme waves in cubic nonlinear systems, if the amplitude of the "seed" waves is large enough. In this case, nonlinear effects amplify the wave much weaker. It is even possible that, as a result of the influence of nonlinearities, the resulting wave inside the resonant band may even weaken inside the resonant band relative to the "seed" wave.

The above makes us think that the influence of the above phase shift between the waves φ_1 and φ_2 can be quite significant. It must be taken into account in calculations, but we do not have a nonlinear theory that allows us to determine the magnitude and change of this phase. Therefore, here we will limit ourselves to only showing how significantly this phase can influence the shape of the wave entering, or passing through, or leaving the resonant region. To this end, we rewrite Eq. (2.92) by introducing a phase shift into it, and then carry out a series of calculations for different phase values.

$$\begin{aligned}&\varphi_2^3 + 3A\cos(kr+\varphi^*)\varphi_2^2 + \left[3A^2\cos^2(kr+\varphi^*) - \lambda^{-1}m^2\right]\varphi_2\\&+A^3\cos^3(kr+\varphi^*) + C = 0.\end{aligned} \tag{2.95}$$

Here it must be emphasized that the linear term of Eq. (2.95) depends on k, r, amplitude A, and magnitude $\lambda^{-1}m^2$. These quantities determine how the linear term passes and whether it passes through the zero value. In the general case, the linear term of Eq. (2.95) can pass through resonance several times (pass through the zero value), and the larger the value of r, the smaller the extent of the corresponding resonant region. The above shows that the phase can greatly influence the shape of the wave as it propagates. During the propagation, the wave can repeatedly pass through resonant regions.

When choosing the phase value, we will proceed from the results of the linear theory of oscillations, namely, we will carry out calculations for three phase values;

let it be equal to 0, $\frac{1}{2}\pi$, or π. Despite the crudeness of this assumption, we will accept, for at least some certainty, that the case of a phase equal to 0 corresponds to the entry of a wave into the resonant band, the case when the wave passes through the middle of the band corresponds $\frac{1}{2}\pi$, and the case π corresponds to the exit of the wave from the resonant band.

We started our calculations under the assumption that $\varphi^* = 0$ in Eq. (2.95). In addition, we set $C = 0$ and $\lambda^{-1}m^2 = 2$. We then looked at the cases $\varphi^* = \frac{1}{2}\pi$ and $\varphi^* = \pi$. The calculation results for these parameters are presented in Fig. 2.39, where a certain value proportional to is plotted along the axis r.

From the calculations, it follows that the maximum amplitude of the resulting waves lies in the vicinity of value 2. Moreover, the maximum amplitude is practically independent of the excitation amplitude. However, the wave profile strongly depends on the driving force. If $\varphi^* = 0$, then a wave appears similar to the Euler

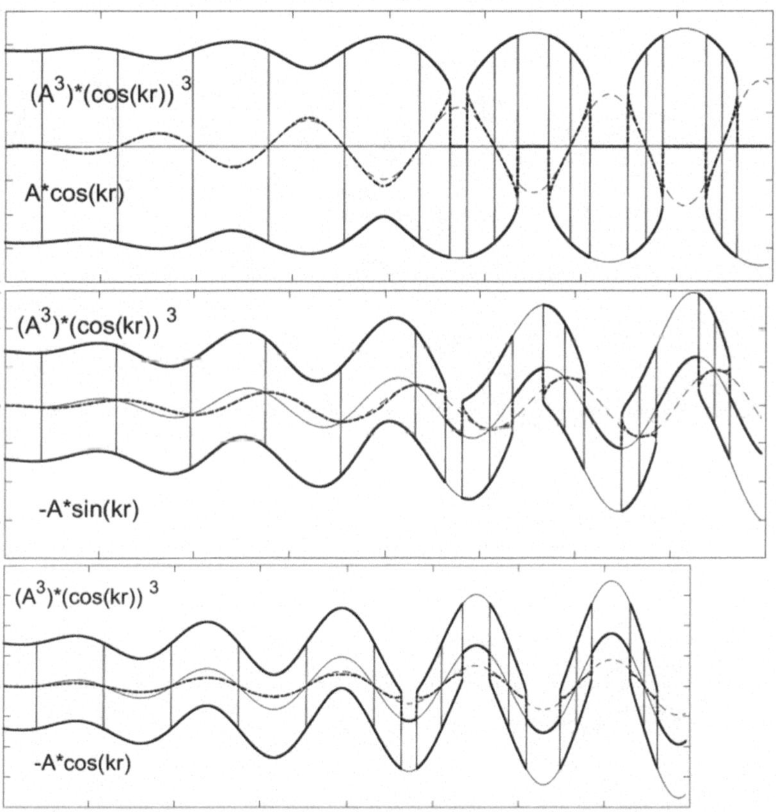

Fig. 2.39 The top picture corresponds to $\varphi^* = 0$, the middle picture corresponds to $\varphi^* = \frac{1}{2}\pi$, and the bottom picture corresponds to $\varphi^* = \pi$. The thick lines correspond to the solution (2.95). The broken harmonic lines determine the linear part $\varphi^{(1)}$ of the solution φ. The thin solid lines determine the forcing term in (2.95)

figure 7 (see Fig. 2.1); if $\varphi^* = \frac{1}{2}\pi$, then a wave configuration appears that sometimes appears at the beginning of the onset of turbulence (see Fig. 2.34). Chester found that a similar configuration occurs at the beginning of the development of turbulence during resonances in gas-filled pipes (Fig. 2.24) [38]. If $\varphi^* = \pi$, then waves appear that can be called mushroom shaped. I□t can be seen that during the considered interval of change in the variable, the system passed through five wave amplification zones, which we identify as resonance zones. The amplification at the last two resonances is so great that extreme multivalued waves (EMW) arise. Their forms are different.

Of course, waveforms depend not only on phase. Calculations have shown that, depending on A, k, $\lambda^{-1}m^2$ and C, a diverse spectrum of EMW can arise. In particular, in Fig. 2.40 shows the influence of a constant C on the evolution of a mushroom-like EMW into a particle-wave.

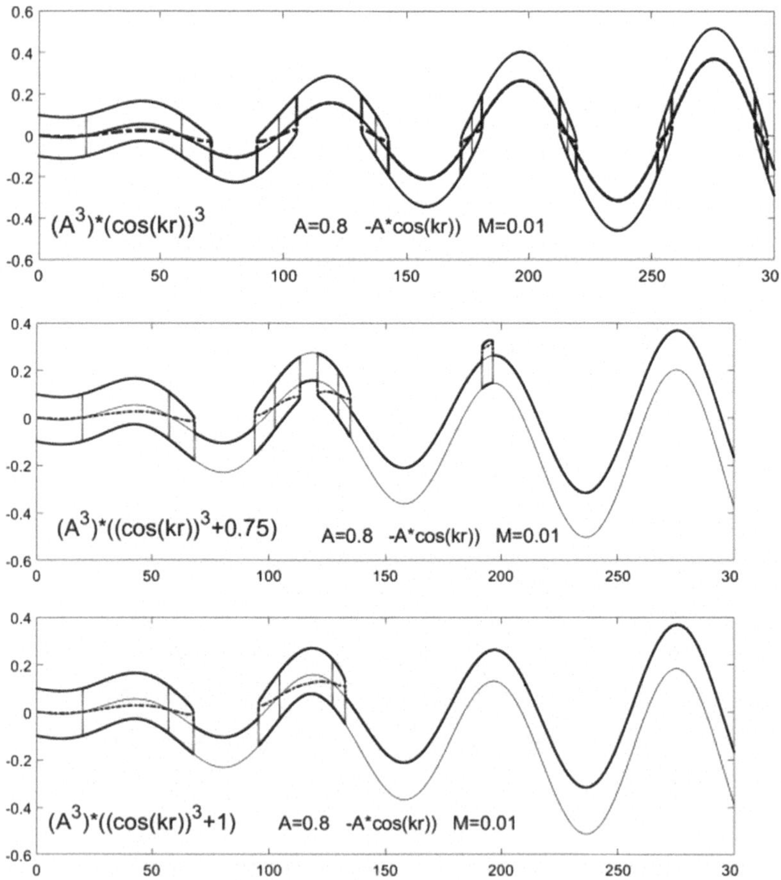

Fig. 2.40 The top picture corresponds to $C = 0$, the middle picture corresponds to $C = 0.75$ const, and the bottom picture corresponds to $C =$ const. Here $M = \lambda^{-1}m^2$

The calculation results for $A = 0.8$, $\lambda^{-1}m^2 = 0.01$, and three values C are presented in Fig. 2.39. The formation of a particle above the top of the wave is shown. We call a system that includes both a wave and a particle above its top particle-wave.

2.5.2 *Multivalued Waves Which May Be Understand as Containing Jumps, "Hair" or "Wings," and Vortices*

We considered some options for solving Eq. (2.95) assuming the continuity of the curves describing the solutions. However, this condition can be considered as some limitation of nature's capabilities. Indeed, this chapter examines many waves whose profile contains folds. A particle moving or located on such folds can be considered as being in an unstable state, well known in catastrophe theory. In principle, a particle can fall on the lower branch of the solution, or in the opposite case, jump to the upper branch of the solution.

More than 10 years ago, books were published [30, 34] where the concept of catastrophic waves was introduced, that is, waves whose profile may contain folds. Here we must remember that the wave profile can have three values at each point according to (2.95). The jumps further increase and complicate the possible wave profiles.

Taking this into account, we continue our study of the influence of the magnitude of the initial (seed) disturbance on φ_2. The following equations were used in the calculations:

$$\varphi_2^3 - 0.0011\varphi_2 + A\sin(5\pi r/L) = 0. \tag{2.96}$$

Here r changes from 0 to L. The calculation results are presented in Fig. 2.41. When constructing the curves, we used either two solutions or one of the generally existing three solutions to clearly illustrate the possibility of jumps in the system.

At A = 0.02412, there are two independent continuous wave profiles resembling cnoidal waves. With a very small increase in amplitude to 0.024126, local jumps appear connecting these solutions. With a further increase in amplitude, these local jumps break up into two jumps. As a result, a very peculiar wave structure appears, which is clearly visible in the case of A = 0.04.

If we accept the possibility of the existence of waves with discontinuities, then the waves presented above can be interpreted as traveling waves carrying vortices or waves with hair (waves with wings). The latest waves are presented in Fig. 2.42.

The profiles calculated for $A = 0.0245$ (top) and $A = 0.04$ (bottom) are shown. For both cases, two options for their interpretation are presented. In the first version, $A = 0.0245$, we illustrated the possibility of the simultaneous existence of two

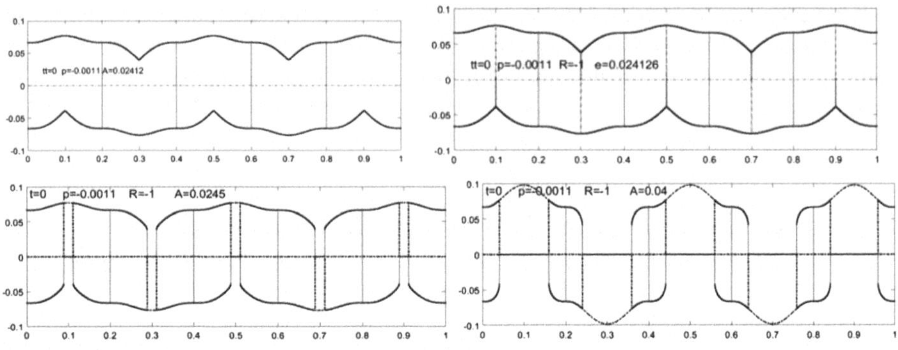

Fig. 2.41 Wave profiles calculated for four values of A

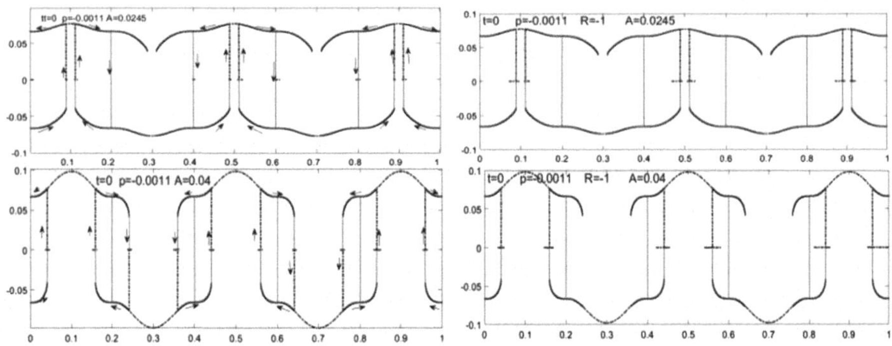

Fig. 2.42 Wave profiles illustrating the possibility of the emergence of vortices and jets ("hairs" or "wings") as part of traveling waves

differently directed vortices concentrated near the wave centers, as well as jets ("hair" or "wings") bordering the vortices (left). In the second interpretation of the calculations (on the right), only jets ("hair" or "wings") are shown. Similar options are presented for the case $A = 0.04$. However, in this case, we assumed that the emergence of vortices excludes the simultaneous existence of a jet ("hair" or "wings") in the wave formations (bottom, left).

It can be seen that, qualitatively, all the waves presented in Figs. 2.41, 2.42 and 2.43 can be called waves with hair (waves with wings). They generated during the Richtmyer-Meshkov instability from sinusoidal, initial perturbation [59, 60]. In conclusion, we note that we have considered waves φ_2. In the case considered, their amplitude is several times greater than the amplitude of the initial perturbation A. Therefore, the wave φ_1 will not change the presented results.

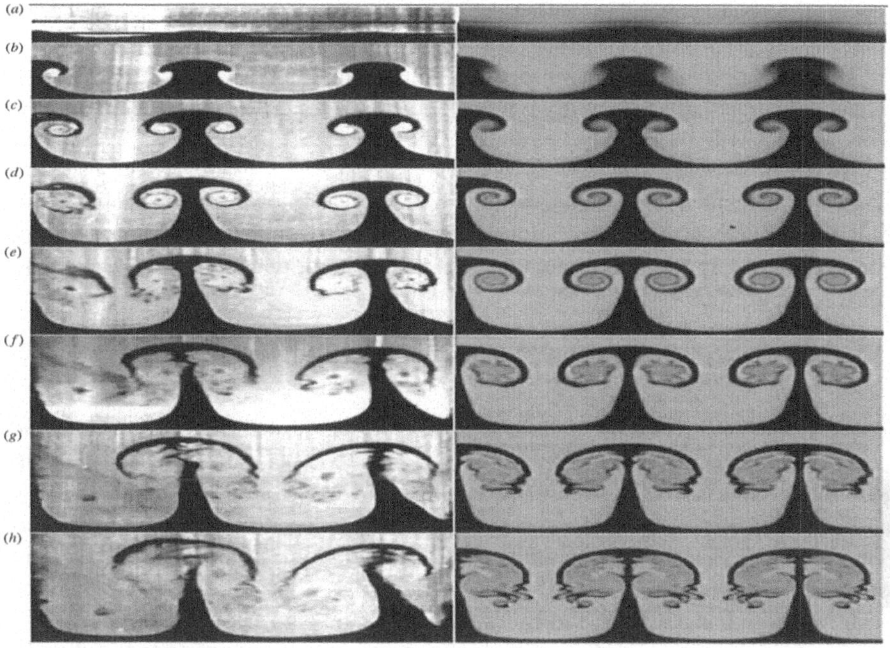

Fig. 2.43 Comparisons of experimental results (left) with numerical calculations (right) performed for different time points (**a**) t = −0.08 ms, (**b**) t = 1.42 ms, (**c**) t = 2.92 ms, (**d**) t = 4.42 ms, (**e**) t = 5.92 ms, (**f**) t = 7.42 ms, (**g**) t = 8.92 ms, and (**h**) t = 10.42 ms. Good qualitative agreement is observed. There is also a qualitative agreement between the presented data and the results of analytical studies presented in Figs. 2.41 and 2.42

2.5.3 Transresonance Evolution of Waves into Particle-Waves and Quanta (One-Frequency Seed Perturbation)

We showed the strong dependence of waves on the parameters of the original equation. In the previous section. In particular, a change in the forcing amplitude A or constant C can fundamentally change the configuration of the excited wave. We must keep in mind that the parameters of Eq. (2.95) can be interdependent. Thus, we do not have a general understanding of how strongly a nonlinear wave system will respond to a particular influence if it is in a resonance zone. In our opinion, the equation resembles a kind of "black box" (see Fig. 2.38) whose response to any change in parameters cannot be predicted. However, the simplest calculations similar carried out above something can give useful information.

Let us consider again Fig. 2.40 as an example of the fruitfulness of such guesses. It shows the possibility of forming particle-waves in the system if $C = 1$.

The formation of particle-waves is of fundamental interest to science. We have already illustrated this possibility with Figs. 2.20, 2.21, and 2.22 and other drawings. This section will be entirely devoted to the emergence of particle-waves during transresonant processes.

Let us look at transresonant evolution of waves following the book [19]. We illustrate the transresonant evolution of forced waves from harmonic forms through Euler's elastica figures and cnoidal waves to again harmonic forms (Fig. 2.44). First, we rewrite Eq. (2.95) in the form

$$\varphi_2^3 + r_*\varphi_2^2 + \mathrm{R}\varphi_2 + A\cos kr = 0. \tag{2.97}$$

Single-frequency excitation is considered. Let us see how the excitation amplitude affects the waveform, provided that $R = -\lambda^{-1}m^2$ is the transresonant parameter. Results of model calculations are presented in Fig. 2.44 for different values of k, r_*, and R. Parameter R varies from negative values to positive values through zero. Thus, using (2.97), we try to simulate the strongly nonlinear transresonant evolution of waves.

In certain changes of R, the Euler figures are transformed into particles-waves (curves C). The corresponding particle-wave curves can be interpreted in different ways. In particular, if one of the roots of the solution to Eq. (2.97) is not taken into account, then this equation describes the motion in space of an isolated particle (i.e., an isolated wave particle) (see curves D). We will call it a quantum of wave. The results of the calculations, qualitatively, agree with the Euler data (see Fig. 8 in Fig. 2.1).

Let us recall that we carried out calculations using model equation (2.92) and (2.95), (2.96), and (2.97). We showed above that various wave equations of mathematical physics can be reduced to these model equations. Therefore, the data obtained can and should be compared with data from various physical experiments. They are most simply interpreted as surface waves excited in resonators.

One can easily imagine that the evolution shown in the last figures could occur on the surface of the liquid in the case of the existence of significant capillary forces or in the absence of the influence of gravity.

In the general case, the vertical motion of liquid surface can become chaotic.

The appearance and evolution of various extreme multivalued surface waves are shown in Fig. 2.45. The appearance of these waves and the atomization of the surface is determined by the capillary effect, which in this case is much stronger than the effect of gravity (see also Figs. 2.14 and 2.16). In general, the experimental data are consistent with the calculation (Fig. 2.44), and it is clear that the results of our theory presented above are in qualitative agreement with the experiment.

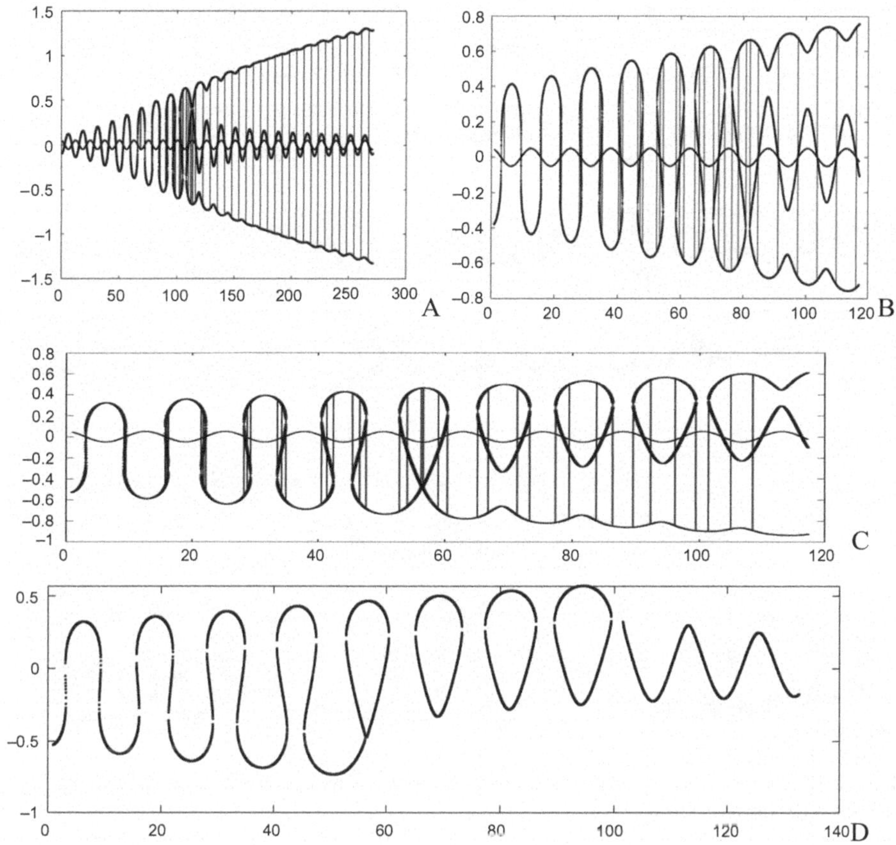

Fig. 2.44 Examples of transresonant evolution of waves described by Eq. (2.97). The thick strongly nonlinear lines correspond to the solution $\varphi = \varphi_1 + \varphi_2$. The thinner harmonic lines determine the linear part $\varphi^{(1)}$ of the solution φ. Curves **A** show the transresonant evolution when R was reduced slowly from 0.008 to -0.296 if $r^* = 0$. Curves **B** show details of the transresonant wave evolution for R ≈ 0. Curves **C** show details of the transresonant wave evolution for R ≈ 0 and $r^* = 0.355$ in (2.97). Curve **D** is the modification of curve **C**

2.5.4 *Transresonance Evolution of Waves into Particle-Waves and Quanta (Two-Frequency Seed Perturbation)*

Let us study a possible transresonant change in the waveform within resonant bands. To do this, we use Eq. (2.95) but rewrite it in the following form, assuming $C = 0$, $k = 0.5$ and $\varphi^* = 0$.

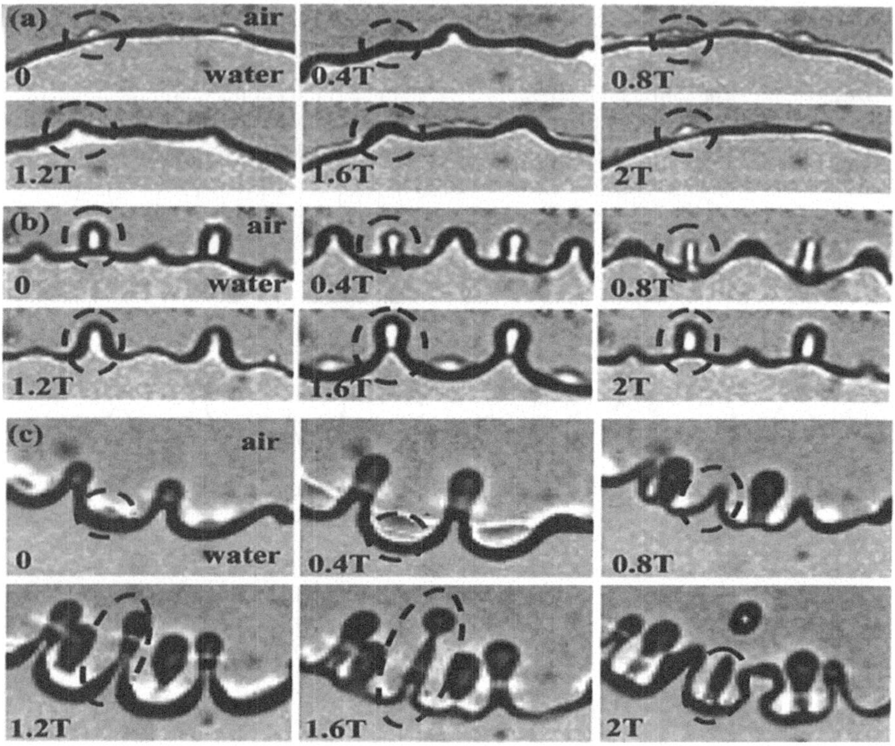

Fig. 2.45 Surface instabilities resembling Faraday waves generated at a driving frequency = 100 kHz [91, 92]

$$\varphi_2^3 + 0.345\varphi_2^2 + \mathrm{R}(r)\varphi_2 + 0.05\cos^3 kr = 0. \quad (2.98)$$

Here R(r) changes, therefore (2.98) describes the transresonant evolution of the wave. The amplitude of the disturbance decreased significantly compared to the cases considered above. The amplitude did not change during the calculation process. The calculation results are presented in Fig. 2.46.

We assume that R(r) changes slowly from 0.008 to −0.296. Calculations were also performed for different but constant R(r).

Calculations show that during the transresonance transition, waves evolve from almost harmonic to those described approximately by Euler figures, which are replaced by particle-waves and quanta. We mean a quantum which is the smallest discrete unit of strongly nonlinear wave (see Fig. 2.44). At the end of the resonant band, cnoidal waves appear and then harmonic waves. In general, it seems to us that the described process resembles the process of droplet atomization shown in Fig. 2.14. We emphasize that (2.98) describes two-frequency excitation since $\cos^3 kr = \frac{1}{4}(3\cos kr + \cos 3kr)$ (see thin line in Fig. 2.46). Therefore, we believe

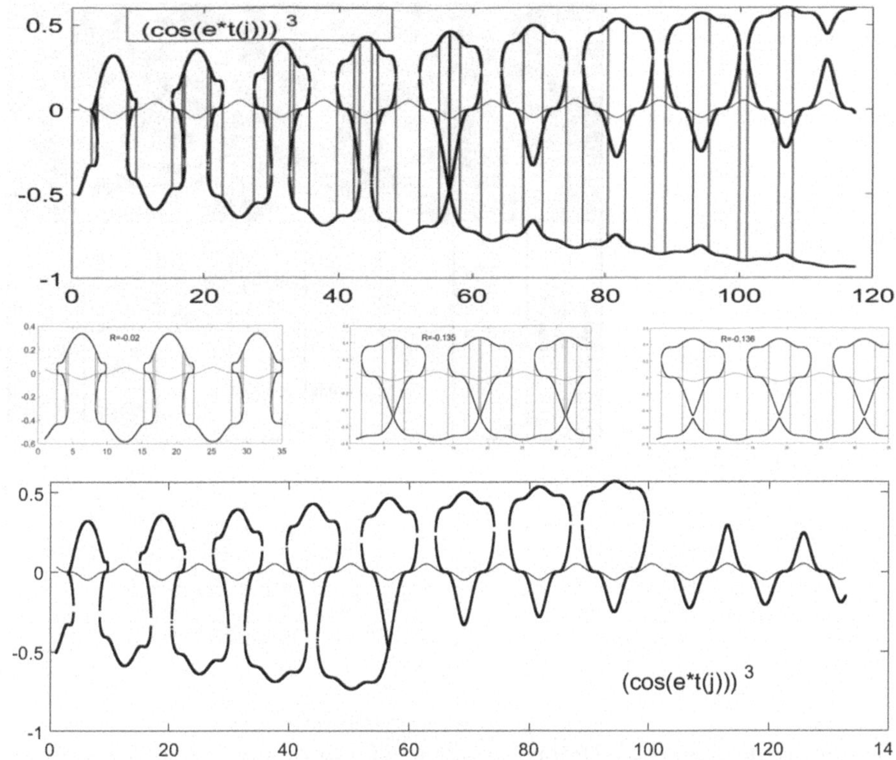

Fig. 2.46 Above is the case when R(r) decreases from 0.008 to −0.296. Below are calculations performed for fixed R(r) (R(r) = − 0.02, R(r) = − 0.135, and R(r) = − 0.136). Bottom is the case of generation of quanta (R(r) decreases from 0.008 to −0.296)

that Fig. 2.46 models, qualitatively, the results of the experiments shown in Fig. 2.47.

We here present experimental results for the Faraday waves excited by vertical oscillations having two frequencies. In general, it is clear that the results of our theory presented above are in qualitative agreement with the experiment.

Harmonic Excitation Above, we examined cases when the seed force in algebraic equations was specified $\cos^3 kr$ as or $\cos kr$. From a comparison of the results obtained, it follows that the calculation results for these two cases may differ significantly. This especially applies to cases when particle-waves appear in the system. We associate this difference with the emergence of new harmonics in the case $\cos^3 kr$. Let us evaluate here the influence of additional exciting harmonics on the occurrence of extreme multivalued waves.

Let us consider the case of two-frequency excitation. The calculation results are presented in Fig. 2.48. We used Eq. (2.98) where the excitatory function was specified by one of the following expressions: 0.05coskr or 0.05coskr - 0.025 cos 2kr or 0.05coskr - 0.05 cos 2kr.

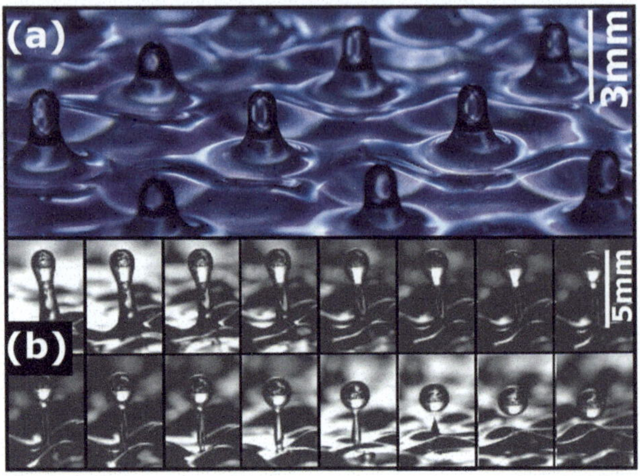

Fig. 2.47 (**a**) A side view of the Faraday waves at forcing acceleration composed of a combination of two commensurable frequencies (2.54 and 4.47). (**b**) Time sequence of a droplet ejection event. Successful images are separated by 1/800 s [61, 62]

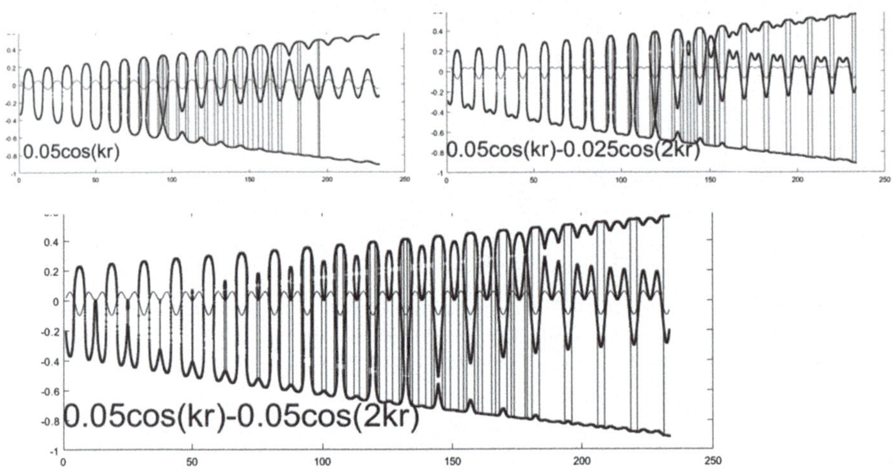

Fig. 2.48 Evolution of multivalued waves into particle-waves under three excitation options

It can be seen that the second harmonic can greatly change the nature of wave formation in the system. In particular, it becomes possible to form a complex system of particle-waves in a certain transresonant region, which is shown in Fig. 2.44.

Let us look at some more experimental data presented in a very interesting article [62, 63]. We will refer to this article several times below. Here we only emphasize that in the article, when studying waves on the contact surface of two liquids, two-frequency excitation was used. We modeled it with Eq. (2.98) written in the following form:

$$\varphi_2^3 + [1 - 0.0006(J - 1)]\varphi_2 + 0.9\left(\cos kr + \frac{1}{2}\cos 3kr\right) = 0. \tag{2.99}$$

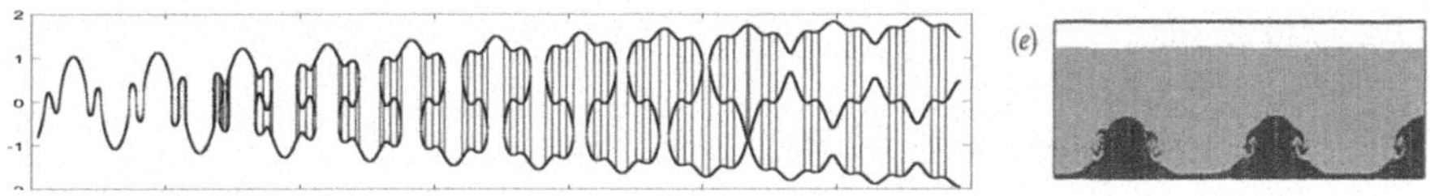

Fig. 2.49 Evolution of extreme multivalued waves during transresonance transition (left). Waves on the contact surface of two liquids corresponding to one of the values of the transresonant transition (right) [60, 62, 63]

Here J varies 1, 2, 3,. . . .., 3050. The calculation results are presented in Fig. 2.49. The results of the experiments are also given there.

The experimental study of extreme multivalued waves is a difficult problem. This may be the reason why we did not find many articles devoted to the experimental study of extreme multivalued waves. We will continue to refer to them in the future. These articles leave no doubt that the calculations presented above cover a huge range of wave resonance phenomena, from surface waves to quantum phenomena.

2.5.5 Stepwise Excitation

We constructed above solutions for the nonlinear component of waves under the assumption that the linear component is determined by harmonics, in particular one harmonic $\varphi^{(1)} = A \cos kr$. However, the linear solution of the wave equation can be represented by various functions. Below, we will consider the case when the seed wave is determined by the expression $A \tanh(\gamma \sin kr)$. It is remarkable that if $\gamma \ll 1$, then $\tanh(\gamma \sin kr)$ is well described by a harmonic function.

If γ increases the function, $\tanh(\gamma \sin kr)$ describes the transition from almost harmonic excitation to stepwise periodic disturbance of the system.

According to calculations (see Fig. 2.50), if $\gamma \ll 1$, then the results obtained for $A \tanh(\gamma \sin kr)$ are similar to those obtained for $\cos kr$. As γ increases, the discrepancy increases and if $\gamma > 1$ significant, step-like waves appear in the system. But in the most important respects, the situation has not changed qualitatively.

In both cases under consideration, there are some regions that can be called transresonant, where particle-waves can be excited.

Moreover, the shape and number of emerging particles can be quite diverse.

Indeed, nature often demonstrates the possibility of the existence of ambiguous solutions to various problems, in particular through the example of breaking and multivalued waves. In particular, even in the case of relatively simple cubic nonlinear systems, it can create very complex, for example, discontinuous and quantum processes.

As examples, we give wave processes that occur during vertical excitation of a granular medium (Fig. 2.51). The formation of bubbles (cavitation zones) in the vertical tubes was studied in [64]. The mechanics of forming bubbles was investigated by using glass beads lying on a vertical vibrating base of the tube. The aim of

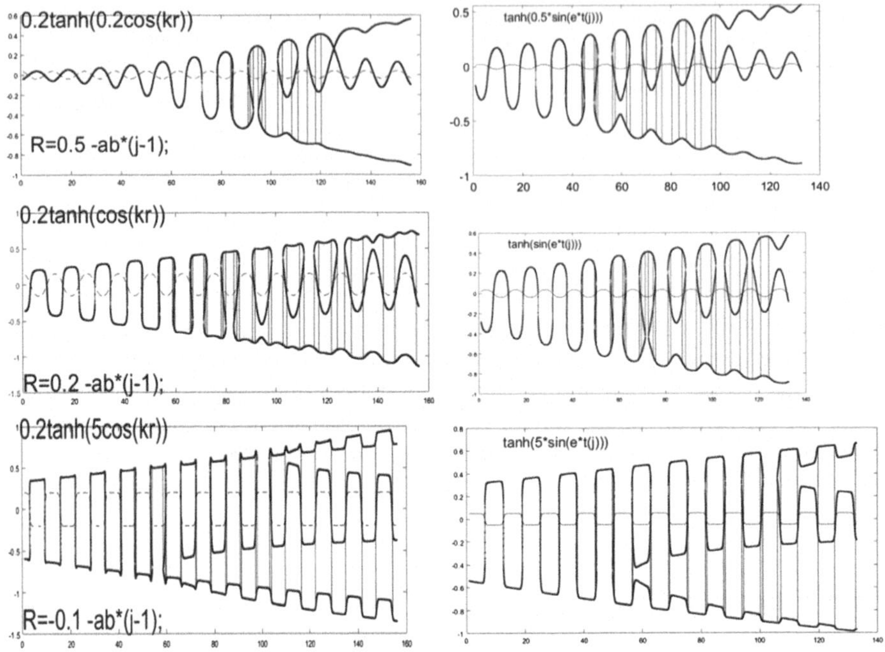

Fig. 2.50 Evolution of waves with an increase γ from 0.5 to 5

the experiments was to determine whether bubbling is possible in the granular beds which are subjected to vertical vibrations. Figure 2.51 (left) shows an image of the bed in which a bubble is created near the base, a second bubble is located in the middle of the bed, and a third bubble is approaching the free surface of the system. As the vertical acceleration increases, the bubbles become larger as shown in Fig. 2.49 (right). Bubbles close to each other may coalesce and form a larger bubble.

Figure 2.51 shows that continuous harmonic excitation is capable of exciting waves in the system that are separated from each other.

To illustrate this new idea, I assumed [19, 20] to use the results of Natanzon's experiments [52] with a water column. Figure 2.36 (left) shows the sketch of the experimental apparatus which Natanzon used. During these experiments, the water column was excited by the oscillating piston. The length of the column and the piston frequency were being changed. Figure 2.52 summarizes some results of the experiment. Oscillograms of the pressure, which is measured near the piston, are presented [52]. The sinusoidal curves show the piston position. During the experiments, the piston acceleration was slowly increased from 0.2 to 2 g, and then it was slowly reduced to 0.37 g. One can see that in the frequency range extending from 7.7 to 17 Hz, there are large variations of the pressure waves. When the down acceleration increases above g, then the harmonic oscillations transform into peaks which correspond to collisions of the water column and the piston. The smooth parts of the pressure curves correspond to the free flight of the column. During these

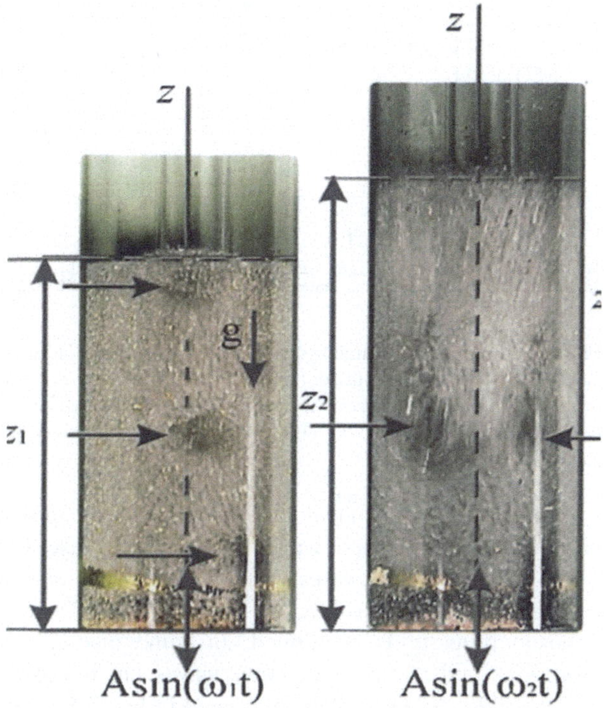

Fig. 2.51 Results of experiments in which bubbles are produced in vertically shaken granular materials [19, 20, 64]

moments in the water near the piston, the rarefaction (cavitation) zone is formed. When the acceleration reduces, then a strong hysteretic effect takes place. For this case, discontinuous waves exist in the system up to 0.5 *g* and then instantly transform into harmonic waves of very small amplitude.

In general, Natanzon's results resemble those presented in Fig. 2.51. Thus, continuous harmonic excitation is capable of exciting waves separated from each other in the system. Cavitation waves propagating up the pipe were not studied in Natanzon's experiments, but perhaps they were similar to the rarefaction waves shown in Fig. 2.51.

Thus, according to the experiments (see Figs. 2.51 and 2.52), the strongly nonlinear discontinuous waves may be generated in the upper-lying material, when the down acceleration of the base exceeds the gravity acceleration *g*. The linear continuous waves are excited if the vertical acceleration is smaller than 0.5 *g*. Thus, if the forced acceleration increases above some threshold level, then the linear prediction breaks down. The nonlinearity changes the amplitude and form of the forced waves. In particular, the rarefaction (cavitation) wave is periodically generated above the vibrating base. However, according to Fig. 2.52, the amplitude of the forced waves is bounded and depends on the hysteretic effect. Because of this effect, the maximal amplification takes place when the forced acceleration approximately equals 0.8 *g*.

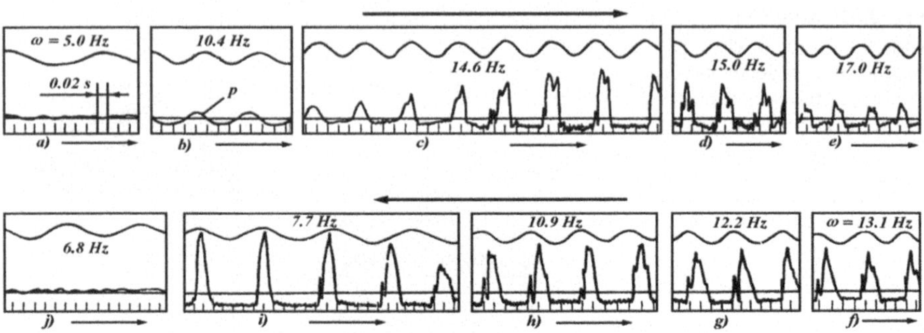

Fig. 2.52 Hysteretic dynamics of strongly nonlinear oscillations of the water column (the exciting amplitude is 0.002 m and the column length is 7 m). The vertical acceleration is slowly increased from 0.2 *g* to 2 *g* (forced frequency from 5 Hz to 17 Hz) and then is slowly reduced to 0.37 *g* (forced frequency from 13.1 Hz to 6.8 Hz) [19, 20, 30, 52]. Figures **a**–**j** determine the wave dynamics for change of frequency

We think that the considered possibility of the emergence of a chain of isolated, but seemingly connected, waves will appear in cases of interaction of several scalar fields, for example, in cases where the existence of waves (or particle-waves) of a certain property is possible, and the existence of waves (or particle-waves) of the opposite property is prohibited. In some media, only compression waves can propagate, and strong rarefaction waves are impossible. The opposite cases are also possible. An example is presented in Fig. 2.51.

Another possibility is offered in Figs. 2.44, 2.46, 2.47, 2.49, and 2.50, where a few transresonant wave evolutions are shown. Let us consider the simplest of them. We used all the real roots of the corresponding cubic algebraic equations during their construction. We followed Figs. 2.17 and 2.18. This allowed us to show the continuity of wave patterns describing, in particular, the evolution of Euler figures number 7 (Fig. 2.1) from harmonics to particle-waves. However, possibly, sometimes Nature cannot realize cases of one real root (see Figs. 2.41, 2.42, 2.43). In this case, the continuous sequence of evolving Euler figures breaks up into a chain of individual fragments. We named them quanta (Figs. 2.44 and 2.46). Such a scenario is shown in Fig. 2.53.

These fragments can be imagined as individual particle-waves, quanta, or vortices that arise in a certain resonant region when a seed harmonic wave enters it. It is remarkable that upon leaving this region, the harmonic form of the seed wave was restored. Thus, it is possible for the seed wave to reenter another resonant region and again decay into fragments. Perhaps this possibility corresponds to the so-called intermittent turbulence (a periodic change of turbulent and laminar regimes of wave fluid flows).

As another possibility, we point out the case of ultrashort distances. One can imagine that with a continuous decrease in the distance between the peaks of the waves, a continuous wave will split into a sequence of isolated clots of energy.

This possibility opens up even for harmonic waves if the distance between the vertices approaches, the corresponding Planck constant. This understanding will be

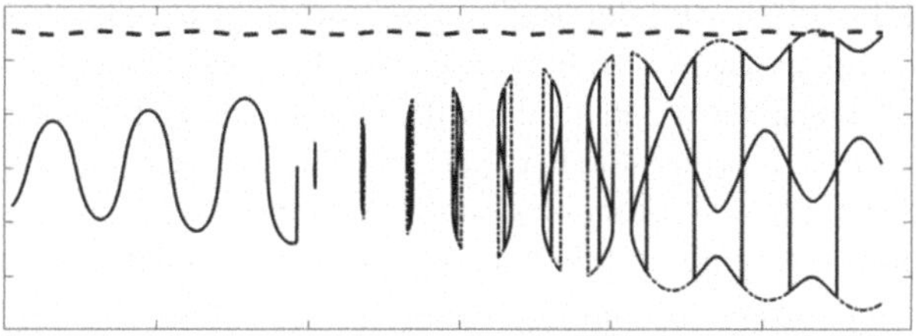

Fig. 2.53 Decay of Euler figures (Fig. 2.1) into fragments in the resonance zone

helped by Figs. 2.17, 2.18, 2.19, 2.46, and many other figures presented in this chapter. As an illustration of this possibility, consider Fig. 2.17 (curves **c**). We emphasize that when constructing these curves, all solutions of the cubic nonlinear equation are used. However, it is possible to use only solutions defining curves **a** (Fig. 2.17). That is, to assume that at Planck distances the weakly changing branch of the solution is not realized, and the wave process breaks up into a chain of energy clusters, that is, into a chain of isolated waves.

We emphasize that the issues of transitions of continuous wave processes into discontinuous ones, their fragmentation, and the emergence of particle-waves are very interesting to us, and we will return to them constantly throughout the book. These transitions connect our study of EMWs with the problems of quantum mechanics. Indeed, the oddities of quantum physics will seem less surprising if you stop thinking about electrons and energy quanta as point masses, but rather as the EMWs studied in this book, or at least as particle-waves. The idea of the wave nature of quantum mechanics is certainly not new, but apparently, the emphasis on the resonant and nonlinear nature of quantum mechanics is made most strongly in the author's works [19, 20]. This approach avoids many of the difficulties of traditional approaches.

Remark The problem of creating very short pulses can be solved by superposing (summing) many harmonics with their appropriate synchronization. The high value of the peak intensity usually leads to the need to take into account nonlinear harmonic interactions even in this approach. Implementation of the highly nonlinear process shown in Fig. 2.51 opens up the possibility of generating ultrashort pulses of Planck duration.

Conclusion In this section, we shown as harmonic waves entering a certain resonant region transform into multivalued waves, particle-waves, quanta, or vortices.

2.6 Transresonance Evolution and Effects of Fifth-Order Nonlinearities

So far, we have limited ourselves to studying the effect of cubic nonlinearity on waves. Here, we will carry out model calculations that can qualitatively describe the contribution of fifth-order nonlinearities to the evolution of waves. In particular, we are particularly interested in the process of evolution of Euler figures into particle-waves and wave-vortex structures. We emphasize that wave equations containing nonlinear terms up to the fifth degree are well known, although they are poorly studied. It is possible to expect that this view can be fruitful in the case of research of different wave processes, since wave processes are described by the similar equations. Developing this thought, we will consider the Airy-type Eq. (1.49):

$$u_{tt} - a_0^2 u_{xx}\left(1 - \frac{3}{2}u_x^2 + \frac{15}{8}u_x^4 u_x^2\right) + k u_{xxxx} = 0. \tag{2.100}$$

Recall that this equation describes highly nonlinear vibrations of strings (1.63). The Klein-Gordon's equation (1.12) may contain the same type of nonlinearity:

$$\varphi_{tt} - c_0^2 \varphi_{xx} = m^2 \varphi - \lambda \varphi^3 - \lambda_5 \varphi^5. \tag{2.101}$$

For three-dimensional wave processes, Klein-Gordon's equation may be written as

$$\varphi_{tt} - c_0^2 \nabla^2 \varphi = m^2 \varphi - \lambda \varphi^3 - \lambda_5 \varphi^5. \tag{2.102}$$

It is possible to continue the list of the wave equations containing explicitly the d'Alembertian operator and highly nonlinear terms. We will study cases when this operator can be equal to 0 or very small value (a "small" divider). We will consider unidirectional waves. Using the procedure described in Sects. 2.2.1 and 3.1, we can reduce Eqs. (2.100), (2.101), and (2.102) to a fifth-order algebraic equation containing a forcing (seed) term. We will write this model equation in the following form (see also (2.35), Sect. 2.2.2, and Chap. 4)

$$-0.11\varphi_2^5 + \varphi_2^3 + r_* \varphi_2^2 + 3R/2^{2/3}\varphi_2 + \cos L^{-1} r = 0. \tag{2.103}$$

Here L has the dimension of the length and L equals 1. Equation (2.103) is the algebraic equation which may be solved numerically. We introduce in Eq. (2.103) the term $r_* \varphi_2^2$ to appreciate the effect of quadratic nonlinearity on wave profiles.

Highly Nonlinear Effects Let us assume $r_* = -0.1$ or $r_* = 0.1$ in Eq. (2.103). The transresonant evolution surface waves calculated for $r_* = -0.1$ are shown in Fig. 2.54. The transformation of the initial harmonic waves (0.3) into the shock-like waves (−0.05) and elastica-like waves (profiles −0.55, −0.61) is seen. Then, we study the influence of r_*. For $R = -0.82$ and $r_* = -0.15$, the elastica-like profiles bifurcate into the elastica-like waves + the periodical ellipsoidal structures. These structures may be considered as bubbles periodically generated under the interface. These bubbles disappear, and the loops form on the interface (−0.252) (see also Euler figure 11 in Fig. 2.1). The loops transform into ellipsoidal structures (drops or particles (see Figs. 2.48 and 2.58)) above the saw-like wave surface (−0.26). Finally, the harmonic waves again form on the liquid surface (−1).

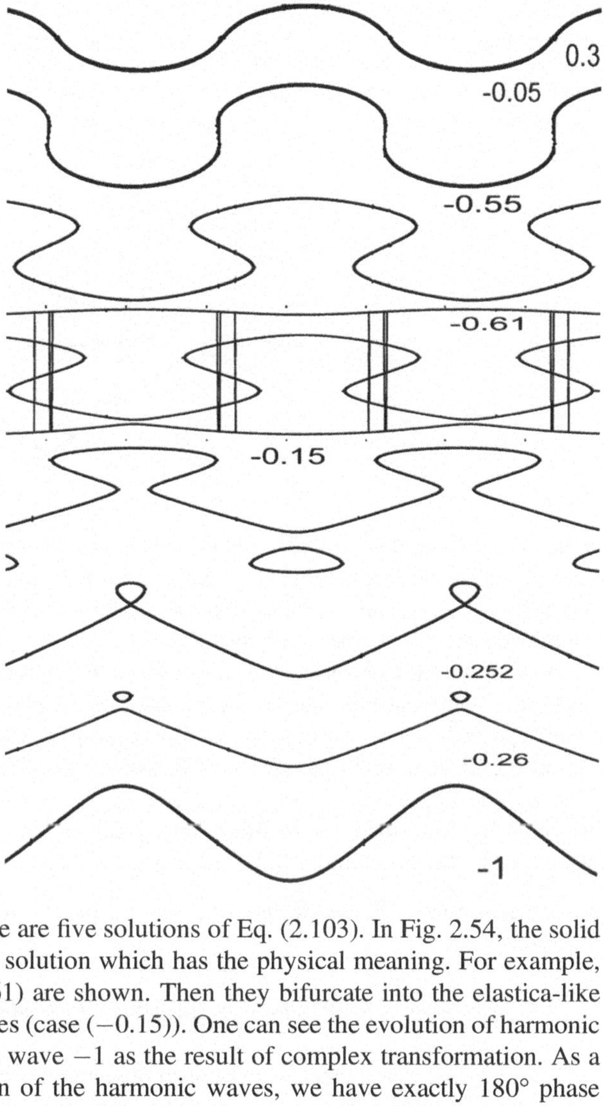

Fig. 2.54 The effect of strong nonlinearity. The evolution of the waves into the drop (particles)-, bubble-, and vortex-like structures. The parameter R is varied from 0.3 to -0.61 when $r_* = -0.1$. Then R is the constant ($R = -0.82$), but r_* is varied from -0.15 to -1

Generally speaking, there are five solutions of Eq. (2.103). In Fig. 2.54, the solid line corresponds to the real solution which has the physical meaning. For example, three curves for case (-0.61) are shown. Then they bifurcate into the elastica-like waves + the elliptic structures (case (-0.15)). One can see the evolution of harmonic wave 0.3 into the harmonic wave -1 as the result of complex transformation. As a result of this transformation of the harmonic waves, we have exactly 180° phase difference between them.

The transresonant evolution waves calculated for $r_* = 0.1$ are shown in Fig. 2.55. The bifurcation of the elastica-like waves (-0.6) into the elastica-like waves + the periodical ellipsoidal structures (-0.82) is seen. These structures may be considered as the drops (particles) periodically generated above the interface having the craters. The craters evolve into the ellipsoidal structures (bubbles) under the sawlike surface (-0.3).

We emphasize that, at the whole, the results of Figs. 2.54 and 2.55 correspond to the Euler data, data of experiments (Figs. 2.45, 2.47, and 2.58), and data for cubic nonlinear models (Sect. 2.2.2). It is found that the generation of the elastica-like

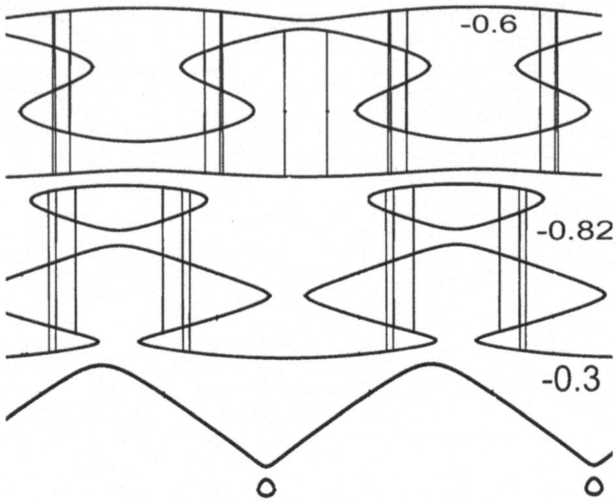

Fig. 2.55 The effect of strong nonlinearity. The parameter R is varied from 0.3 to -0.82 when $r_* = -0.1$. Then R is the constant ($R = -0.82$), but r_* is -0.3

waves is connected with cubic nonlinearity. The drops (particles) and bubbles are formed due to both quadratic and cubic nonlinearities. For the liquid surface, the generation of drops and bubbles may be explained by both the capillary effect and nonlinear elasticity of the liquid surface.

In the latest calculations, we found only one example of the strong influence of fifth-order nonlinearities on cubic nonlinear calculations. This is the case of -0.82 in Fig. 2.55 and -0.61 in Fig. 2.54. The examples contradicted our calculations presented in other sections, so we will study it here additionally focusing on the fundamental experiments [59, 63]. There highly nonlinear waves parametrically excited at the interface of two fluids were studied. These waves have already been discussed in this book and will be discussed further in special sections.

Transresonance Bifurcations The governing equation was taken in the following form:

$$-0.11\varphi_2^5 + \varphi_2^3 - 1.6875\varphi_2 + A\left[\sin\left(1.5L^{-1}r\right) - \sin\left(4.5L^{-1}r\right)\right] = 0. \quad (2.104)$$

The equation parameters were chosen absolutely arbitrarily. The purpose of the calculations is to trace the influence of the amplitude of the seed disturbance on the wave shape. The harmonics of the seed disturbance are equal in amplitude. Their frequencies differ by a factor of three. The calculation results are presented in Fig. 2.56.

Calculations were carried out for amplitudes A of the seed disturbance varying from 0.4 to 2.4. For those values of A, the bifurcation of one type of wave into another type occurs. Let us note the following bifurcation: when passing from 0.4 to 0.5, the two-frequency harmonic curve transforms into a one-story mushroom-shaped structure reminiscent of the experimentally recorded strongly nonlinear evolution of initially harmonic waves [42, 66]. We emphasize that on

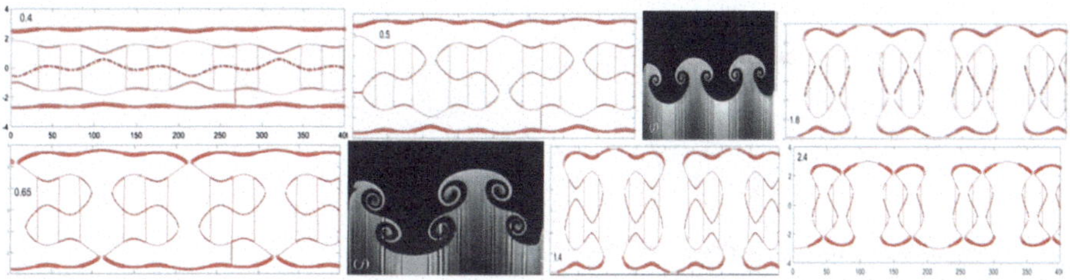

Fig. 2.56 Evolution and bifurcation of strongly nonlinear waves with an increase in the amplitude of the seed disturbance from 0.4 to 2.4. The calculation pictures are supplemented with experimental data on studies of the emergence and evolution of multivalued wave-vortex structures

the right side, there are calculated curves obtained for $A = 1.8$, which also qualitatively resemble the data of the indicated experiment. With this, we finish our consideration of the upper wave patterns shown in Fig. 2.56. The bottom row of pictures shows the result of the bifurcation of the calculation for $A = 0.5$ into the patterns obtained for $A = 0.65$ and $A = 1.4$. Thus, the one-story mushroom structure evolves into a two-story wave structure. We emphasize that the two-story wave structures found in the calculation qualitatively describe the evolution of two-story multivalued wave-vortex structures observed in the experiments [42, 66].

The Transresonant Evolution In the above sections, we examined the emergence and evolution of extreme multivalued waves when passing through resonant zones. The cubic nonlinear theory was used. Many extraordinary wave processes have been discovered. Now we have the opportunity to study the influence of fifth-order nonlinearities on the transresonant process.

Calculations were carried out using equation

$$-0.11\varphi_2^5 + \varphi_2^3 - 2.5(-0.3 + 0.005r)\varphi_2 - 1.47\cos\left(0.25L^{-1}r\right) = 0. \qquad (2.105)$$

Here r decreases from 30 to -210. Thus, at $r = 60$, the linear term in (2.105) vanishes.

In general, the calculation in Fig. 2.57 resembles cases obtained for cubic nonlinearity. As before, the possibility of the emergence of particle-waves during the passage of a resonant band was found. However, in the latter case, particles appear not only above the top of the waves and under the troughs. The process of transition from multivalued waves to particle-waves has now become significantly more complicated. Thus, first, a one-story mushroom-shaped structure appears, which is then transformed into a two-story mushroom-shaped structure through complex bifurcation, and only then wave formations arise, which we call particle-waves!

We got an interesting result that can be interpreted in completely different ways. In particular, it can be associated with processes that arise during surface cavitation.

In Fig. 2.58, results are shown of acoustic cavitation generated by ultrasonic vibrations. It is found that strong forcing leads to the excitation of nonlinear surface

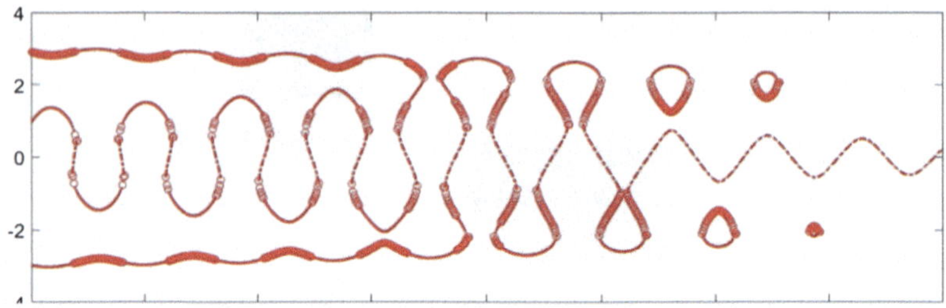

Fig. 2.57 Curves modeling of atomization of a string (surface) or evolution of multivalued waves into a structure reminds the Karman vortex street

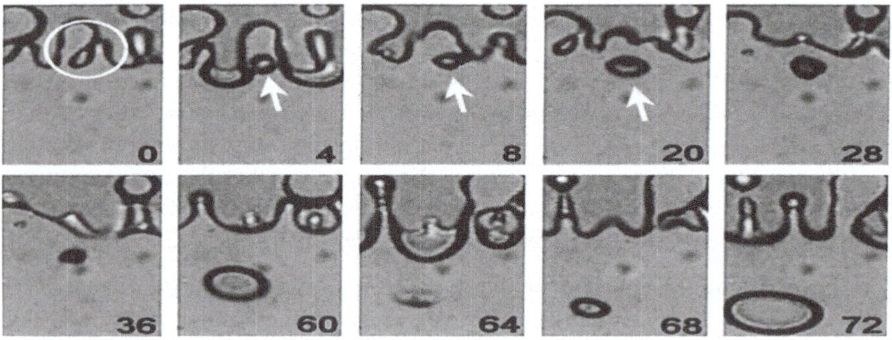

Fig. 2.58 Selected frames showing the entrapment of gas bubbles and separation of drops during surface cavitation [60, 65]

waves on gas-liquid interfaces. If the excitation is strong enough, then both the appearance of droplets and the generation of air bubbles occur simultaneously. Qualitatively, these processes correspond to the scheme described by Eqs. 2.35 and 2.103 and shown in Fig. 2.58.

The study of the influence of the fifth-order nonlinearity on extreme multivalued waves began in a 2001 paper [28].

Extreme waves can change in time and space so that they break up into waves, vortices, and particles, or into a chain of isolated particle-waves (Fig. 2.59) [62].

Calculations were carried out according to the equation

$$-0.11\varphi_2^5 + \varphi_2^3 + \left(3R/2^{2/3}\right)\varphi_2 + \cos(kr) = 0. \tag{2.106}$$

It was accepted that $R = R(r) = 0.0025kr + \tanh(0.025kr)$, where kr during calculations varied from 20 to -980. Therefore, the parameter $R = R(r)$ varies from 0.4 to -1.25. The harmonic wave grows and transforms into shock-like wave and the Euler figures, when $R(r)$ reduces from 0.4. The solutions bifurcate when $R(r) \approx -1$. As a result, two rows of particles (vortices) and the saw-like curve are generated. The particles reduce and disappear when $R(r) \approx -1.2$. Moreover, the

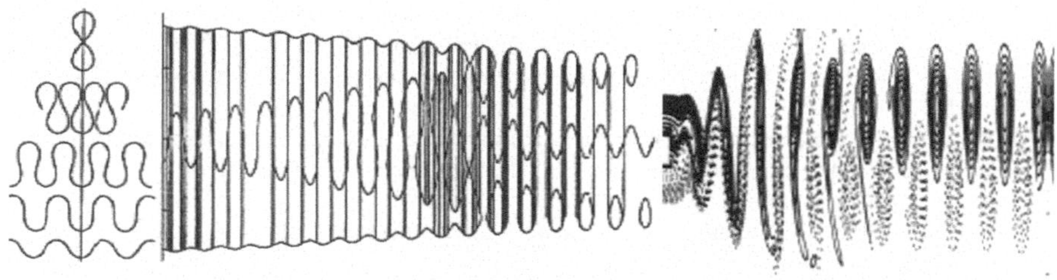

Fig. 2.59 Elastica figures of Euler (left) and examples of the evolutions of harmonic waves into elastica-like waves, particle-waves, and von Karman chain (center and right (experiment data))

saw-like curve transforms into the harmonic curve. It is possible to give different interpretations of Figs. 2.57 and 2.59 [67].

As a whole, Fig. 2.59 can be treated as some model of the processes accompanying formation of the Kármán vortex street. On the other hand, Fig. 2.59 may be treated as some "pitchfork" bifurcation diagram, which describes the evolution of initial harmonious waves into "elastica" forms, ellipse structures, and sawtooth curve. First, the harmonious wave evolves into shock waves and further into the "elastica" form. In some sense, this site corresponds to the "handle" of "the pitchfork." Then the transitive zone follows, where may be the jumps from curves (solutions) to another curves (another solutions). The zone corresponds to the bifurcating zone of the pitchfork. Behind this zone, the sawtooth curve and two chains of the ellipse structures are being formed. In some sense, they form the "pitchfork teeth." Certainly, the described scheme is only very rough diagram of possible wave bifurcations. We underline that this diagram corresponds to the transresonant wave evolution.

Thus, we have shown that taking into account fifth-order nonlinearities in wave equations in some cases allows us to detect bifurcations that are not described by cubic nonlinear equations. Apparently taking into account fifth-order nonlinearities can be very important. Indeed, the choice of coefficients in Eqs. (2.103), (2.104), (2.105), and (2.106) was absolutely random. Nevertheless, these equations described qualitatively such wave structures and bifurcations that were observed during experiments but cannot be described by cubic nonlinearity.

References

1. Bernoulli JJ (1999) Quadratura curvae, e cujus evolutione describitur inflexae laminae curvatura, Die Werke von Jakob Bernoulli: Bd. 5: Differentialgeometrie, Birkhäuser, p 223–227 [1692. Med. CLXX; Ref. UB: L Ia 3, p. 211–212]
2. Euler L (1744) Methodus inveniendi lineas curvas... sive solutio problematis isoperimetrici. (cum tabulis aeneis), vol 14. Marcus-Michael Bousquet
3. Laplace PS (1880) Œuvres complètes de Laplace, vol 4. Gauthier-Villars, p 349–401

4. Faraday M (1831) On a peculiar class of acoustical figures and on certain forms assumed by groups of particles upon vibrating elastic surface. Phil Trans R Soc Lond 52:299
5. Kirchhoff GR (1859) On the equilibrium and the movements of an infinitely thin bar. Crelles J Math 56
6. Born M (1906) Untersuchungen über die Stabilität der elastischen Linie in Ebene und Raum, unter verschiedenen Grenzbedingungen. PhD thesis, University of Göttingen
7. Born M (1968) My life and views. Scribner
8. Love AEH (1944) A treatise on the mathematical theory of elasticity, 4th edn. Dover Publications
9. Taylor GI (1969) Instability of jets, threads, and sheets of viscous fluid. In: Applied mechanics: proceedings of the twelfth international congress of applied mechanics. Stanford University, Springer, Berlin, p 382–388
10. Truesdell C (1987) Der Briefwechsel von Jacob Bernoulli, chapter Mechanics, especially Elasticity, in the Correspondence of Jacob Bernoulli with Leibniz
11. Jones N (1989) Structural impact. Cambridge University Press, Cambridge
12. Raragiozova D, Jones N (2002) On dynamical buckling phenomena in axially loaded elastic-plastic cylindrical shells. Int J NonLin Mech 37:1223
13. Goss VGA (2003) Snap buckling, writhing and loop formation in twisted rods. Thesis of Doctor of Philosophy, The University of London
14. Pocheau A, Roman B (2004) Uniqueness of solutions for constrained Elastica. Phys D 192:161
15. Fraser CG (2007) Mathematical technique and physical conception in Euler's investigation of the Elastica. Centaurus 34(3):211
16. Levien R (2008) The elastica: a mathematical theory. http://levien.com/phd/elastica_hist.pdf
17. Lamb H (1932) Hydrodynamics, 6th edn. Dover Publications, New York
18. Linde A (1990) Particle physics and inflationary cosmology. CRC Press, Boca Raton
19. Galiev SH (2020) Evolution of extreme waves and resonances. CRC Press, p 508
20. Galiev SU (2015) Darwin, geodynamics and extreme waves. Springer
21. Сибгатуллин НР (1972) О нелинейных поперечных колебаниях при резоансе ы упругом слое и слое идеальной проводящцкй жмдкостю ПММ, том 36. Стр, p 79
22. Сибгатуллин НР (1984) Колебания и волны в сильных гравитационных и электромагнитных полях. Наука, Москва. (Sibgatullin NR (1991) Oscillations and waves in strong gravitational and electromagnetic fields. Springer)
23. Филиппов АТ (1986) Многоликий солитон. М., Наука, 1986 (in Russian) (Filippov AT (2010) The versatile soliton, Birkh user. Boston)
24. Nishinari K (1997) Nonlinear dynamics of solitary waves in an extensible rod. Proc R Soc Lond A 453:817–833
25. Fukumoto Y (2007) Analogy between a vortex-jet filament and the Kirchhoff elastic rod. Fluid Dyn Res 39:511–520
26. Parkes EJ (2008) Some periodic and solitary travelling-wave solutions of the short-pulse equation. Chaos, Solitons Fractals 38(1):154–159. Parkes EJ (2010) A note on loop-soliton solutions of the short-pulse equation. Phys Lett A 374:4321–4323
27. Ilgamov MA, Galiev ShU, Mace BR, Galiyev TSH (2015) Modelling of resonant and transresonant waves in natural resonators: from gravity waves to the origin of the Universe. Известия Уфимского Научного Центра Российской Академии Наук (Herald of Ufa Scientific Center, Russian Academy of Sciences (RAS)) 4:26–37. http://journal.ufaras.ru/wp-content/uploads/2021/12/izvestiya_4_2015_1-126.pdf
28. Galiev SU, Galiyev TS (2001) Nonlinear trans-resonant waves, vortices and patterns: from microresonators to the early universe. Chaos: an interdisciplinary. Journal of Nonlinear Science 11(3):686–704
29. Waitz IA, Greitzer EM, Tam CS (1995) Aero–propulsion systems. In: Green SL (ed) Fluid vortices. Kluwer Academic Publishers
30. Galiev SU (2011) Charles Darwin's geophysical reports as models of the theory of catastrophic waves. Moscow, pp. 655 (in Russian)

31. James AJ, Vukasinovich B, Smith MK, Glezer A (2003) Vibration-induced drop atomization and bursting. J Fluid Mech 476:1–28
32. Vilenkin A (2006) Many worlds in one. Hill and Wang, New York
33. Galiev SU (2003) The theory of non-linear transresonant wave phenomena and an examination of Charles Darwin's earthquake reports. Geophys J Int 154:300–354
34. Galiev SU (2009) Modelling of Charles Darwin's earthquake reports as catastrophic wave phenomena. The University of Auckland
35. Ostrovsky LA, Potapov AI (1999) Modulated waves. Theory and applications. Johns Hopkins University Press, Baltimore and London
36. Bullough RK, Jack PM, Kitchenside PW, Saunders R (1979) Soliton in laser physics. Phys Scr 20:364
37. Kneubhl FK (1997) Oscillations and waves. Springer
38. Chester W, Moser A (1982) On an equation which describes the resonant oscillations of a gas in an open-ended tube. Proc R Soc Lond A 380:409–415
39. Ilgamov MA, Zaripov RG, Galiullin RG, Repin VB (1996) Nonlinear oscillations of a gas in a tube. Appl Mech Rev 49:138–154
40. Chester W (1981) Resonant oscillations of a gas in an open-ended tube. Proc R Soc Lond A 377:449–467
41. Merkli P, Thomann H (1975) Transient to turbulence in oscillation pipe flow. J Fluid Mech 68: 567–575
42. Niederhaus CE, Jacobs JW (2003) Experimental study of the Richtmyer-Meshkov instability of incompressible fluids. J Fluid Mech 485:243
43. Galiev SU (2003) Transresonant evolution of wave singularities and vortices. Phys Lett A 311: 192–199
44. Sarpkaya T (2002) Experiments on the stability of sinusoidal flow over a circular cylinder. J Fluid Mech 457:157–180
45. Galiev SU, Galiev TS (1994) Linear and nonlinear forced oscillations of the flow of bubble liquid in a deformable pipe. Problems Strength 26(9):633–654
46. Schowalter DG, van Atta CW, Lasheras JC (1994) A study of streamwise vortex structure in a stratified shear layer. J Fluid Mech 281:247–291
47. Suplee C (1999) Physics in the 20th century. Harry N. Abrams, New York
48. Poincaré H (1899) Les méthodes nouvelles de la mécanique céleste, vol 3. Gauthier-Villars et fils
49. Пригожин И (2001) Конец определенности. РХД, Москва (Prigogine I (1997) The end of certainty. The Free Press)
50. Пригожин И, Николис Г (2008) Познание сложного. Издательство ЛКИ, Москва (Nicolis G (1989) Prigogine I Exploring complexity. W. H. Freeman, New York)
51. Hawking SW (2011) The illustrated theory of everything: the origin and fate of the universe. Phoenix Books
52. Natanzon MS (1965) Forced discontinuous oscillations of liquid in tubes. Izv. Akad. Nauk USSR, Mechanics, 2, 33–42 (in Russian); Natanzon MS (1977) Продольные Автоколебания Жидкостной Ракеты (Longitudinal selfexcited oscillations of a liqquid-fuel rocket). Mashinostroenie, Moscow
53. Van der Pol B (1926) On oscillation hysteresis in a simple triode generator. Philos Mag 43:700–719
54. Nayfeh AH (1973) Perturbation Mechods. Wiley, New York
55. Thompson JMT, Soliman MS (1991) Indeterminate jumps to resonance from a tangled saddle-node bifurcation. Proc R Soc Lond A 432:101–111
56. Galiev SU (1999) Topographic effect in a Faraday experiment. J Phys A Math Gen 32:6963–7000
57. Galiev SU, Ilgamov MA, Sadikov AV (1970) Periodic shock waves in a gas. Izv. Akad. Nauk USSR, Mech. Fluid and Gas 2:57–66

58. Galiev SU (1988) Nonlinear Waves in Bounded Continua. Naukova Dumka: Kiev (in Russian); Galiev SU (1972) Forced oscillations of a nonlinear elastic layer. Izv. Acad. Nauk USSR, Mech. Solid Body 1:58–63; Galiev SU (1972) Forced longitudinal oscillations of a nonlinear elastic body. Izv. Acad. Nauk USSR, Mech. Solid Body 4:80–87; Galiev SU (1973) Nonlinear gas oscillations maintained in a closed pipe by a heat source. Soviet Aeronautics 16(2):7–12; Galiev SU, Galiyev TS (1994) Linear and discontinuous forced oscillations of flow of bubbly liquid in a Deformable Pipe (survey). Strength Mater 9:633–65
59. Morgan RV et al (2012) On the late-time of the two-dimensional Richtmyer-Meshkov instability in shock tube experiments. JFM 712:354–383
60. Galiev SU (2024) Multivalued waves and the Richtmyer-Meshkov instability as the causes of the formation of galaxies, vol 1. Herald of Ufa Scientific Center Russian Academy of Sciences, p 5–17. https://doi.org/10.31040/2222-8349-2024-0-1-5-17
61. Sampara N, Gilet T (2016) Two-frequency forcing of droplet rebounds on a liquid bath. Phys Rev E 94(5):053112
62. Ilgamov MA (2023) Annotation book Galiev S.U. "Extreme multivalued waves in scalar fields" planned by Springer for publication in early 2024. Herald of Ufa Scientific Center, Russian Academy of Sciences. No 3, p 115–120
63. Liu D, Lin P (2022) Two-layer liquids Interface instabilities in Faraday waves of two-layer liquids with free surface. J Fluid Mech 941:A33. https://doi.org/10.1017/jfm.2022.259
64. Zamankhan P (2012) Solid structures in a highly agitated bed of granular materials. Appl Math Model 36:414–429
65. Ohl SW, Ow DSW, Klaseboer E, Wong VV, Camattari A, Ohl CD (2010) Creation of cavitation activity in a microfluidic device through acoustically driven capillary waves. Lab Chip 10(14): 1848–1855
66. Cavelier M, Gréa B-J, Briard A, Gostiaux L (2022) The subcritical transition to turbulence of Faraday waves in miscible fluids. J Fluid Mech 934:A34. https://doi.org/10.1017/jfm.2021.1124
67. Newell AC, Rumpf D (2011) Wave turbulence. Annu Rev Fluid Mech 43:59–78

Chapter 3
Waves in Resonators: d'Alembert Method and Multivalued Faraday Waves

At the beginning of the chapter, we will present some results of publications that indicate the possibility of forming multivalued waves in resonators. We have already discussed the results of Chester [1, 2]. Another remarkable study was published in 1968 [3]. Two-dimensional internal gravity waves in a rectangular container are examined theoretically and experimentally. The possibility of the formation of waves with stable folds on the profile in the case of resonant excitation of the liquid is shown. The resulting folds and wave profiles are presented in Fig. 3.1.

In a paper [5], the dynamics of standing periodic waves at the interface between two inviscid fluids with different densities, subject to monochromatic oscillations normal to the unperturbed interface, is studied under normal- and low-gravity conditions. Calculation results are shown in Fig. 3.2 (left).

We showed in Fig. 3.2 (right) the wave evolution of the liquid layer laying on another liquid. It is easy to see that the presented wave profiles resemble the Euler figures shown in Fig. 2.5. In particular, the curves $\lambda = 2, 1, 0.5, 0.4, 0.35$ in Fig. 2.5 qualitatively describe the instability of the interface of two liquids with different densities, which was shown in Fig. 3.2 (right).

The evolution of the perturbations of interface of two environments into vortices was studied in [4, 6, 7, 9]. They are shown by Fig. 3.3 (a, at the left). The interface changes as a result of growth of the perturbations (t = 100, 300 and 1700 сек). Figure 3.3 (a, on the right) shows positions of vortices at the same moments of time. Thus, Fig. 3.3a demonstrates the evolution of the originally smooth interface into mushroom-like waves. Further, they are transformed into a chaotic mix of two environments and vortices. Figure 3.3b shows a similar development of instability of an interface calculated according to Ginzburg-Landau equation.

Multivalued waves, similar presented Figs. 3.1, 3.2, and 3.3, are possible in the most various environments and fields. Everywhere, where there are enough strong interaction of media and high—nonlinear effects. In certain cases, the surface tension is responsible for appearance of such waves. Thus, a behavior of the systems consisting of two or more adjoining environments (phases) can be determined by a thin layer of the contact between these environments (phases). For example, in the

S. U. Galiev, *Extreme Multivalued Waves as a Model of Everything*,
https://doi.org/10.1007/978-3-031-67027-5_3

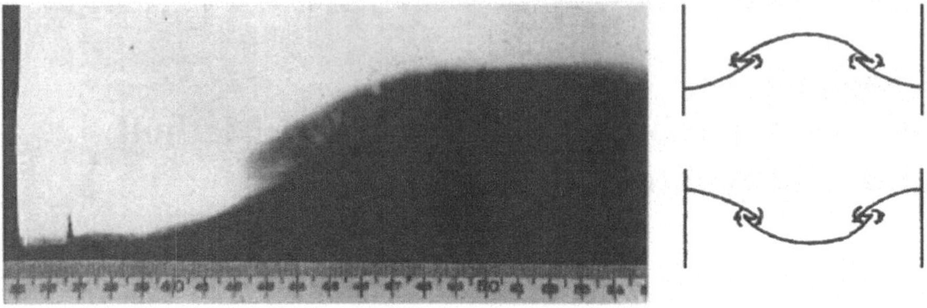

Fig. 3.1 An element of the wave multivalued profile observed in [3] (**a**). The profiles of wave multivalued wave for two moments of time (**b**) [3, 4]

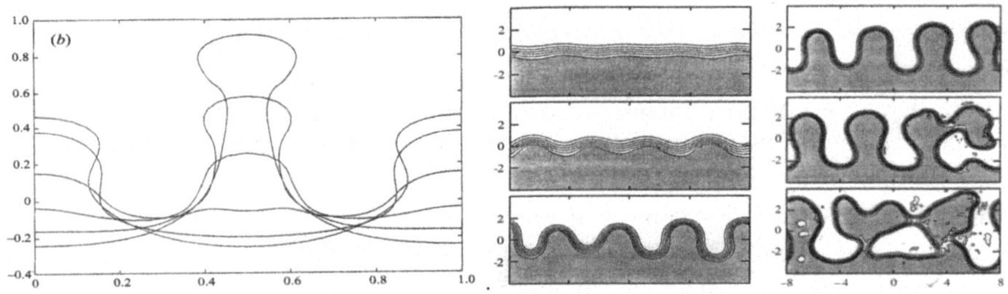

Fig. 3.2 Formation and evolution of the mushroom-like standing waves on an interface of two liquids during Faraday's type experiments [5–7] (left). Results of numerical simulation of forms of the thin surface layer [6–8] (right)

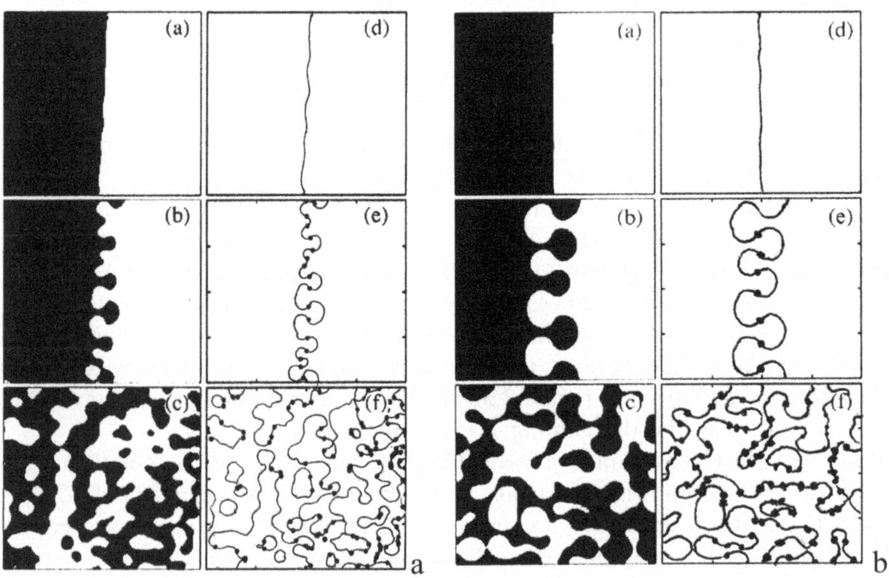

Fig. 3.3 Evolution of the initial perturbations of some interface into the "elasica" forms, vortices, and turbulence. The occurrence of the vortices, which is shown by spots, during the Belousov-Zhabotinsky reaction (**a**). The occurrence of the vortices and turbulence according to the numerical solution of the Ginzburg-Landau equation (**b**) [9]

case of very short surface waves, mechanical properties of such surface are characterized, first of all, by the surface tension. This tension defines properties of a surface similar to properties of a thread (string).

In the previous chapter, we looked at numerous examples of the emergence and propagation of unidirectional waves. Resonant amplification of them was associated with the approach of the wave speed to a certain characteristic value. Of course, the examples considered do not even to a small extent describe the wealth of resonant processes existing in natural and artificial environments. In this chapter, we will include in our consideration cases of the occurrence of extreme waves in limited media (in resonators). The theory of such waves was proposed in 1988 [10, 11] and developed in [12–14] for multivalued resonant waves. Here we will use the approaches developed there, with the main attention being paid to modeling of recently published experimental data.

D'Alembert proposed to describe the wave field by two oppositely traveling waves. We are developing this approach for extreme waves in finite resonators. In particular, we study the wide spectrum of Faraday waves.

3.1 Generalization of the d'Alembert's Solution for Nonlinear Long Waves

Let us consider Eq. (1.31). If we put $q = 1$ and $T = t$, we get the Airy equation,

$$u_{tt} = a_0^2 u_{aa}(1 + u_a)^{-3}. \tag{3.1}$$

Airy [15] predicted that a nonlinear wave could not propagate without change of form. Generally speaking, (3.1) can describe as continuous so discontinuous, multivalued waves. Let us rewrite this equation by discarding the very high nonlinear terms and introducing the dispersion term ku_{xxx} and the viscosity term μu_{xxt}. For further use, we introduce in (3.1) a new notation for the Lagrangian variable a. In what follows, we will denote this variable in the same way as the Euler variable, that is, we will assume that $a = x$.

Then Eq. (3.1) will take the form

$$u_{tt} - a_0^2 u_{xx} = \left(\beta u_x + \beta_1 u_x^2 + \beta_2 u_x^3 + \beta_3 u_x^4\right) u_{xx} + \mu u_{xxt} + k u_{xxxx} \tag{3.2}$$

Equation (3.1) may be considered as the generalization of Airy's theory. If $a_0^2 = gh$ $\beta = -3a_0^2$, $\beta_1 = 6a_0^2$, $\beta_2 = \beta_3 = 0$, $\mu = 0$, and $k = 0$ in (3.1), we have the Airy equation written for cubic nonlinear waves. Equations (3.1) and (3.2) describe longitudinal waves [13]. In the case of the surface waves, we can calculate vertical displacements using the formula (1.40) rewritten in the form

$$\eta = h\left[(1 + u_x)^{-1} - 1\right] - h_x u. \tag{3.3}$$

The solution to Eq. (3.2) will be constructed by the perturbation method. Equation (3.2) is now rewritten in terms of new variables:

$$r = a(t) - x\ , \quad s = a(t) + x\ , \tag{3.4}$$

where $a(t)$ is an unknown function that will be determined later. It is found that $u_x = u_s - u_r$, $u_t = a_t(u_s + u_r)$, $u_{xx} = u_{rr} - 2u_{rs} + u_{ss}$,

$$\begin{aligned} &u_{tt} = a_t^2(u_{rr} + 2u_{rs} + u_{ss}) + a_{tt}(u_r + u_s)\ , \\ &u_{xxxx} = u_{rrrr} - 2u_{rrss} + u_{ssss}\ , u_{xxt} = a_t(u_{rrr} - u_{rss} - u_{rrs} + u_{sss})\ . \end{aligned} \tag{3.5}$$

Here $a = a(t)$ and the subscripts r and s refer to partial derivatives with respect to r and s, respectively. Then, using (3.4), we rewrite Eq. (3.2) so that

$$\begin{aligned} &\left(a_t^2 - a_0^2\right)(u_{rr} + u_{ss}) + 2\left(a_t^2 + a_0^2\right)u_{rs} + a_{tt}^2(u_r + u_s) = \\ &+ \left[\beta - \beta_1(u_r - u_s) + \beta_2(u_r - u_s)^2 - \beta_3(u_r - u_s)^3\right](u_s - u_r)(u_{rr} - 2u_{rs} + u_{ss}) \\ &+ \mu a_t(u_{rrr} - u_{rss} - u_{rrs} + u_{sss}) + k(u_{rrrr} - 2u_{rrss} + u_{ssss}) \end{aligned} \tag{3.6}$$

It is seen that (3.6) yields the nonlinear diffusion equation if $u_{rs} \approx 0$. We assume site resonant conditions

$$|R_e| < < 1\ , \quad |a_{tt}| < < 1\ , \quad u_{rs} \approx 0\ , \tag{3.6a}$$

where a transresonant parameter $R_e = a_t^2 - a_0^2$. At the exact site resonance, $a_t^2 = a_0^2$. We assumed that a_0 and a_t are functions of t. Therefore, approximately,

$$a = a(t) = \int a_0 dt + C. \tag{3.7}$$

Here C is an integration constant. We also assume that $a = a(t)$ is a smooth function, while the unknown u is a fast variable function. Therefore, we will often neglect by changing of $a = a(t)$ compared to changing u! When considering parametrically excited waves, a_t can be considered as a seed perturbation.

The solution of (3.6) is sought using the perturbation method:

$$u = u^{(1)} + u^{(2)} + u^{(3)} + u^{(4)} + u^{(5)} + \ldots, \tag{3.8}$$

where $u^{(1)} > > u^{(2)} > > u^{(3)} > > u^{(4)} > > u^{(5)}$. Expansion (3.8) is used widely in nonlinear physics (see Sect. 2.4.3). We assume that at the resonance, an amplitude of the waves becomes significantly larger than an amplitude of forcing oscillations.

Substituting sum (3.8) into (3.6) and equating terms of the same order, we obtain the following linear differential equations:

$$u_{rs}^{(1)} = 0, \tag{3.9}$$

$$\begin{aligned} 4a_0^2 u_{rs}^{(2)} = & -R_e\left(u_{rr}^{(1)} + 2u_{rs}^{(1)} + u_{ss}^{(1)}\right) - a_{tt}\left(u_r^{(1)} + u_s^{(1)}\right) \\ & + \mu a_0\left(u_{rrr}^{(1)} - u_{rss}^{(1)} - u_{rrs}^{(1)} + u_{sss}^{(1)}\right) + k\left(u_{rrrr}^{(1)} - 2u_{rrss}^{(1)} + u_{ssss}^{(1)}\right) \\ & + \beta\left(u_s^{(1)} - u_r^{(1)}\right)\left(u_{rr}^{(1)} - 2u_{rs}^{(1)} + u_{ss}^{(1)}\right), \end{aligned} \tag{3.10}$$

$$\begin{aligned} 4a_0^2 u_{rs}^{(3)} = & -R_e\left(u_{rr}^{(2)} + 2u_{rs}^{(2)} + u_{ss}^{(2)}\right) - a_{tt}\left(u_r^{(2)} + u_s^{(2)}\right) \\ & + \beta\left(u_s^{(2)} - u_r^{(2)}\right)\left(u_{rr}^{(1)} - 2u_{rs}^{(1)} + u_{ss}^{(1)}\right) \\ & + \beta\left(u_s^{(1)} - u_r^{(1)}\right)\left(u_{rr}^{(2)} - 2u_{rs}^{(2)} + u_{ss}^{(2)}\right) \\ & + \beta_1\left(u_s^{(1)} - u_r^{(1)}\right)^2\left(u_{rr}^{(1)} - 2u_{rs}^{(1)} + u_{ss}^{(1)}\right), \end{aligned} \tag{3.11}$$

$$4a_0^2 u_{rs}^{(4)} = \beta_2\left(u_s^{(1)} - u_r^{(1)}\right)^3\left(u_{rr}^{(1)} - 2u_{rs}^{(1)} + u_{ss}^{(1)}\right), \tag{3.12}$$

$$4a_0^2 u_{rs}^{(5)} = \beta_3\left(u_s^{(1)} - u_r^{(1)}\right)^4\left(u_{rr}^{(1)} - 2u_{rs}^{(1)} + u_{ss}^{(1)}\right). \tag{3.13}$$

Equations for $u^{(4)}$ and $u^{(5)}$ following from (3.6) were very long and complex. We retained only a few terms in (3.12) and (3.13) to simplify the problem. It is assumed that $\mu \approx R_e \approx 0$и $k \approx a_{tt} \approx 0$,in (3.11).

Thus, we reduced the solution of the original nonlinear Eq. (3.2) to the solution of the homogeneous linear Eq. (3.9) and inhomogeneous linear Eqs. (3.10)–(3.13). These simplified equations will allow us to evaluate qualitatively the influence of high nonlinearity on the waves.

3.1.1 Generalization of the Perturbation Method

The solution of (3.9) is

$$u_1^{(1)} = J(r) + j(s)\ . \tag{3.14}$$

The solution (3.14) resembles the d'Alembert-type solution, but the velocity of waves $J(a-x)$ and $j(a+x)$ can vary and depends on t [13]. We emphasize that $J(a-x)$ and $j(a+x)$ are unknown functions (see Sect. 2.4.3). They will be unknown up to the end of the solution of Eq. (3.2). Now we shall correct solution (3.14) taking into account $u^{(2)}$, and then $u^{(i)}$ (i = 3; 4; 5). We assume that $R_e \approx const$, $a_{tt} \approx const$. Substituting (3.14) into (3.10) and integrating, one can find that

$$
\begin{aligned}
u^{(2)} = & J_2(r) + j_2(s) - \frac{1}{2}a_0^{-2}[s(R_e J' + a_{tt}J) + r(R_e j' + a_{tt}j)] \\
& + \frac{1}{8}\beta a_0^{-2}\left[r(j')^2 - s(J')^2 + 2jJ' - 2j'J\right] + \frac{1}{4}k\left(sJ''' + rj'''\right) \\
& + \frac{1}{4}\mu a_0^{-1}(sJ'' + rj''),
\end{aligned}
\tag{3.15}
$$

where $J_2(r)$ and $j_2(s)$ are functions of integration. In (3.15), $J = J(r)$ and $j = j(s)$, and the prime denotes differentiation with respect to the appropriate variables r or s. The influence of d and d_1 on the high-order values will not be considered. Using (3.14) and (3.15), we can rewrite Eq. (3.11) in the following form:

$$
\begin{aligned}
u_{rs}^{(3)} = & \beta^2 a_0^{-4}\{1/2(j^2)'J''' + 1/2(J^2)'j''' \\
& - Jj'j''' - jJ'J''' - 5/6\left[(j')^3\right]' - 5/6\left[(J')^3\right]' - JJ''j'' - jj''J'' - 7/2(j')^2J'' - 7/2(J')^2j'' \\
& + 7/2\left[(J')^2\right]'j' + 7/2\left[(j')^2\right]'J' - j(J'')^2 - J(j'')^2\}/16 \\
& - 1/32\beta^2 a_0^{-4}s\left\{j'\left[(J')^2\right]'' - j''\left[(J')^2\right]' - 2/3\left[(J')^3\right]''\right\} + 1/32\beta^2 a_0^{-4}r\{-J'\left[(j')^2\right]'' + J''\left[(j')^2\right]' \\
& + \frac{2}{3}\left[(j')^3\right]''\} + \frac{1}{4}\beta_1 a_0^{-2}\left[J''(j')^2 - 2J'j'J'' + J''(J')^2 + j''(j')^2 - 2J'j'j'' + j''(J')\right] \\
& + \frac{1}{4}\beta a_0^{-2}[(J_2'' + j_2'')(j' - J') - (J_2' - j_2')(J'' + j'')].
\end{aligned}
\tag{3.16}
$$

Here $J_2 = J_2(r)$ and $j_2 = j_2(s)$. Equation (3.16) yields the expression for $u^{(3)}$:

$$
\begin{aligned}
u^{(3)} = & J_3(r) + j_3(s) + \frac{1}{32}\beta^2 a_0^{-4}\Big(J''j^2 + j''J^2 - 2j'J'J - 2j'J'j - 2sjJ'J'' \\
& + sj'J'^2 - 2rJj'j'' - rJ'j'^2 + s^2J'^2J'' + r^2j'^2j''\Big) \\
& - \frac{1}{16}a_0^{-2}(4\beta_1 - 3\beta^2 a_0^{-2})\left(jJ'^2 + Jj'^2\right) \\
& + \frac{1}{12}a_0^{-2}\left(\beta_1 - \frac{5}{8}\beta^2 a_0^{-2}\right)\left(sJ'^3 + rj'\right) \\
& + \frac{1}{4}a_0^{-2}\left(\beta_1 - \frac{5}{8}\beta^2 a_0^{-2}\right)\left(j'\int J'^2 dr + J'\int j'^2 ds\right) \\
& + \frac{1}{4}\beta a_0^{-2}(J'j_2 - j'J_2 + jJ_2' - Jj_2' + rj_2'j' - sJ_2'J') + \frac{1}{2}(s - r)d + d_1.
\end{aligned}
\tag{3.17}
$$

Here d and $d1$ are constants of integration. Now we can approximately take into account highly nonlinear effects. We shall neglect the interaction of the waves J' and j' in (3.12) and (3.13). In addition, we will take into account only the most important singular (resonant) terms. In this case, the previous equations yield

$$u_1^{(4)} = J_4(r) + j_4(s) + \beta_2 a_0^{-2}\left[r(j')^4 - s(J')^4\right]/16, \tag{3.18}$$

$$u_1^{(5)} = J_5(r) + j_5(s) + 0.05\beta_3 a_0^{-2}\left[r(j')^5 - s(J')^5\right] \tag{3.19}$$

The approximate general solution of the Eq. (3.6) is the sum (3.8):

$$\begin{aligned}
u = {} & J + j + J_2(r) + j_2(s) + J_3(r) + j_3(s) + J_4(r) + j_4(s) + J_5(r) + j_5(s) \\
& - \frac{1}{4}a_0^{-2}[s(R_eJ' + a_{tt}J) + r(R_ej' + a_{tt}j)] \\
& + \frac{1}{8}\beta a_0^{-2}\left[r(j')^2 - s(J')^2 + 2jJ' - 2j'J\right] + \frac{1}{4}k\left(sJ''' + rj'''\right) \\
& + \frac{1}{4}\mu a_0^{-1}(sJ'' + rj'') + + \frac{1}{32}\beta^2 a_0^{-4}\Big(J''j^2 + j''J^2 - 2j'J'J - 2j'J'j - 2sjJ'J'' \\
& + sj'J'^2 - 2rJj'j' - rJ'j'^2 + s^2J'^2J'' + r^2j'^2j''\Big) \\
& - \frac{1}{16}a_0^{-2}(4\beta_1 - 3\beta^2 a_0^{-2})\left(jJ'^2 + Jj'^2\right) + \frac{1}{12}a_0^{-2}\left(\beta_1 - \frac{5}{8}\beta^2 a_0^{-2}\right)\left(sJ'^3 + rj'\right) \\
& + \frac{1}{4}a_0^{\ 2}\left(\beta_1 - \frac{5}{8}\beta^2 a_0^{\ 2}\right)\left(j'\int J'^2 dr + J'\int j'^2 ds\right) \\
& + \frac{1}{4}\beta a_0^{-2}(J'j_2 - j'J_2 + jJ_2' - Jj_2' + rj_2'j' - sJ_2'J') + J_5(r) + j_5(s) \\
& + \beta_2 a_0^{-2}\left[r(j')^4 - s(J')^4\right]/16 + 0.05\beta_3 a_0^{-2}\left[r(j')^5 - s(J')^5\right] \\
& + \frac{1}{2}(s - r)d + d_1.
\end{aligned} \tag{3.20}$$

Here unknown functions $J(r)$, $j(s)$, $J_2^*(s)$, $j_2^*(s)$, $J_3^*(s)$, and $j_3^*(s)$ are determined by initial and boundary conditions.

3.1.2 Bounded Solution of (3.2) for Resonators

In the expression (3.20), there are terms whose coefficients explicitly contain coordinates r or s. Thus, they contain terms that increase without limit if the coordinate $x$or time t. increases. The noted fact significantly limits the scope of application of the solution. Indeed, with increasing x or t, or with their simultaneous growth, a moment inevitably comes when the condition $u^{(1)} >> u^{(2)} >> u^{(3)} >> u^{(4)} >> u^{(5)}$ is violated and the presented solution becomes inapplicable.

However, there is a case of applicability of the expression (3.20) that is very important for practice. It is the case when wave processes in finite systems are studied. Of course, the solution (3.20) needs to be somewhat transformed. Thus, expression (3.20) should be modified to exclude the secular terms depending on t.

First, we modify expression (3.15). The secular (resonant) terms in time will be eliminated there if

$$J_2(r) = 0.25a_0^{-2}r(R_eJ' + a_{tt}J) + \beta a_0^{-2}r(J')^2/8 + \Psi_2(r),$$
$$j_2(s) = 0.25a_0^{-2}s(R_ej' + a_{tt}j) - \beta a_0^{-2}s(j')^2/8 + \psi_2(s). \tag{3.21}$$

As a result, we have

$$\begin{aligned} u^{(2)} = {} & \Psi_2(r) + \psi_2(s) - \frac{1}{4}a_0^{-2}(s-r)[R_e(J'-j') + a_{tt}(J-j)] \\ & + \beta a_0^{-2}\left[(r-s)(j')^2 - (s-r)(J')^2 + 2jJ' - 2j'J\right]/8. \end{aligned} \tag{3.22}$$

Using (3.14) and (3.22), and following the procedure described for (3.15) and the author's 2003 article [13], we rewrite (3.17) entering the quantities $\frac{1}{2}(s-r)$ there. As a result, we find an approximate solution to Eq. (3.11)

$$\begin{aligned} u^{(3)} = {} & \Psi_3 + \psi_3 - \frac{1}{8}(s-r)\beta a_0^{-2}\left(j'^2 + J'^2\right) + \frac{1}{4}\beta a_0^{-2}(jJ' - j'J) \\ & - \frac{1}{4}(s-r)k\left(j''' - J'''\right) + \frac{1}{4}(s-r)\mu a_0^{-1}(J'' - j'') \\ & + \frac{1}{4}a_0^{-2}\left\{\frac{1}{8}\beta^2 a_0^{-2}\left[-5J'\int j'^2 ds - 5j'\int J'^2 dr - \frac{16}{6}(s-r)\left(J'^3 - j'^3\right)\right.\right. \\ & + \frac{1}{2}(s-r)^2 j'\left(j'^2\right)' + \frac{1}{2}(s-r)^2 J'\left(J'^2\right)' \\ & + (s-r)\left(J'^2 j' - J'j'^2 - 2jJ'J'' + 2j'j''J\right) + 7J'^2 j + 7Jj'^2 \\ & \left. + j''J'^2 + j^2 J'' - 2jj'J'' - 2j'JJ'\right] + \beta\left[\psi_2 J' - \Psi_2 j' + j\Psi_2' - J\psi_2' - (s-r)\left(\psi_2' j' + \Psi_2' J'\right)\right] \\ & \left. + \beta_1\left(J'\int j'^2 ds - jJ'^2 + \frac{1}{3}(s-r)\left(J'^3 - j'^3\right) - j'^2 J + j'\int J'^2 dr\right)\right\} \\ & + \frac{1}{2}(s-r)d + d_1. \end{aligned} \tag{3.23}$$

Here $\Psi_2 = \Psi_2(r)$, $\Psi_3 = \Psi_3(r)$ and $\psi_2 = \psi_2(s)$, $\psi_3 = \psi_3(s)$. For simplicity we assumed $R_e \approx 0$ and $a_{tt} \approx 0$ in (3.23). Finally, in the same way, following to [13] we calculate $u^{(4)}$ and $u^{(5)}$. By summing the quantities determined in this way, we find an approximate solution to Eq. (3.2).

$$
\begin{aligned}
u = {} & J + j + \Psi_2 + \psi_2 + \Psi_3 + \psi_3 + \Psi_4 + \psi_4 + \Psi_5 + \psi_5 - \frac{1}{4}a_0^{-2}(s-r)[R_e(J' - j') + a_{tt}(J - j)] \\
& - \frac{1}{8}(s-r)\beta a_0^{-2}\left(j'^2 + J'^2\right) + \frac{1}{4}\beta a_0^{-2}(jJ' - j'J) - \frac{1}{4}(s-r)k\left(j''' - J'''\right) \\
& + \frac{1}{4}(s-r)\mu a_0^{-1}(J'' - j'') + \frac{1}{4}a_0^{-2}\left\{\frac{1}{8}\beta^2 a_0^{-2}\left[-5J'\int j'^2 ds - 5j'\int J'^2 dr\right.\right. \\
& - \frac{16}{6}(s-r)\left(J'^3 - j'^3\right) + \frac{1}{2}(s-r)^2 j'\left(j'^2\right)' + \frac{1}{2}(s-r)^2 J'\left(J'^2\right)' \\
& + (s-r)\left(J'^2 j' - J'j'^2 - 2jJ'J'' + 2j'j''J\right) + 7J'^2 j + 7Jj'^2 + j''J^2 \\
& \left. + j^2 J'' - 2jj'J' - 2j'JJ'\right] + \beta\left[\psi_2 J' - \Psi_2 j' + j\Psi_2' - J\psi_2'\right. \\
& \left. - (s-r)\left(\psi_2' j' + \Psi_2' J'\right)\right] + \beta_1\left(J'\int j'^2 ds - jJ'^2 + \frac{1}{3}(s-r)\left(J'^3 - j'^3\right)\right. \\
& \left.\left. - j'^2 J + j'\int J'^2 dr\right)\right\} - \frac{1}{16}\beta_2 a_0^{-2}(s-r)\left(j'^4 + J'^4\right) \\
& - 0.05\beta_3 a_0^{-2}(s-r)\left(j'^5 + J'^5\right) + \frac{1}{2}(s-r)d + d_1.
\end{aligned}
\tag{3.24}
$$

Here $\frac{1}{2}(s-r) = x$. It is emphasized that some coefficients in (3.24) explicitly depend on x. Expression (3.24) explicitly demonstrates the nonlinear effect. Because of the nonlinear terms in (79), the wave shape deforms as the coordinate x varies. In particular, harmonics are generated within the wave. For example, if we have at a given point, say $x = 0$, a sinusoidal wave with frequency ω, then, according to (3.24), harmonics with frequencies 2ω, 3ω, 4ω, and 5ω appear in $x > 0$.

Here unknown functions $J(r)$,$j(s)$, and Ψ_2, ψ_2, Ψ_3, ψ_3, Ψ_4, ψ_4, Ψ_5, ψ_5 d, d_1 are determined by initial and boundary conditions. The principal difficulty is connected with the integrals $\int J'^2 dr$, $\int j'^2 ds$. Their expressions cannot be simplified or changed. However, these integrals are eliminated in important special cases.

We emphasize that solution (3.24) is valid in the finite intervals of the variation of r and s, while the nonlinear terms are smaller than J and j. Thus, the solutions are valid only if conditions $u^{(1)} >> u^{(2)} >> u^{(3)}$ in (3.8) take place.

1. The above limitation is fully consistent with the traditional perturbation method and thus, at first glance, differs strikingly from the approach developed and used in the second chapter. But this is not entirely true! The fact is that in the solution (3.24), all quantities $J(r)$, $j(s)$, and Ψ_2, ψ_2, Ψ_3, ψ_3, Ψ_4, ψ_4, Ψ_5, ψ_5 d, d_1 are unknown. At the same time, they can be specified within the limits of the above assumptions completely arbitrarily and at the same time remain a solution to Eq. (3.2). In this, they are quite consistent with the decision recorded by d'Alembert. Of course, the unknown functions J and j are, by definition, the most important, and these will be determined in the future when solving boundary value problems. At the same time, in some cases, we will not take into account the influence on the wave process of the remaining unknowns. In other cases, those unknown functions can be considered as seed quantities.

2. It follows from the solutions that it is very difficult to excite purely harmonic wave. As the solution is nonlinear, the formulas show that together with the first primary (leading) harmonic wave the second harmonic wave is being generated. Then the third harmonic wave is excited and so on. These waves travel from the source and interact with the primary wave. Near the source, the amplitudes of the high harmonic waves are small and can only slightly change the wave form. However, the amplitudes resonant grow with the coordinate x and can strongly change the wave form far from the source. The form variation is also connected with the dispersive and viscous effects.

3. It is very important to notice that the cubic nonlinearity in (3.24) can generate the harmonious wave similar to the initial wave, but which grows during the propagation. In particular, this mechanism is responsible for the amplification of unidirectional waves. For example, it is true for the ocean waves.

4. There is a great variety of natural, technical, and other processes at which strongly nonlinear resonant oscillations and waves appear. Taking into account solution (3.24), it would be possible to assume that in the case of the finite resonators, the research of the resonant waves will lead to a more complex analysis and to very large expressions for unknown values. However, paradoxically, it is not absolutely so. A growth of a quantity of harmonics and their amplitudes during the approach to a resonance are being replaced in an immediate proximity from the exact resonance by the simplification of the governing equations. In particular, the research of nonlinear waves can be simplified at the vicinity of the resonance up to the solution of the algebraic equation. Examples of this are given in the Chap. 2, and there will be many more in the following chapters.

Thus, following [13], we wrote the general approximate solutions for the nonlinear wave equation. In particular, the solutions describe both the nonlinear standing waves and the traveling waves in the finite long resonators. The coefficients contain the time t and the coordinate x, or only the coordinate x (3.24). Thus, the received solutions obviously show the occurrence of the quickly growing resonant components in the nonlinear waves. These terms define the development of the wave system. In comparison with them, other terms of the solutions, in some limit, become ignorable. Thus, thanks to the resonant terms, the description of the wave dynamics can be essentially simplified in certain cases. Of course, it is necessary to remember that the solutions are fair, while/if the amendments to the d'Alembert solution are small.

Summary Above we have suggested to seek for the solution for the long waves in the form of the sum (3.24). We emphasise, that such presentation essentially excludes occurrence during the analysis of the "small" dividers in the cases of resonators.

In the general case, since we have a one-dimensional equation, it corresponds to two boundary conditions. Using one of them, you can determine the relationship between J and j functions. The second boundary condition will then determine the equation for, for example, the function J. In this case, the functions Ψ_i and $\varPsi_i$ $(i = 2, 3, 4, 5)$ can be considered arbitrary functions or seed perturbations. As a result, after satisfying the boundary and initial conditions, the problem is reduced to solving an ordinary nonlinear differential equation. Of course, the described process can be reduced to very complex mathematics and, apparently, can only be realized in the simplest cases. Moreover, even in this case, it can lead to a nonintegrable differential equation. Let us emphasize once again that in the general case, in the expressions given above, and are arbitrary functions. At the same time, on the one hand, they are solutions of the linear equation (3.9). On the other hand, when solving specific problems, they are determined from solutions of nonlinear differential equations following from the corresponding boundary and initial conditions.

3.2 Counterintuitive Singlevalued Faraday Waves

Faraday about 200 years ago began the experiments with vertically excited layers of liquid or granular materials [17]. In recent decades, interest in such research has increased greatly, in particular, to the study of surface waves in layers of liquid or weakly bound media. To some extent, this is due to the fact that in these experiments we can control the mechanical properties of the excited medium. It is known that the undisturbed mechanical properties of many geomaterials may often be described by the theory of elasticity. However, due to the forced vibrations, the pore volume in geomaterials can increase, and then they begin to move like a liquid. On the other hand, we can control the characteristics of vertical acceleration and therefore influence the speed of surface waves (see Figs. 1.5, 1.6, 1.7, and 1.8). For example, in the case of long surface waves, their speed is approximately defined by the expression $\sqrt{gh}$, where $g = g_0 + g_d$ [13, 18]. Here g_0 is the gravitational acceleration, and g_d is the forcing acceleration. According to the last expressions, if at some instant $g_d = -g_0$, then the wave stops. During other moments of time, the speed of the wave depends on current value g_d. If the speed becomes imaginary, it may be considered as equal to zero [6, 19, 20].

It is also very important for us that by changing the excitation characteristics, we can study the nonlinear properties of waves in a very wide range of their changes. In fact, this chapter is based on this study.

The experimental research of parametric-excited strongly nonlinear waves is connected to the significant difficulties, and the results received are sometimes unexpected.

3.2.1 Main Ideas About Strong Parametric Excitation and Nonlinear Eigenfunctions

Here we present the main ideas which will be used for simulation of experiments. In particular, we focus our attention on the surface waves for which there are a lot of experimental data. Thus, we distinguish waves excited in cases of standing and traveling waves. Cases of simultaneous existence of both standing and traveling waves are also studied.

All these cases differ from each other by governing equations and methods of modeling vertical (parametric) excitation.

1. Basic relations and equations for long waves. We shall use here the long wave theory. Thin layers of liquid or granular media under the action of vertical excitation will be considered. For this case, Eq. (3.2) yields

$$u_{tt} - a_0^2 u_{xx} = \beta u_x u_{xx} + \beta_1 u_{xx} u_x^2 + \mu u_{txx} + k u_{ttxx}. \tag{3.25}$$

Here $a_0^2 = gh$ and $g = g_0 + g_d$, where $g_d = g_d$ (t) is the exciting vertical acceleration. Eq. (3.25) describes the granular waves on the surface thin granular layer after the "fluidization" of the material.

Let $h = const$ and

$$g_d = \delta \cos \omega t \tag{3.26}$$

In this case, the wave velocity is

$$a_0(t) = \sqrt{(g_0 + \delta \cos \omega t)h} \tag{3.27}$$

The wave velocity a_0 changes with δ cos ωt. It can increase or reduce from $a_0 = \sqrt{g_0 h}$. The maximal velocity is $a_0 = \sqrt{2gh}$. The minimal velocity is zero. If $\delta > g_0$, the wave stops at the moments when a_0 is imagery.

If $\delta > g_0$, then the layer can separate from the bottom and the fly time can appear (Fig. 1.8). If $\delta > > g_0$, then the fly time becomes much larger than the contact time [13, 18].

2. Long-standing waves and modeling of vertical excitation. The standing wave method can describe both linear and nonlinear waves in resonators. We must also distinguish parametrically excited waves from other surface waves. In the case of parametrically excited surface waves, it is convenient to use variables $r = a(t) - x$ and $s = a(t) + x$ (3.4). Here $a(t)$ value is different from $a_0(t)$.

To illustrate this case, we will consider a case of the purely harmonic excitation, when $g_d = \delta$ cos ωt.

The case $\delta = g_0$ is considered in [4, 6, 18]. For it $a_0^2 = g_0 h(1 + \cos \omega t)$ and

$$a_0(t) = (2g_0 h)^{0.5} \cos \omega t/2 \text{ and } a(t) = \int a_0 dt = \left(8 g_0 h \omega^{-2}\right)^{0.5} \cos \omega t/2. \tag{3.28}$$

If $\delta < g_0$, then according to (3.27)

$$a_0(t) \approx \sqrt{hg_0}\left(1 + \frac{1}{2}\delta g_0^{-1}\cos\omega t - \frac{1}{8}\delta^2 g_0^{-2}\cos^2\omega t + \ldots\right) \text{ and}$$
$$a(t) \approx \sqrt{h_0 g_0}\left(t + \frac{1}{2}\delta\omega^{-1}g_0^{-1}\sin\omega t - \frac{1}{16}\delta^2 g_0^{-2}t - \frac{1}{32}\delta^2\omega^{-1}g_0^{-2}\sin 2\omega t\ldots\right). \tag{3.29}$$

The convergence of the last expansion depends on δg_0^{-1}. It is seen that the wave velocity can vary according the complex law. Thus, in cases of not too strong excitation, the vertical force or displacement acting on the resonator may be presented by a harmonic law.

3. Traveling waves and modeling of vertical excitation. Let us assume we are dealing with a resonator containing a layer of a certain medium. At a certain amplitude of vertical excitation, the medium begins to detach from the base [4]. As a result, the initial harmonic excitation begins to change. In particular, peaks appear on it, the duration of which may even be comparable with the period of the excitation. This markedly complicates the study of the problems considered here.

In general, if $\delta > g_0$, a pulsed vertical load begins to act on the material layer. The model of the standing wave becomes incorrect. However, if the amplitude of the vertical acceleration is not too large, we can use the standing wave model, if the occurrence of traveling waves is also taken into account. If $\delta \approx g_0$, then the pulses lead to the appearance of traveling waves of small amplitude (ripple waves) that move along the surface of the standing waves. In this case, the wave excited in the resonator is the sum of standing and traveling waves. If $\delta >> g_0$, then the model of the standing wave is not fully suitable, and the results of the experiments should be modeled on the basis of the traveling wave model.

If approximately $\delta > g_0$, the layer can separate periodically from the base. This case was modeled in [4]. The acceleration was considered as a fast variable function of the contact and the separation of the material from the base. In this case, the traveling wave velocity $a_0(t)$ may be approximated by the expression.

$$a_0(t) = \delta_* \sin^n m^{-1}\omega t\ , \tag{3.30}$$

where δ_* is an unknown constant, $n = 1, 2, 3, \ldots$ and $m = 1, 2, 3\ldots.$ Constants δ_*, n, and m can be defined experimentally. In (3.30), the power n is defined by the forcing excitation and by the contact time. The amplitude δ_* is defined by the speed of the collision of the layer and the base. The value m is defined by the flight time of the layer.

Thus, the constants in (3.30) are defined by the excitation. However, it is underlined that (3.30) is not the container acceleration. The expression (3.30) is defined by the layer acceleration which can be absolutely distinct from the base acceleration. In particular, (3.30) gives the subharmonic oscillations excited by the simple harmonic oscillations of the base, if $m = 2, 3, \ldots.$

Let us consider approximately the case when (3.28) takes place and at the same time the peak of the load acting on the layer has already appeared (3.30). In this case, we used the expression (3.28) written with this peak

$$a_0(t) = \delta_* \left[\sin \omega t/2 - k_1 \sin^3(\omega t/k_2)\right]. \tag{3.31}$$

Here, k_1 and k_2 are arbitrary constants which are determined during corresponding calculations. We emphasize that the expressions (3.30) and (3.31) will be further used to describe ripple waves in resonator caused by impulse loadings.

Thus, we have determined the basic equations and assumptions that we will use during the calculations. Now we will start modeling the results of experiments.

4. Eigenwave functions of a parametrically excited resonator. We will consider a vertically harmonically excited container containing a liquid with a free surface [4, 13, 18]. The side walls of the container are motionless in the horizontal direction. Boundary conditions are written:

$$u = 0 \text{ at } x = 0 \text{ and } x = L\ , \tag{3.32}$$

The periodic solution is sought. In this case, the condition at $x = 0$ gives

$$J = -j. \tag{3.33}$$

Taking into account the last requirements, we rewrite the solution (3.23) in the form, which includes only the unknown J'. After the solution is substituted into the condition at $x = L$ and after satisfying some resonant requirements, the following Eq. [4, 12] is given:

$$J''' + D_2 J' + D_4 J'' + D_1 (J')^2 + D_3 (J')^3 + D_* J' J'' = d_*. \tag{3.34}$$

Here D_1, D_2, D_3, D_4 и D_* are constants. Apparently, by varying the constants in (3.34), we can precisely describe the results of different experiments. However, carrying out calculations in which the constants fully correspond to the experiments is not our task in this section. We will strive to identify general trends. We emphasize that a number of nonlinear wave equations written for a travelling wave has the form (3.34). Therefore, our further consideration does not apply only to surface long waves. The consideration is more general, although to illustrate the results, we will turn to surface waves.

On the other hand, Eq. (3.34) can only qualitatively describe some features of the waves and wave patterns that we are interested in. We think that this is largely due to the fact that during the excitation process, the mechanical properties of the excited layers change. However, we leave aside these doubts and deal with the solution of Eq. (3.34). Of course, in the general case, it can be solved analytically only approximately.

Eq. (3.34) takes into account the effect on the wave of nonlinearities, dissipative dispersion effects, and even the interaction of waves. Following [24, 25, 27], we shall seek the approximate solution of the last equation as a sum of traveling periodical waves:

$$J'(r) = B\tanh(l\sin\zeta_-)\cos^k\zeta_- + A\,\mathrm{sec}\,h^2(l\sin\zeta_-)\cos^K\zeta_- + C\cos^z\zeta_-, \quad (3.35)$$

where $\zeta_- = \pi L^{-1}M^{-1}(a(t) - x + c_-)$; A, B, C, and l are unknown constants; k, K, and z are integers; and c_- is an arbitrary constant. Expression (3.35) describes the interaction and the competition between the nonlinear, dissipative, and dispersive effects. In the limit cases, (3.35) simulates shock-, soliton-, cnoidal-, and harmonic-type waves. Thus, expression (3.35) describes all well-studied waves. Therefore, we shall consider (3.35) as universal wavefunctions. They define infinite spectra of waves because $M^{-1} = 1, 2, 3,\ldots$ It will be shown that (3.35) describes many interesting results of recent experiments.

Expression (3.35) is substituted into (3.34). Next, equating to zero nonlocalized terms, and terms containing $\mathrm{sech}^2(l \sin \zeta_-)$, or $\mathrm{sech}^4(l \sin \zeta_-)$ or $\tanh(l \sin \zeta_-)$, we obtain

$$\begin{aligned} &C\left(D_2 - z\pi^2L^{-2}M^{-2}\right)\cos^z\zeta_- + D_1B^2\cos^{2k}\zeta_- + D_1C^2\cos^{2z}\zeta_- \\ &\quad + 3D_3CB^2\cos^{2k+z}\zeta_- + D_3C^3\cos^{3z}\zeta_- = d_*, \end{aligned} \quad (3.36)$$

$$\begin{aligned} &A\left(D_2 - K\pi^2L^{-2}M^{-2}\right)\cos^K\zeta_- - 2Al^2\pi^2L^{-2}M^{-2}\cos^{K+2}\zeta_- \\ &\quad - D_1B^2\cos^{2k}\zeta_- + 2D_1AC\cos^{K+z}\zeta_- + D_4Bl\pi L^{-1}M^{-1}\cos^{k+1}\zeta_- \end{aligned} \quad (3.37)$$

$$- 3D_3CB^2\cos^{2k+z}\zeta_- + 3D_3AC^2\cos^{K+2z}\zeta_- + D_sBCl\pi L^{-1}M^{-1}\cos^{z+k+1}\zeta_- = 0,$$

$$\begin{aligned} &D_3A^3\cos^{3K}\zeta_- + 3D_3CA^2\cos^{2K+z}\zeta_- + D_1A^2\cos^{2K}\zeta_- \\ &\quad + D_*ABl\,\pi L^{-1}M^{-1}\cos^{K+k+1}\zeta_- = 0, \end{aligned} \quad (3.38)$$

$$\begin{aligned} &B\left(D_2 - k\pi^2L^{-2}M^{-2} + 2D_1C\cos^2\zeta_-\right)\cos^k\zeta_- + D_3B^3\cos^{3k}\zeta_- \\ &\quad + 3D_3BC^2\cos^{2z+k}\zeta_- = 0. \end{aligned} \quad (3.39)$$

For fixed k, K, and z, one can obtain algebraic equations for A, B, C, and l from (3.36) to (3.39). Here, as an example, parametric surface waves will be considered. Following [4, 12], we consider a few cases when coefficients in (3.35) may be easily found.

(I). Let $A = C = 0$, $k = 1$ and $d_* = D_1B^2/2$. In this case, $B^2 = 4(\pi^2L^{-2}M^{-2} - D_2)/3D_3$ and $l = BMLD_1/\pi D_4$. Thus, quadratic nonlinear and viscous effects are taken into account. If $D_3 = (\pi^2L^{-2}M^{-2} - D_2)/3$, then we have

$$u_x \approx -2\tanh(l\sin\zeta_-)\cos\zeta_- - 2\tanh(l\sin\zeta_+)\cos\zeta_+. \quad (3.40)$$

Here $\zeta_+ = \pi L^{-1}M^{-1}(a(t) + x + c_+)$.

Remark Solution (3.40) is some modified expression for the shock waves running in weakly cohesive media (e.g., gas or water).

(II). Let A = 0, k = 0, z = 1, and $D_1 \approx D_* \approx 0$, $d_* = 0$. In this case, one can find that $B^2 = (3\pi^2L^{-2}M^{-2} - 2D_2)/8D_3$, $C^2 = -(2D_2 + \pi^2L^{-2}M^{-2})/4D_3$, and $l = 3BCMLD_3/\pi D_4$. If $D_3 = (3\pi^2L^{-2}M^{-2} - 2D_2)/32$, then $B = \pm 2$, and we have the next expression

$$u_x \approx -2\tanh(l\sin\zeta_-) - C\cos\zeta_- - 2\tanh(l\sin\zeta_+) - C\cos\zeta_+. \quad (3.41)$$

(III). Let B = 0, K = z = 2, and $d_* = 0$. In this case, one can find $A = -3C - 6D_1/5D_3$ and $l^2 = \pi^{-2}M^2L^2(2D_2/3 - 4\pi^2M^{-2}L^{-2}/3 + D_1C + 5D_3C^2/4)$, where $C = 0$ or C is defined by the equation $5D_3C^2 + 6D_1C + 8D_2 - 16\pi^2M^{-2}L^{-2} = 0$. Now we can write

$$u_x \approx -\left[A\sec h^2(l\sin\zeta_-) + C\right]\cos^2\zeta_- - \left[A\sec h^2(l\sin\zeta_+) + C\right]\cos^2\zeta_+. \quad (3.42)$$

Let us consider a case when $C = 0$ and $l \to \infty$. Then the wave transforms into a particle (jet) and we have approximately

$$u_a \approx -A\left[\sec h^2(l\sin\zeta_-)\cos^2\zeta_- + \sec h^2(l\sin\zeta_+)\cos^2\zeta_+\right]. \quad (3.43)$$

(IV). Let $D_1 = 0$, $B = 0$, $K = z = 1$, and $d_* = 0$. In this case, one can find $A = -3C$, $l^2 = 2 - 2\pi^{-2}M^2L^2D_2$, and $C^2 = 2\left(\pi^2M^{-2}L^{-2} - D_2\right)D_3^{-1}$. Now we can write

$$u_x \approx C\left[3\mathrm{sech}^2(l\sin\zeta_-) - 1\right]\cos\zeta_- + C\left[3\mathrm{sech}h^2(l\sin\zeta_+) - 1\right]\cos\zeta_+. \quad (3.44)$$

For liquid layers, we found [4] that $C = 1.7(\mathrm{h/L})1/3$ in (3.44).

We presented several approximate solutions of Eq. (3.34), which can be valid under the above assumptions. Using these approaches, it is possible to calculate the surface wave. During calculations of the elevation, we used

$$\eta = -hu_x \quad (3.45)$$

It is the linear approximation of the formula (3.3).

In the conclusion of this section, we emphasize that the solution (3.35) describes only unambiguous wave profiles, while profiles may have folds.

3.2.2 Oscillons, Stand-Traveling Waves, and Wave Patterns

Let us check the accuracy of the developed approach by comparing its results with experimental data [21–24].

1. **Vertical oscillons. 1.** At first, we consider abnormal parametrically raised surface waves observed in the long rectangular container. Two isolated waves were excited in certain cases, which oscillated relatively the container middle (Fig. 3.4).

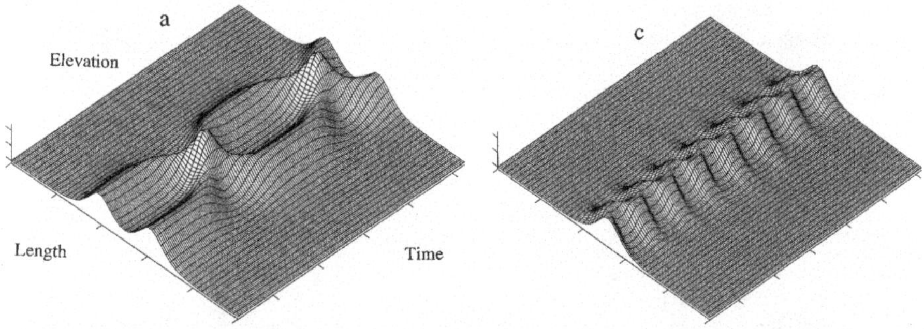

Fig. 3.4 Simulation of parametrically excited surface water waves [4, 6, 12] with the help of solution (3.43) ($c_+ = c_- = L/2$): oscillations and the interaction of a pair of soliton-like waves (left); the localization of the parametric harmonic waves(right) [12]

During calculations, we used $\zeta_\pm = \pi L^{-1}(a(t) \pm x + c_\pm)$ and $a(t) = 2[\sin(\omega t/2) - 0.333\sin^3(\omega t/2)]$ or $a(t) = \frac{1}{3}\left[\sin(3\omega t) - 0.333\sin^3(3\omega t)\right]$ in (3.43).

2. In the experimental system [21], a layer of a suspension was vibrated vertically. A cubical container with 20 cm sides was used, with Plexiglas lateral boundaries. The container was mounted on a mechanical shaker providing vertical acceleration from 0 to 30 g_0. The range of driving frequencies was limited from 10 to 60 Hz. The maximum amplitude was 1.25 cm. The working suspensions were a mixture of water with commercial clay powder. At a critical value of acceleration, the initial spatially uniform surface loses its stability, and localized vertically oscillating waves (oscillons) can appear (Fig. 3.5).

The experiments are illustrated by Fig. 3.6.

We used (3.41) written in the form $u_x \approx -2\tanh(l\sin\zeta_-) - 2\tanh(l\sin\zeta_+)$ for the calculation of the oscillon. Here $\zeta_\pm = \pi L^{-1}(a_0\omega^{-1}\sin(\omega t/2) \pm x + c_\pm)$ and $c_- = L/2$, $c_+ = -L/2$. For the calculation of two oscillons and the oscillon triad observed in [4, 12], we used (3.40) и $\zeta_\pm = \pi L^{-1}(a_0\omega^{-1}\sin(\omega t/2) \pm 2x + c_\pm)$ (two oscillons) or $\zeta_\pm = \pi L^{-1}(a_0\omega^{-1}\sin(\omega t/2) \pm 3x + c_\pm)$ (the oscillon triad).

Thus, the approach developed above, which takes into account the influence of vertical disturbance on the speed of Faraday waves, made it possible to describe qualitatively the experimental data.

In the next section, we will present the calculation results taken from [4].

2. **Stand-traveling oscillons.** Let us continue our research using the following expression to describe Faraday waves.

$$\begin{aligned}\eta =& A\sec h^2\left[l\sin N\pi L^{-1}(a(t)-x+c_-)\right]\cos^M N\pi L^{-1}(a(t)-x+c_-)\\ &-A\sec h^2\left[l\sin N\pi L^{-1}(a(t)+x+c_+)\right]\cos^M N\pi L^{-1}(a(t)+x+c_+).\end{aligned} \tag{3.46}$$

This expression follows from (3.35) and (3.45). Apparently, by varying constants in (3.46), we can describe the results of different experiments. However, carrying out calculations in which the constants fully correspond to the experiments is not our

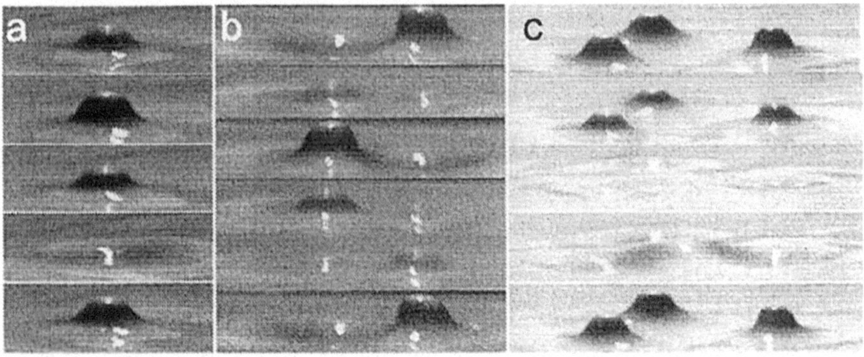

Fig. 3.5 Typical oscillons excited on a suspension surface. One oscillon (**a**), two oscillons (**b**), and three oscillons (**c**) [4, 6, 21]

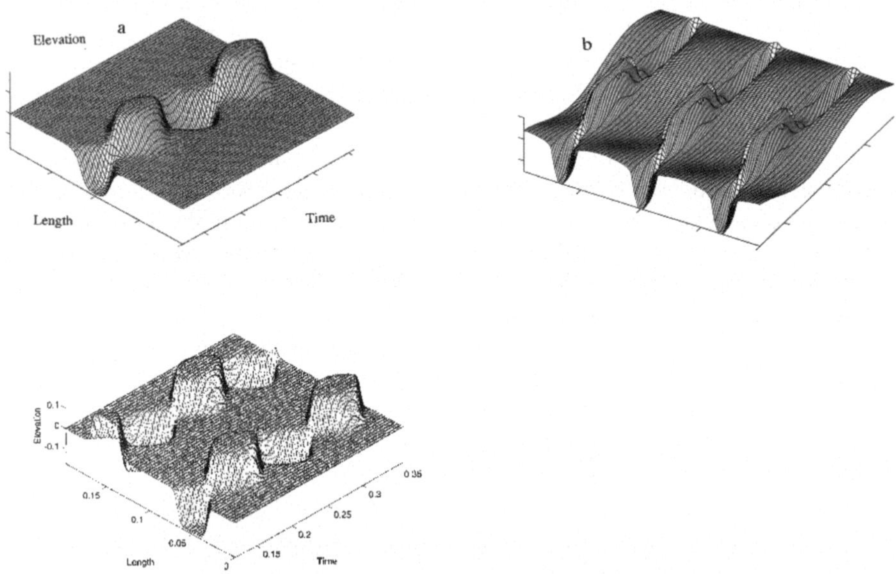

Fig. 3.6 Simulation of step-like standing surface oscillons excited in suspension: single oscillon (**a**), and oscillon triad (**b**), and two oscillons (**c**) [4, 12]

task in this section. We will strive to identify general trends. For example, during the calculation of ripple waves, we restricted ourselves by using of two expressions for vertical acceleration (3.30) and (3.31). It turned out that the expression (3.30) models situations when a ripple wave stops for a considerable period of time. This case corresponds to the experiments, when the layer is detached from the bottom of the container and is in flight for a considerable time. Therefore, waves, even moving in opposite directions, do not have time to diverge significantly (Fig. 3.7). In this case, the vertical acceleration described by the expression (3.30) corresponds to the

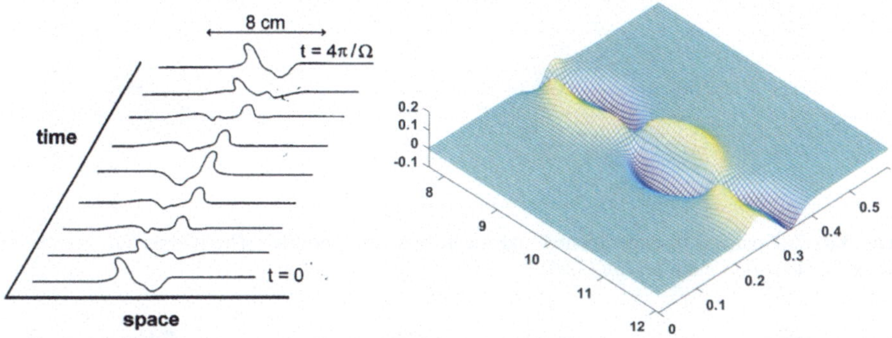

Fig. 3.7 Overlapping two ripple waves. Results of the experiment are on the left and the calculation results are on the right [4, 23]

experiments in which the flight time of the layer is small. At the same time the ripple waves have time for collisions. As a result, complex surface wave patents cannot form on the surface of the medium. In this case, ripple waves form some dynamic bands in the *x*- *t* plane.

We emphasize that in the calculations, we were not much worried about the coincidence of the results of the experiments and calculations in terms of amplitudes. We were mainly interested in the influence$a(t)$ on the dynamics of ripple waves. To illustrate this effect, we carried out a number of calculations in which expression (3.46) was used. In this case, we change only constant in $a(t)$, while the other constants in (3.46) did not change.

The results of this test are presented in Figs. 3.8 and 3.9.

The picture on the left is taken from [4, 22]. It shows a point of time when the oppositely moving waves collide and the peak appears in the middle of the resonator. The possibility of such temporarily stationary peak (also the left) is illustrated by calculations located next to the experiment under discussion. The next picture shows the experiment results taken from 15, 105]. There, the dynamics of overlay of two oppositely moving waves and the periodic appearance of a peak are well represented. This dynamic is illustrated by calculations (on the right).

Let us stop on Fig. 3.9 where four longitudinal oscillating oscillons are shown [4, 7, 22]. The length of the resonator was 0.2 m. The width was 0.025 m. The depth of water was 0.02 m. According to the developed above theory, the wave patterns on the water surface are defined by the expression (3.46). Results of calculations for $\delta = g_0$ and different M-1 and $c \pm$ are presented (bottom). We used a-(t) $= (\delta_* h_0)^{0.5} \sin \omega t/2$ and assumed that $\omega = 63$ Hz, $M^{-1} = 0.5$, c- $= -7\ L/8$, and $c + = 5\ L/8$. Figure 3.9 demonstrates the large amplitude oscillations of the ripples (solitons). They oscillate independently without collisions.

Thus, it is shown that by varying the constants in (3.46), we can qualitatively describe the experimental results for the vertically excited waves on water.

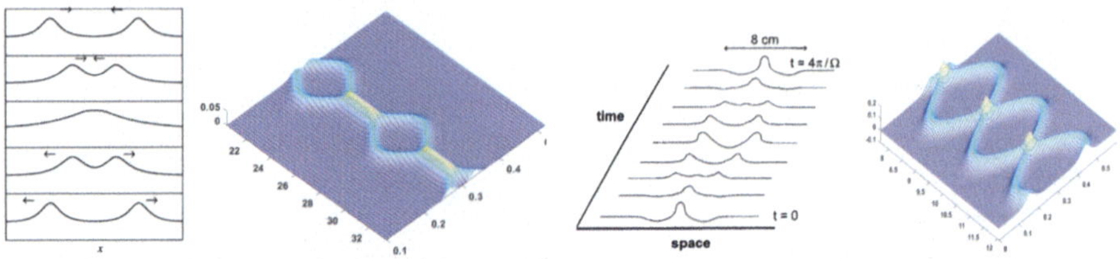

Fig. 3.8 Comparison of experimental and theoretical wave profiles. The profiles are calculated according to $a(t) = \delta_* \sin^n m^{-1} \omega t$ (3.30)

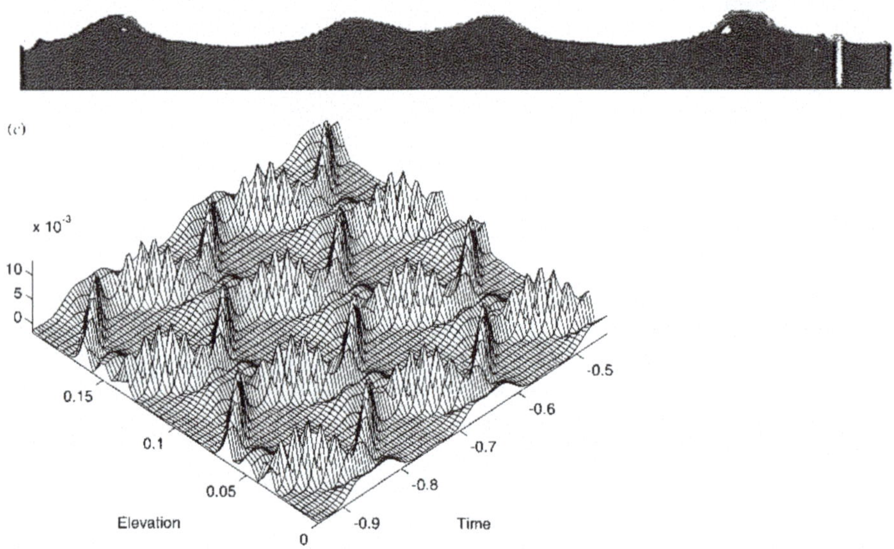

Fig. 3.9 Comparison of experimental and theoretical wave profiles. Four longitudinally oscillating solitons (experiment, upper) and corresponding calculations (bottom)

The picture of weak interaction of two oscillons is shown in Fig. 3.10a. The picture (a) qualitatively describes two weakly interacting waves. Figure 3.10b describes the formation of the oscillon.

Calculations were fulfilled according to (3.46) for h0 / $L = 0.05$, $L = 1$ м, $m = 2$, $c \pm = L/4$, $\delta^* = 2.5\ g_0$, and $a(t) = (\delta_* h_0)^{0.5} m\omega^{-1}(\sin m^{-1}\omega t - 0.333 \sin^3 m^{-1}\omega t)$. Figure 3.10 corresponds to $\omega = 25$ Hz. Figure 3.11 represents the wave patterns, calculated for different $\omega = 15$ (a), 7 (b), and 5 Hz (c).

The oscillon appearance may be considered as the following stage of the evolution of oscillating solitons shown in Fig. 3.9. They contact if the amplitude of their oscillations becomes so big that they begin to collide. Thus, the oscillons arise as a result of the interaction of two oscillating solitons. Thus, the peak arises at the moment of the collision, and the crater is formed as a result of a reflection of the solitons.

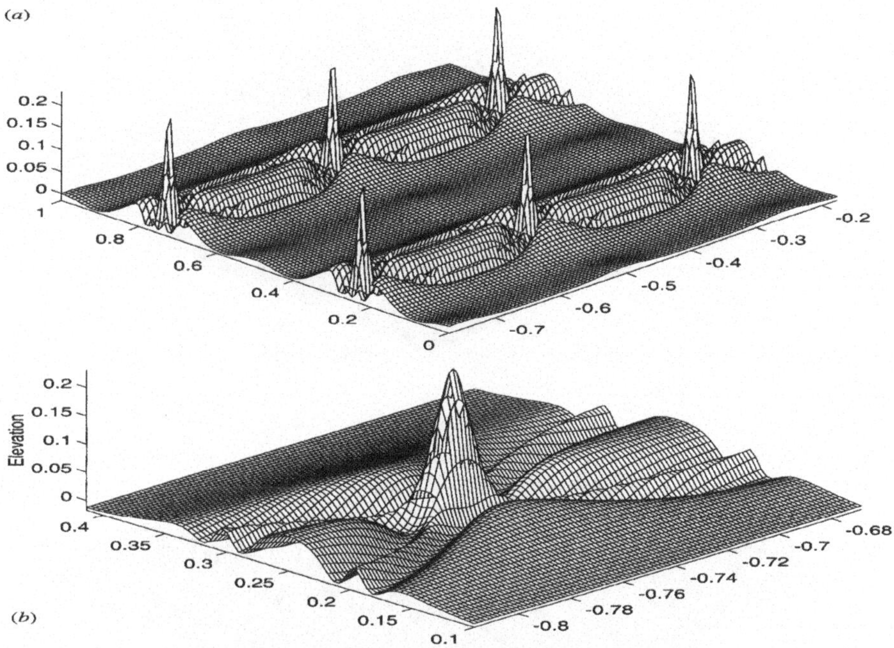

Fig. 3.10 Periodic eruptions of material and the formation of the surface craters (**a**). The detailed dynamics of the crater and the eruption (the formation of the oscillon) (**b**) [4, 6, 7]

According to the calculations, the speed of the ripples forming the oscillons strongly depends on the forcing gravitation and on bottom/layer collision. The ripples do not practically move during the free flight (after separation from the bottom). The wave dynamics takes place small enough time during the collision of the layer with the base.

The surface patterns change with the forcing frequency. The corresponding examples are given in Fig. 3.11.

When the forcing frequency is great and vertical acceleration much exceeds g_0, the ripples practically stay on the place. In this case, the interaction of oscillons is absent. If the frequency decreases, but the vertical acceleration continues and exceeds g_0, the effects of layer/base collision increase. The possibility of oscillons coupling may also increase. At a certain frequency, they start to interact strongly.

This interaction and an influence of the boundaries explain the formation of square, rectangular, and hexagonal patterns shown in Fig. 3.11. According to calculations, different patterns can be formed by highly nonlinear waves of Faraday [4, 6, 7, 18].

We should not forget that we used the model of deep water determining the above solutions and the results of calculations. It is difficult to accept this model in the case of thin layers of granular material or suspension. Indeed, the heights of oscillons (Fig. 3.11) or oscillating soliton on the surface of the water (Fig. 3.9) are quite comparable with the thickness of the layers on which they were excited.

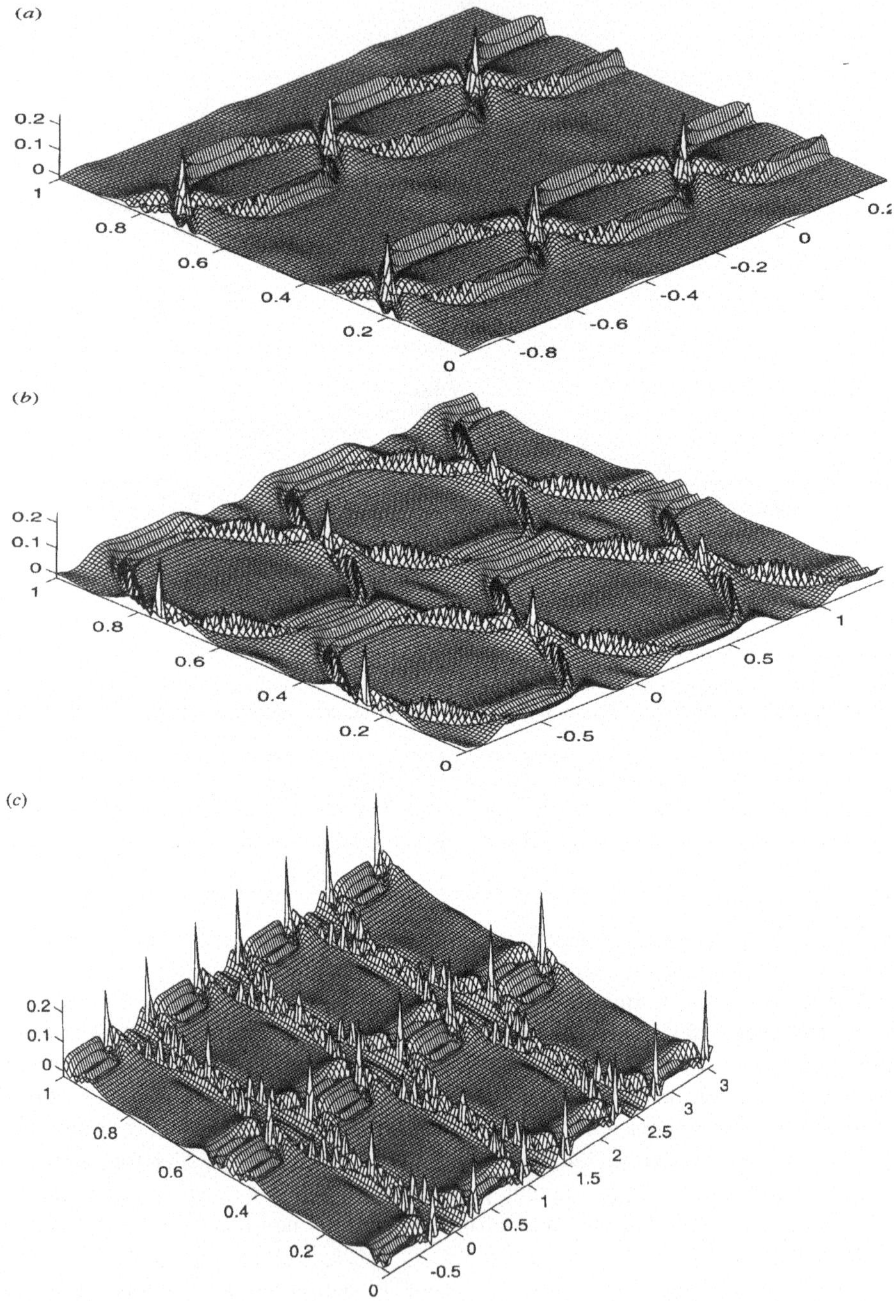

Fig. 3.11 The wave patterns, calculated for different ω =15 (**a**), 7 (**b**), and 5 Hz (**c**). The evolution of the oscillons into ellipsoidal, hexagonal, and other complex periodic structures [6, 7]

We explain this amazing qualitative agreement between the results by the fact that we are considering Faraday waves. They are characterized by both vertical and horizontal movements. However, we underline that it is possible to ignore often the horizontal movement of waves. Such behavior reminds the behavior of ball on the vibrating base. Therefore, with strong vertical resonant excitation, the error in using the theory of long surface gravitational waves when studying Faraday waves may not be very significant.

3. **Surface patterns.** The possibility of the wave coupling may be increased when the frequency decreases. At certain frequency, they start to interact strongly. As a result, the complex surface patterns may be formed (Fig. 3.12).

It is seen that the four oscillating solitons excited when the frequency equals 15 or 7. These cases qualitatively correspond to stripes showed in Fig. 3.9. Oscillating solitons begin strongly interact when the frequency is 6.2. As a result of the collisions in the plane x-t, the square pattern of oscillons is formed. After each peak, the crater follows. Near each peak, there are craters and vice versa. In general, the resulting patent corresponds to the patents presented in Figs. 1.6 and 1.7. During the calculations, we used (3.44) where $a(t) = A \sin \omega t/2$. Here A is some constant.

Changing the constants in the above solutions (3.40)–(3.44), we can get a variety of patterns and describe some experiments. Figure 3.13 (upper) shows two experimentally obtained patterns [7, 24]. We modeled them. Results are presented in Fig. 3.13 (bottom). We used solutions (3.43) and (3.31), when the left patent was calculated. The resulting pattern (left), in general, satisfactorily describes the data of the experiments (upper, left). The same can be said about the other pattern (bottom, right). During the calculations, we used (3.40) where $a(t) = (g_0 h)^{0/5} t$ .The squarely located peaks showed there describe the data of the experiment (upper, right).

We considered ripple waves excited on layers of various media. It turned out that with a corresponding change in excitation, they can form surface patterns. In general, these patterns correspond to experiments. We emphasize that the diversity of these patents depends on the frequency and amplitude of forcing vertical oscillations.

If the forcing frequency is decreased, but its amplitude is increased, the ripples start to interact.

This interaction explains the formation of square, rectangular, and hexagonal patterns shown in Figs. 3.12, 3.13, and 3.14. These patterns correspond theoretical

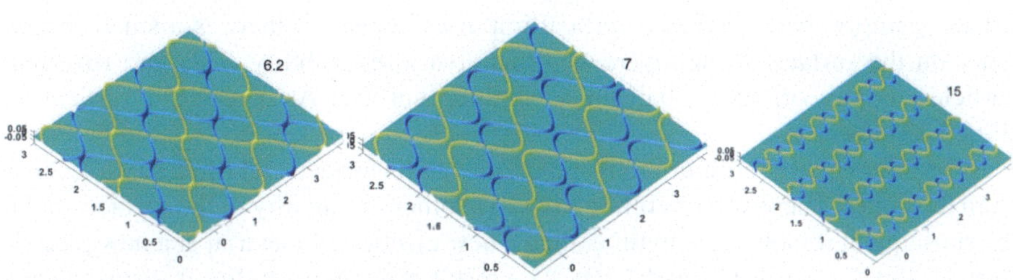

Fig. 3.12 Example of evolution of stripes into squares when frequency of vertical oscillations changes from 15 to 6.2

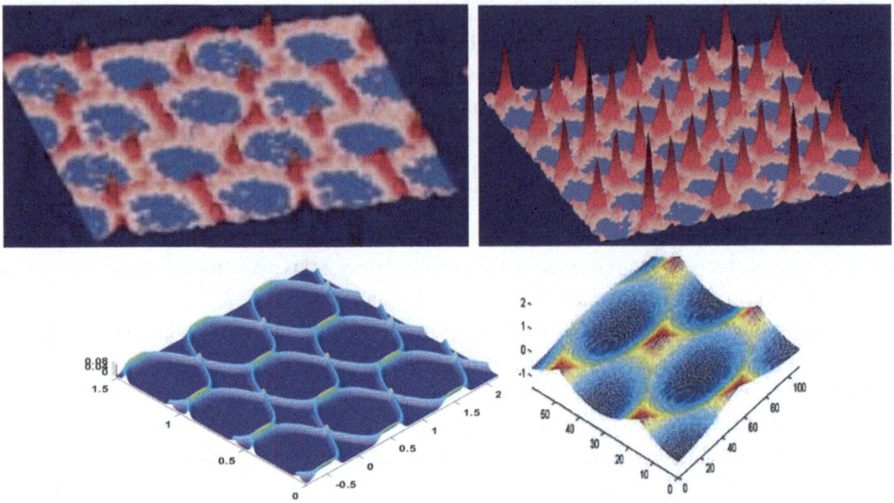

Fig. 3.13 Examples of the surface pattern. The upper row—experiments [7, 24] and the bottom row—results of calculations

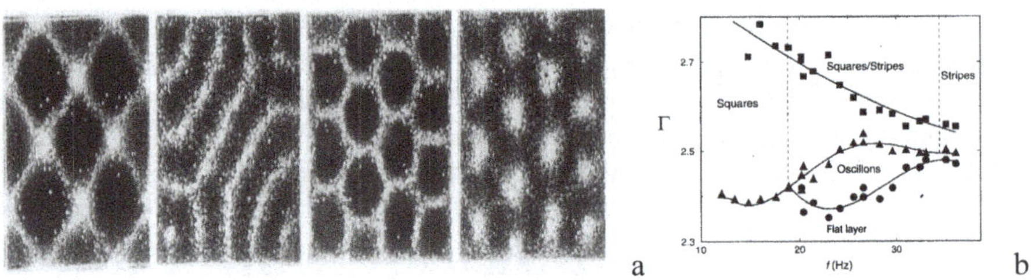

Fig. 3.14 Examples of the surface patterns of standing waves (**a**). The areas of the various surface patterns in a plane of the dimensionless vertical acceleration/forcing frequency (**b**) [6, 25]

results (Figs. 3.9, 3.10, and 3.11). According to calculations, different patterns can be formed by highly nonlinear waves of Faraday.

Remark. It is important that oscillons occur if the vertical acceleration is greater than the gravitational acceleration. In this case, the layer begins to periodically detach from the base. Periodic vertical impulses appear. If there is a small irregularity on the surface of the layer, then this area loses stability. This part (medium particles) comes off like a ball on a vibrating surface. As a result, a vertical jet (Fig. 3.13) occurs.

When these particles return to its place, a crater is formed and ripples (waves) run from the crater. These waves collide to form another irregularity of the surface, and if the subsequent impulse acts in time, then a new ejection of material at a new place is formed (Figs. 1.6, 1.8, 1.9, and 1.10). As a result, a strongly nonlinear surface patent arises that is clearly visible in the example of oscillons.

3.3 Faraday Waves and Counterintuitive Results of Experiments [26, 27]

We emphasize that nonlinear longitudinally excited waves in resonators have been studied for more than 50 years [1, 2, 10, 11, 13, 28, 29]. However, this cannot be said about highly nonlinear parametrically excited waves. The interesting results for these waves were obtained in the interval of 1996–2000. In particular, spatially localized, oscillating excitations (oscillons) were observed [30–32] on the surface of a layer of vertically vibrated brass balls (see also, Sect. 1.5). Physicists suggested these excitations possibly relevant to areas of physics as remote as semiconductors, earthquakes, and clustering of galaxies. A few models were proposed useful for the interpretation of the experimental data. Here we will look at the model proposed in 1999 [18] and developed in [4, 6]. There it was proposed that in some cases, Faraday waves consist of standing and traveling, oscillating waves.

3.3.1 Faraday Waves as a Sum of Nonlinear Standing Waves and Nonlinear Eigenfunctions

The standing wave may be presented as

$$u = u_1 + u_2, \tag{3.47}$$

where we assume that

$$u_1 = U_1(x) \cos \frac{1}{2}\omega t\ , \tag{3.48}$$

and

$$u_2 = U_2(x) \cos \omega t \tag{3.49}$$

We substitute (3.47) into the equation for long waves (3.25). Waves u_1 are considered primary, and waves u_2 are secondary. Separating these waves, we get two equations:

$$u_{1,tt} - a_0^2 u_{1,xx} = 6a_0^2 u_{1,xx} u_{1,x}^2 + \frac{1}{3} h^2 u_{1,ttxx}, \tag{3.50}$$

$$\begin{aligned} u_{2,tt} - a_0^2 u_{2,xx} = & -3a_0^2 \left[u_{1,x} u_{1,xx} + u_{1,x} u_{2,xx} + u_{2,x} (u_1 + u_2)_{xx} \right] \\ & + 6a_0^2 \left(u_{1,x}^2 + 2u_{1,x} u_{2,x} + u_{2,x}^2 \right) u_{2,xx} \\ & + 6a_0^2 \left(2u_{1,x} u_{2,x} + u_{2,x}^2 \right) u_{1,xx} + \frac{1}{3} h^2 u_{2,ttxx}. \end{aligned} \tag{3.51}$$

This shows that even in the case of purely vertical excitation, the forcing and parametric terms arise in Eq. (3.51) due to the nonlinearity of the original expression

(3.25). It is possible that the influence of the waves u_2 on the results of the experiments discussed below is not great. However, the parametric resonance frequency can coincide with one of the natural frequencies of the longitudinal oscillations of the resonator. In the case of such synchronization, the problem of analyzing parametrically induced oscillations becomes complicate.

On the other hand, from the simplest analysis of Eq. (3.50), it follows that its solution can be represented as

$$u_1 = \overline{u}_1(x)\left(\cos\frac{1}{2}\omega t + A\cos\frac{3}{2}\omega t + B\cos\frac{1}{6}\omega t\right) \tag{3.52}$$

We do not know how significant the influence of the components $A\cos\frac{3}{2}\omega t$ and $B\cos\frac{1}{6}\omega t$ on the resulting waves is. However, taking it into account complicates the problem in the case of the study of extreme waves.

We have given the reasons characterizing the difficulties of studying highly nonlinear vertically excited waves.

Let us consider solution in the form of harmonic modes. We will mainly consider waves vertically excited in resonators with fixed boundaries. For this case, we look for the solution of Eq. (3.50) in the form

$$u_1 = U\sin\left(N\pi aL^{-1}x\right)\cos\frac{1}{2}\omega t. \tag{3.53}$$

Thus, at this stage of the study, we do not take into account the contribution of oscillations $\cos\frac{3}{2}\omega t$ and $\cos\frac{1}{6}\omega t$ (3.52) to the waves in the resonator. Expression (3.53) is substituted into (3.50). Then, after a simple, albeit lengthy calculation, as well as dropping high harmonics, Eq. (3.50) is reduced to the following quadratic algebraic equation:

$$\frac{1}{4}\left[\left(N\pi L^{-1}\right)^{-2} + \frac{1}{3}h^2\right]\omega^2 - hg_0 - \frac{1}{2}h\delta \approx \left(\frac{9}{8}hg_0 + \frac{3}{4}h\delta\right)\left(N\pi L^{-1}\right)^2 U^2. \tag{3.54}$$

We find from (3.54)

$$U \approx \pm\frac{2}{\sqrt{3}}\left(N\pi L^{-1}\right)^{-1}\left\{\frac{1}{4}\left[h^{-1}\left(N\pi L^{-1}\right)^{-2} + \frac{1}{3}h\right]\omega^2 - g_0 - \frac{1}{2}\delta\right\}^{0.5}\left(\frac{3}{2}g_0 + \delta\right)^{-0.5}. \tag{3.55}$$

Now we can find u_1 (3.53) using (3.55). The expression (3.53) can also be written as traveling waves (modes):

$$.u_1 = \frac{1}{2}U\left[\sin\left(\frac{1}{2}\omega t - N\pi aL^{-1}x\right) - \sin\left(\frac{1}{2}\omega t + N\pi aL^{-1}x\right)\right]. \tag{3.56}$$

Using (3.53) we can find that

$$u_{1,x} = u_A\cos\left(N\pi aL^{-1}x\right)\cos\frac{1}{2}\,\omega t. \tag{3.57}$$

Here

$$u_A = N\pi L^{-1} U \ . \tag{3.58}$$

The expression (3.57) can also be represented as

$$u_{1,x} = \frac{1}{2} u_A \left[\cos\left(\frac{1}{2}\omega t - N\pi a L^{-1} x \right) + \cos\left(\frac{1}{2}\omega t + N\pi a L^{-1} x \right) \right]. \tag{3.59}$$

After that, you can find a nonlinear wave profile (3.3)

$$\eta = h u_{1,x} \left[-1 + u_{1,x} - (u_{1,x})^2 \right] . \tag{3.60}$$

We looked at Eq. (3.50). Now let us turn to Eq. (3.51). When considering it, we will neglect the interaction of waves u_1 and u_2. In this case, (3.51) is written as Eq. (3.25), the solution of which we obtained above (see 3.8, 3.14, 3.22 and 3.23). In the last solution, the sought functions are defined in the form (3.35). Thus, the expression for u_2 is found. The general expression for Faraday waves, written as the sum of standing and traveling waves $\eta = -h\left[u_A \cos\left(N\pi a L^{-1} x \right) \cos \frac{1}{2}\ \omega t - J'(r) + j'(s) \right]$, can be written as follows:

$$\begin{aligned} \eta = \ & -h u_A \cos\left(N\pi a L^{-1} x \right) \cos \frac{1}{2}\omega t - h\left[B \tanh(l \sin \zeta_-) \cos^k \zeta_- \right. \\ & \left. + A \sec h^2 (l \sin \zeta_-) \cos^K \zeta_- + C \cos^z \zeta \right] - h\left[B \tanh(l \sin \zeta_+) \cos^k \zeta_+ \right. \\ & \left. + A \sec h^2 (l \sin \zeta_+) \cos^K \zeta_+ + C \cos^z \zeta_+ \right]. \end{aligned} \tag{3.61}$$

Here $\zeta_- = \pi L^{-1} M^{-1}(a(t) - x + c_-)$, $\zeta_+ = \pi L^{-1} M^{-1}(a(t) + x + c_+)$, A, B, C, and l are unknown constants; k, K, and z are integers; and c_-, c_- are arbitrary constants. We now turn to modeling the results of some remarkable experiments. To do this, we will use expression (3.61) written in a somewhat simplified form. Assuming $B = C = 0$ and $K = z = 2$, it is possible to write the sum

$$\begin{aligned} \eta = \ & -h u_A \cos\left(N\pi a L^{-1} a \right) \cos \frac{1}{2}\ \omega t + \\ & h A \sec h^2 \left[l \sin N\pi L^{-1}(a(t) - x + c_-) \right] \cos^2 N\pi L^{-1}(a(t) - x + c_-) \\ & \pm\ h A \sec h^2 \left[l \sin N\pi L^{-1}(a(t) + x + c_+) \right] \cos^2 N\pi L^{-1}(a(t) + x + c_+). \end{aligned} \tag{3.62}$$

Here $a(t)$ may be determined by the expression (3.30) or (3.31). The constants in (3.62) were chosen according to the experiments or so that the calculations fit the experiment qualitatively.

3.3.2 Experimentally Observed Faraday Waves as a Sum of Nonlinear Standing and Traveling Waves

Here we will continue to study the extraordinary Faraday waves accompanied by jets and craters on a free surface.

The waves presented in Fig. 3.15 were excited in the vertically oscillating container. Its length was 0.6 m and width was 0.06 m. The depth of water was 0.3 m. All the wave profiles of Fig. 3.15 were observed during one experiment. The forcing frequency was 1.6 Hz and the amplitude was 4.6 mm. It is possible for qualitative analysis of the arisen waves to present them as a sum of the main standing wave and traveling nonlinear perturbations. They are the periodical waves. It is visible that intervals of time from 0 till 0.44, from 0.52 till 1.4, and from 1.16 till 1.66 seconds define approximately a period T of oscillations of the main standing wave.

However, each interval is characterized by the local singularities. The first interval is determined by the eruptions of water from the wave crest. The wave with almost plane or slightly concave peak and the folds on the crest edges appears during the second interval. The wave with the dome-like crest forms within the third interval. Subsequently, there were repeats of the described sequence. Thus, the period of nonlinear traveling waves was equal to three period T of oscillations of the main standing wave. The extraordinary waves appeared at the center and near container edges (Fig. 3.15). However, the waves at edge had a delay on 1.5 T in comparison with waves at center (T is the period of forcing oscillations). These singularities in [26] have been paid special attention.

For modelling of the experiments with the surface waves in the container were used described above theory. The surface waves were represented by the sum of the standing wave and travelling ripples.

On the basis of this approach the patterns presented in Fig. 3.16 are received. all profiles of the waves of Fig. 3.15 are presented on the upper row of Fig. 3.16.

Modelling of the extraordinary waves of Fig. 3.15 is based on the assumption that the speed of ripples is some function of vertically excitation. Changing the frequency and the amplitude, we can vary the acceleration and the speed. As a result the wave direction and the number of the waves can be changed. Two examples of the ripples spread on the surface of the smooth main wave are presented in Fig. 3.16 (bottom row).

The left waves were calculated using $a(t)=\sqrt{\delta h m \omega}^{-1}(\sin m^{-1}\omega t-\frac{1}{3}\sin^3 m^{-1}\omega t)$. The right waves were calculated using $a(t) = \delta_* \sin^n m^{-1}\omega t$. In the left option, it was assumed that the forcing vertical acceleration was slightly less than the gravitational acceleration. In the right option the forcing acceleration was slightly higher of the gravitational acceleration.

According to Figs. 3.15 and 3.16 the maximum amplitudes and the water eruptions arise when the opposite travelling ripples meet each other at the standing wave peak. The ripples form the plane crests in cases when these meetings take place on a slope of the main wave at the moment of its growth. At last, if this meeting takes place on a slope of the main wave at the moment of its recession, the ripples form the dome-like crest.

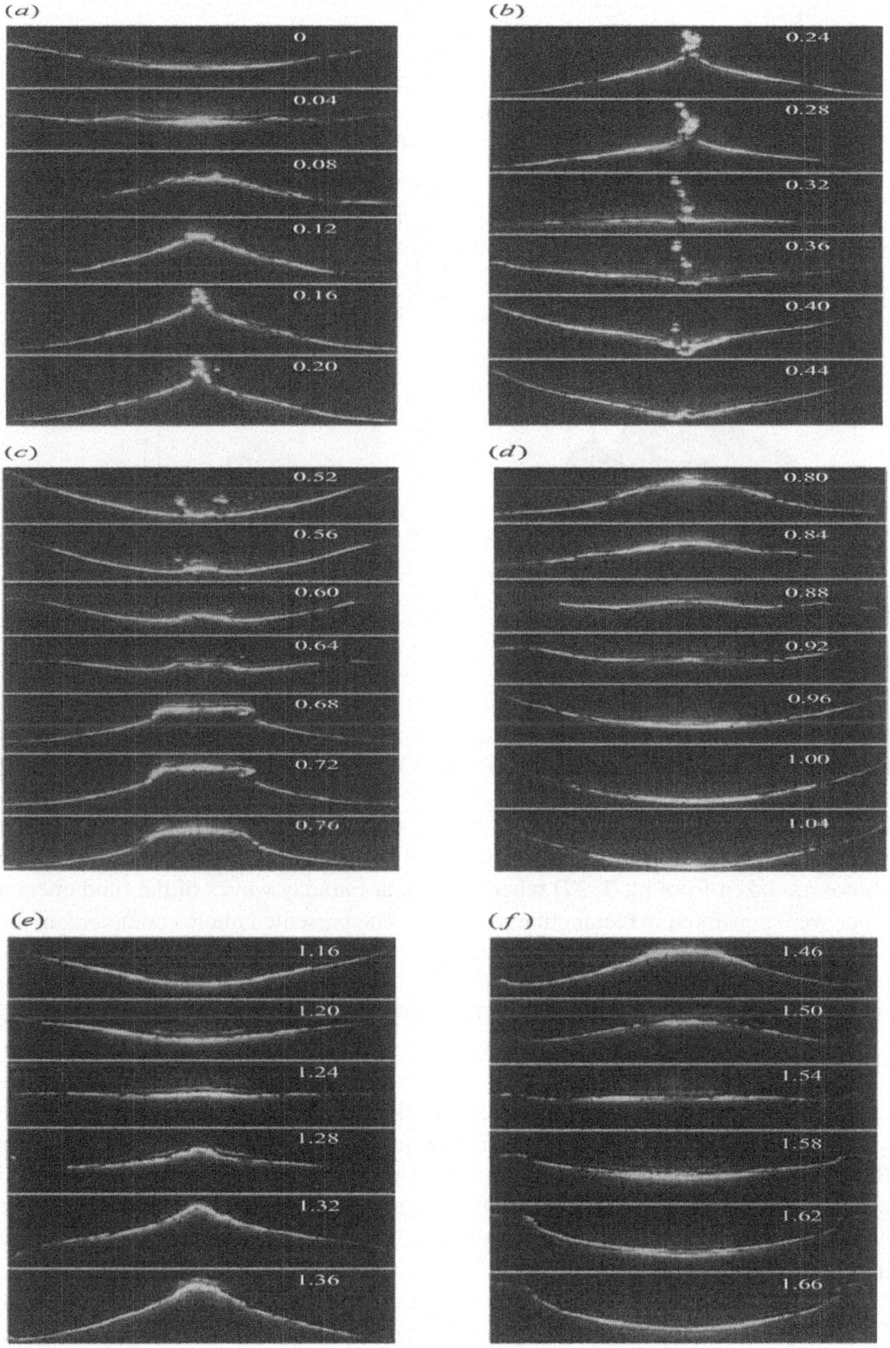

Fig. 3.15 The complex evolution of surface waves accompanied by eruption from wave crests. Digits define time moments [4, 6, 7, 26]

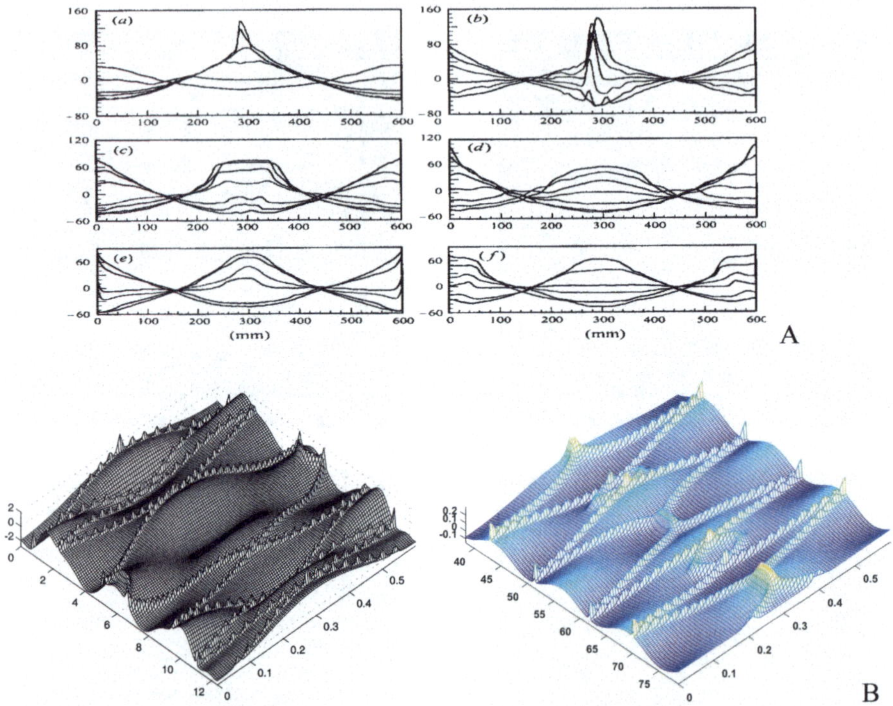

Fig. 3.16 Wave profiles (**a**) corresponding to waves in Fig. 3.15. Result (**b**) of modeling of these profiles and the waves presented in photos of Fig. 3.15. The picture (**b**) is calculated using linear formula for η

Let us consider the data of calculations and experiments showed in Fig. 3.17. The photos are taken from [4, 7, 27] where nonlinear Faraday waves of the fundamental mode were generated in rectangular resonator. The presented photos complement the photos shown in Fig. 3.15. The wave ripples moving on the surface of the water are more clearly visible. The resonator has internal dimensions 600 * 60* 483 mm (length, width, and height) [4, 7, 27]. The water depth is 0.3 m. The forcing amplitude is 4.04 mm and the forcing frequency is 3.17 Hz. The wave height is about 130 mm. The photos are not very clear but show some important particularities of wave profiles. Aside from the small breaking that occurs at the sharp crest at $t = 0.528$ s, the wave profile is similar to the non-breaking standing waves. Ripple-like wave to each side of the centerline are also present with the dimpled crest at the beginning of the series. The waveform is spatially symmetric about the centerline, and the same breaking events occur at the two ends of the container with a half-period lag (see time 0.372 s).

In Fig. 3.17, the results of experiments are accompanied by the calculations. In the calculations, we used the expression (3.62) and $a(t) = \sqrt{\delta_* h} m\omega^{-1}\left(\sin m^{-1}\omega t - \frac{1}{3}\sin^3 m^{-1}\omega t\right)$.

We considered cases when the forced acceleration was close to gravitational acceleration. Therefore, the time of separation of the liquid layer from the bottom

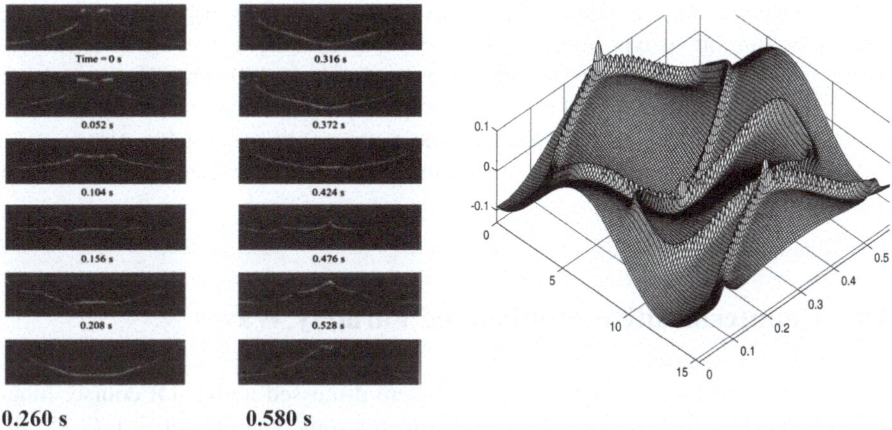

Fig. 3.17 The traveling resonant ripples on the surface of the standing harmonic wave. Comparison of experimental (left) and theoretical (right) results

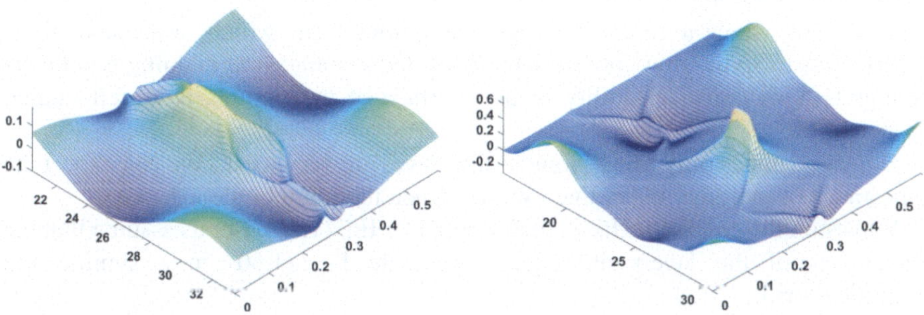

Fig. 3.18 The traveling resonant ripples on a surface of the standing harmonic wave. It is assumed that $a_0(t) = \delta_* \sin'' m^{-1} \omega t$

was short and the time when the traveling waves stopped was also not long. Cases when the flight time is significant and it is accompanied by the collision of oppositely traveling waves are shown in Fig. 3.18 [4].

Of course, the appearance of the traveling waves discussed above is not entirely obvious. They cannot arise in the case of linear analysis. However, doubts about these waves disappear if nonlinear effects are taken into account. It is necessary to use a nonlinear theory. According to it, a linear parametrically excited wave, if nonlinear terms are taken into account in the governing equations, generates forcing terms in them. The latter excite traveling longitudinal waves. See (3.47) and Eqs. (3.50) and (3.51). Thanks to what has been said, on the free surface, in addition to the main standing waves, traveling waves arise. They are usually much smaller than a standing wave, but with resonance this is not the case. Let the exciting frequency is close to one of the resonant frequencies of longitudinal vibrations of the water layer.

The resonant amplification of the wave in this case is especially strong if oppositely running waves meet at the top of a standing wave. Such a case is shown in Fig. 3.15. In this case, the wave amplitude is 120 mm. Thus, we have almost twofold amplification of the wave compared to other maximum amplitudes as shown in Figs. 3.15 and 3.16. Thus, extreme wave amplification takes place.

We will call the resonant wave arising in this case, as an extreme Faraday wave [18, 33].

3.4 Counterintuitive Multivalued Faraday Waves

Waves described by eigenfunctions (3.35) were discussed above. Of course, functions (3.35) are only approximate expressions for some solutions to Eq. (2.25). In particular, the adopted approximation (3.35) rejects multivalued solutions from the very beginning.

Let us correct this shortcoming. Let us consider and explore the possibility of multivalued solutions.

To do this, we write (2.25) in a highly simplified form, which, at the same time, corresponds to some physical problem. Consider a container containing two layers of liquid. A dynamic instability of an interface of two liquids is a well-studied phenomenon.

However, this is only true as long as the waves are very weak. We can expect the appearance of extreme multivalued waves, when their amplitude increases.

Waves on the contact surface are described by Eq. (1.60). Let us assume that the thicknesses of the layers are equal and write Eq. (1.60) in a significantly simplified form:

$$N_{tt} - a_0^2 N_{xx} = \frac{1}{3} a_0^2 h_1 h_2 N_{xxx} + \beta_1 a_0^2 N_x^2 N_{xx}. \tag{3.63}$$

Here according to (1.59) we have

$$\beta_1 = -3 h_1^{-2} a_0^{-2} c_0^{-2}. \tag{3.64}$$

We emphasize that the entered value β_1 differs in writing from expression (1.59). Let us recall that after the definition N, the wave profile (elevation) is determined as follows:

$$\eta = N_x\ . \tag{3.65}$$

At the boundaries of the container, we accept that

$$N = 0 \quad (x = 0\ ;\ \ x = L) \tag{3.66}$$

It is easy to see that (3.63) is a special case of Eq. (3.2). Therefore, we can use solution (3.23). Using it, we write the solution to Eq. (3.63) in the form

$$N = J + j + \Psi_3 + \psi_3 + \frac{1}{4} a_0^{-2} \beta_1 \left(J' \int j'^2 ds - jJ'^2 + \frac{2}{3} x \left(J'^3 - j'^3 \right) - j'^2 J + j' \int J'^2 dr \right) + d_1. \tag{3.67}$$

We will look for periodic solutions having a period 2 L. As a condition for exact resonances, we accept the equalities

$$J(a(t) - L) = J(a(t) + L) \quad \text{and} \quad j(a(t) - L) = j(a(t) + L) \tag{3.68}$$

and similar equalities for the functions Ψ_3 and ψ_3. Now consider the boundary condition at $x = 0$ ($N = 0$). It is easy to see that it is satisfied if

$$J = -j\,. \tag{3.69}$$

Now consider the boundary condition at $x = L$. There, the following Taylor series expansions are approximately valid in the vicinity of resonances:

$$J(s) = J\left(r + \overline{\Delta}\right) = J(r) + \overline{\Delta} J'(r) + \ldots \tag{3.70}$$

Here $\overline{\Delta}$ is a deviation from the exact resonance. In this case, the expression (3.67) and the condition $x = L$ give the following algebraic equation:

$$J'^3 - 3a_0^2 \beta_1^{-1} L^{-1} \left(\overline{\Delta} J' - \Psi_3 - d_1 \right) = 0. \tag{3.71}$$

The function Ψ_3 is determined by the seed function (seed wave). Equation (3.71) has a similar form for an oppositely traveling wave. Equations (3.71) and (3.65) are the subject of our further study. It contains several parameters. Therefore, it describes a wide variety of unidirectional waves. However, our research goal is to study waves in a resonator, where the wave pattern can be much more complex than that determined by unidirectional waves. Therefore, when studying waves in a resonator, we will be guided by the results of remarkable studies of extreme multivalued Faraday waves presented in [34]. We have already discussed these results in Chap. 2. To an even greater extent, these results will be useful in this chapter.

3.4.1 Method for Calculating Extreme Multivalued Traveling Waves Arising in Resonators

We find approximately using experimental data [34, 35] the quantity

$$a_1(t) = 1 + 0.8 \sin\left(7.65 \sqrt{g h_1} t\right), \quad a_2(t) = \sqrt{a_1(t) g h_1}. \tag{3.72}$$

Then Eq. (3.71) yields

$$J'^3 - \Delta a_1(t)J' + 0.0001a_1(t)[\sin 7.65(a_2t - x) - 0.4 \sin 22.3(a_2t - x)] = 0. \tag{3.73}$$

Here Δ is the transresonant parameter. We emphasize that Δ is an arbitrary quantity, which includes the quantities $2.127\beta_1^{-1}L^{-1}\overline{\Delta}$ from (3.71). The terms sin $7.65(a_2t - x)$ and 0.4 sin $22.3(a_2t - x)$ in (3.73) are determined by the seed function Ψ_3. The equation for j' is written in a similar way.

The purpose of the calculations presented in this section is to create the calculation method itself, suitable for studying multivalued waves arising in resonators. The calculation of resonant waves, despite the apparent simplicity of the original Eq. (3.63), turned out to be not an easy task for the author of the book. To some extent, this is due to the novelty of the transresonant transition problem itself, and on the other hand, as the author believes, with the algorithm itself used in the calculations, described in Chap. 2.

After determining J' and j' from the indicated equations, the wave profile is found as a sum $\eta = J'(r) + j'(s)$ or difference $\eta = J'(r) - j'(s)$. Note that the difference $\eta = J'(r) - j'(s)$ corresponds to the solution of the wave equation under the boundary conditions $u_x = 0$.

The solution of Eq. (3.73) was carried out according to the algorithm presented in Sect. 2.2.1. "Strongly nonlinear effects and unification of their description and analysis." According to this algorithm, when calculating the four roots of the equation, we used the expression

$$J_i = -2D \cos\left[\tfrac{1}{3}\arccos\left(0.5q_* D^{-3}\right) + 2N\pi/3\right], \tag{3.74}$$

(where N = 0; 1; and 2) and the expression

$$J = -2D \sinh\left[\tfrac{1}{3}\operatorname{arcsin} h\left(0.5q_* D^{-3}\right)\right], \tag{3.75}$$

which defines the single-valued discontinuous part of the general wave profile. Therefore, in many cases, when calculating complex wave profiles, some refinements were introduced into the algorithm connecting three-valued sections of the wave profile with single-valued sections.

To test the suitability of the algorithm, we decided to use the experimental data presented in [4, 7, 34, 35]. In the calculations, Eq. (3.73) was used where $\Delta = 0.0005$. The results are presented below (Fig. 3.19).

Figure 3.20 (left) shows the results of calculations using Eq. (3.73). On the right are data from an experimental study of Faraday waves excited on the contact surface of two liquids having different mechanical parameters [34]. The figure shows waves calculated as the difference of traveling waves $\eta = J' - j'$ (top) and their sum $\eta = J' + j'$ (bottom). The solution to Eq. (3.73) depends on how close the problem parameters are to the exact resonance. Deviations from resonance are determined by Δ. When calculating, Fig. 3.20 accepted $\Delta = -0.0005$. The exception is the second (central) curve at the top, which is calculated for $\Delta = -0.001$.

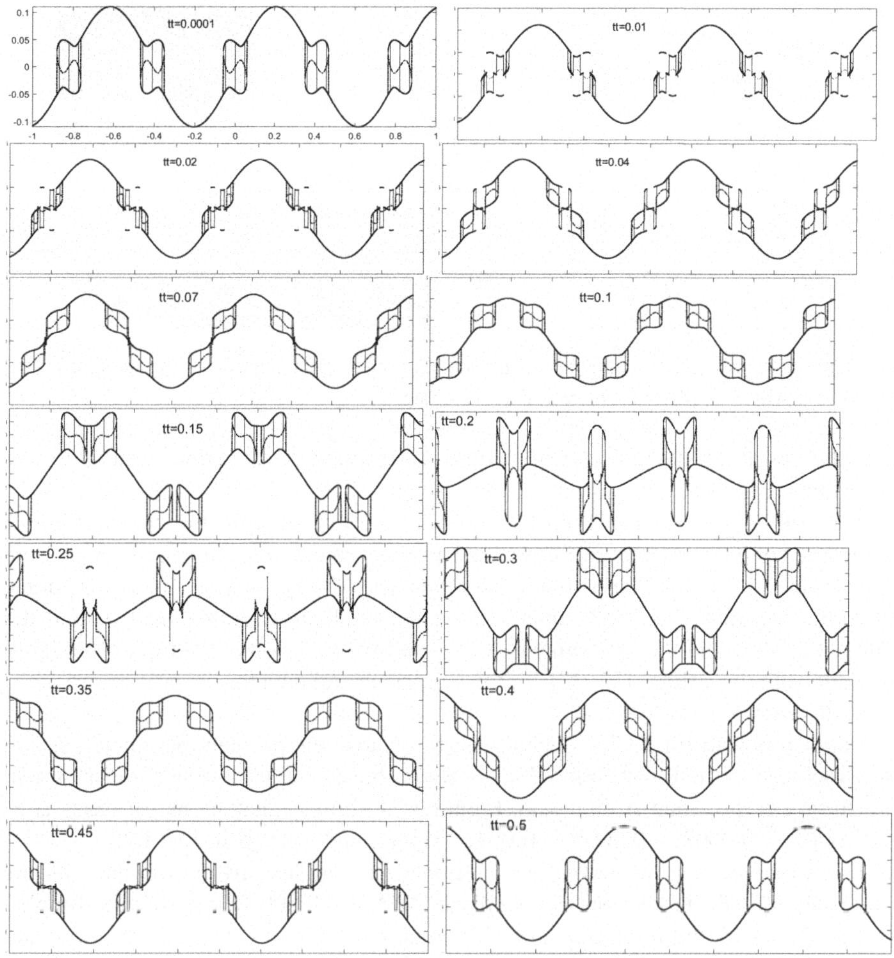

Fig. 3.19 The change in the wave profile $\eta = J' - j'$ calculated according to (3.73). The quantity *tt* determines the dimensionless time. The curves correspond to experimental data (see Fig. 3.25 and [34])

Let us recall that Eq. (3.73) was written using the solution (3.23) of the wave Eq. (3.25). This wave equation qualitatively corresponds to the equation describing waves on the surface of contact of two liquids (1.60). We emphasize once again that the last equation is also derived from approximate equations that describe various models of the propagation of nonlinear waves on the contact surface of two liquids. Finally, it is necessary to say something about the parameters of the algebraic Eq. (3.73). There the quantities and 5.41 were calculated using experimental data [34, 35]. Perhaps they do not fully correspond to the experimental conditions. The amplitude of the seed disturbance and its mathematical representation are taken completely arbitrarily. True, the amplitude of the seed disturbance was chosen so

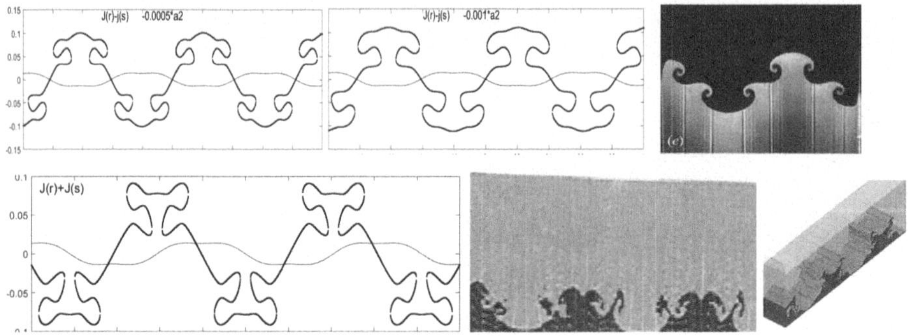

Fig. 3.20 Comparison of theoretical results and experimental data. The thin line defines a seed disturbance whose amplitude is increased by 100 times

that the result of the calculation would correspond to a certain extent to the experiment [34–36].

The curves presented in Fig. 3.20 do not take into account the contribution of nonlinear components that are contained in the general solution (3.23). Thus, the curves Fig. 3.19 and 3.20 are obtained on the basis of very, very strong assumptions. Nevertheless, from Fig. 3.20, one can see the qualitative agreement between the calculation data and the experiments (the experimental result at the top of the figure is taken from the articles [35, 37], and the experimental data for the bottom are taken from the articles [34–36]).

The solution to Eq. (3.73) depends on how close the problem parameters are to the exact resonance. It was found that the wave shape changes during a transresonant transition (as the value Δ changes). Figure 3.21 shows calculations for cases $\Delta = -0.00002$; -0.0003; -0.0005; -0.001; -0.003; and -0.005 in Eq. 3.73.

It can be seen that inside the resonant band, a complex process of changing the wave pattern in the resonator can take place. At $\Delta = -0.00002$, a wave of a complex profile is excited, but it remains unambiguous. Further, with changing Δ, more and more complex multivalued wave profiles appear, composed of solutions to the algebraic Eq. (3.73). In particular, we highlighted with a thick line the wave profile, which changes in the considered interval from a continuous form corresponding $\Delta = -0.00002$ to a continuous form corresponding $\Delta = -0.005$. Within the specified interval, the shape of the profile changes very strongly and, generally speaking, it becomes multivalued. The calculation results for $\Delta = -0.0003$; -0.0005; and -0.001, qualitatively correspond to the profiles presented in Fig. 3.20 (see the top row of waves there). The situation with calculations at $\Delta = -0.003$ and -0.005 requires separate consideration. Apparently, it corresponds to the system leaving the resonance zone. In this case, according to the calculation for $\Delta = -0.005$, two options are possible. The first corresponds to the appearance of a continuous wave whose amplitude is much less than the amplitude of the waves excited inside the resonant band. The second corresponds to the emergence of discontinuous stepped multivalued waves whose amplitude continues to increase compared to the waves arising in the resonant band.

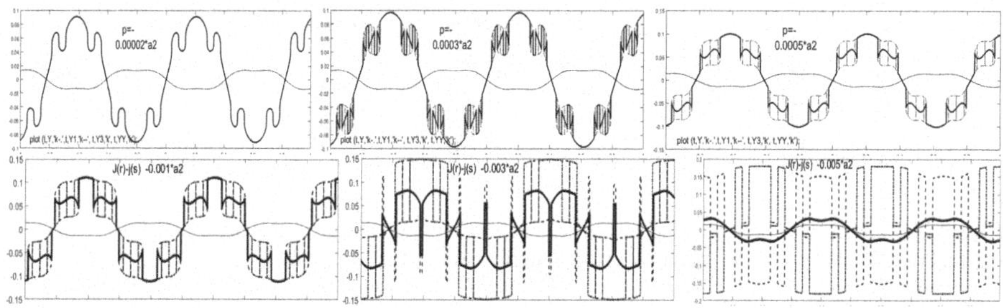

Fig. 3.21 Changes in wave profiles during transresonant evolution. The thin line defines a seed disturbance whose amplitude is increased by 100 times

This is an amazing result! It requires further research. However, it corresponds very well to the analysis carried out in Chaps. 1 and 2 (see Figs. 1.4, 2.33, 2.37, and 2.38). If we take into account the results presented there, the result obtained can be illustrated by the results of Natanzon's experiments (Fig. 2.48) or Figs. 2.17 and 2.18. At small oscillation amplitudes, the acceleration acting on the fluid is much less than gravitational acceleration. Therefore, even in the resonant region, the liquid remains in full contact with the bottom of the container, and when leaving the resonant region, wave formation on its surface becomes completely linear. Another situation arises if the liquid begins to come off the bottom of the container. In this case, the surface of the liquid begins to experience shocks, and the wave process takes on a completely new character. In particular, the waves in the medium become discontinuous, and their amplitude increases sharply. I would like to emphasize that all this was well demonstrated by Natanzon's experiments [38], in particular Fig. 2.48.

The stepped waves corresponding to $\Delta = -0.005$ in Eq. 3.73 presented in Fig. 3.21. Let us isolate the step waveforms from noted case. We found two forms. They are shown in Fig. 3.22.

We believe that the calculation data presented in Fig. 3.22 show the possibility of evolution of continuous extreme multivalued waves shown in Fig. 3.21 in isolated stepped waves, which to a certain extent agrees with experiment [34, 35].

Of course, the result presented above requires verification, although it occurred in many of my studies of waves excited in various resonators. Their peculiarity is that they are constructed using only two branches of the solution out of three determined by solutions (2.32).

Another possibility for interpreting some of the results presented is related to the waves we called in Sect. 2.5 "waves with hair" (Fig. 3.23). These waves resemble, qualitatively, the experimental data presented in [34, 35].

In Chap. 2, it was shown that the shape of the seed perturbation does not greatly affect the calculations. The main thing is that the disturbance is harmonic. The same result was obtained for waves in the resonator (Fig. 3.24).

In the last calculations, we assumed $\Delta = -0.0005$ and $t = 0.1$ in (3.73).

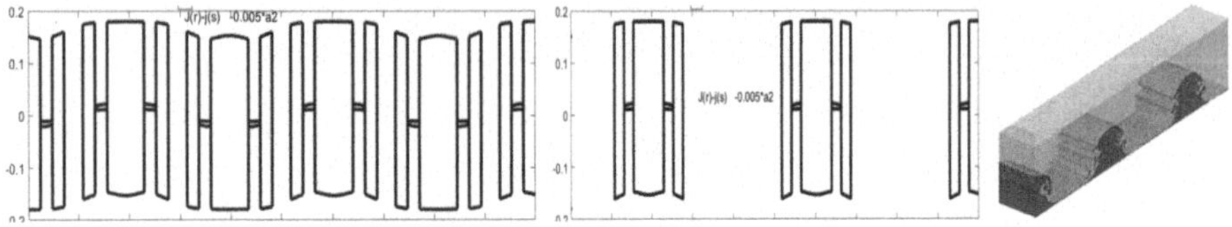

Fig. 3.22 Variants of stepped isolated waves (particle-waves) that can arise near the boundary of the resonant band (left). On the right are isolated extreme Faraday waves

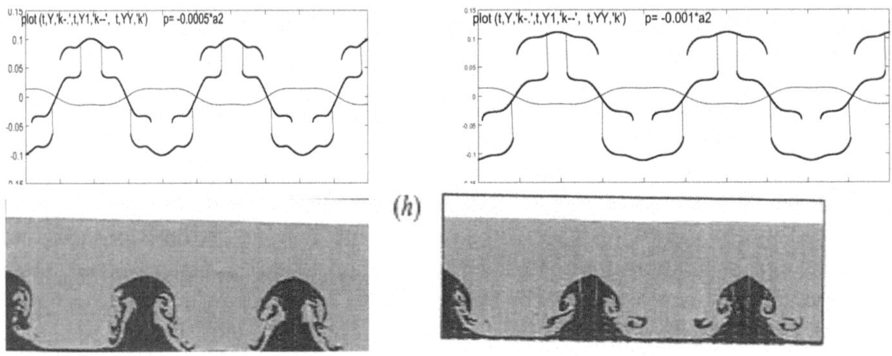

Fig. 3.23 Comparison of theoretical results, which can be interpreted as the appearance of the wave with hair, and experimental data

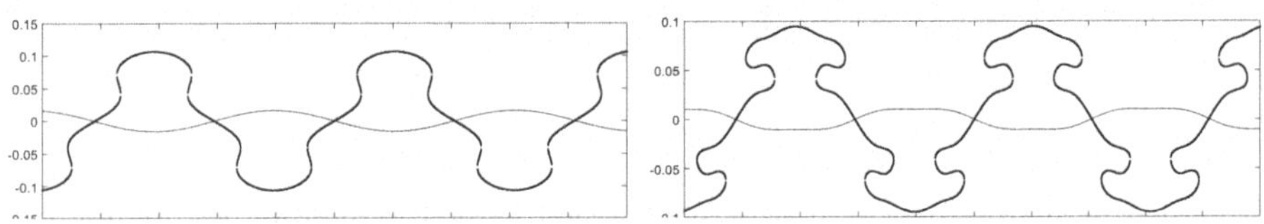

Fig. 3.24 Wave profiles obtained for cases of seed disturbances containing $\sin 7.65(a_0 t - x)$ (left) or $\sin^3 7.65(a_0 t - x)$ (right)

3.4.2 Application of Method of "Resonant" Perturbations to Counterintuitive Faraday Waves

We considered above the option of describing resonant Faraday waves as traveling waves. The speed of these waves changes periodically in accordance with the law of parametric excitation. This approach apparently works if the seed function is much smaller than the excited Faraday waves. However, this approach is apparently unfair in cases where the amplitude of parametric excitation is large, for example, when the amplitude of the forcing oscillations is comparable to g. In this case, both a standing wave and two oppositely moving waves can coexist in the system simultaneously. Of course, with strong excitation, the last waves can be almost motionless, as our calculations in Sect. 3.4.1 showed. But this does not negate the fact that they must be considered as traveling waves.

This is the case that we believe is realized in article [34].

There a tank containing a two-layer liquid with a free surface was considered. During vertical excitations, the layer interface and the free surface can be excited separately or simultaneously. Of particular interest to us are resonant waves at the interface between layers. Analysis of these waves showed that their surprising multivalued is associated with the influence both standing and traveling waves.

Naturally, one must use the algorithm implemented in Sect. 3.3 to describe the case under consideration.

In accordance with it, we will look for solutions to Eq. (3.63) as the sum

$$N = N^{(1)} + N^{(2)}. \tag{3.76}$$

The conditions at the ends of the resonator do not change (3.66).

Let us define the wave speed in the form (3.27)

$$a_0(t) = a_0\sqrt{(1 + A_0 \cos \omega t)}. \tag{3.77}$$

Then Eq. (3.63) can be approximately rewritten as follows:

$$a_0^{-2}(1 - A_0 \cos \omega t)N_{tt} - N_{xx} = \frac{1}{3}h_1 h_2 N_{xxxx} + \alpha_2^* N_x^2 N_{xx}. \tag{3.78}$$

We can write taking into account (3.76):

$$\begin{aligned} &a_0^{-2}(1 - A_0 \cos \omega t)\left(N^{(1)} + N^{(2)}\right)_{tt} - \left(N^{(1)} + N^{(2)}\right)_{xx} = \\ &\frac{1}{3}h_1 h_2\left(N^{(1)} + N^{(2)}\right)_{xxx} + \alpha_2^*\left(N^{(1)} + N^{(2)}\right)_x^2\left(N^{(1)} + N^{(2)}\right)_{xx}. \end{aligned} \tag{3.79}$$

We divide the last equation into two equations

$$a_0^{-2}(1 - A_0 \cos \omega t){N^{(1)}}_{tt} - {N^{(1)}}_{xx} - \alpha_2^* {N^{(1)}}_x {N^{(1)}}_{xx} = 0, \tag{3.80}$$

$$\begin{aligned} a_0^{-2}{N^{(2)}}_{tt} - {N^{(2)}}_{xx} = &\frac{1}{3}h_1 h_2\left(N^{(1)} + N^{(2)}\right)_{xxx} + \alpha_2^*\left(N^{(1)}\right)_x^2\left(N^{(1)}\right)_{xx} \\ &+ \alpha_2^*\left(N^{(1)}\right)_x^2 {N^{(2)}}_{xx} + \alpha_2^*\left(N^{(2)}\right)_x^2 {N^{(2)}}_{xx}. \end{aligned} \tag{3.81}$$

In the second equation, we do not take into account the interaction of waves $N^{(1)}$ and $N^{(2)}$. We approximately find from the first equations, taking into account the boundary condition (3.66) that

$$N^{(1)} = A\cos\frac{1}{2}\omega t \sin\pi NL^{-1}x. \quad (N = \pm 1, \pm 2, \pm 3, \ldots.) \tag{3.82}$$

Here

$$A^2 = \frac{16}{3\alpha_2^*}\left[\frac{1}{8}a_0^{-2}(2 - A_0)\omega^2 + \left(\pi NL^{-1}\right)^2\right]\left(\pi NL^{-1}\right)^{-4}. \tag{3.83}$$

Taking (3.82) and (3.83) into account, we rewrite (3.81) in the form

$$\begin{aligned}
&a_0^{-2}\left(N^{(2)}\right)_{tt} - \left(N^{(2)}\right)_{xx} + \alpha_2^*\frac{1}{16}A^3\left(\pi NL^{-1}\right)^4 \\
&\quad\times\frac{1}{2}\left[\sin 3\left(\frac{1}{2}\omega t + \pi NL^{-1}x\right) - \sin 3\left(\frac{1}{2}\omega t - \pi NL^{-1}x\right)\right] \\
&\quad = \frac{1}{3}h_1h_2A\left(\pi NL^{-1}\right)^4\frac{1}{2}\left[\sin\left(\frac{1}{2}\omega t + \pi NL^{-1}x\right) - \sin\left(\frac{1}{2}\omega t - \pi NL^{-1}x\right)\right] \\
&\quad + \frac{1}{3}h_1h_2N^{(2)}_{xxx} + \alpha_2^*\left[\frac{1}{4}A^2\left(\pi NL^{-1}\right)^2N^{(2)}_{xx} + N^{(2)2}_{x}N^{(2)}_{xx}\right]
\end{aligned} \tag{3.84}$$

Considering this equation, we introduce new variables. We assume that

$$r = (a_0 + \Delta)t - x \text{ and } s = (a_0 + \Delta)t + x. \tag{3.85}$$

Moreover, the unknown $N^{(2)}$ is sought in the form of the sum of traveling waves

$$N^{(2)} = J(r) + j(s)\ . \tag{3.86}$$

Then, taking (3.85) and (3.86) into account, we rewrite Eq. (3.84). We do not take into account the interaction of oppositely moving waves and assume $\frac{1}{3}h_1h_2N^{(2)}{}_{xxx} \approx 0$. In this case, the described equation is divided into two equations

$$\begin{aligned}
&2a_0^{-1}J_{rr}\Delta - \alpha_2^*\frac{1}{16}A^3\left(\pi NL^{-1}\right)^4\ \frac{1}{2}\left[\sin 3\left(\frac{1}{2}\omega t - \pi NL^{-1}x\right)\right] \\
&\quad = -\frac{1}{3}h_1h_2A\left(\pi NL^{-1}\right)^4\ \frac{1}{2}\left[\sin\left(\frac{1}{2}\omega t - \pi NL^{-1}x\right)\right] + \alpha_2^*\left[\frac{1}{4}A^2\left(\pi NL^{-1}\right)^2(J_{rr}) + J_r^2J_{rr}\right],
\end{aligned} \tag{3.87}$$

$$\begin{aligned}
&2a_0^{-1}(j_{ss})\Delta + \alpha_2^*\frac{1}{16}A^3\left(\pi NL^{-1}\right)^4\frac{1}{2}\left[\sin 3\left(\frac{1}{2}\omega t + \pi NL^{-1}x\right)\right] \\
&\quad = \frac{1}{3}h_1h_2A\left(\pi NL^{-1}\right)^4\ \frac{1}{2}\left[\sin\left(\frac{1}{2}\omega t - \pi NL^{-1}x\right)\right] + \alpha_2^*\left[\frac{1}{4}A^2\left(\pi NL^{-1}\right)^2(j_{ss}) + j_s^2J_{ss}\right].
\end{aligned} \tag{3.88}$$

Let us consider Eq. (3.87) assuming that $\frac{1}{2}\omega\left(\pi NL^{-1}\right)^{-1}=g_*^{0.5}$. After integration, (3.87) takes the form

$$J_r^3-\left[6\Delta\alpha_0^{-1}\left(\alpha_2^*\right)^{-1}-\frac{3}{4}A^2\left(\pi NL^{-1}\right)^2\right]J_r$$
$$-\frac{1}{2}h_1h_2A\left(\alpha_2^*\right)^{-1}\left(\pi NL^{-1}\right)^3\cos\pi NL^{-1}r$$
$$+\frac{3}{32}A^3\left(\pi NL^{-1}\right)^3\cos 3\pi NL^{-1}r=0. \qquad (3.89)$$

We assume that $\left[6\Delta\alpha_0^{-1}\left(\alpha_2^*\right)^{-1}-\frac{3}{4}A^2\left(\pi NL^{-1}\right)^2\right]=0.0005$, $-\frac{1}{2}h_1h_2A\left(\alpha_2^*\right)^{-1}\left(\pi NL^{-1}\right)^3=0.0001$, $\frac{3}{32}A^3\left(\pi NL^{-1}\right)^3=-0.00004$, $a_0=0.72$, $L=1$ and $\pi NL^{-1}=7.65$. Then

$$J_r^3-0.0005J_r+0.0001\sin 7.65r-0.00004\sin 22.95r=0. \qquad (3.90)$$

Here $r=\sqrt{gh_1}t-x$. The equation for j_s has a similar form.

The sum J_r+j_s will be used to describe the dynamics of parametrically excited waves on the contact surface of two different layers of liquid.

We emphasize that the values of the coefficients in (3.90) are taken arbitrarily. However, the coefficient for the linear term in (3.90) was chosen so that the solution was close to the resonant one, and the amplitude of the sum J_r+j_s was close to the amplitude of the waves recorded in the experiment.

Calculations have shown that the influence of propagation speed on the wave pattern is insignificant. Therefore, most of the following calculations were carried out without taking into account the vertical forced acceleration.

3.4.3 Counterintuitive Faraday Waves on the Interface of Two Liquids

Let us describe analytically certain results from article [34]. There a tank containing a two-layer liquid with a free surface was considered. During vertical excitations, the layer interface and the free surface can be oscillated separately or simultaneously. Of particular interest to us are multivalued waves at the interface between layers. Analysis of these waves showed that their behavior is a result of the combined effects of vertical acceleration of the fluid and external excitation. This contribution comes down to the fact that the upper layer of liquid in some cases becomes significantly "heavier" than the lower layer.

The experiments used a container 1 m long, 0.15 m wide, and 0.3 m high. The thickness of the lower, heavier layer of liquid was 0.05 m. The thickness of the upper layer of liquid was 0.2 m. The layer was excited according to the harmonic law and lasted approximately 15 s. The acceleration amplitude of vertical external excitation is approximately 0.85 g. The snapshots of the free surface and interface at the mid-plane of the tank width at different moments of time are shown in Fig. 3.25 [34–36].

We will not discuss the results of the experiment [34] here. The reader can find the discussion in the article [36]. Let us note only the most important for us. It can be seen that from the beginning of parametric excitation, there is a very strong change in the profile of Faraday waves from almost harmonic to multivalued. They resemble the waves with hair discussed earlier in Sects. 2.5.2 and 3.4.1. It is important for us that, albeit very roughly, the wave evolution retains some semblance of periodicity over several cycles of excitation. Afterward, the flow becomes chaotic as can be seen from Figs. 3.25 (i–l).

Of course, based on the experimental data (a)—(d) presented in Fig. 3.25, it is difficult to imagine how a wave with hair (d) arises from the initial harmonics.

We carried out calculations for the case when the amplitude of the seed disturbance increased but the time did not change. The results are presented in Fig. 3.26. It can be seen how, as the amplitude increases, the harmonics begin to interact, and at an amplitude of 0.01, they quite well describe the shape (e) in Fig. 3.25.

The results of more detailed calculations are presented in Fig. 3.27. There the numbers determine dimensionless time.

In this case, it is assumed that we are at some fixed point of the resonant band.

There is a qualitative consent of the calculated profiles and the wave profiles in Fig. 3.25. In particular, Fig. 3.27 describes the experimental data almost on all lengths of the container. In certain cases, the discrepancies are only near the boundaries.

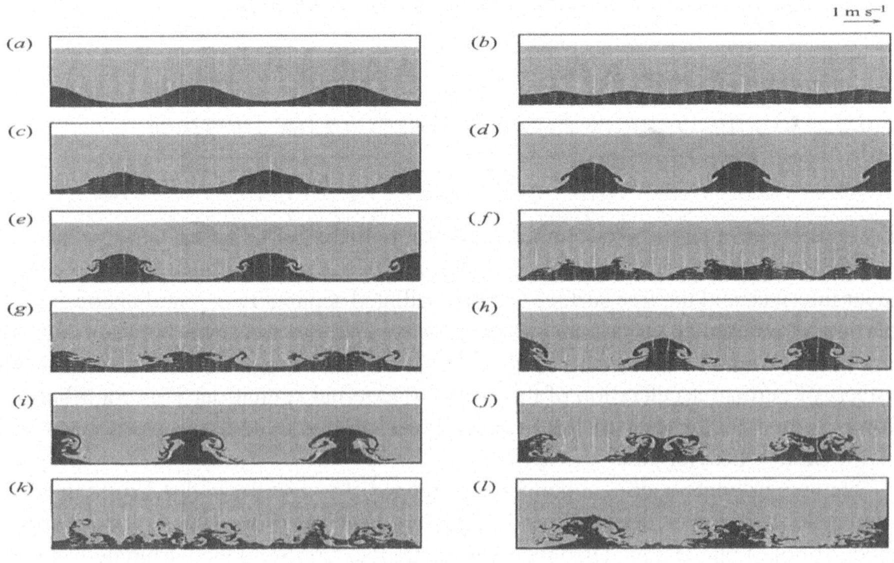

Fig. 3.25 Snapshots of Faraday waves on the free surface and the interface

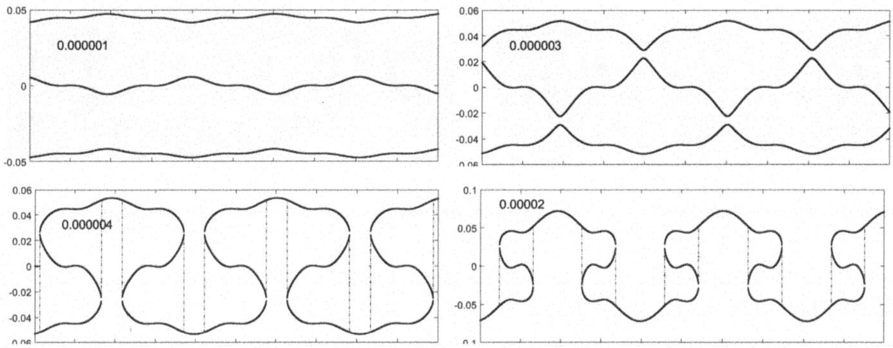

Fig. 3.26 The beginning of the formation of the waves in Fig. 3.25

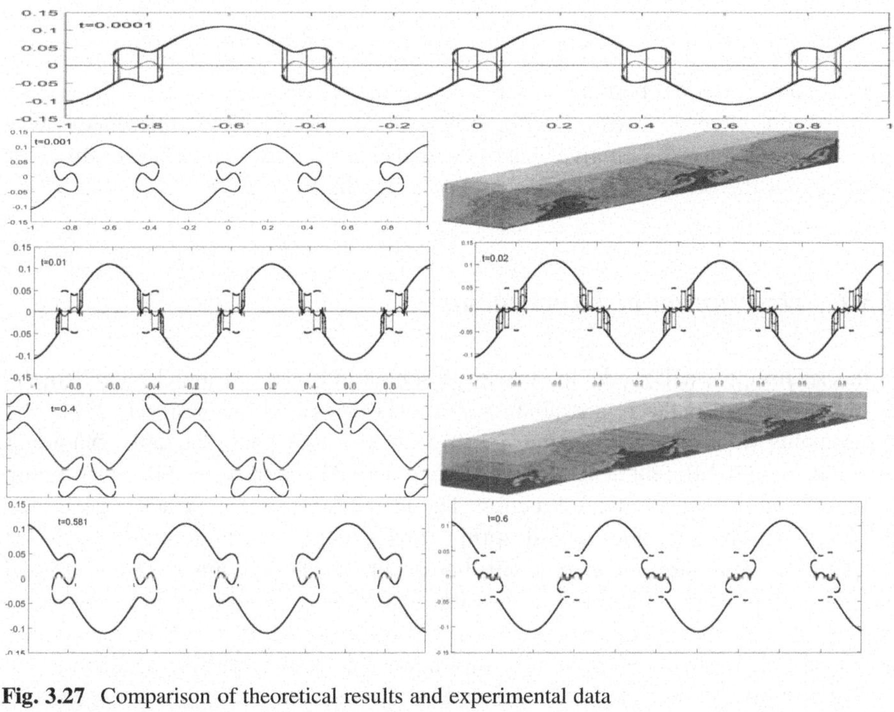

Fig. 3.27 Comparison of theoretical results and experimental data

It is surprising, but the results of Figs. 3.27 and 3.19 are consistent with each other and do not contradict experimental data (Fig. 3.25).

The generation of surprising multivalued waves (Fig. 3.25) may be associated with the fact that the upper layer of liquid is limited by the waves of the lower layer. As a result, wave forms arise that resemble the shapes of strings deformed when movement is limited from above (see Fig. 3.28).

The string which is submitted to compression at both ends is shown in Fig. 3.28. Upward movements are limited. Therefore, the string takes on a shape reminiscent of

Fig. 3.28 Flat-topped wave form (left) and mushroom-like wave form (right) of a string [39]

one of Euler figures (Fig. 2.1), as well as some Faraday waves on the contact surface of two layers of liquid (Fig. 3.25).

3.5 The Richtmyer-Meshkov Instability

Vertical sea earthquakes can create giant internal ocean waves. In particular, these waves can occur at the boundary between cold and warm ocean water. The process of formation of these waves is very complex. In particular, giant multivalued waves can arise there. Some qualitative idea of the waves arising at the interface of different media under the action of a vertical pulse is given by Figs. 3.29, 3.30, and 3.34.

3.5.1 Description of Experiments

In this section, we will model the waves presented in Figs. 3.31 and 3.35, where the results are given of the remarkable experiments described in the article [37]. The experiments were conducted using a 3 m drop tower. A plexiglas tank containing two liquids with different densities was mounted on a linear rail system, constraining its main motion to the vertical direction. The instability of the interface is generated by dropping the sled onto a coil spring, producing a nearly impulsive upward acceleration. The subsequent free fall that occurs as the container travels upward and then downward on the rail allows the instability to evolve in the absence of gravity. The interface initially has a well-defined sinusoidal perturbation that quickly inverts and then grows in amplitude, after undergoing the impulsive acceleration.

We will now give a very brief description of the main points of the experiment and how we understand them.

For several moments from $t = -0.025$ s to approximately $t = -0$, harmonic disturbances existed on the interface surface. At $t = 0$, this surface wave began to be affected by a short-term pressure pulse propagating from the bottom of the container with a duration of about 0.02 s (Fig. 3.29).

At first, pressure acts on the downward-facing wave antinodes of the harmonic initial disturbance. Something similar can happen with axial impact of a cylindrical tube falling under gravity and filled with a liquid wetting the tube wall. The curvature of the spherical meniscus (crater) is reversed violently. As a result, a rapidly ascending jet is formed [40] (Fig. 3.30).

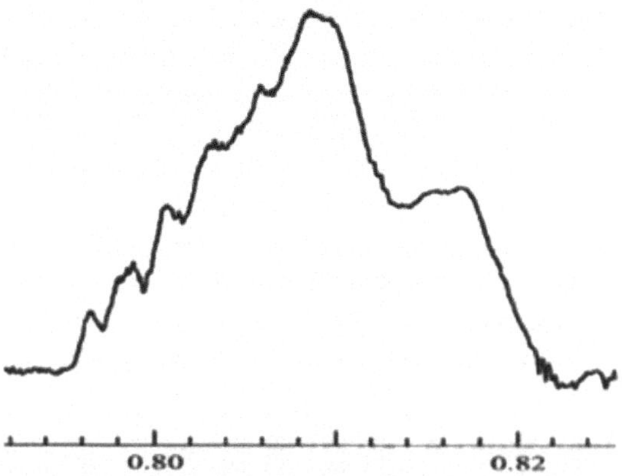

Fig. 3.29 A plot of container acceleration versus time for a typical experiment. The plot is described as $\gamma_* \sec h^2\gamma(t - t_*)$.

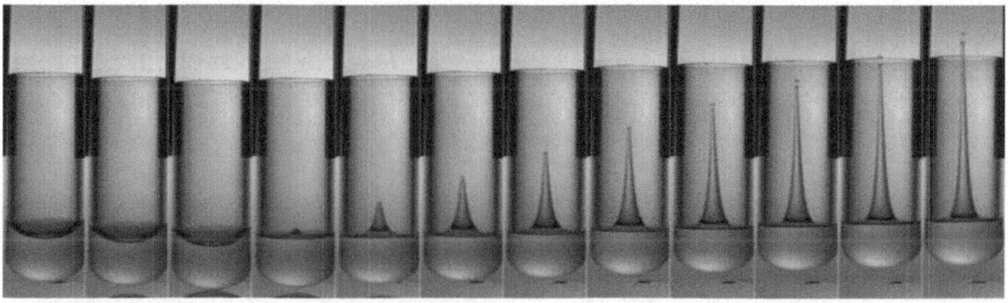

Fig. 3.30 Axial impact of a cylindrical tube falling under gravity and filled with a liquid wetting the tube wall. The curvature of the spherical meniscus (crater) is reversed violently. As a result, a rapidly ascending jet is formed [7, 40]

As a result of this, each downward-facing antinode begins to break up into two traveling waves. The speed of these waves is very small, so during the pulse action, the surface wave pattern does not change very much. However, due to the action of the impulse, the initially standing harmonic wave began to represent the sum of traveling waves, the arguments of which differ by a very small amount. This value is determined by the distance that each wave traveled during the pulse, that is, in 0.02 s.

After the action of the impulse, the container goes into a state of free flight. During this flight, the waves again behave as if they were stationary, namely, they practically do not move relative to each other.

Thus, the evolution of the waves under consideration in time can be divided into three stages.

First stage, up to approximately $t — -0$. In this case, a standing harmonic wave exists on the surface. It is described by the initial condition. This standing wave can be represented as the sum of two traveling waves, the velocities of which depend only on gravitational acceleration.

The impact acts on the container during the second stage. At the same time, the gravity continuous to act. We do not know how these accelerations affect the speeds of the emerging traveling waves. Obviously, they are formed under the influence of the impulse, which we present as $\gamma_* \sec h^2\gamma(t - t_*)$ (Fig. 3.29). Thus, the impulse reaches its maximum value γ_* at $t = t_*$. This occurs in the time interval from $t \approx 0$ to $t \approx 0.02$ s (see Fig. 3.31).

The third stage begins from $t > 0.02$ s. This is the free flight time of the container. This stage takes up almost the entire time of the experiment! In this stage, all the dynamics of the formation of extreme multivalued waves observed in the experiment and shown in Fig. 3.31 take place.

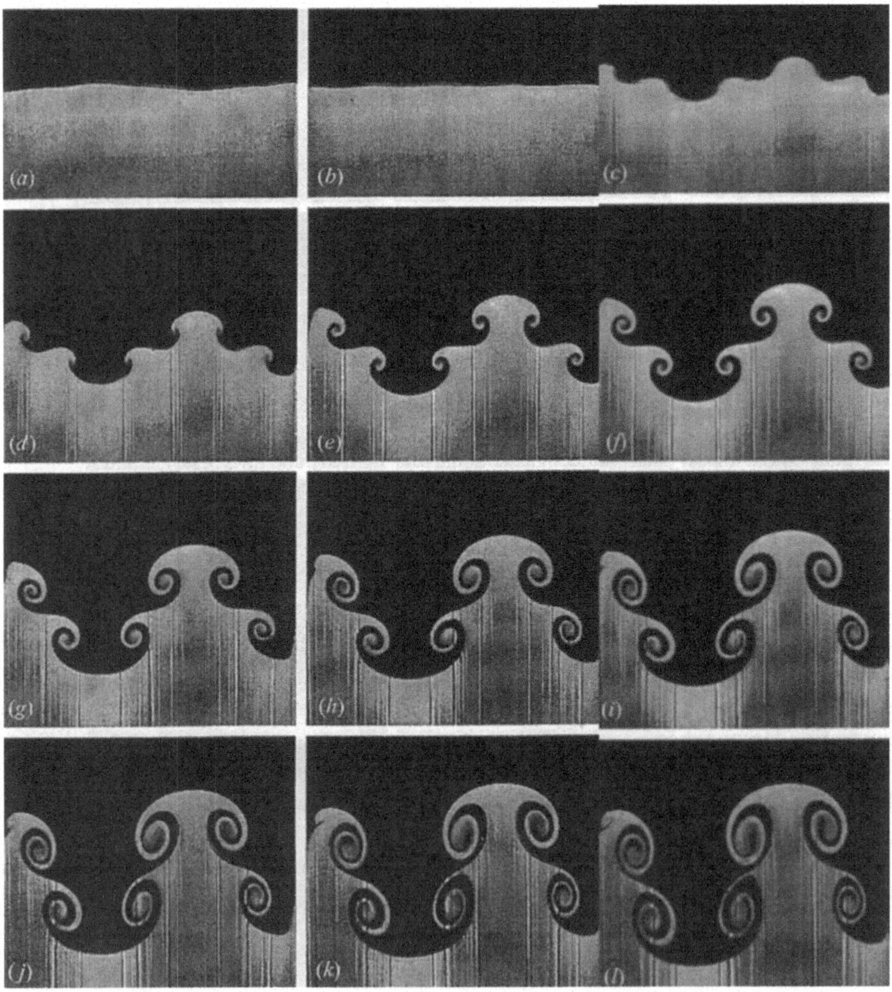

Fig. 3.31 An example of highly nonlinear waves appearing on the interface of two liquids with different densities, under harmonious perturbation [4, 6, 37]

The above description of the experiment shows how complex its mathematical modeling can be. The problem is aggravated by the fact that we are going to study the mathematical model of the experiment analytically. Therefore, our model should be simple enough for mathematical research. Therefore, when studying the model, in some cases, we will go for radical simplifications of mathematical expressions, believing that if we manage to describe, at least qualitatively, the experimental data, then this will to a certain extent justify the adopted simplifications.

Let us look at Fig. 3.31. We can assume based on the stages introduced above that image (a) corresponds to stage 1. Images (b) and (c) correspond to stage 2. At this stage, there is a strong change in both the amplitude and the waveform. However, during the third stage (images (d)—(l)), the change in time of both the wave profile and its amplitude slows down very much. However, there are very strong, but local, changes in the wave profile along the coordinate. In this case, there is a very weak tendency for the lower lateral surfaces of the waves to converge and their upper lateral surfaces to diverge. Thus, according to what has been said, during the main time of the experiment, the wave amplitude changes weakly and is approximately equal to 0.1 within a fairly wide time interval.

Based on the above, we accept the following assumptions for subsequent mathematical modeling. First, the seed perturbation can change greatly both in time and in coordinate, but only during a very short, initial stage in time. Second, the profile of the main wave changes very slow in time, but very fast in coordinate.

3.5.2 *Modeling of the Richtmyer-Meshkov Instability*

Formulation of the problem. Waves on the contact surface are described by Eq. (3.63) where $\frac{1}{3}a_0^2 h_1 h_2 N_{xxx} \approx 0$. In this case, we have

$$N_{tt} - a_0^2 N_{xx} = \beta_1 a_0^2 N_x^2 N_{xx}\ , \tag{3.91}$$

$$a_0^2 = \left[g_0 + \gamma_* \sec h^2 \gamma (t - t_*)\right] \frac{h_1 h_2 (\rho_2 - \rho_1)}{(h_1 + h_2)(\rho_2 + \rho_1)}\ , \quad \beta_1 = \frac{3}{(h_1 h_2)^2} \left(\frac{h_1^3 + h_2^3}{h_1 + h_2}\right) c_0^{-1}. \tag{3.92}$$

Here t_* is the moment of the maximum magnitude of the impulse, and γ and γ_* are experimentally determined constants. The quantities g_0 and γ_* sec $h^2\gamma(t - t_*)$ determine the gravitational acceleration and the acceleration caused by the impulse, respectively. In the experiment under consideration, the influence of the impulse and flight time of the container is so significant that the influence g_0 can be ignored. Therefore, in Eq. (3.91), we take

$$a_0^2 = g_* \sec h^2 \gamma (t - t_*)\ , g_* = \frac{\gamma_* h_1 h_2}{h_1 + h_2} (\rho_2 - \rho_1)/(\rho_2 + \rho_1) \tag{3.93}$$

At the boundaries of the container, we accept that

$$N = 0 \ \ (x = 0\ ;\ \ x = L) \tag{3.94}$$

At the first stage $t = -0$, we have the initial condition

$$\eta = N_x = A_0 \cos N\pi x/L + B\cos 3N\pi x/L. \tag{3.95}$$

The initial condition $\eta = N_x = A_0 \cos N\pi x/L$ will also be considered. It follows from (3.95) as a special case when $B = 0$.

Solution method. We will look for N as the sum of the seed perturbation and the nonlinear part of the wave

$$N = N_1 + N_2\ . \tag{3.96}$$

We define the seed perturbation in the form

$$N_1 = A_1(t)\left[\left(\bar{J}(r) + \bar{j}(s)\right].\right. \tag{3.97}$$

Here $r = a_0 t - x$ and $s = a_0 t + x$. The functions $A_1(t)$, $\bar{J}(r)$, and $\bar{j}(r)$ are determined subsequently based on the initial conditions and impulse loading. In particular, we accept that

$$A_1(t) = A_0 \tanh\gamma(t - t_*)\ . \tag{3.98}$$

Here A_0 is a constant. The magnitude of the impulse is maximum at $t = t_*$. The shape of the waves changes rapidly only during the action of the pulse. We described this change by introducing a factor $\tanh\gamma(t - t_*)$ into the expression for the seed wave.

We believe that what was said about N_1 completely defined it. Now let us look at N_2. Let us represent this function like this:

$$N_2 = \tanh\gamma(t - t_*)[(J(r) + j(s)], \quad \eta = \tanh\gamma(t - t_*)[(J_x(r) + j_x(s)]. \tag{3.99}$$

We introduce expressions (3.96), (3.97), (3.98), and (3.99) into the wave Eq. (3.91). It has already been emphasized that, according to experiments, waves N_2 vary slightly over time. Changes in function N_2 over time are not taken into account in (3.91). In this case, (3.91) gives

$$\begin{aligned}
&-2\gamma^2 \sec h^2\gamma(t - t_*)\tanh\gamma(t - t_*)A_0 \int \left[\bar{J}(r) + \bar{j}(s)\right]_{tt} dx \\
&- g_* \sec h^2\gamma(t - t_*)\tanh\gamma(t - t_*)\left[A_0\bar{J}(r) + A_0\bar{j}(s) + J(r) + j(s)\right]_x \\
&= \frac{1}{3}\alpha_2^* g_* \sec h^2\gamma(t - t_*)\tanh^3\gamma(t - t_*)\left[\left(A_0\bar{J}(r) + A_0\bar{j}(s) + J(r) + j(s)\right]_x^3.\right.
\end{aligned} \tag{3.100}$$

Further, the interaction of oppositely moving waves is not taken into account. Grouping the terms containing $\tanh\gamma(t - t_*)$ and taking into account the above, we obtain from (3.100) an equation whose variables depend only on r:

$$-2\gamma^2 A_0 \int \bar{J}_{tt} dx - g_* \left(A_0\bar{J} + J\right)_x = \frac{1}{3}\alpha_2^* g_* \left(A_0\bar{J} + J\right)_x^3. \tag{3.101}$$

The equation for $A_0\bar{j}(s)$ and $j(s)$ is written similarly to (3.101). Finally, we assume that the seed function $A_0\bar{J}(r)$ is described by a harmonic traveling wave following from expression (3.95). Let us assume that the amplitude of this wave is very small and this wave does not interact with $J(r)$. As a result, we rewrite Eq. (3.101) as follows:

$$-2\gamma^2 A_0 \int \bar{J}_n dx - g_* \left(A_0\bar{J} + J\right)_x = \frac{1}{3}\alpha_2^* g_* \left(A_0^3\bar{J}_x^3 + 3A_0^2\bar{J}_x^2 J_x + J_x^3\right). \tag{3.102}$$

We have obtained the basic equation for the mathematical description of the experiments under consideration. The specific type depends on the seed wave and the coefficients of the equation determined by a particular experiment. We will further carry out two series of calculations. We first consider case (3.95). After expression (3.95), we take $B = 0$.

Both of these cases correspond to the two main experiments carried out in [37]. The coefficients in the equation will be specified arbitrarily, but in such a way that the calculation results are qualitatively consistent with the experiments.

We write Eq. (3.102) in the following form:

$$J_x{}^3 - 0.0011 J_x + 0.0001\left[\sin\Omega(a_0 t - x) + B\sin 3\Omega(a_0 t - x)\right] = 0. \tag{3.103}$$

The equation for j is written similarly. The results of calculating the sum $J_x + j_x$ for several initial times are presented in Fig. 3.32.

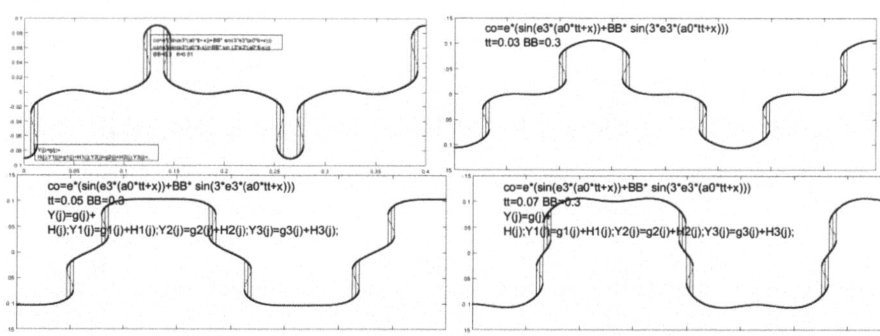

Fig. 3.32 Calculations performed for four points in time: $t = 0.01$, $t = 0.03$, $t = 0.05$, and $t = 0.07$

It can be seen that the calculations are in qualitative agreement with the experiment. They show a picture of a mushroom wave forming. Moreover, a tendency is visible for the lower side surfaces of the wave to run inside and the upper side surfaces to run outside. Note that, according to this approximation, the wave amplitude changes weakly and is approximately equal to 0.1 within a fairly wide time interval.

Let us continue modeling experiments. We emphasize that it continues to be of high quality. Now let me remind readers that the study of the sum $J + j$ is an intermediate stage of our research. Obviously, next we must more accurately take into account the change in wave amplitude with time.

According to the developed approach, the change in wave amplitude over time is determined by the factor $\tanh\gamma(t - t_*)$ (3.99). Therefore, further clarification comes down to calculation $\tanh\gamma(t - t_*)(J_x + j_x)$. In the calculations, we will accept a very small value t_* and γ such that $\tanh\gamma(t - t_*) \approx 1$ at $t = 0.07$. Let $t_* = 0.001$ and $\gamma = 50$. The calculation results are presented below (Fig. 3.33).

The refinement carried out, which corresponds to the theory developed above, significantly improved their agreement with experiment. In particular, during the considered time, the amplitude of the wave increased by three times. This is consistent with the experiment. It is also important that the greatest increase in amplitude occurs in the first moments of evolution, which is also consistent with the experiment. Note that for t greater than 0.07, the discrepancy between the calculations and the experiment increases.

Now consider the case of the one side traveling wave when $j(s) = 0$. Let us substitute into (3.103) $B = 0$. The results of the corresponding calculations are given below (Fig. 3.34).

It is easy to see that the latest calculations are consistent with the experimental data presented in Fig. 3.35. The amplitude of the waves grows very quickly at the beginning of the process. At the end of the considered time interval, growth stops. As in the experiment, in the process of evolution, extreme multivalued waves retain their mushroom shape.

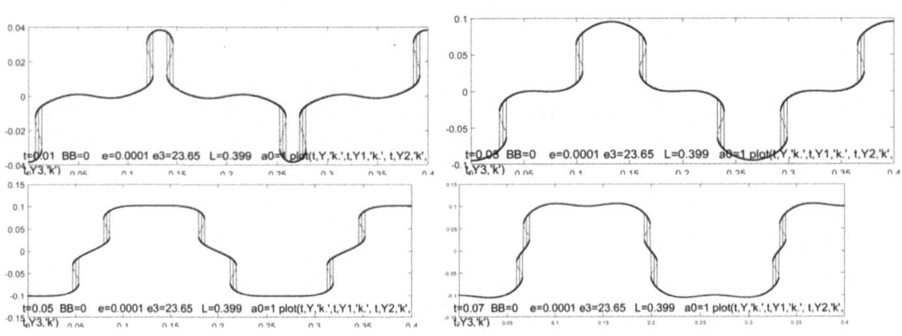

Fig. 3.33 Calculations performed for $B = 0$ in (3.103) and four points in time: $t = 0.01$, $t = 0.03$, $t = 0.05$, and $t = 0.07$

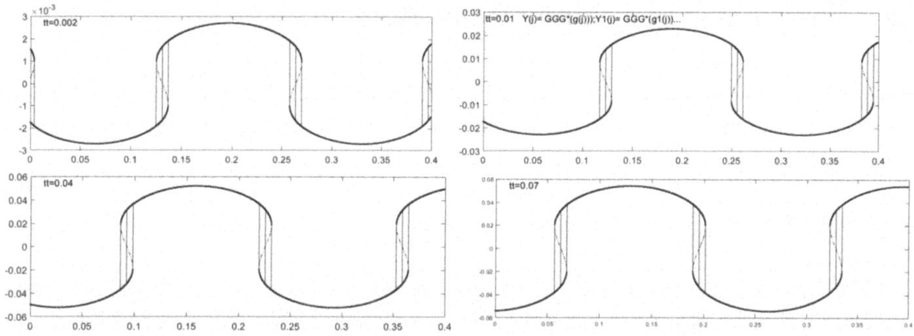

Fig. 3.34 Calculations performed for four points in time: $t = 0.01$, $t = 0.03$, $t = 0.05$, and $t = 0.07$

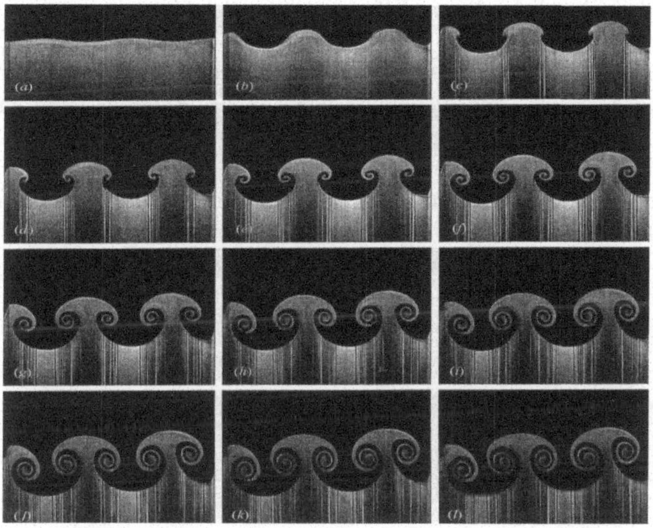

Fig. 3.35 The result of the Richtmyer-Meshkov instability [4, 6, 37]

It is easy to see that the latest calculations are consistent with the experimental data presented in Fig. 3.35, where an example of highly nonlinear standing waves appearing on the interface of two liquids with different densities as the result of the Richtmyer-Meshkov instability. The interface has the low-frequency sinusoidal perturbation.

The amplitude of the waves grows very quickly at the beginning of the process. At the end of the considered time interval, growth stops. As in the experiment, in the process of evolution, extreme multivalued waves retain their mushroom shape.

3.6 Forced Waves in Finite Resonators

Many artificial and natural objects may be considered as certain resonators. When an object is stimulated at its resonant frequency, the amplitude of the vibrations is greatly increased.

As we have repeatedly shown with the example of waves, not only does the amplitude increase, but there can also be a very large change in the shape of the wave.

This form is influenced by many circumstances. Of these, perhaps the most interesting is how close the excitation is to resonance. No less interesting is the influence of nonlinearities on the transresonant evolution of the wave, as well as the influence of viscosity and dispersion. Of course, we have noted a tangle of complex and interdependent effects, which is not unraveled even for the case of resonant oscillations described by ordinary differential equations (see Sect. 2.4 and Figs. 2.32 and 2.34).

At the same time, I know some results of studying the mutual influence of nonlinearities, viscosity, and dispersion in resonance regions. These are given in this section.

We consider experimental and theoretical results obtained for surface waves in a liquid excited harmonically in a rectangular container, and for waves in a gas column excited at one end according to a periodic law.

Thus, to simplify the problem, we shall limit ourselves to resonators which may be approximated by simple one-dimensional geometries (Fig. 3.36).

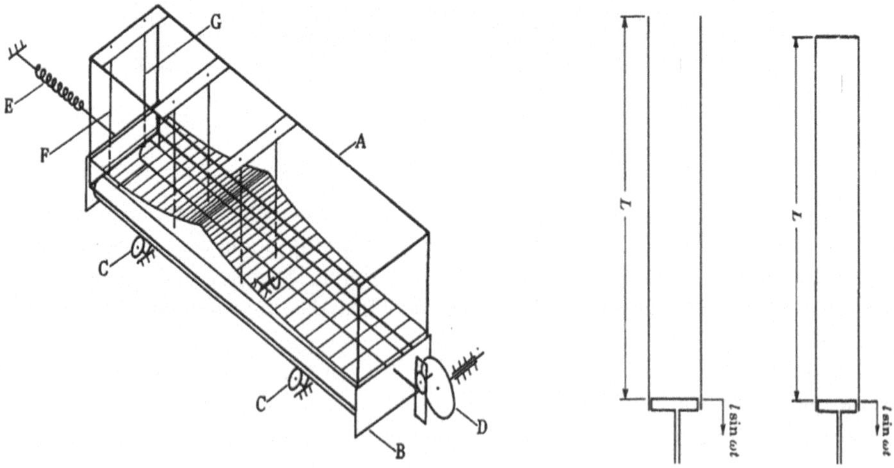

Fig. 3.36 One-dimensional schemes modeling some resonators: the tank with liquid and the open and closed vertical tubes having a gas and the oscillating piston (right)

3.6.1 *Effect of Quadratic Nonlinear and Dispersion on Water Waves*

Below we will compare our theoretical results determined in [4, 6, 7] with the experimental data of Chester and Moser [41]. They excited water waves in a rectangular tank, with length 60 cm and width 15 cm. The forcing amplitude and the depth were varied during the experiments. The amplitude defined a contribution of the nonlinearity on the wave formation. Variation of depth changes the dispersive effect.

The theory predicts that in the case of small depth and strong excitation, waves should occur with steep front resembling shock waves in gas. On the other hand, periodic solitons should be excited if the depth is large enough. An unexpected result was found in the case of strong excitation with depth not very small—extreme waves with a height about equal to the water depth were observed.

1. Shallow water. For thin layer, the measured profiles of surface waves are given in Fig. 3.37 (left). It is easy to see that they correspond to the calculations (Fig. 3.37, right). Quadratic nonlinearity and viscous effects are taken into account during calculations. Thus, if the dispersion is very small (very thin layer), surface waves can resemble shock waves well known in gas (see Fig. 3.40).

Effect of weak dispersion. We found [4, 6] that weak internal viscosity and the very weak dispersion cannot strongly change the steep front and the wave profile

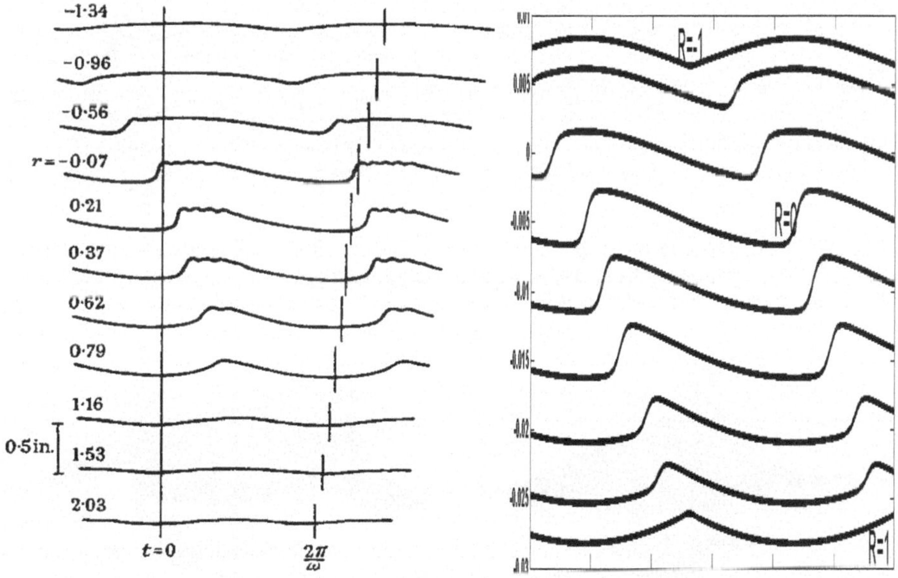

Fig. 3.37 Experimental recordings of the evolution of the surface waves within the resonance band (left) [4, 6, 20, 41] and theoretical results (right). The thickness of the water layer $h \approx 1.25$ cm, the forcing amplitude is 0.31 cm

formed by the quadratic nonlinearity. However, the influence of the dispersion increases when the depth increases. Data of experiments and calculations are presented in Fig. 3.38 for this case. The transresonant evolution of water waves differed strongly from the evolution shown in Fig. 3.37.

The left and central curves are experimental, but the right curves are theoretical. The thickness of the water layer was increased (5 cm); however, the forcing amplitude was reduced (0.0775 cm) in respect of the data in Fig. 3.37.

Some periodical solitons appear instead of the shock-like waves in the tank. The appearance of similar waves was also demonstrated in [7] (Fig. 3.38, central). The right curves were calculated.

Spatial resonant oscillations of the water waves are shown in Fig. 3.39. One solitary wave is excited near the first fundamental resonance of the container (left). Near the second fundamental resonance, two solitary waves are excited (right). These waves are quite different from the waves excited in very shallow water (Fig. 3.37).

Because of the dispersion, a sharp crest is formed in the wave top. Figures 3.37, 3.38, and 3.39 show that strongly nonlinear waves can be excited in water tanks. Generally speaking, a few oscillations with resonant frequencies might be enough to form similar strongly nonlinear waves in tanks. After these oscillations, the strongly

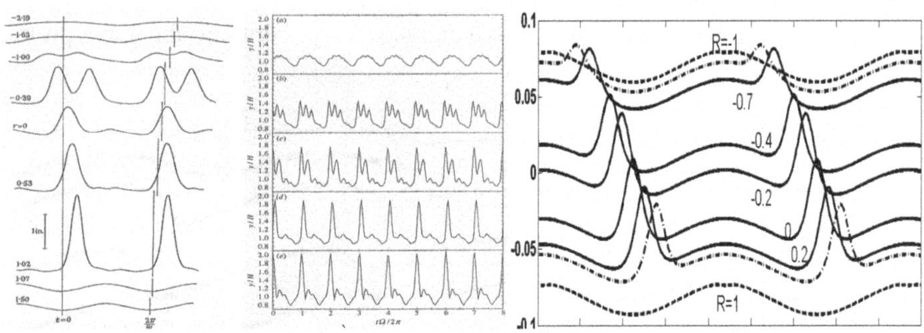

Fig. 3.38 2. **The effect of the dispersion.** Evolution and catastrophic amplification of the surface waves crossing the resonant band. The left curves and the central curves are given in [4, 6, 20, 41, 42]. The right curves were calculated

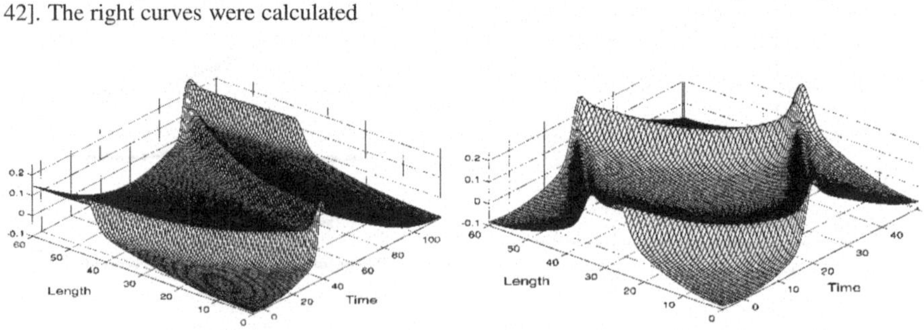

Fig. 3.39 Solitary waves travel back and forth in the tank, being repeatedly reflected from the boundaries

nonlinear waves can proceed for some time, since dissipative losses of energy in water are usually small.

The dispersive property of the layers may be very important. The solitary wave amplitude can greatly exceed the amplitude of the resonant waves excited in thinner layers. In particular, near a resonance, the amplitude of the solitary waves can exceed ten and more times the amplitude of nonresonant waves (see Fig. 3.38, left). For example, if the height of a harmonic ocean wave is 7 m, then, according to our research, the height of the resonant waves can be about 70 m.

3.6.2 Resonant Quadratic Nonlinear Waves in Closed, Gas-Filled Tubes

Physical reasons for the generation and transformation of resonant waves can differ dramatically. Nevertheless, the equations and analytical solutions describing these processes are often similar. That was not surprising for us. Many years ago, Lamb [43] stressed the similarity between the one-dimensional wave equations for shallow water and gas. Therefore, the resonant waves excited in gas-filled tubes can resemble nonlinear ocean waves. On the other hand, waves in gassy magma of volcano conduits can be modeled as resonant waves in closed or semi-open gas-filled tubes.

We will consider in this section the waves produced in a closed, gas-filled tube by oscillations of a piston at one end (Fig. 3.36, right).

1. **The linear resonance.** Forced waves excited in the vicinity of the first linear resonance of the closed tube are shown in Fig. 3.40 (left). Waves were excited by a piston in a pipe in the cases of $R = 2, 1, 0.97, 0.6, 0, 0.5$ [18]. R is the transresonant parameter. The transresonant evolution of a harmonic wave into shock waves is shown. Three upper curves of the figure are taken from [7], which show completely the evolution of the waves in the case of change of the forcing frequency near a resonance. Those waves differ from the waves measured for $R = 2, 1, 0.97, 0.6, 0, 0.5$, because those waves are influenced also by some average current existing in the pipe [7].

We simulate the experimental data shown in Fig. 3.40 (left). Results of the calculations are presented in Fig. 3.40 (right).

The weakly nonlinear harmonic waves begin to evolve into cnoidal waves (Fig. 3.40, $R = 1$), when the transresonant parameter R is reduced to $R = 1$. Then shock waves can appear traveling to and fro in the tube, being repeatedly reflected from the piston and from the closed end.

Effects of the nonlinearity and viscosity on the transresonant evolution of the pressure waves are shown in Fig. 3.40 (right). The discontinuous lines are calculated for very small viscosity. The smooth lines are calculated for two different viscosities. The discontinuous lines transform into the harmonic curves, if the viscosity of the gas is large enough.

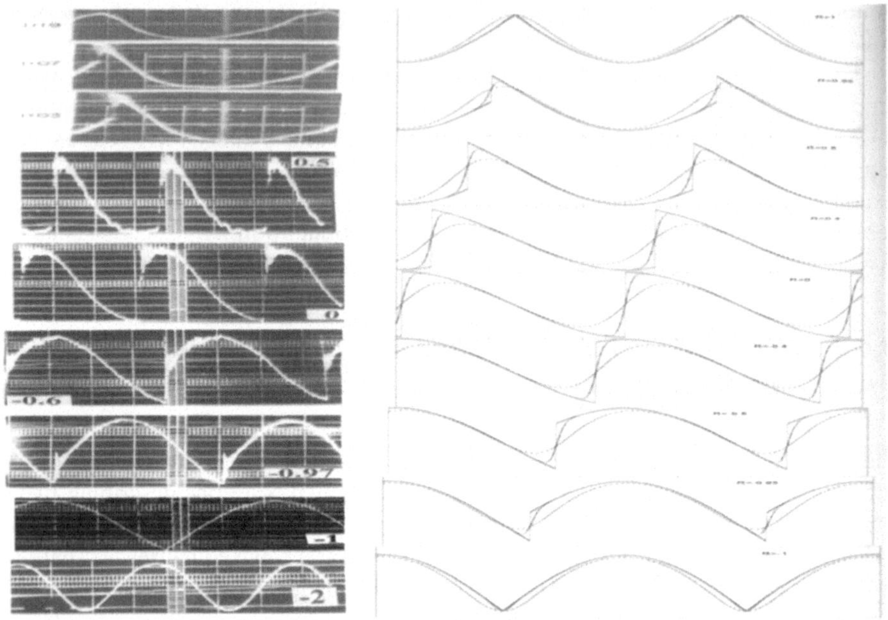

Fig. 3.40 Comparison of experimental pressure profiles at the closed end (left) and theoretical (right) results

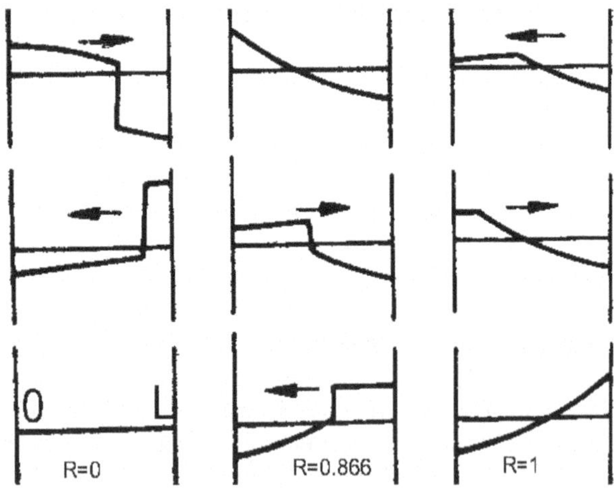

Fig. 3.41 Transresonant evolution of the wave excited in the tube when $R = 0$, 0.866 and $R = 1$. The shock wave travels in the tube, being repeatedly reflected from the piston and the closed end, if $R < 1$. There is an amplification of the pressure when the shock wave reflects from the boundaries

The spatial nonlinear oscillations of the wave in the tube calculated for the fundamental resonance are shown in Fig. 3.41. Waves are calculated for $R = 0$, 0.866 and 1. The transresonant evolution of the wave profile is demonstrated. It is seen that the shock amplitude is increased when R is reduced to zero.

Near the second resonant frequency, two shock waves appear in the tube (Fig. 3.42). The collision of shock waves explains the extreme amplification of the pressure at the point L/2 .

2. Nonlinear resonance. It is found [6, 7] that shock waves may be also generated in the tube at the vicinity of nonlinear resonances (Fig. 3.43).

The curves in Fig. 3.43 (left) are symmetrical around $R = 0$. The principal difference from Fig. 3.40 is that there are discontinuities with smaller jump, located between the strong discontinuities in each piston period. Moreover, the entire wave

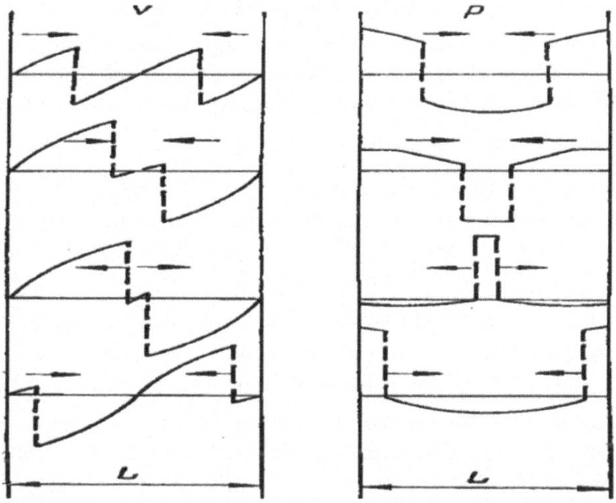

Fig. 3.42 The resonant shock waves of the velocity (left) and the pressure (right) excited at the second fundamental resonance of the gas column [6]

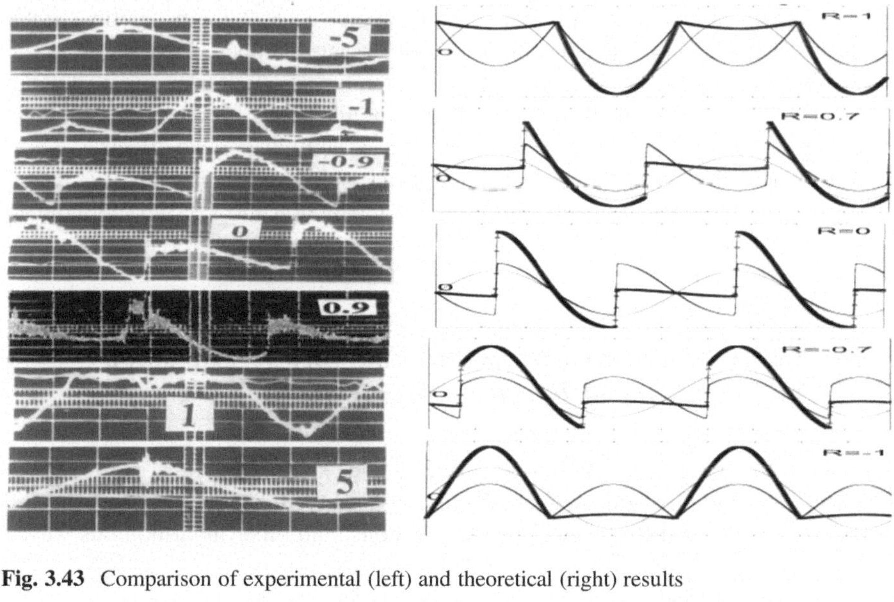

Fig. 3.43 Comparison of experimental (left) and theoretical (right) results

structures are less intense. Results of calculations are in Fig. 3.43 (right). The very thin lines describe the linear harmonic waves. The thicker lines are defined by the sum of linear and nonlinear waves. The transresonant evolution of waves is more complex in this case than the evolution of waves in Fig. 3.40.

We considered the resonant waves in the closed pipe. The experiments and the theory show the strong amplification of the waves within the resonant bands. In the case of quadratic resonance, the amplification is two times; in case of the linear resonance, the amplification is three and more times. Therefore, we can claim that extreme waves have appeared in the pipe within the resonant bands. We think that the experiments with gas-filled tubes can qualitatively describe resonant oscillations of gassy magma in volcano conduits. If a frequency of the earthquake-induced oscillations of a volcano base coincides with the fundamental resonance of the magma column, then a strong amplification of oscillations both the magma and the volcano can occur.

Remark The competition between the nonlinearity, the dispersion, and the dissipation determines wave processes. For example, in the case of initially harmonic wave, the nonlinearity leads to the appearance of high harmonics during wave propagation. As a result, the profile of the wave is distorted. The nonlinearity is stronger, and the distortion of an initial profile is stronger. Finally, the nonlinearity can generate an overturning wave. However, it is hindered by the dispersion. Because of it, the originated harmonicas have different speeds. Therefore, the harmonics cannot accumulate and form the steep wave fronts. Thus, the nonlinearity may be counterbalanced by the dispersion, if the cumulation effect equals, approximately, to the scatter effect. In this case, the stable running waves may appear. However, if the dispersion is small, only speeds of the high harmonics are changed. Speeds of the lowest harmonicas practically match each other. Therefore, they can form the rather steep wave fronts. Here we dare to give a parallel with an effect of the viscosity. If it is small only, the high harmonicas are damped. Possibility of the formation of very steep wave fronts in this case remains. Something similar can take place in the case of weak dispersion. Thus, enough steep ocean waves may be generated due to the interaction of nonlinear and weakly dispersive effects.

Apparently, the interaction of the nonlinear and dispersive effects increases sharply within the resonant band. But how this interaction shows itself? Of course, we cannot give here the complete response to this question.

3.6.3 *Cubic Nonlinear Effects and First Step to Transition to Turbulence of Faraday Waves*

We have shown a wide variety of waves that can be excited in resonators, if we take into account quadratic nonlinearity. Waves arise that differ significantly from the predictions of linear theory. Shock waves, solitons, and other inharmonious waves can arise inside the resonant zones. From Figs. 3.37, 3.38, 3.39, 3.40, 3.41, 3.42, and 3.43, it is clear that wave modes appear inside the resonant band that are very different from those prescribed by the linear theory.

It is obvious that even more striking differences can be expected if the influence of cubic nonlinear effects is taken into account in calculations, and cases of very strong excitation of resonators are studied in experiments. We have demonstrated these differences for Faraday waves.

There are not many experiments that show the excitation of multivalued waves in resonators. Sibgatullin was the first to study excitations of multivalued waves in resonators theoretically [44, 45]. Then the same effect was studied by Chester (see Fig. 2.24). Here we present results that significantly develop the results of these authors. Let us compare the results of linear theory and calculations that take into account the influence of cubic nonlinearity on resonant waves in one-dimensional resonators.

Figures 3.44, 3.45, and 3.46 illustrate the cubic nonlinear waves [4, 6, 7, 20]. The waves were formed by two opposite nonlinear waves excited in the vicinity of the resonant frequencies. Each of these waves can be very different from harmonic. Thus, very surprising waves can be excited in the resonator. These waves may be quite different from quadratic nonlinear waves showed in Figs. 3.39, 3.41, and 3.42.

Figures 3.44 gives an example of the evolution of cubic nonlinear waves for first mode ($N = 1$).

Those profiles were not observed in resonators known to us. They are destroyed due to the instability of the curved parts of the profiles for surface waves. This instability should disappear if the influence of gravity is reduced. This is possible in

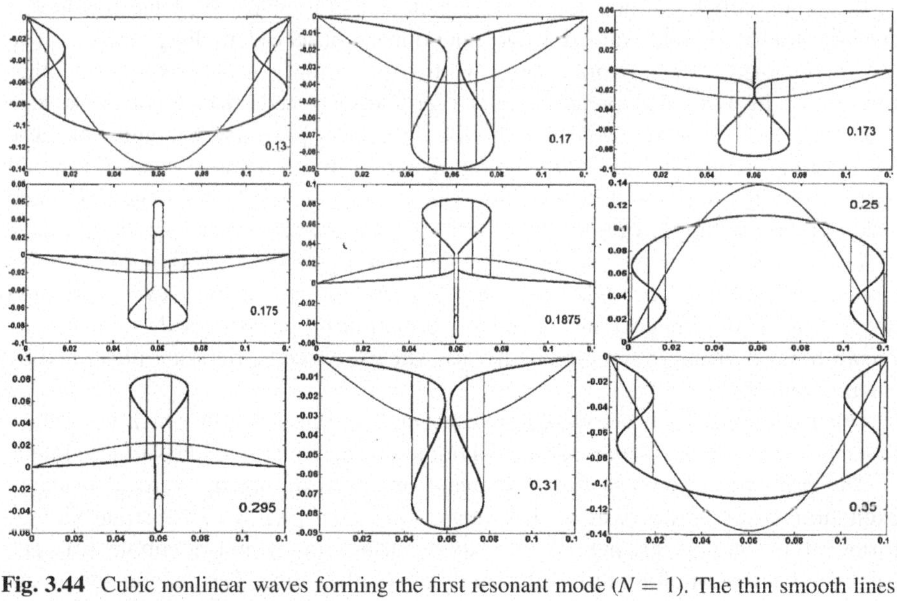

Fig. 3.44 Cubic nonlinear waves forming the first resonant mode ($N = 1$). The thin smooth lines correspond to the linear oscillations [4, 6, 7]

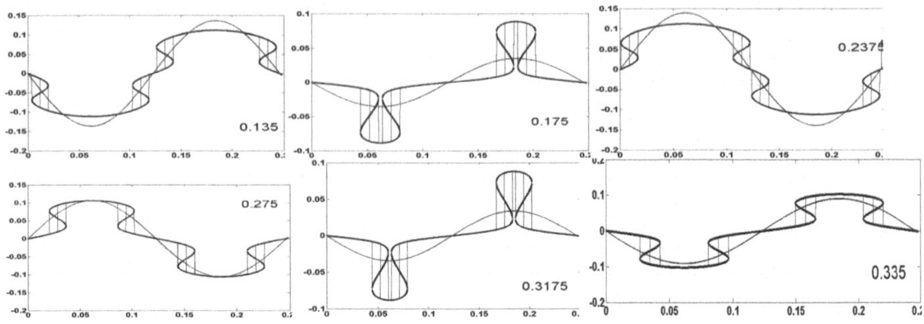

Fig. 3.45 Resonant continuous multivalued waves corresponding to the second resonance ($N = 2$). The thin smooth lines correspond to the linear oscillations [4, 6, 7]

the space. Under terrestrial conditions, the influence of instability decreases under conditions of high surface tension of the medium or on the contact surface of two liquids. In the latter case, some of the profiles are shown in Sect. 3.4.

To give more complete picture of cubically nonlinear standing waves excited in the resonator during of the harmonic excitation, we present the figures calculated for the case $N = 2$ (Fig. 3.45) and $N = 3$ (Fig. 3.46). The shapes of the nonlinear waves are quite different from the linear forms. In particular, the nonlinear shapes have folds. Within and near these folds, the curve can have jumps of discontinuity.

The above results, to a certain extent, agree with those obtained in [46].

Several examples illustrating and explaining the calculations of Figs. 3.44, 3.45, and 3.46 are presented in Figs. 3.47 and 3.48.

Resonant particles, drops, jets, surface craters, and bubbles. We have considered the appearance of multivalued wave phenomena observed in long objects. The results obtained, on the whole, agree with extraordinary waves excited in other conditions. In particular, parametrically (vertically) forced surface waves, known as Faraday waves, were studied in circular containers [47], and examples of these waves are presented in Fig. 3.47. The figure shows the formation of jets of a liquid during the collapse of the surface depression or cavity caused by the large-amplitude axisymmetric standing waves.

The experiments were conducted in circular cylindrical Plexiglas containers having radius $R = 5$ cm filled to a depth 6 cm. Images of the wave depression (lower part of the images) and of the jet formation (upper part of the images) are shown for two different forcing conditions. The experiments (left) are illustrated by calculations (right).

Other examples of these waves are presented in Fig. 3.48, in which the photomontages show the formation of jets of a liquid during collapse of the surface craters.

The processes shown in Fig. 3.48 were observed during the vertical resonant excitation of a circular basin with a liquid [47]. The photos demonstrate various forms of the craters arising on the liquid surface (a–c) and a bubble (c). The calculated curves are found for different dimensionless time. The general agreement of experiments and calculations is visible.

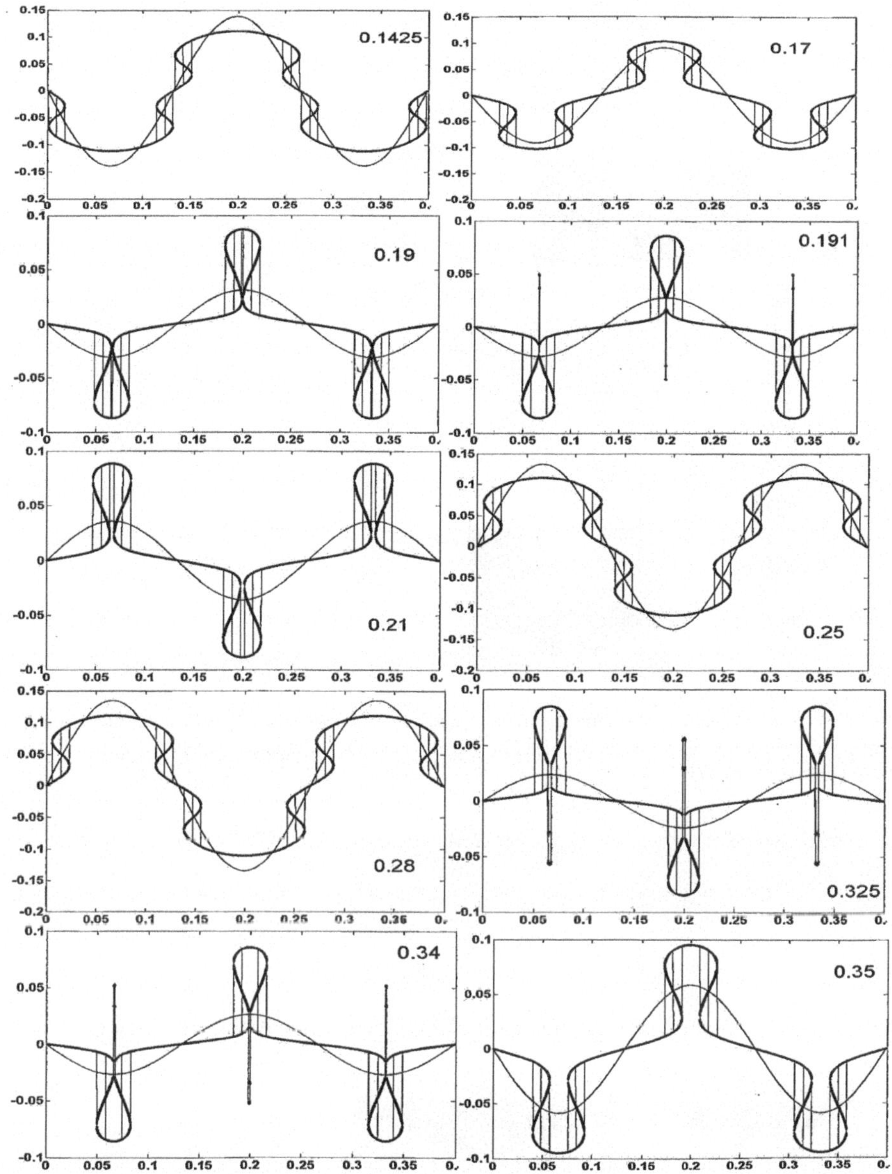

Fig. 3.46 Resonant oscillations corresponding to the third mode ($N = 3$). The thin smooth lines correspond to the linear oscillations [4, 7]

Conclusion From joint consideration of Figs. 3.44, 3.45, 3.46, 3.47, and 3.48, it is clear that the appearance of jets and particles is associated with the moments of convergence (collision) of multivalued waves. At these moments, the ambiguity of these waves manifests itself especially strongly, which leads to unusual phenomena. Perhaps, the results of such collisions correspond to the emergence of particles from "nowhere" (see Chap. 5).

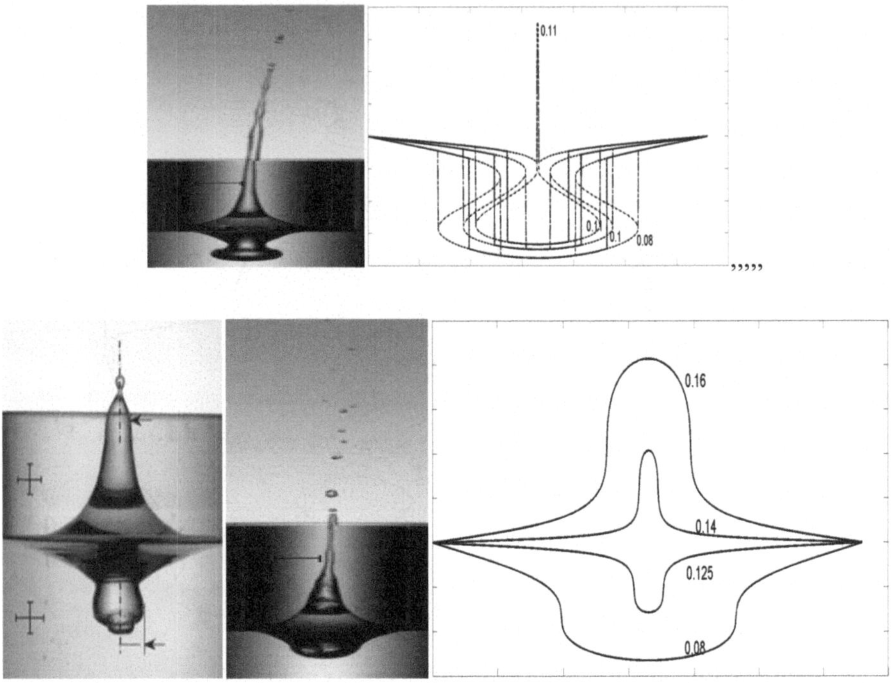

Fig. 3.47 Photomontages showing the axisymmetric wave depressions (surface cavities) and jets (left), and results of calculations (right). The figures near the calculated curves determine dimensionless time [4, 6, 47]

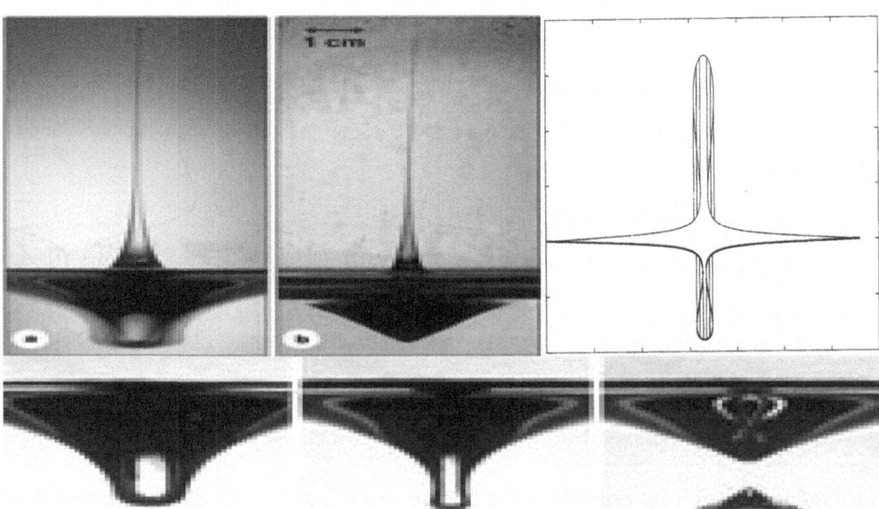

Fig. 3.48 Photomontages of the surface collapsing craters, a bubble near the liquid surface and surface jets. The calculated curves illustrate the photomontages [4, 6, 7, 47]

3.7 Historical Comments

Apparently, Galileo Galilei [48] first studied surface waves experimentally. In particular, he noted a nonlinear event when he excited the surface waves in a glass of the water. He wrote "that a glass of water may be made to emit a tone ..... sometimes happens, the tone of the glass jumps an octave higher .... each of the aforesaid waves divides in two." Galileo also discussed a row of parallel, equidistant streaks excited on a brass plate. Thus, Galileo studied the wave patterns on water and solid body surfaces and described the nonlinear wave effect. Ernst Chladni [49] presented his technique of sprinkling sand on vibrating plates to reveal interesting patterns (Chladni figures). In 1809, Chladni gave a live demonstration of these patterns to Napoleon [50]. It is interesting that, under the impression of the Chladni's experiments, Emperor Napoleon allocated 6000 francs to translate its book "Acoustic."

Faraday [17] performed a number of experiments with liquids and sand-like materials. In particular, he described the "Chladni heaps" phenomenon for weakly cohesive materials and examined the Galileo experiments with a drinking glass [51]. Perhaps, his most popular finding in the summer of 1831 was that the surface waves on a vertically vibrated fluid vibrate at precisely half the forcing frequency. In particularly, Faraday observed a square pattern which was reminiscent of one from the Chladni figures. Rayleigh [52] recognized that the waves were the result of parametric resonance. For inviscid liquid, this idea may be transformed into the Mathieu Eq. [53]. Experimental and theoretical studies of vertically vibrated weakly viscous liquid demonstrated a variety of regular patterns (e.g., rolls, squares, or hexagons) [54]. The richness of this variety increases in the case of highly viscous liquid [55]. The variety depends on the depth of the liquid and frequency of the excitation. Patterns reminiscent of the Chladni figures were described for different systems and circumstances [56–60].

The Chladni and Faraday experiments with weakly cohesive materials have led to a very interesting development. In experiments [30, 32], the surface wave patterns in vibrated initially flat granular thin layers were excited if an amplitude of forced acceleration was larger than 2 g_0 (g_0 is the acceleration due to gravity). This behavior stands in sharp contrast to that of fluid systems, where the wave patterns exist under more weak excitation. Apparently, nonlinear and dissipative effects are more important for granular media than for liquids. In particular, we do not know of any investigations using a Mathieu-type equation for studying of the granular waves. For the last time, a few theoretical models (see, for instance [4, 6, 7, 18, 20]) were proposed, which are useful for the interpretation of the experimental data for granular material. These theories described nonpropagating parametrically excited solitary-type waves and wave patterns.

References

1. Chester W, Moser A (1982) On an equation which describes the resonant oscillations of a gas in an open-ended tube. Proc R Soc Lond A 380:409–415
2. Chester W (1981) Resonant oscillations of a gas in an open-ended tube. Proc R Soc Lond A 377:449–467
3. Thorpe SA (1968) On standing internal gravity waves of finite amplitude. J Fluid Mech 32(3): 489–528
4. Galiev SH (2020) Evolution of extreme waves and resonances. CRC Press, p 508
5. Wright J, Yon S, Pozrikidis C (2007) Numerical studies of two-dimensional Faraday oscillations of inviscid fluids. J Fluid Mech 588:279–308
6. Galiev SU (2010) Charles Darwin's geophysical reports as models of the theory of catastrophic waves. The University of Auckland, p 655
7. Galiev SU (2015) Darwin, geodynamics and extreme waves. Springer
8. Karsten RH, Swaters GE (2000) Nonlinear effects in two-layer large-amplitude geostrophic dynamics. Part 2 the weak-beta case. J Fluid Mech 412:161–196
9. Marts B, Hagberg A, Meron E, Lin AL (2004) Bloch-front turbulence in a periodically forced Belousov-Zhabotinsky reaction. Phys Rev Lett 93(10):108305; Galiev SU (2008) Strongly-nonlinear wave phenomena in restricted liquids and their mathematical description. In: Perlidze T (ed) New nonlinear phenomena research. Nova Science Publishers, New York, pp 109–300
10. Galiev SU, Ilgamov MA, Sadikov AV (1970) Periodic shock waves in a gas. Izv. Akad. Nauk USSR, Mech. Fluid and Gas 2:57–66
11. Galiev SU (1988) Nonlinear Waves in Bounded Continua. Naukova Dumka: Kiev (in Russian); Galiev SU (1972) Forced oscillations of a nonlinear elastic layer. Izv. Acad. Nauk USSR, Mech. Solid Body 1:58–63; Galiev SU (1972) Forced longitudinal oscillations of a nonlinear elastic body. Izv. Acad. Nauk USSR, Mech. Solid Body 4:80–87; Galiev SU (1973) Nonlinear gas oscillations maintained in a closed pipe by a heat source. Soviet Aeronautics 16(2):7–12; Galiev SU, Galiyev TS (1994) Linear and discontinuous forced oscillations of flow of bubbly liquid in a Deformable Pipe (survey). Strength Mater 9:633–65
12. Galiev SU, Galiyev TS (2001) Nonlinear trans-resonant waves, vortices and patterns: from microresonators to the early universe. Chaos: an interdisciplinary. Journal of Nonlinear Science 11(3):686–704
13. Galiev SU (2003) The theory of non-linear transresonant wave phenomena and an examination of Charles Darwin's earthquake reports. Geophys J Int 154:300–354
14. Galiev SU (2003) Transresonant evolution of wave singularities and vortices. Phys Lett A 311: 192–199
15. Airy GB (1845) Tides and waves. In: Encyclopaedia metropolitana, London
16. Poincaré H (1899) Les méthodes nouvelles de la mécanique céleste, vol 3. Gauthier-Villars et fils
17. Faraday M (1831) On a peculiar class of acoustical figures and on certain forms assumed by groups of particles upon vibrating elastic surface. Phil Trans R Soc Lond 52:299
18. Galiev SU (1999) Topographic effect in a Faraday experiment. J Phys A Math Gen 32:6963–7000
19. Kneubhl FK (1997) Oscillations and waves. Springer
20. Galiev SU (2009) Modelling of Charles Darwin's earthquake reports as catastrophic wave phenomena. The University of Auckland
21. Lioubashevski O, Hamiel Y, Agnon A, Reches Z, Fineberg J (1999) Oscillons and propagating solitary waves in a vertically vibrated colloidal suspension. Phys Rev Lett 83(16):3190–3193
22. Wang X, Wei R (1997) Dynamics of multisoliton interactions in parametrically resonant systems. Phys Rev Lett 78(14):2744–2747
23. Rajcnenbach J, Leroux A, Clamond D (2011) New standing solitary waves in water. PRL 107: 024502
24. Shinbrot T, Muzzio FJ (2001) Noise to order. Nature 410:251–258

25. Bizon C, Shattuck MD, Swift JB, McCormick WD, Swinney HL (1998) Patterns in 3D vertically oscillated granular layers: simulation and experiment. Phys Rev Lett 80(1):57–60
26. Jiang L, Perlin M, Schultz WW (1998) Period tripling and energy dissipation of breaking standing waves. J Fluid Mech 369:273–299
27. Jiang L, Ting C-L, Perlin M, Schultz WW (1996) Moderate and steep Faraday waves: instabilities, modulation and temporal asymmetries. J Fluid Mech 329:275–307
28. Ilgamov MA, Zaripov RG, Galiullin RG, Repin VB (1996) Nonlinear oscillations of a gas in a tube. Appl Mech Rev 49:138–154
29. Merkli P, Thomann H (1975) Transient to turbulence in oscillation pipe flow. J Fluid Mech 68: 567–575
30. Umbanhowar PB, Melo F, Swinney HL (1996) Localized excitations in a vertically vibrated granular layer. Nature 382:793–796
31. Cerda E, Melo F, Rica S (1997) Model for subharmonic waves in granular materials. Phys Rev Lett 79(23):4570. https://doi.org/10.1103/PhysRevLett.79.4570
32. Clement E, Labous L (2000) Pattern formation in a vibrated granular layer: the pattern selection issue. Phys Rev E 62(6):8314–8323. Clement E, Vanel L, Rajchenbach J, Duran J (1996) Pattern formation in a vibrated granular layer. Phys Rev E 53:2972–2975
33. Galiev SU (2000) Unfamiliar vertically excited water waves. Phys Lett A 266:41–52
34. Liu D, Lin P (2022) Two-layer liquids Interface instabilities in Faraday waves of two-layer liquids with free surface. J Fluid Mech 941:A33. https://doi.org/10.1017/jfm.2022.259
35. Ilgamov MA (2023) Annotation book Galiev S.U. "Extreme multivalued waves in scalar fields" planned by Springer for publication in early 2024. Herald of Ufa Scientific Center, Russian Academy of Sciences. No 3, p 115–120
36. Galiev SU (2024) Multivalued waves and the Richtmyer-Meshkov instability as the causes of the formation of galaxies, vol 1. Herald of Ufa Scientific Center Russian Academy of Sciences, p 5–17. https://doi.org/10.31040/2222-8349-2024-0-1-5-17
37. Niederhaus CE, Jacobs JW (2003) Experimental study of the Richtmyer-Meshkov instability of incompressible fluids. J Fluid Mech 485:243
38. Natanzon MS (1965) Forced discontinuous oscillations of liquid in tubes. Izv. Akad. Nauk USSR, Mechanics, 2, 33–42 (in Russian); Natanzon MS (1977) Продольные Автоколебания Жидкостной Ракеты (Longitudinal selfexcited oscillations of a liqquid-fuel rocket). Mashinostroenie, Moscow
39. Pocheau A, Roman B (2004) Uniqueness of solutions for constrained Elastica. Phys D 192:161
40. Antkowiak A, Bremond N, Le Dizes S, Villermaux E (2007) Short-term dynamics of a density interface following an impact. J Fluid Mech 577:241–250
41. Chester W, Bones JA (1968) Resonant oscillations of water waves. II experiment. Proc R Soc A 306:23–39
42. Kobine JJ (2008) Nonlinear resonant characteristics of shallow fluid layers. Philos Trans R Soc A 366:1331–1346
43. Lamb H (1932) Hydrodynamics, 6th edn. Dover Publications, New York
44. Сибгатуллин НР (1972) О нелинейных поперечных колебаниях при резоансе ы упругом слое и слое идеальной проводящцкй жмдкостю ПММ, том 36. Стр, p 79
45. Сибгатуллин НР (1984) Колебания и волны в сильных гравитационных и электромагнитных полях. Наука, Москва. (Sibgatullin NR (1991) Oscillations and waves in strong gravitational and electromagnetic fields. Springer)
46. Cavelier M, Gréa B-J, Briard A, Gostiaux L (2022) The subcritical transition to turbulence of Faraday waves in miscible fluids. J Fluid Mech 934:A34. https://doi.org/10.1017/jfm.2021.1124
47. Das SP, Hopfinger EJ (2008) Parametrical forced gravity waves in a circular cylinder and finite-time singularity. J Fluid Mech 599:205–228. Royon-Lebeaud A, Hopfinger EJ, Cartellier A (2007) Liquid sloshing and wave breaking in circular and square-base cylindrical containers. J Fluid Mech 577:467–494

48. Galilei G (1946) Discorisi e Dimostrazioni Matematiche Intorne a due Nuove Scienze, Elfevirii, 1638 (Dialogues concerning two new sciences (1638) translated by H. Crew and A. De Salvio, Northwestern University Press, Evanston and Chicago
49. Chladni EFF (1802) Die Akustic. Breitkopf und Ha..rtel, Leipzig
50. Fenny B, Guran A, Hinrichs N, Popp K (1998) A historical review on dry friction and stick-slip phenomena. Appl Mech Rev 51:321–341
51. Faraday M (1932) In: Martin T, London G (eds) Entry in Faraday's diary. Bell
52. Rayleigh L (1883) On the crispations of fluid resting upon a vibrating support. Philos Mag 16: 50–58
53. Benjamin TB, Ursell F (1954) The stability of the plane free surface of a liquid in vertical periodic motion. Proc R Soc Lond A 225:505–515
54. Miles J, Henderson D (1990) Parametrically forced surface waves. Ann Rev Fluid Mech 22: 143–165
55. Cross MC, Hohenberg PC (1993) Pattern formation outside of equilibrium. Rev Mod Phys 65: 851–1088
56. Ibrahim RA (2015) Recent advanced in physics of fluid parametric sloshing ans related problems. J Fluid Eng 135:1–52
57. Edwards WS, Fauve S (1994) Patterns and quasi-patterns in the Faraday experiments. Fluid Mech 278:123–148
58. Bechhoefer J, Ego V, Manneville S, Johnson B (1995) An experimental study of the onset of parametrically pumped surface waves in viscous fluids. J Fluid Mech 288:325–350
59. Cerda EA, Tirapegui EL (1998) Faraday's instability in viscous fluid. J Fluid Mech 368: 195–228
60. Bowman C, Newell AC (1998) Natural patterns and wavelets. Rev Mod Phys 70:289–301

Chapter 4
Particle-Waves and Quanta in Resonant Scalar Fields, and Classification of Multivalued Waves

Extreme waves can have profiles that resemble harmonic waves, shock waves, solitons, and oscillons. On the other hand, they can have multivalued profiles, that is, profiles that have folds.

In previous chapters, we examined a wide range of problems and a wide range of multivalued waves corresponding to different problems. It must be emphasized that considering them, we remain within the framework of the theory of nonlinear acoustics. This determined the difficulties of our research and some of its successes. The fact is that the limitations of acoustics at the same time open up the possibility of using fundamental acoustic ideas and analytical methods for studying multivalued waves. At the same time, the acoustic approach becomes obviously unacceptable in areas of very high amplitudes and in cases of strong effects of interaction of multivalued waves. In the latter cases, it is necessary to use other equations and numerical methods for calculating multivalued waves. Finally, the area in which the results of the acoustic approach are valid is not yet entirely clear.

It seems to the author that his approach is already on the verge, beyond which it is necessary to use more general equations and more powerful research methods.

The use of these methods is known [1, 2]. They encounter enormous difficulties in their implementation. The problem of the existence and study of multivalued solutions in Nature, apparently, will be a problem for many centuries of the development of science.

Calculated multivalued waves [3] were so different from the widely known ones that the possibility of their existence in reality had raised doubts in the author's mind. Doubts were to some extent overcome after the publication of a number of remarkable articles with experimental data.

Doubts were caused by the fact that the developed approaches were based on the old fundamental idea of d'Alembert about the possibility of a general description of waves using one or several arbitrary functions, each of which propagates in a certain direction. The question arises: how applicable is such a description for multivalued waves, which in principle cannot be approximated by a set of harmonic functions? Indeed, since the time of d'Alembert, an arbitrary function very often meant the

S. U. Galiev, *Extreme Multivalued Waves as a Model of Everything*,
https://doi.org/10.1007/978-3-031-67027-5_4

possibility of its description by one or several harmonics. In the case of multivalued waves in one-dimensional resonators, to describe them, it was proposed, following d'Alembert, to use two functions that define oppositely traveling waves. One may wonder how acceptable this approximation is in the case of highly nonlinear waves. Experiments more than 50 years ago showed that this approach is acceptable in cases of propagation and interaction of weak shock waves and solitons in resonators. But how this approximation would behave in cases of multivalued waves was completely unclear! This ambiguity continues to this day. It may well be that the appearance of multivalued waves is associated with the transition of the wave system to the beginning of an unstable state. Is it possible that the theory and approach to the problems proposed above only work at the beginning of the appearance of folds, jets, and breaks on the surface of the wave? Is it possible that the theory and approach to the problems proposed above only work at the beginning of collisions of multivalued waves in resonators until the effect of their collisions has accumulated?

These thoughts are, to a certain extent, confirmed by experiments. Thus, in the case of a unidirectional wave, its propagation is accompanied by the emergence of vortices and turbulization of the wave process. Collisions of multivalued waves.

perhaps eliminates the long-time existence of periodic wave patterns in the resonators.

On the other hand, in some cases, for example, in the case of Faraday waves, experiments often excite waves that are more similar to standing ones than to traveling ones. The use of the mathematical apparatus of traveling waves in such cases may not be justified.

The questions posed above will be further discussed in this chapter. In particular, multidimensional multivalued waves will be considered. Thus, to a certain extent, the study of multivalued waves arising in various environments and under various circumstances is completed.

This will allow us to give a certain classification of multivalued waves which the author encountered during calculations, or which were observed in experiments or nature.

It was shown in Chaps. 2 and 3 that surface waves in the process of resonant oscillations in finite resonators can have have profiles corresponding to Euler figures. In particular, in the resonator, Euler fig. 11 appears (see Figs. 3.44, 3.45, and 3.46), which Laplace interpreted as capillary waves. This Euler figure resembles a loop. This figure can be interpreted as a description of the critical moment of separation of a particle of matter from the top of a wave (waves) propagating in the medium. In particular, according to Laplace, the surface tension of a liquid can make such configuration stable.

In some approximation, such configurations can be considered as composed of two parts. The bottom of the configuration can be interpreted as a cnoidal wave, and the loops (upper parts) can be interpreted as particles. In general, the configuration can be considered as a critical stage of formation of wave-particle configuration. In this section, we show that these configurations can exist in scalar fields.

4.1 Forced and Eigenmultivalued Waves in One-Dimensional Scalar Fields

At the beginning of the chapter, we present the results of the calculations.

We are interested in the following questions:

1. To what extent do calculations of waves in scalar fields correspond to the experimental data discussed in the second and third chapters of the book?
2. To what extent does the nonlinear scalar field in the resonators exhibit the properties of standing and/or travelling waves?
3. How the initial field parameters affect multivalued waves in resonators?

It is remarkable that the study of these waves is greatly simplified in regions where the d'Alembert operator in nonlinear wave equations becomes negligibly small compared to the nonlinear terms. In this book, we are looking at different wave fields. We call the indicated areas in these fields resonant.

Let us recall that in the first chapter the equations were introduced to describe the scalar field:

$$\varphi_{tt} - c_0^2\nabla^2\varphi = -\partial V(\varphi)/\partial\varphi + \mu\nabla^2\varphi_t + k\nabla^4\varphi. \tag{4.1}$$

Here

$$V(\varphi) = V_0 + \frac{1}{2}m^2\varphi^2 - \frac{1}{4}\lambda\varphi^4 + \lambda_6\Lambda^{-2}\varphi^6 + \lambda_8\Lambda^{-4}\varphi^8 + \ldots . C^*\varphi. \tag{4.2}$$

Of course, it is very difficult to study waves based on these equations. We discussed above several options for finding a solution. In the case of unidirectional waves, the problem was reduced to solving an equation of the form (1.74). In the case of finite resonators, the results of Sect. 3.1 were used. However, in both cases, the wave problem was reduced to solving a nonlinear algebraic equation. We also emphasize that Chaps. 2 and 3 considered mainly surface waves.

In the following chapters of the book, we will study mainly waves in scalar fields; however, the research methods developed above when considering surface waves will not change. Very roughly speaking, problems will be reduced to solving algebraic equations. Their appearance, qualitatively, can be represented as follows:

Therefore, as the development of the main ideas used above, we will reduce our research, practically, to solving an equation of the form

$$\partial V(\varphi)/\partial\varphi = f(x_i, t), . \tag{4.3}$$

Here $f(x_i, t)$ is some initial perturbation and x_i are coordinates. We have formu lated the main idea of simplifying the tasks discussed below:

4.1.1 The Nonlinear Klein-Gordon Equation and Forced Resonant Waves

The nonlinear Klein-Gordon equation (NKGE) is written as follows:

$$\varphi_{tt} - c_*^2\varphi_{xx} - m^2\varphi - m_2\varphi^2 + \lambda\varphi^3 = C. \tag{4.4}$$

The conditions are written at the ends of the resonator

$$\varphi_x = 0 \quad \text{at} \quad x = 0 \quad \text{and} \quad \varphi_x = f(t), \quad x = L \tag{4.5}$$

Here *f*(*t*) is a periodical function. The question of the occurrence of periodic waves in a resonator is investigated. We are looking for waves with a period *L*/*N*, where $N = 1, 2, 3, \ldots.$

New coordinates *r* and *s* are introduced:

$$r = ct - x, \quad s = ct + x, \tag{4.6}$$

Here *c* is a constant. In new variables, Eq. (4.4) is transformed to form

$$2\left(c^2 + c_*^2\right)\varphi_{rs} + \left(c^2 - c_*^2\right)(\varphi_{rr} + \varphi_{ss}) - m^2\varphi - m_2\varphi^2 + \lambda\varphi^3 = C. \tag{4.7}$$

We assume that

$$c = c_* + \overline{c}, \tag{4.8}$$

where $\overline{c}$ is the perturbation of the speed *c*. Thus, we accepted that Eq. (4.4) describes the resonant scalar field. The scalar field function φ is presented as a sum:

$$\varphi = \varphi^{(1)} + \varphi^{(2)}, \tag{4.9}$$

where $\varphi^{(1)} > > \varphi^{(2)}$.Taking into account (4.8) and (4.9), we split (4.4) into two equations:

$$4c^2\varphi^{(1)}{}_{rs} = 0, \tag{4.10}$$

$$\begin{aligned} &4c^2\varphi^{(2)}{}_{rs} + \left(2c_*\overline{c} + \overline{c}^2\right)\left(\varphi^{(1)}{}_{rr} + \varphi^{(1)}{}_{ss}\right) \\ &- m^2\varphi^{(1)} - m_2\varphi^{(1)^2} + \lambda\varphi^{(1)^3} = C. \end{aligned} \tag{4.11}$$

Here $c^2 \approx c_*^2$ and $2c_*\overline{c} + \overline{c}^{-2} < < 1$. Considering (4.10), we assume following expression:

$$\varphi^{(1)} = J(r) + j(s), \tag{4.12}$$

We emphasize that this simple expression takes place only if the resonance condition (4.8) occurs. Then Eq. (4.11) is considered. After the integration we found that

$$4c^2 \varphi^{(2)}{}_r = J_*(r) + m^2\left(sJ + \int jds\right) + m_2\int (J+j)^2\, ds \\ -\lambda\left(sJ^3 + 3J^2\int jds + 3J\int j^2 ds + \int j^3 ds\right) + C_1. \tag{4.13}$$

Here $J_*(r)$ is an arbitrary function. The expression for $\varphi^{(2)}r$ is written in a similar way. Now we can write the expression $\varphi_x = \varphi_s - \varphi_r$ defining the boundary conditions (4.5):

$$4c^2(j_s - J_r) + j_*(s) - J_*(r) + m^2\left(\int Jdr + rj - sJ - \int jds\right) \\ +m_2\int (J+j)^2 dr - m_2\int (J+j)^2 ds \\ -\lambda\left(rJ^3 + 3j^2\int Jdr + 3j\int J^2 dr + \int J^3 dr - sJ^3 - 3J^2\int jds - 3J\int j^2 ds\right. \\ \left.-\int j^3 ds\right) + C_1 + C_2. = 0. \tag{4.14}$$

Here we will not take into account the interaction of waves of oppositely traveling waves. Therefore, (4.14) gives

$$4c^2(j_s - J_r) + j_*(s) - J_*(r) + m^2\left(\int Jdr + rj - sJ - \int jds\right) \\ +m_2\int (J^2 + j^2)dr - m_2\int (J^2 + j^2)ds \\ -\lambda\left(rJ^3 + \int J^3 dr - sJ^3 - \int j^3 ds\right) + C_1 + C_2. = 0. \tag{4.15}$$

We obtain from (4.15) accordingly selecting arbitrary functions $j_*(s)$ and $J_*(r)$ (see also, (4.43) and (4.60)):

$$4c^2(j_s - J_r) + j_{1*}(s) - J_{1*}(r) + m^2[(r-s)j - (s-r)J] \\ +m_2 j^2(r-s) - m_2 J^2(s-r) - \lambda[(r-s)J^3 + -(s-r)J^3] + C_1 + C_2 = 0. \tag{4.16}$$

Next, we take into account (4.6) and obtain the following equation defining the boundary conditions:

$$\varphi_x = 4c^2(j_s - J_r) + j_{1*}(s) - J_{1*}(r) - 2x[m^2(j+J) \\ +m_2(j^2 + J^2) - \lambda(J^3 + j^3)] + C_1 + C_2 = 0 \tag{4.17}$$

Let j_s and J_r be equal to each other at $x = 0$ and $x = L$. Then we consider arbitrary functions $j_{1*}(s)$ and $j_{1*}(s)$ Let them be equal at $x = 0$. Then the boundary condition at $x = 0$ is satisfied. Now let us accept that at $x = L$, we have

$$j_{1*}(s) = \frac{1}{2} l \sin\left(N + \frac{1}{2}\right)\pi s \quad \text{и} \quad J_{1*}(r) = \frac{1}{2} l \sin\left(N + \frac{1}{2}\right)\pi r. \tag{4.18}$$

Here l is an arbitrary value. Now we return to the condition $\varphi_x = f(t)$ at $x = L$ (4.5). Using (4.17) and (4.18) in this condition, we get the equation

$$J^3 - m_2\lambda^{-1}J^2 - m^2\lambda^{-1}J - \frac{1}{4} l\lambda^{-1}L^{-1} \sin\left(N + \frac{1}{2}\right)\pi r$$
$$+ \frac{1}{2}\lambda^{-1}L^{-1}(C_1 + C_2) = \frac{1}{2}\lambda^{-1}f(t) \tag{4.19}$$

This equation can be written as

$$J^3(r) + r_* J^2(r) + RJ(r) + A \sin \omega c^{-1} r = \frac{1}{2}C + \frac{1}{2}\lambda^{-1}f(t). \tag{4.20}$$

Here $r_* = -m_2\lambda^{-1}$, $R = -m^2\lambda^{-1}$, $A = -\frac{1}{4}l\lambda^{-1}L^{-1}$ and $C = -\lambda^{-1}L^{-1}(C_1 + C_2)$. The equation for $j(s)$ is written as (4.20). Then we find the sum $\varphi = J(r) + j(s)$. Thus, we have obtained an expression defining the scalar field in the resonator. Both continuous and multivalued waves can occur there.

4.1.2 Particle-Waves and Quantum-Like Configurations in Resonators

We emphasize that A, C, and $f(t)$ are arbitrary constants and the function in Eq. (4.20). By changing them, we can consider various cases of forced and natural extreme waves in the resonator.

The subsequent material of the section will be devoted to different resonant extreme waves in the resonant scalar field. We will explore how the transresonant parameter R, the amplitude of the disturbance, and the choice of function $f(t)$ affect the waves. The main attention will be paid to the formation of particle-waves in the resonator. We studied this question in Chap. 2 for one-direction waves.

Calculations of the profiles were carried out for $f(t) = 0$, $r_* = 0$, $A = 5$, $R = -3$, $\omega = 25.3$, $L = 0.25$, and $C = 0$. The dynamics of the second mode is presented in Fig. 4.1 (see also Fig. 3.44).

It is very interesting that the profiles φ seem to describe the dynamics of Euler figures when Euler fig. 7 evolves into fig. 11 and then the figures evolve back (Fig. 2.1). The profile does not cease to remind harmonic forms. But on these profiles folds appear. But more surprising is that at some point, the wave peaks eject the

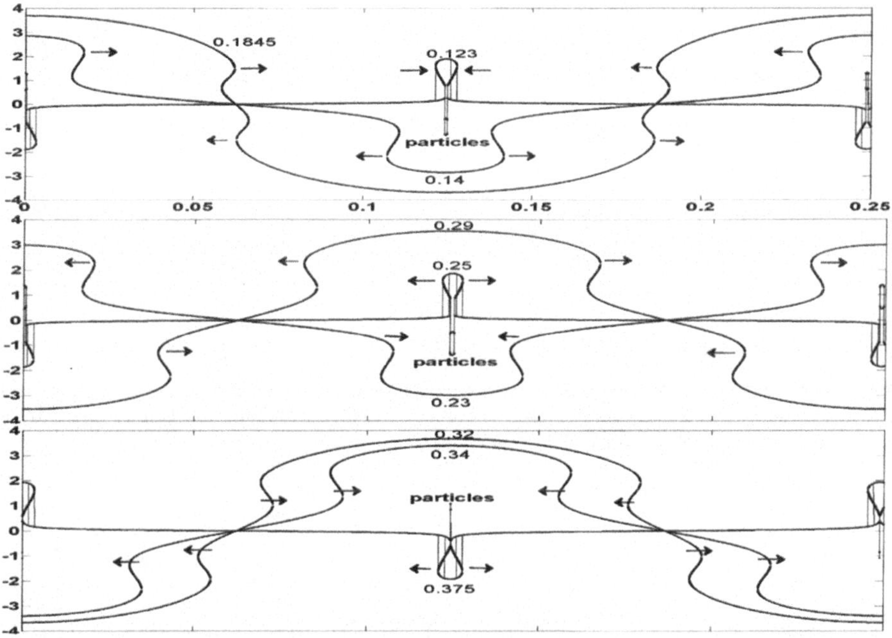

Fig. 4.1 Second mode evolution. It was calculated when $f(t) = 0$ in (4.20)

particles. We are especially interested in the possibility of the appearance in the scalar field of these particles.

Let us explore the question under which parameters the particles appear and disappear. Consider two options. Figure 4.2 presents the evolution determined by the change in the amplitude of the forcing oscillations. In Fig. 4.3, the evolution is shown connected with changing the coefficient of the linear term R.

First profiles of Fig. 4.2 correspond to a small amplitude of the forcing oscillations and small negative R ($R = -0.14$). In this case, the oscillations remain harmonic; however, there are three independent harmonics. With an increase in the forcing amplitude, the amplitudes of the oscillations increase and the two harmonics merge to form particles. Each particle is located above the corresponding top of the remaining harmonic.

With further growth of the forcing amplitude, the harmonic tops and the particle growing in size approach each other. Moreover, the harmonic is distorted into a cnoidal wave. Euler fig. 11 occurs (Fig. 2.1) at the moment of contact of these curves. During further growth of the amplitude A, we have Euler fig. 7.

A very similar evolution also occurs when R changes.

In Fig. 4.3, the value R varied from 0.3 to 0.625. As a result, the initially almost harmonic wave transforms into a stepped wave. The latter splits into a particle-wave configuration when R is close to -0.44. These configurations break down into three independent curves. Each of them, to a certain extent, resembles cnoidal waves. We

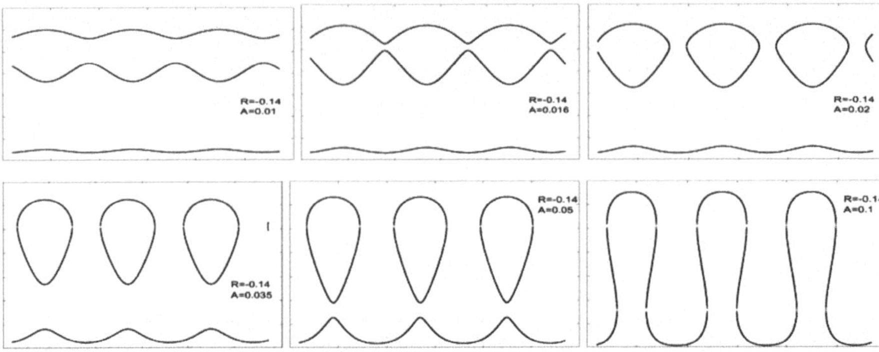

Fig. 4.2 The origin, change, and disappearance of a particle-wave configuration with an increase in the amplitude of the forcing oscillations. The amplitude of the forcing oscillations varied from 0.01 to 0.1, the value $R = -0.14$, $r_* = 0.345$, and $C = 0$

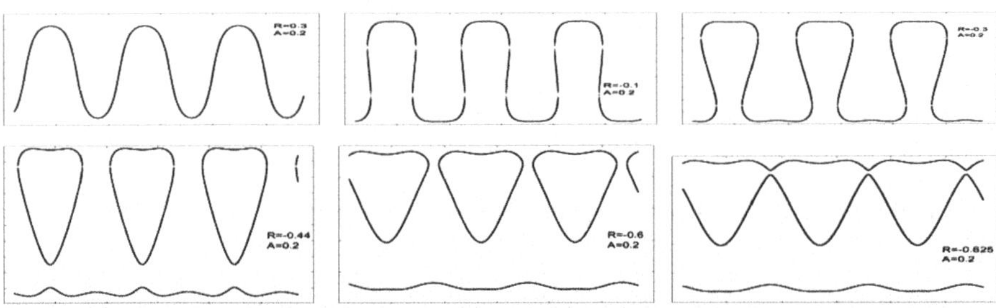

Fig. 4.3 The origin, change, and disappearance of a particle-wave configuration as R changes. The amplitude of the forcing oscillations $A = 0.2$, $r_* = 0.345$, $N = 1$, and $C = 0$

can see how an almost harmonic wave ($R = 0.3$) evolves into the Euler figs. ($R = -0.1$ and $R = -0.3$) which yield particle-wave configurations ($R = -0.44$ and $R = -0.6$) and which split further into saw-tooth waves ($R = -0.625$).

4.1.3 Formation of Quanta of Resonant Scalar Field

The quantities $A \sin \omega c^{-1} r$ and $f(t)$ in Eq. (4.20) can be considered as seed perturbations.

Let us emphasize that so far we have not discussed the question of why we have the right to introduce seed perturbations into consideration. At first glance, there is some arbitrariness in their use. In reality, there is no such arbitrariness. Even the ancient Greeks came to the conclusion that all bodies constantly vibrate according to an infinite set of their own modes. The amplitude of these oscillations is generally very small, although the energy of oscillations can be significant. The Greeks came to the conclusion by observing the vibrations of strings. Indeed, an ideally elastic thin string, according to the equations of acoustics, can have an infinite number of vibration modes and, according to theory, all modes are continuous. This contradicts the postulate of quantum mechanics that as soon as the wavelength on the string approaches the Planck length, the continuity of the waves on the string should be lost.

The above applies to all media, including scalar fields.

In light of the above, Figs. 4.2 and 4.3, as well as many of the results of Chap. 2, acquire special interest, since they show that in the process of transresonant evolution, an initially continuous scalar wave field can decay into isolated parts.

We are not aware of studies where such an effect is interpreted as a mathematical description of the main postulate of quantum mechanics.

Let us explore those quanta in relation to resonant scalar fields. Let us change the constants $\boldsymbol{A}$, $\boldsymbol{C}$, ω, and r_* in Eq. (4.20) and write it in the following two forms:

$$J^3(r) + RJ(r) + A\sin^N \omega c^{-1} r = 0, \tag{4.21}$$

$$J^3(r) + RJ(r) + A\left(\cos\omega c^{-1} r + C\right) = 0. \tag{4.21a}$$

Using these equations and comparing their solutions, we will try to extend the results of Figs. 4.2 and 4.3 on natural waves arising in finite resonators. The equationfor $j(s)$ is written as (4.21) or (4.21a). Then we find the sum $\varphi = J(r) + j(s)$ for each of the cases discussed below.

Let us consider the indicated effect of the appearance of quantization of an initially continuous field on the basis of Eqs. (4.21) and (4.21a).

In Fig. 4.4, calculations for $N = 1$ (left) and $N = 2$ (right) in Eq. (4.21) are presented.

We assumed $A = 0.000025$, $R = -0.0007$, $c = 1$, and $\omega = 26.1$. It can be seen that the change only of the forcing oscillations causes appearance particle-waves **in** the **resonator**.

Then we investigated how the wave profiles change over time. We strongly increased the amplitude of the forcing oscillations $A = 0.1$ and substitute $R = -0.001$ and $\omega \approx 26.1$.

The top of Fig. 4.5 shows the dynamics of wave profiles calculated for $N = 1$ (4.21); the bottom of the figure shows the dynamics calculated for $N = 2$ (4.21). The top of the figure showing dynamics is consistent with Fig. 3.43. According to the

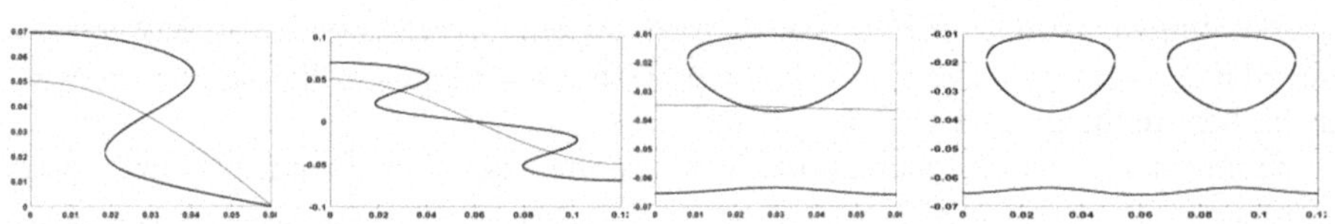

Fig. 4.4 Results of calculations for $N = 1$, A sin $\omega c^{-1} r$ (4.21). Multifold curves are the result of nonlinear modeling (left). Results of nonlinear calculations for $N = 2$, $A\sin^2\omega c^{-1} r$ (4.21). Multifold curves transform into particle-waves (right). Smooth harmonic lines are the result of linear modeling

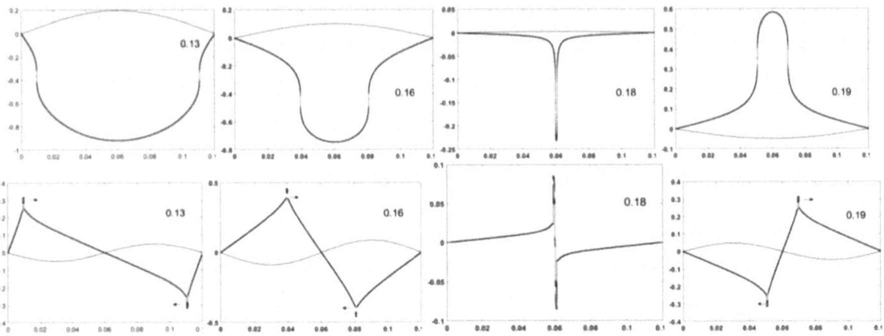

Fig. 4.5 Resonant oscillations of nonlinear wave (top). Resonant oscillations of wave configurations having particles (drops and bubbles) (bottom). Thin lines define the linear oscillations

bottom dynamics, the particles move, converging and diverging. In particular, they pass half the length of the resonator and are encountered at $t \approx 0.18$. At this moment, the particles destroy each other and then arise again. Subsequent dynamics can be assessed differently. These are either oscillations of particles along the entire length of the resonator, or there is periodic particle interactions in the middle of the resonator at which a positive particle is converted into a negative particle and vice versa.

Figures 4.2, 4.3, 4.4, and 4.5 showed particles moving with an accompanying continuous wave. This corresponds to a particle-wave, which was repeatedly demonstrated in Chap. 2. However, the possibility of the appearance of isolated particles moving in space was also emphasized there (see Figs. 2.40, 2.42, and 2.49).

Let's focus on the last case using Eq. (4.21a), where $C = 0$ or $C = 1$. The results of calculations carried out for various points in time are given in Figs. 4.6 and 4.7.

We assumed $A = 5$, $R = -0.7$, $c = 1$, and $\omega = 26.1$. The upper curves of the figure are calculated for the case $C = 0$. Curves located in the middle strip in Fig. 4.6 are calculated for the case $C = 1$. They determine the movement of the particle-wave. The lower curves show the movement of only the particle (quantum of the resonant scalar field).

Note that in experiments, continuous curves are located in the middle strip Fig. 4.6 may not be fixed, since they change weakly. Even the particles shown below Fig. 4.6 can be recorded as a point if the measuring device is capable of measuring only zones of rapid change of function, that is, the ends of closed curves.

We emphasize that in Fig. 4.6, there is no direct correspondence between the calculations for $C = 0$ and $C = 1$. In the first case, waves arise with frequencies twice as high as in the case of $C = 1$.

Figure 4.7 provides calculations only for the case $C = 1$ and different points in time.

On the left are particle-waves, and on the right are the corresponding quanta of the scalar field.

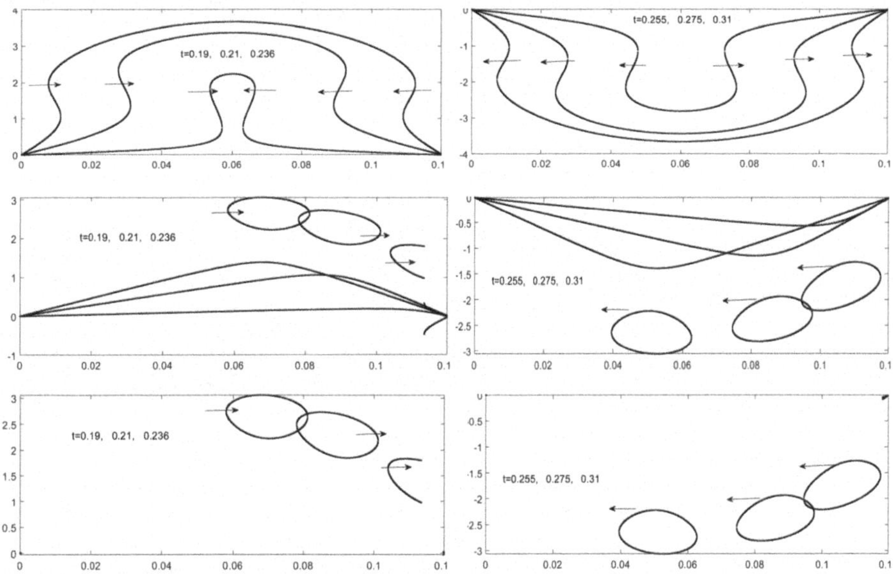

Fig. 4.6 Curves illustrating the evolution of the scalar field in the resonator, calculated for various values of C in Eq. (4.21a). Each figure shows calculations for three points in time. These values are shown there. It can be seen that the particle is located above the top of the wave and moves along with the wave top

The results presented in Figs. 4.6 and 4.7 are obtained using Eq.(4.21a) in which it is assumed that the amplitude of the seed perturbation is equal to the stationary energy of the scalar field C (see 4.2). Thus, if we limit ourselves to considering bare disturbances in the form of simple harmonics of the form $A \sin \omega c^{-1} r$, then according to (4.21a), particle-waves and quanta arise only when $A = C$. In this case, waves appear in the resonator whose frequency is two times lower.

At the same time, if we assume that the seed perturbation can have the form.

$\sin^N \omega c^{-1} r$, where $N > 1$, then particle-waves and quanta can arise in other cases. Let's consider one of them. Figures 4.7a and 4.7b show calculations obtained using Eq. (4.21), where $N = 2$.

This figure corresponds to Figs. 4.7a and 4.7b. Thin lines correspond to $A\sin^2 \omega c^{-1} r$.

The upper curves of the figure are calculated for the case $N = 1$ ($A \sin \omega c^{-1} r$). The curves located in the middle strip of Fig. 4.7b are calculated for the case $N = 2$ ($A\sin^2 \omega c^{-1} r$). They determine the movement of the particle-wave. The lower curves show the movement of only the particle (quantum of the resonant scalar field) over time. It can be seen that the particle is located above the top of the wave and moves along with the wave.

It's interesting to learn what the relationship between the seed perturbation and C is. Before this, we assumed that they were completely independent. Let us

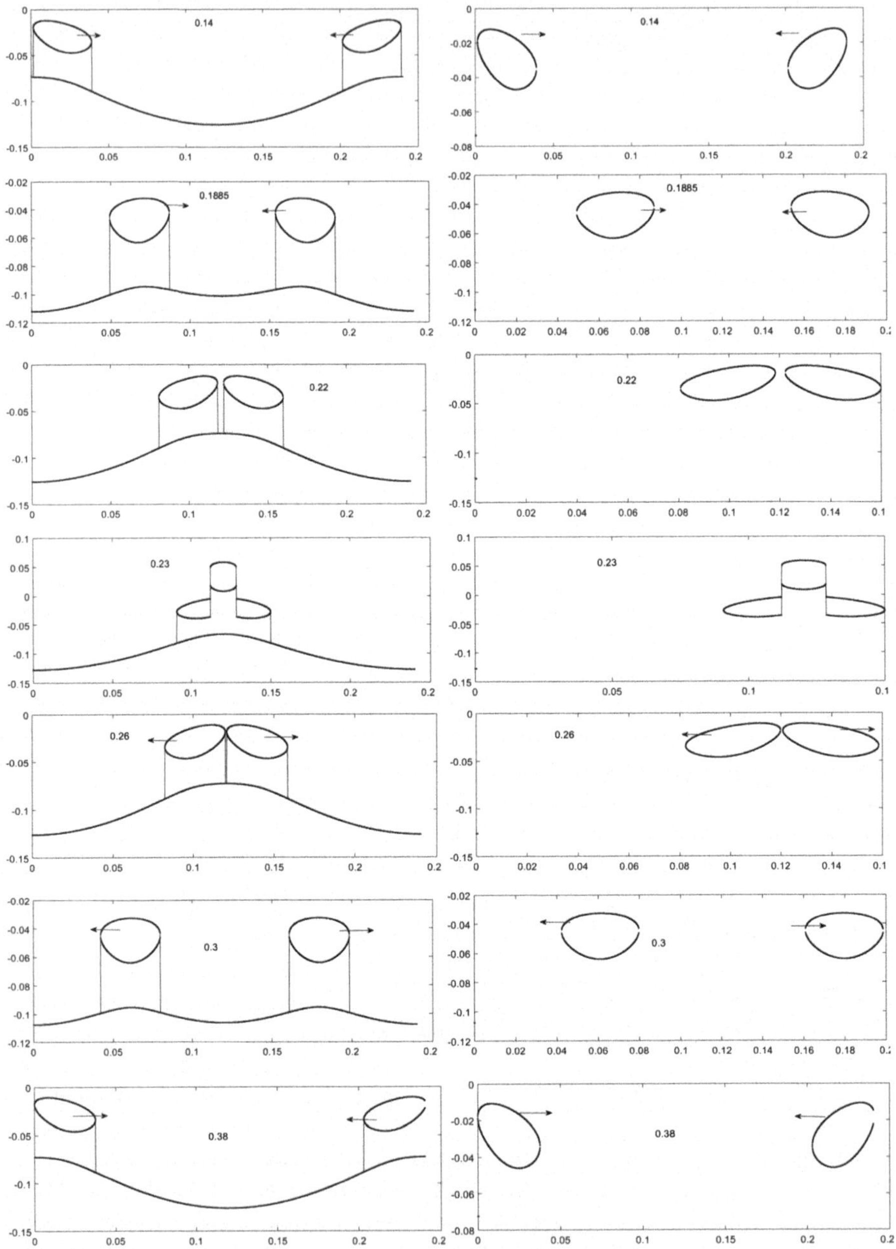

Fig. 4.7 Dynamics of particle-waves and corresponding scalar field quanta. It can be seen that the particle is located above the top of the wave and moves along with the wave top

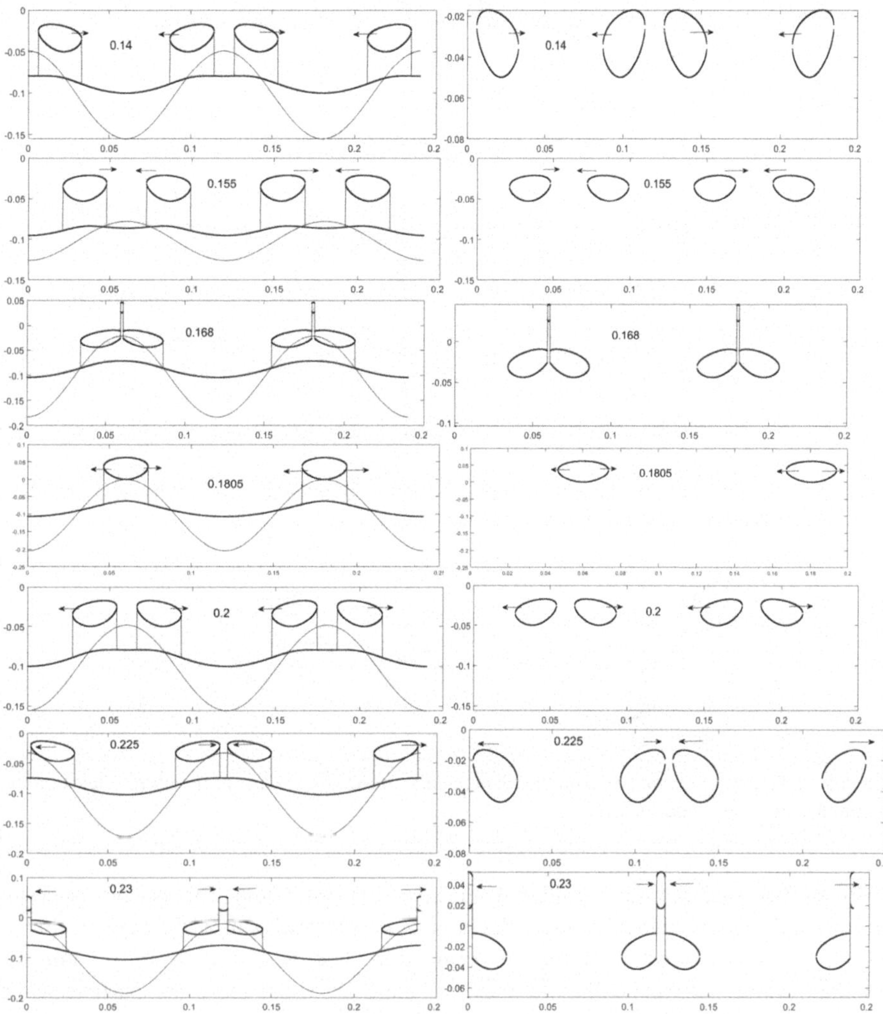

Fig. 4.7a Dynamics of particle-waves and corresponding scalar field quanta calculated according to (4.21) for $A\sin^2\omega c^{-1}r$

investigate what will happen if the amplitude of the harmonic seed perturbation begins to approach the value C.

Figure 4.8 shows the calculation results for the case when in Eq. (4.21a), the value $C = 0.9$, 0.95, and 0.99. Calculations were performed for the same point in time. On the right, the calculated curves are superimposed on each other. It can be seen how, as C increases, a peculiar loop on the wave profile transforms into a particle.

The same evolution is illustrated by the curves on the right in Fig. 4.8. There, the calculated curves are presented separately from each other.

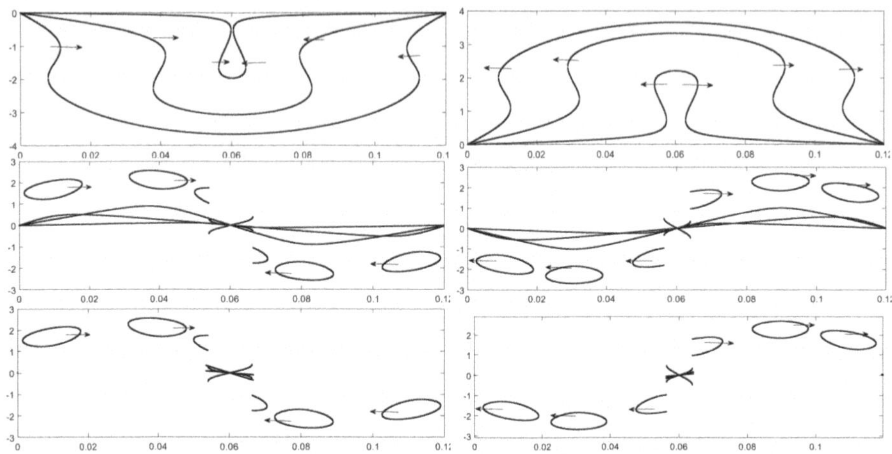

Fig. 4.7b Curves illustrating the evolution of the scalar field in the resonator, calculated according to (4.21) for $A\sin^N\omega c^{-1}r$. This figure corresponds to Fig. 4.6

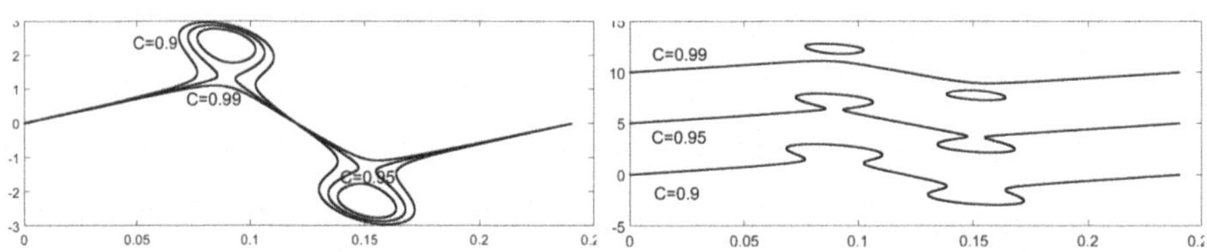

Fig. 4.8 Calculation results showing the dependence of the occurrence of quantum effects on the magnitude of the seed perturbation

From the calculations presented in this section, some connection between multivalued waves and quantum effects is more clearly visible. Of course, this is only the first step in establishing this connection. The picture of strongly nonlinear resonant waves unexpectedly includes the stage at which there are particle-wave configurations. The emergence of this configuration is rather unexpected, but generally corresponds to the latest experiments [10, 11].

In connection with the above, we present experimental data in which a remarkable attempt is made at the macro level to simulate quantum effects.

Remark. Above is a broad picture of the possible evolution paths of Euler figures in relation to wave dynamics. At the same time, the question under what conditions these new, multivalued waveforms can be realized is not entirely clear. The existence of a wave-particle configuration for a very short time, generally speaking, is not surprising. It was surprisingly, when in 1978, Walker discovered that a droplet of soapy water could bounce for several minutes on a surface wave excited on the water [4].

In 2005 Couder, Protiere, Fort, and Baddoud started the systematic study of this phenomenon using droplets of silicone oil [5]. Couder and Fort were struck by the

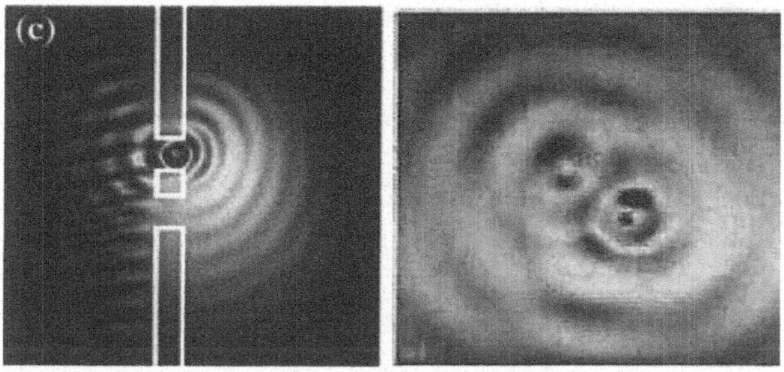

Fig. 4.9 Imitation of the famous double-slit experiment (left). Particles traveling on top of a surface wave attracted such attention because they can be interpreted as some analog of quantum waves [7, 8]

similarity between the droplets (walkers) and de Broglie's pilot waves. They explored the strength of this analogy and how closely the walker could imitate the quantum behavior of the elementary particle. In particular, Couder and Fort imitated the famous double-slit experiment [6] (Fig. 4.9).

Thus, some analogy is shown between macro- and microphenomena [4–6, 8–11]. In this book, we emphasize this possible analogy many times. When we talk about particle physics, we do not usually emphasize that wave field physics is considered. Thus, we do not deal with particles and their trajectories. In reality, we have systems of waves and, possibly, the resonant effects of their amplification up to the emergence of field clusters, that is, particles.

4.1.4 Eigenmultivalued Waves in Resonators

Let us consider free nonlinear oscillations of waves in a finite resonator (potential well).

Equation (4.1) is written as follows:

$$\varphi_{tt} - c_*^2\varphi_{xx} = m^2\varphi - \lambda\varphi^3. \tag{4.22}$$

We write zero conditions at the ends of the resonator:

$$\varphi = 0 \quad \text{at} \quad x = 0 \quad \text{and} \quad x = L. \tag{4.23}$$

We look for an expression for the potential in the form of the sum of initial perturbation $\varphi^{(1)}$ and evolution component (part) $\varphi^{(2)}$,

$$\varphi = \varphi^{(1)} + \varphi^{(2)}. \tag{4.24}$$

Then we can write

$$\left(\varphi^{(1)} + \varphi^{(2)}\right)_{tt} - c_*^2\left(\varphi^{(1)} + \varphi^{(2)}\right)_{xx} = m^2\left(\varphi^{(1)} + \varphi^{(2)}\right) - \lambda\left(\varphi^{(1)} + \varphi^{(2)}\right)^3. \tag{4.25}$$

Then we divide the resulting equation into the following two equations:

$$\begin{aligned} {\varphi^{(1)}}_{tt} - c_*^2 {\varphi^{(1)}}_{xx} &= m^2 \varphi^{(1)}, \\ {\varphi^{(2)}}_{tt} - c_*^2 {\varphi^{(2)}}_{xx} &= m^2 \varphi^{(2)} - \lambda \left(\varphi^{(1)} + \varphi^{(2)} \right)^3. \end{aligned} \tag{4.26}$$

Next we accept

$$\begin{aligned} &\varphi^{(1)} = A \sin \omega t \sin \pi N L^{-1} x, \quad c_* = \pm \pi^{-1} N^{-1} L \sqrt{m^2 + \omega^2}, \\ &\omega L c^{-1} = 2\pi N \quad (N = \pm 1, \pm 2, \pm 3, \ldots.) \end{aligned} \tag{4.27}$$

In this case, $\varphi^{(1)}$ satisfies the corresponding boundary conditions.

Now consider Eq. (4.26). We are looking for a solution in traveling waves. We accept that

$$r = ct - x, \quad s = ct + x \quad \text{and} \quad c = c_* + \overline{c} \tag{4.28}$$

where $\overline{c}$ is the perturbation of the speed c. Taking into account what has been said, we rewrite (4.26) in new variables:

$$4c_*^2 {\varphi^{(2)}}_{rs} + \left(2c_* \overline{c} + \overline{c}^2 \right) \left[{\varphi^{(2)}}_{rr} + 2{\varphi^{(2)}}_{rs} + {\varphi^{(2)}}_{ss} \right] = m^2 \varphi^{(2)} - \lambda \left(\varphi^{(1)} + \varphi^{(2)} \right)^3. \tag{4.29}$$

We will look for a solution to this equation in the form

$$\varphi^{(2)} = J_1(r) - J_1(s). \tag{4.30}$$

In this case, (4.30) immediately satisfies the boundary condition $\varphi^{(2)} = 0$ at $x = 0$. Next we assume that $\left(2c_* \overline{c} + \overline{c}^2 \right) \lambda^{-1} (J_{1rr} + J_{1ss}) \approx 0$. The functions $J_1(r)$ and $J_1(s)$ have a period L, and oppositely propagating waves do not interact with each other. In this case, the functions $J_1(r)$ and $J_1(s)$ are described by the equation

$$J_1^3 + 3JJ_1^2 + \left(3J_1^2 - m^2 \lambda^{-1} \right) J_1 + J^3 = 0. \tag{4.31}$$

Let us look at the calculation results. We emphasize that our research is qualitative in nature.

A sequence of two calculations was performed. Their results are presented in Figs. 4.10 and 4.11. They show the sum $J_1(s)$ + $J_1(r)$. The seed waves $J(r)$ and $J(s)$ were not taken into account due to their smallness. In each series of calculations, the time and coordinate changed, but each series of calculations was carried out with the other parameters of Eq. (4.31) being constant.

First, we carry out the calculation for the case $J = 0.1 \cos(5\pi(0.01t + x))$ and $m^2\lambda^{-1} = 0.005$.

Figure 4.10 shows the complex wave evolution during the half-cycle of oscillations of the seed wave functions. This evolution can be represented as consisting of three main stages. The first stage is transitional. It is determined by the transition from a single-fold wave profile to a double-folded profile (the stage of the existence of waves with hair). In the second stage, the peaks and troughs change places. In this case, there are moments of existence of particle-waves. At the third stage, the two-fold configuration that emerged at the second stage again returns to the single-fold configuration. Thus, Fig. 4.10 models the half period of oscillations of a scalar field.

Now we carry out the calculation for the case $J = 0.0216 \cos(5\pi(t + x))$ and $m^2\lambda^{-1} = 0.001$.

Joint consideration of Figs. 4.10 and 4.11 allows us to draw the following conclusions:

1. The figures show peculiar harmonic waves with folds. The half-cycle of their oscillations is shown. Overall, their behavior resembles oscillations of ossilons (see Figs. 1.6, 1.7, and 1.8) or linear standing waves in resonators.
2. A careful examination reveals small amplitude waves running up or down the surface of each large wave. When these traveling waves converge (collide) at the peaks and troughs of a standing wave, a portion of the field can be separated from the main standing wave. In this case, parts of the standing wave change sign.
3. There are two surprising features of the presented evolution. The first is the existence of folds in the wave profile. The second is the emergence of particles when these folds converge at the peaks and troughs of the waves.

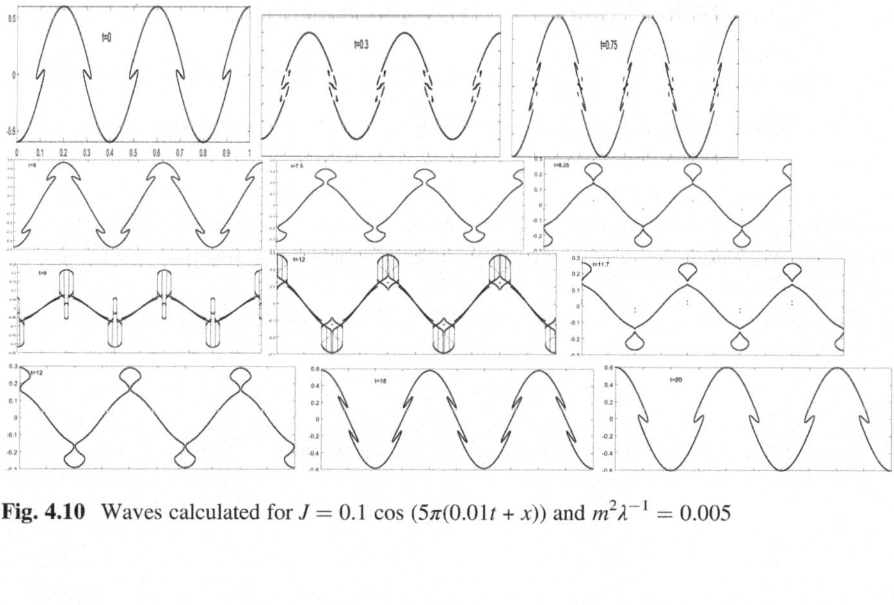

Fig. 4.10 Waves calculated for $J = 0.1 \cos(5\pi(0.01t + x))$ and $m^2\lambda^{-1} = 0.005$

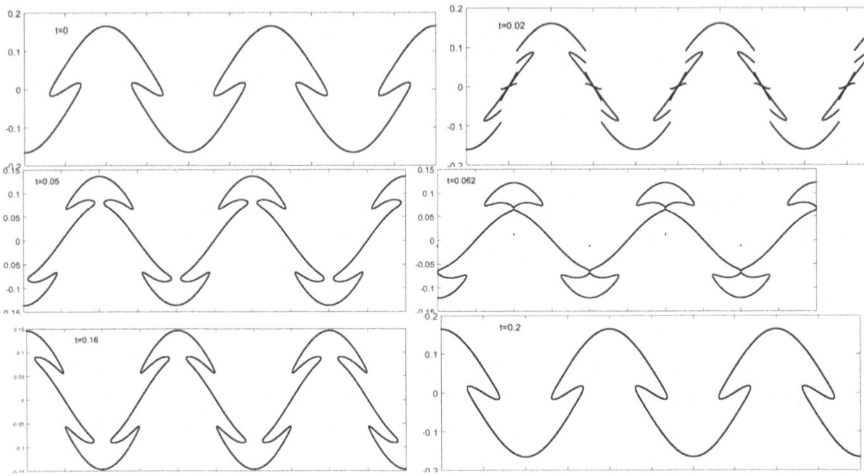

Fig. 4.11 Waves calculated for $J = 0.0216 \cos (5\pi(t + x))$ and $m^2 \lambda^{-1} = 0.001$

4. Finally, the calculations reveal an interesting wave pattern in the x-t plane, reminiscent of those shown in Figs. 3.16, 3.17, and 3.18. This pattern is formed by smooth standing waves on the surface of which multivalued ("overturning") waves of small amplitude, which can be called ripple waves, perform periodic movements.
5. At the same time, the curves presented in Figs. 4.10 and 4.11 can be interpreted as waves traveling to the right or left. During the movement, they change their form, which is periodically restored.

4.2 Theory of Parametrically Excited Nonlinear Scalar Waves

Let us consider a parametrically excited wave φ. We assume that the speed c_* changes according to some periodic law. For example, if we mean the speed of light c_*, then this means that the properties of the medium in which the wave φ propagates periodically change and the speed c_* changes accordingly. Moreover, the frequency of changes in the properties of the medium is comparable to the frequency of the light wave.

The nonlinear Klein-Gordon equation (NKGE) is written as follows:

$$\varphi_{tt} - c_*^2(t)\varphi_{xx} = m^2\varphi - \lambda\varphi^3 \tag{4.32}$$

The conditions are written at the ends of the resonator:

$$\varphi_x = 0 \quad \text{at} \quad x = 0 \quad \text{and} \quad x = L. \tag{4.33}$$

We are looking for waves with a period *L*/*N* where $N = 1, 2, 3, \ldots$ We will consider $c_*(t)$ as a slowly variable function such that $(c_*)_{tt} \approx 0$. Let us

$$\varphi = \varphi^{(1)} + \varphi^{(2)}. \tag{4.34}$$

Here $\varphi^{(1)} >> \varphi^{(2)}$. Let us introduce new variables

$$r = c(t) - x, \qquad s = c(t) + x. \tag{4.35}$$

Then we accept that

$$c(t) = c_*(t) + \overline{c}(t), \tag{4.36}$$

where $\overline{c}(t)$ is the perturbation of the speed $c(t)$. Equation (4.32) is rewrited using (4.34) and (4.35):

$$\begin{aligned}&(c_t^2 - c_*^2)\left[\left(\varphi^{(1)} + \varphi^{(2)}\right)_{rr} + \left(\varphi^{(1)} + \varphi^{(2)}\right)_{ss}\right] + 2(c_t^2 + c_*^2)\left(\varphi^{(1)} + \varphi^{(2)}\right)_{rs}\\ &+ c_{tt}\left[\left(\varphi^{(1)} + \varphi^{(2)}\right)_r + \left(\varphi^{(1)} + \varphi^{(2)}\right)_s\right] = m^2\left(\varphi^{(1)} + \varphi^{(2)}\right) - \lambda\left(\varphi^{(1)} + \varphi^{(2)}\right)^3.\end{aligned} \tag{4.37}$$

Next we take into account (4.36). In this case, $\left(c^2 - c_*^2\right)\left({\varphi^{(1)}}_{rr} + {\Phi^{(1)}}_{ss}\right) \approx 0$. In addition, we have $c_{tt} \approx 0$. As a result of these assumptions, Eq. (4.37) can be divided into the following two equations:

$${\varphi^{(1)}}_{rs} = 0, \tag{4.38}$$

$$4c_t^2 {\varphi^{(2)}}_{rs} - m^2\varphi^{(1)} - \lambda\left(\varphi^{(1)}\right)^3. \tag{4.39}$$

Now we have

$$\varphi^{(1)} = J(r) + j(s), \tag{4.40}$$

and

$$4c_t^2 {\varphi^{(2)}}_r = J_*(r) + m^2\left(sJ + \int j ds\right) - \lambda\left(sJ^3 + 3J^2\int j ds + 3J\int j^2 ds + \int j^3 ds\right) + C_1. \tag{4.41}$$

The expression for $4c_t^2\varphi^{(2)}{}_s$ is similar to (4.41). Now let us write

$$\begin{aligned}4c_t^2\left(\varphi_s^{(2)}-\varphi_r^{(2)}\right)=&j_*(s)-J_*(r)+m^2\left((r-s)j+\int Jdr\right)\\&-m^2\left((s-r)J+\int jds\right)-\lambda\left((r-s)j^3+3j^2\int Jdr\right.\\&+3j\int J^2dr+\int J^3dr)+C_2+\lambda\left((s-r)J^3+3J^2\int jds\right.\\&+3J\int j^2ds+\int j^3ds)-C_1.\end{aligned} \tag{4.42}$$

Let us

$$j_*(s)=m^2\int jds-\lambda\int j^3ds,\quad J_*(r)=+m^2\int Jdr-\lambda\int J^3dr. \tag{4.43}$$

We do not take into account the interaction of oppositely traveling waves. Then

$$\begin{aligned}4c_t^2\left(\varphi^{(2)}{}_s-\varphi^{(2)}{}_r\right)=&j_{1*}(s)-J_{1*}(r)+m^2(r-s)j-m^2(s-r)J\\&-\lambda(r-s)j^3+\lambda(s-r)J^3+C_2-C_1.\end{aligned} \tag{4.44}$$

We further assume that the value $4c_t^2\left(\varphi_s^{(2)}-\varphi_r^{(2)}\right)$ is negligible, since it was assumed above that $\varphi^{(1)}\gg\varphi^{(2)}$. As a result, we obtain an equation for finding the unknown function $J(r)$.

$$-J_{1*}(r)-m^2(s-r)J+\lambda(s-r)J^3+C_2-C_1=0. \tag{4.45}$$

The equation for $j(s)$ has the same form. Next we take into account (4.45). As a result, we have the following defining equations $J(r)$ and $j(s)$:

$$2x\left(-\lambda J^3+m^2J\right)+J_{1*}(r)-C_2+C_1=0, \tag{4.46}$$

$$2x\left(-\lambda j^3+m^2j\right)-j_{1*}(s)-C_2+C_1=0. \tag{4.47}$$

Here $J_{1*}(t)=-j_{1*}(t)$ and $J_{1*}(t)=J_{1*}(t+L)$. It means that $J_{1*}(t)$ and $J_{1*}(t)$ are periodic functions with period L. In this case, Eqs. (4.46) and (4.47) have identical periodic solutions, and the expression $\varphi_x=[J(r)+j(s)]_x$ satisfies the boundary conditions (4.33).

Equation (4.46) is written in the following form:

$$J^3 - 0.0005J + 0.0001(\sin(7.65(c-x)) - 0.4\sin(22.95(c-x))) = 0. \quad (4.48)$$

Here $c = 0.1849 \sin(5.4094t)$. Figure 4.12 shows the calculation results.

It is easy to see that the calculation results resemble waves parametrically excited on the contact surface of two liquids (see Figs. 3.19, 3.20, 3.21, 3.23, 3.24, and 3.27). However, there is a significant difference with the calculations in Chap. 3. Now we have obtained an oscillating chain of waves, where each wave component, as a whole, is motionless in space, and oscillates without changing sign. Additional studies have shown that the latter features are related to the amplitude of parametric excitation. After we increased its amplitude by a factor of 4, the elementary waves of the chain began to move relative to each other. As a result, the calculations began to correspond well to Figs. 3.19 and 3.27 throughout the entire period of parametric excitation.

The presented Figs. 4.10, 4.11, and 4.12 allow us to draw the following conclusions:

1. In general, the results of the series correspond to experiments and calculations for a two-layer liquid.
2. A nonlinear scalar field in a resonator can exhibit the properties of both standing and traveling waves. The extent of this manifestation depends on the parameters of the field and and the excitation.

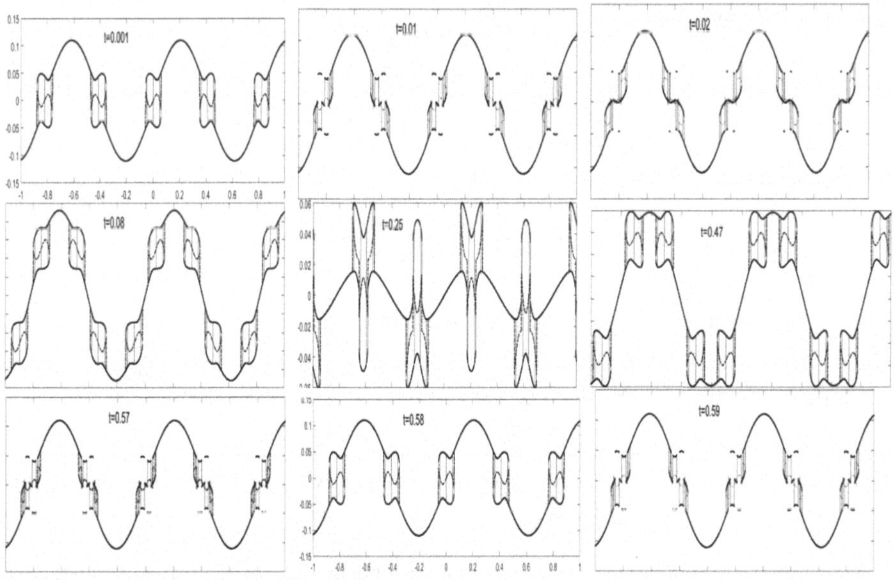

Fig. 4.12 Evolution of extreme multivalued waves φ

This concludes an important part of our research on extreme multivalued waves. We have obtained calculation results that correlate with each other, and, more importantly, with the experiments known to us.

4.3 Example of Multidimensional Multivalued Waves

Above, we considered one-dimensional waves. Of course, the results obtained for them cannot be directly extended to the case of many variables. But there is a case of plane waves when this is, to a certain extent, possible. The simplest version of this case is discussed in this section.

The Klein-Gordon equation is written in the form

$$\varphi_{tt} - c_*^2 \nabla^2 \varphi = m^2 \varphi - \lambda \varphi^3 + \lambda_6 \varphi^5. \tag{4.49}$$

In this book, we consider locally rapidly variable waves. In the interval of the length of the rapid change of the wave, it can be assumed that the coefficients of Eq. (4.49) can change slightly along the coordinates. This assumption is similar to the case of the theory of long waves in shallow water. There, fairly accurate calculations can be carried out under the assumption that the water depth does not change over intervals of the order of the wavelength. In this section, we assume that m^2 is a slightly variable function of coordinates x_n, where $N = 1, 2, 3, 4, \ldots$. This dependence is so weak that we do not take it into account until the start of calculations.

Next we assume that

$$\varphi = \varphi^{(1)} + \varphi^{(2)}. \tag{4.50}$$

Here $\varphi^{(1)} > > \varphi^{(2)}$. The expression (4.50) is substituted in (4.49). As a result, we have approximately

$$\left(\varphi^{(1)} + \varphi^{(2)}\right)_{tt} - c_*^2 \nabla^2 \left(\Phi^{(1)} + \Phi^{(2)}\right) = m^2 \left(\varphi^{(1)} + \varphi^{(2)}\right)$$
$$- \lambda \left(\varphi^{(1)} + \varphi^{(2)}\right)^3 + \lambda_6 \left(\varphi^{(1)} + \varphi^{(2)}\right)^5. \tag{4.51}$$

Further we will consider two contrary moving plane waves. Let.

$$r = ct - \sum_n^N k_n x_n, \qquad s = ct + \sum_n^N k_n x_n. \tag{4.52}$$

We emphasize that we are considering a very particular, apparently simplest, case of spatial plane waves.

Equation (4.51) yields in the variables (4.52),

$$\left(c^2 - c_*^2\sum_n^N k_n^2\right)\left[\left(\varphi^{(1)} + \varphi^{(2)}\right)_{rr} + \left(\varphi^{(1)} + \varphi^{(2)}\right)_{ss}\right]$$
$$+2\left(c^2 + c_*^2\sum_n^N k_n^2\right)\left(\varphi^{(1)} + \varphi^{(2)}\right)_{rs} \tag{4.53}$$
$$= m^2\left(\varphi^{(1)} + \varphi^{(2)}\right) - \lambda\left(\varphi^{(1)} + \varphi^{(2)}\right)^3 + \lambda_6\left(\varphi^{(1)} + \varphi^{(2)}\right)^5.$$

Let

$$c^2 = c_*^2\sum_n^N k_n^2. \tag{4.54}$$

In addition to this, we assume that the coefficients with c, m^2, λ, and λ_6 are such that $4c^2\varphi_{rs}^{(1)} >> -4c^2\varphi_{rs}^{(2)} + m^2(\varphi^{(1)} + \varphi^{(2)}) - \lambda(\varphi^{(1)} + \varphi^{(2)})^3 + \lambda_6(\varphi^{(1)} + \varphi^{(2)})^5$. In this case, Eq. (4.53) can be divided approximately into the following two equations:

$$\varphi^{(1)}{}_{rs} = 0, \tag{4.55}$$

$$4c^2\varphi^{(2)}{}_{rs} = m^2\varphi^{(1)} - \lambda\left(\varphi^{(1)}\right)^3 + \lambda_6\left(\varphi^{(1)}\right)^5. \tag{4.56}$$

Thus, we reduced the solution of the original nonlinear Eq. (4.49) to the solution of the homogeneous linear Eq. (4.55) and the inhomogeneous linear Eq. (4.56). We have repeatedly successfully used this method for solving the nonlinear wave equation in the second and third chapters, starting with Eq. (3.2). This method allowed us to evaluate qualitatively the influence of high nonlinearity on the waves.

Now we have

$$\varphi^{(1)} = J(r) + j(s), \tag{4.57}$$

and

$$4c^2\varphi^{(2)}{}_r = J_*(r) + m^2\left(sJ + \int jds\right)$$
$$-\lambda\left(sJ^3 + 3J^2\int jds + 3J\int j^2ds + \int j^3ds\right) + \lambda_6\int (J + j)^5ds. \tag{4.58}$$

The expression for $4c^2\varphi^{(2)}{}_s$ is written in a similar way. Then we write the following difference:

$$4c^2\left(\varphi^{(2)}{}_s - \varphi^{(2)}{}_r\right) = \begin{aligned} & j_*(s) - J_*(r) + m^2\left((r-s)j + \int Jdr\right) \\ & - m^2\left((s-r)J + \int jds\right) - \lambda\left((r-s)j^3 + 3j^2\int Jdr\right. \\ & + 3j\int J^2dr + \int J^3dr) + \lambda\left((s-r)J^3 + 3J^2\int jds\right. \\ & + 3J\int j^2ds + \int j^3ds) + \lambda_6\int (J+j)^5dr - \lambda_6\int (J \\ & + j)^5ds. \end{aligned} \tag{4.59}$$

We will write arbitrary functions $j_*(s)$ and $J_*(r)$ in the following forms:

$$\begin{aligned} j_*(s) &= sj_{*1}(s) + m^2\int jds - \lambda\int j^3ds + \lambda_6\int j^5ds, \\ J_*(r) &= rJ_{*1}(r) + m^2\int Jdr - \lambda\int\ J^3\,dr + \lambda_6\int j^5dr. \end{aligned} \tag{4.60}$$

Further, the interaction of oppositely traveling waves is not taken into account. Then (4.59) gives

$$\begin{aligned} 4c_*^2\sum_n^N k_n^2(\varphi^{(2)}{}_s - \varphi^{(2)}{}_r) &= 4c^2(\varphi^{(2)}{}_s - \varphi^{(2)}{}_r) \\ &= sj_{1*}(s) - rJ_{1*}(r) + m^2(r-s)\,j - m^2(s-r)J \\ &\quad - \lambda(r-s)\,j^3 + \lambda(s-r)J^3 + \lambda_6(r-s)\,j^5 - \lambda_6(s-r)J^5. \end{aligned} \tag{4.61}$$

We emphasize that despite all the assumptions used to simplify the analysis, we obtained quite complex expressions (4.57) and (4.61) from which we need to find $J(r)$, $j(s)$, $\varphi^{(2)}{}_s$, and $\varphi^{(2)}{}_r$. We do not see the possibility of finding all the indicated quantities. Therefore, we further use the following assumptions that radically simplify the goals of our research.

1. We refuse to search for quantities $\varphi^{(2)}{}_s$ and $\varphi^{(2)}{}_r$. Our goal is to define only $J(r)$ and $j(s)$.
2. We will look for a solution that is fair only at points where

$$J_r = j_s \quad \text{and} \quad \varphi^{(2)}{}_s = \varphi^{(2)}{}_r.. \tag{4.62}$$

In this case, to find $J(r)$ and $j(s)$, we obtain from (4.61)

$$0 = sj_{1*}(s) - rJ_{1*}(r) + m^2(r-s)j - m^2(s-r)J \\ - \lambda(r-s)j^3 + \lambda(s-r)J^3 + \lambda_6(r-s)j^5 - \lambda_6(s-r)J^5. \tag{4.63}$$

Let $j_{1*}(s)$ and $J_{1*}(r)$ be the harmonic functions for which we have $j_{1*}(s) = J_{1*}(r)$ at the points under consideration. Then in these points

$$0 = (j_{1*}(s) - J_{1*}(r)) - m^2 j - m^2 J + \lambda j^3 + \lambda J^3 - \lambda_6 j^5 - \lambda_6 J^5. \tag{4.64}$$

From here we get equations for $J(r)$ and $j(s)$:

$$j_{1*}(s) - m^2 j + \lambda j^3 - \lambda_6 j^5 = 0, \tag{4.65}$$

$$J_{1*}(r) + m^2 J - \lambda J^3 + \lambda_6 J^5 = 0. \tag{4.66}$$

As an example of using the proposed approach, we consider a two-dimensional region. Only the wave $J(r)$ propagates in it. We will accept that $r = c_0 t - x - y$. For this case, Eq. (4.66) is written as follows:

$$-0.11J^5 + J^3 + 3R/2^{2/3}J + \sin \omega c_0^{-1}(c_0 t - x - y) = 0. \tag{4.67}$$

It is assumed that $R = 0.5 - 2 \sec h\{0.01[(25 - x)^2 + y]\}$. Therefore, there is a zone where the coefficient of the linear term smoothly changes from a maximum negative value to a positive value. We assume that in the zone, the wave profile changes so quickly that the fact of the change R to a positive value is more important then the value R..

The evolution of multivalued waves propagating in the negative y-direction is shown in Fig. 4.13 (left). We assume that $\omega c_0^{-1} = 1$ and $t = 0$. Far from center of resonance ($x = 25$ and $y = 0$) the waveform ($y = 11$) slightly reminds the incident harmonic wave although the amplitudes of them strongly different. Then near the line $x = 25$, profiles of the waves curl and mushroom-like waves are generated ($y = 9.9$ and 9.75). As a result of the instability, the mushroom-like waves break, nucleating the closed loops. We may consider these loops as vortices.

It is possible to give a different interpretation of Fig. 4.13 (left). In particular, they qualitatively resemble the formation of shock waves and vortex pairs in a resonator containing the Bose-Einstein condensate (BEC). At first, we stop by an example of the evolution of the linear perturbation in BEC (Fig. 4.13, center) [15]. There the originally direct line is being distorted in the harmonic line. After that, the folds appear. Then the last breaks up into vortices (the points in Fig. 4.13, center). Let us consider another example. The solution of the equation of Ginzburg-Landau has shown that the "elastica" waves with the vortices on the profiles can arise during the evolution of an initial perturbation (Fig. 4.13, right) on an interface of two media [16].

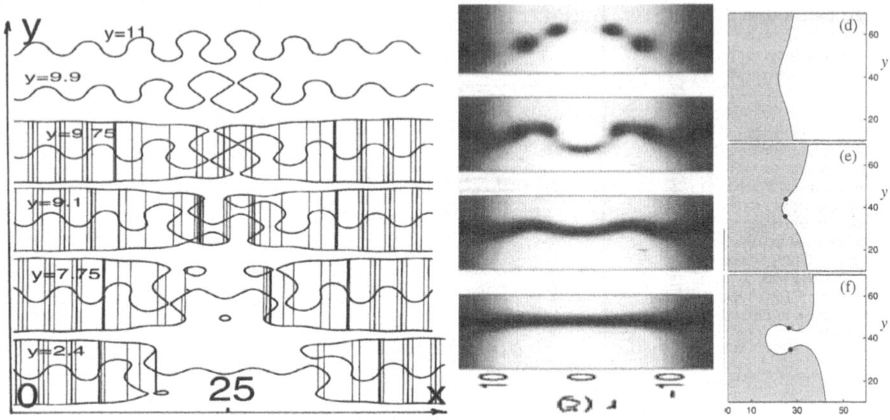

Fig. 4.13 Transresonant evolution of initially harmonic waves into mushroom-like waves and loops (vortices) in the x-y plane (left). Examples of the evolution of originally small perturbations into complex patterns and vortices (center and right) [13–16]

All this in the context of our research should lead to the thought that some "elastica" forms can evolve into vortices (see Sect. 2.3).

Let us consider the circular area near points $x = 25$ and $y = 0$ framed by multivalued waves and vortex zones which is presented in Fig. 4.13 (left). Possibly similar circular areas can originate and exist everywhere, from the "eyes" of a typhoon and a tsunami in the ocean to an unstable interface of the environments and in BEC and, finally, during the origin "black" holes in space and Universes.

Thus, it may be everywhere where the wave evolution is defined by very high nonlinearity.

The first attempt to study multidimensional multivalued waves was made more than 20 years ago. Surface waves were considered [3]. In this section, we extended the previously obtained results to the case of scalar waves. We proceeded from the fact that multivalued waves in the ocean shown in Figs. 4.15, 4.16, 4.17, 4.18, and 4.19 can occur wherever nonlinear scalar field equations are valid.

You don't have to be a particularly attentive reader to notice that the author demonstrated above the existence of multivalued waves in a huge variety of wave systems. The accumulated material allows us to think about introducing a certain classification of multivalued waves. Of course, the introduction of classificatory classification is hampered by the diversity of waves and the unusual nature of their forms.

4.4 Multivalued Ocean Waves

We have considered many multivalued waves. But this is only a small part of those waves that can exist in infinite and diverse scalar fields. It is difficult to imagine, dealing only with surface waves excited in small containers, all the wave diversity that exists on the surface of the ocean. Even if you don't know about the internal ocean waves, an example is waves running onto the shallow shore (Fig. 4.14).

A huge number of multivalued waves appear on the surface of the ocean during storms. But they are not sustainable. Waves that occur near the shore are more stable (Figs. 4.14, 4.15, and 4.16).

Waves are shown in thick lines. Thin straight lines show a change in depth. The thin lines on the left show qualitatively the nature of the change in the transresonant parameter. The curves show a significant effect of small perturbations of the bottom profile. As a result, the initially harmonic waves evolve into an overturning wave (left) or into vertical splashes (right) when approaching the shore. The right curves may be associated with ocean sprays ejected from the top of ocean waves [7, 13].

No less diversity arises when ocean waves pass through underwater topographies.

It is important to emphasize that the maximum resonant amplification of waves may not occur at the top of the topography, but at some distance from the top. On the top of the topography (Figs. 4.16, 4.17, and 4.18) or near a very flat coastal bottom (Figs. 4.14 and 4.15), wave motion may be weak, although already a few tens of meters toward the deep ocean wave amplitudes reach tens of meters.

Fig. 4.14 Ocean islands [13]

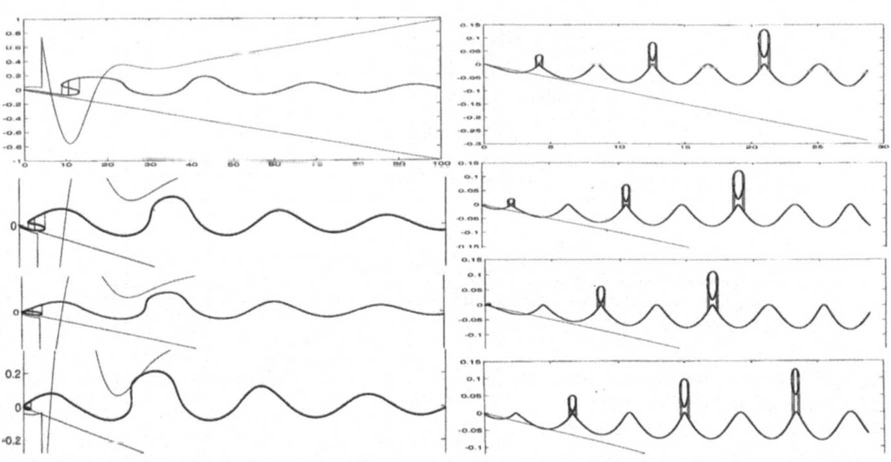

Fig. 4.15 Complexity of the near-shore wave dynamics. Picture of the coastal evolution of a harmonic wave into the shock wave and breaker right up to the formation and collapse of the bubble (left). Picture of the coastal evolution of the harmonic wave into cnoidal waves topped with jet (right)

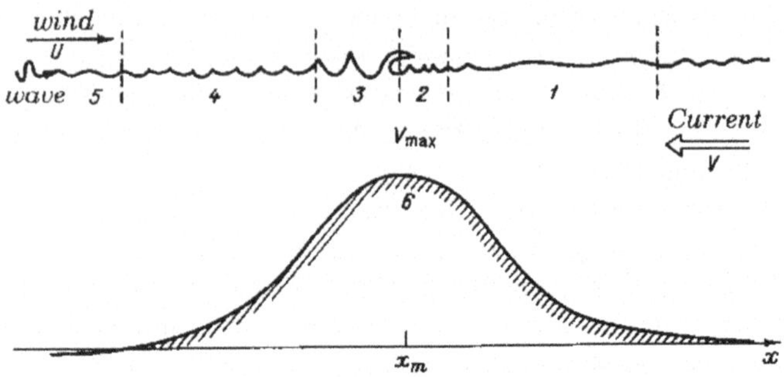

Fig. 4.16 Scheme of zones where different type of waves are generated: 1—calm water; 2—transition zone; 3—breaking zone; 4—whirlpool; 5—background waves; 6—submarine bank; x_m—point of the maximum speed [7, 13]

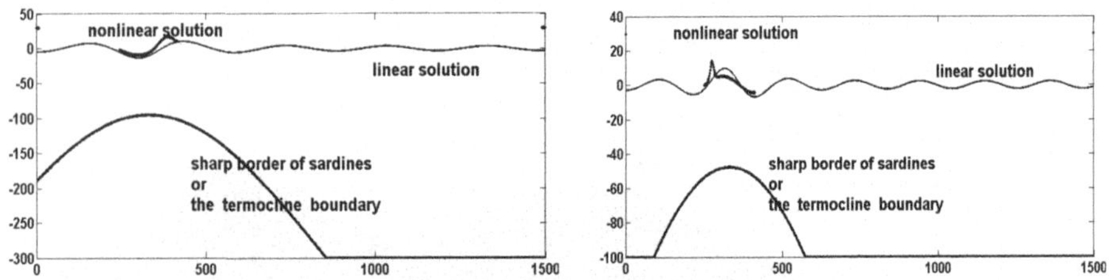

Fig. 4.17 Resonant amplification of ocean wave because of the formation and vertical displacement of a school of sardines or the termocline [7, 13]

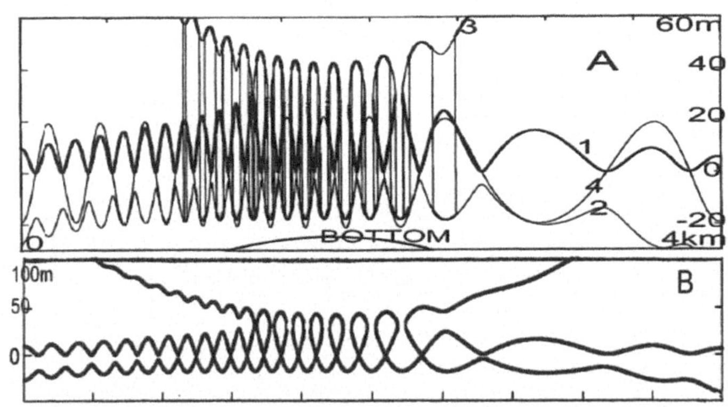

Fig. 4.18 Catastrophic intensification and formation of extreme multivalued ocean waves when a wave passes an underwater hill

The first results of a study of the formation of extreme multivalued waves over underwater topographies were published in 2011 [13]. When modeling these waves, an equation of the form (4.20) was used, written taking into account a slight change in ocean depth. Figure 4.18 A shows curves 1, 2, and 3 corresponding to three analytical solutions of the cubic algebraic equation. Curve 4 defines a linear wave coming from the ocean. It is clear that these solutions do not interact in any way far from the topography. However, above the topography, they begin to interact to form extreme multivalued waves. Figure 4.18b shows the curves constructed according to the numerical solution of the Eq. (4.20). It can be seen that different calculation methods give similar, amazing wave patterns above the topography. They correspond to the breaking of waves and the destruction of the ocean surface, that is, the appearance of spray domes and huge jets of water-air mixture over the underwater topography.

Of course, the parameters of the harmonic wave used in calculating the curves Fig. 4.18 lie far beyond the amplitudes of real ocean waves. However, such waves can occur in the rarest situations, for example, at some distance from the site of a meteorite impact on the ocean surface. At the same time, the destruction of the water surface and the formation of emissions of an air-water mixture on it, shown in Fig. 4.18, is a completely ordinary phenomenon that is observed by many. The actual amplitudes of the resulting waves and emissions are usually no more than a few meters.

Examples of the diversity of tsunami forms occurring near the coast are given in Fig. 4.19. Apparently, something similar can happen in the case of scalar fields and waves propagating in them. The foregoing allows us to make an attempt to classify the waves we have considered. When thinking about classification, we, of course, had in mind the classification introduced in [12] as a successful example of the classification of strongly nonlinear waves. But it does not cover multivalued waves.

Fig. 4.19 Evolution of the 2004 Indian Ocean tsunami near the coast of Sri Lanka: the form changes from the wall of water (low part), through the breaker to the localized jet (upper part) [7]

4.5 Classification of Almost Multivalued Waves

Vertically excited counterintuitive surface waves were observed in [12]. A rectangular container was used. The water depth was 0.3 m. In these experiments, the authors identified wave modes A, B, and C, which were most characteristic of their experiments (Fig. 4.20).

Indeed, these modes also appeared in experiments performed by other researchers. For example, Fig. 4.20 (bottom) shows the waves obtained in remarkable experiments [18], which fully correspond to the classification.

The waves were excited in a vertically oscillating container. Counterintuitive profiles appeared at the container center. The wave crests erupted thin jets of water, or they were crowned by small breakers. Figure 4.20 (bottom) demonstrates these waves, which can be schematized as modes (a), (e), and (i). Mode (a) is defined by a sharp angle at the crest and by the appearance there of a jet. Mode (e) is characterized by almost-plane or slightly concave peak and by the presence of folds on the crest edges. Mode (i) has a dome-like crest.

It is easy to see that waves (a), (e), and (i) can be schematized as modes A, B, and C.

The appearance of similar modes is possible in many other cases.

In 1901, physicist Bénard discovered a strange effect. It can be observed in the following experiment: vegetable or silicone oil is placed in a shallow vessel and heated evenly from below. A temperature difference occurs between the upper and

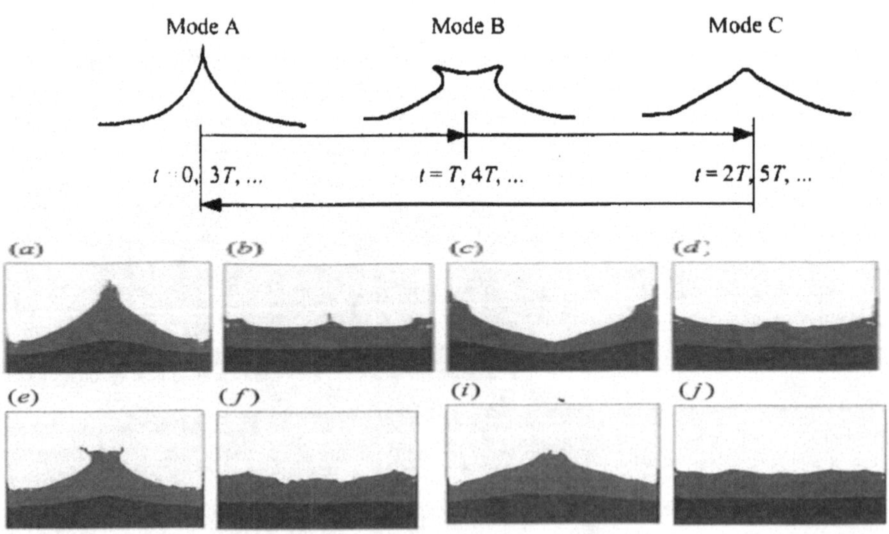

Fig. 4.20 Classification and profiles of counterintuitive wave A, B, and C [7, 12, 13] (up). Waves which can be schematized as wave (a) corresponding to wave A, wave (e) corresponding to wave B, and wave (i) corresponding to wave C [17, 18] (bottom)

lower surfaces. As a result, various phenomena arise in the vessel; in particular, wave dynamics presented in Fig. 4.21 may occur.

Another example of the development of wave structures during Bénard convection in a rectangular container is presented in Fig. 4.22.

It is clearly seen that the wave structures arising from liquid convection can resemble mode B in the classification under consideration. Thus, the classification apparently covers a wide range of wave phenomena. But of course not all.

Quadratic and cubic nonlinear terms in wave equations can form waves with different profiles. We have shown that quadratic terms determine waves with a steep, even discontinuous, front. Cubic nonlinearity can generate catastrophic waves that have folds in their profiles. In the general case, the amplitude of cubic nonlinear waves excited under resonant conditions can significantly exceed the amplitude of quadratic nonlinear waves. We emphasize that the profiles of these waves very much depend on the proximity to the resonance condition and change as the wave passes

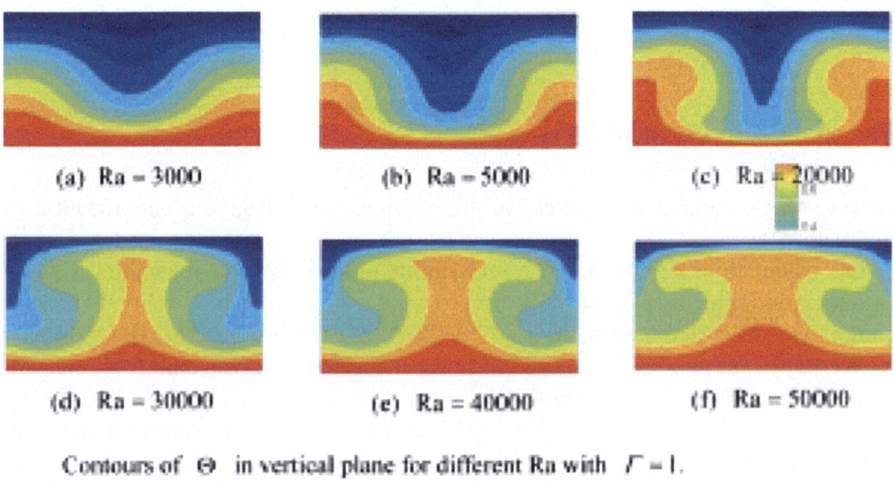

Fig. 4.21 Counterintuitive waves that arise during Rayleigh-Bénard convection in a cylinder with incompressible fluid [19, 20]

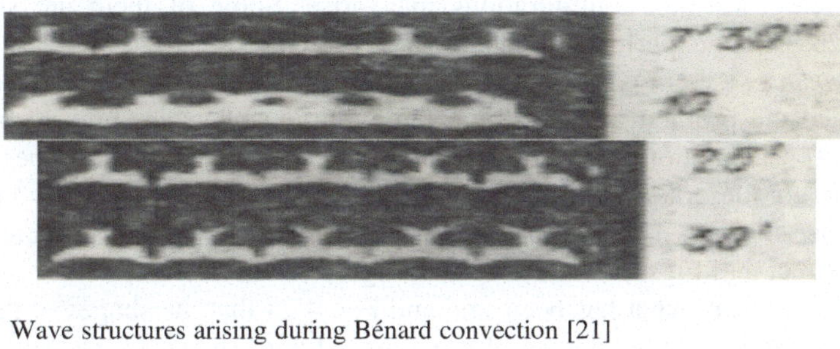

Fig. 4.22 Wave structures arising during Bénard convection [21]

through the resonant region. The classification of waves based on these factors is given in the book [7]. We will not develop it here.

The example of classification of waves based on these factors is given in the book [7]. We will develop this classification below. As the example of the amazing multivalued waves described by this classification we indicate the results presented in Sect. 2.3.4. It was found that the curves of Figs. 2.32, 2.33, 2.34, and 2.35 determune the main wave profiles arising durng the transresonance evolution of waves into the turbulence. These are the mushroom-like waves (Fig. 2.32), the profiles containing the closed curves near the wave hollows (Fig. 2.33), the overturning waves (Fig. 2.34), and the waves framed by "ears" (Fig. 2.35). Below, this classification is developed.

A new wave classification is introduced, which significantly complements the one indicated in Fig. 4.20. As a matter of fact, a classification is introduced for multivalued waves, to which this book is devoted.

This is the first classification known to us. Previously, this could not have happened, since MW were not previously considered as a separate area of wave dynamics.

4.6 Classification of MW

Despite all the doubts about luck, in this section, we will give a classification of multivalued waves.

Our calculations demonstrated the possibility of the existence of a wide variety of extreme multivalued waves (EMW). But to what extent is this possibility realized in nature, or at least in experiments?

We hope that the classification will make it easier to analyze data from calculations and experiments related to EMW.

When explaining classification, we will use illustrations that have some generalized image. At the same time, we will provide links to previous calculations in order to specify each image.

1. *Elastica-like waves* (see Fig. 4.23). The most obvious are the wave profiles which repeat the Euler figures (see Figs. 2.1 and 2.5). Wave shapes can vary from step to mushroom (see Figs. 2.6, 2.8, 2.17, 2.18, 2.19, 2.20, 2.21, and 2.22), and another. In this case, various configurations may arise. Some of them are shown in Fig. 4.23, where experimental data are also presented.
2. *Wave-vortices* (see Figs. 4.24, 4.25, and 4.26). In this book, we have repeatedly addressed the problem of describing the evolution of waves into vortices (see Sect. 2.3). Calculations have shown that multivalued waves in a number of cases qualitatively describe the beginning of the generation of vortices and even turbulence in flows [24]. To be convinced of this, it is enough to compare the vortex street and Figs. 2.9 and 2.11 with figures from Sect. 2.3.

   It is clear from what has been said and Fig. 4.24 that the shapes of the waves can be different, and the degree of proximity of their shapes to the vortex is also different. Here we will additionally consider the case of the Richtmyer-Meshkov

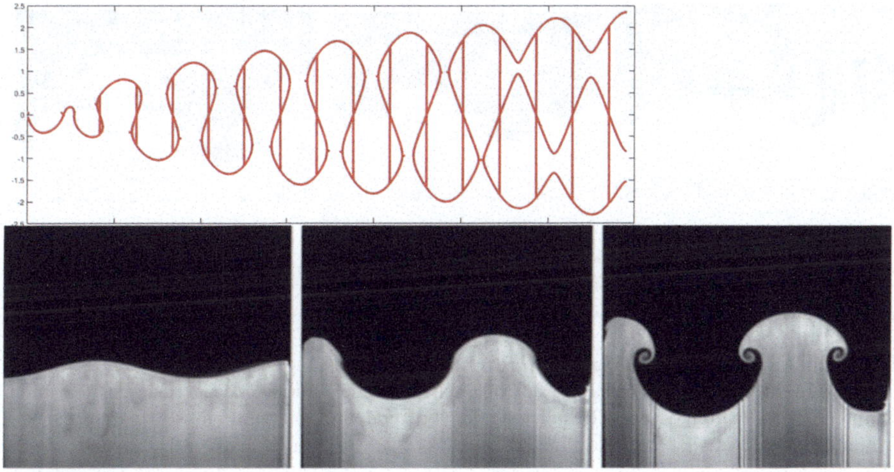

Fig. 4.23 Solutions an algebraic equation (top). A sequence of images of waves generated due to the Richtmyer-Meshkov instability of incompressible liquids [7, 13, 22, 23] (bottom; see also Fig. 3.34)

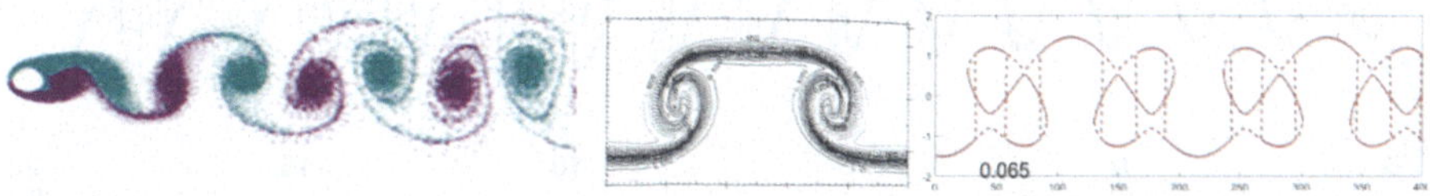

Fig. 4.24 Vortex street and results its theoretical modeling. See also Figs. 4.28 and 4.29

instability of an interface of a two-liquid system [23]. The results of instability are presented in Figs. 4.23 (bottom) and 4.25 (upper rows). The results of our calculations are shown there in the bottom rows.

Figure 4.26 shows one of the possible processes of wave evolution into vortices [25].

It is surprising that the Euler figures presented in Fig. 2.5, numbered 1, 0.5, 0.4, and 0.5, can be easily matched with the chains in Figs. 4.24 and 4.26. This view combines to some extent the mathematics describing Euler figures with the mathematics describing the emergence of the Kármán vortex chain and the theory developed in this book. The possibility of such a unified representation is illustrated in Fig. 2.55.

Wave-vortices are the most complex in our classification. They directly pass into the unstable part of the development of the wave profile. This evolution already enters the area of turbulence and lies beyond the scope of this book.

At the same time, the manifestation of this type of wave is so diverse that the following three types of multivalued waves can, to a certain extent, be associated with the wave-vortices.

3. *Waves with hair* (see Fig. 4.27). Another possibility for interpreting some of the results presented in this book related to the waves we called in Chap. 2 "waves

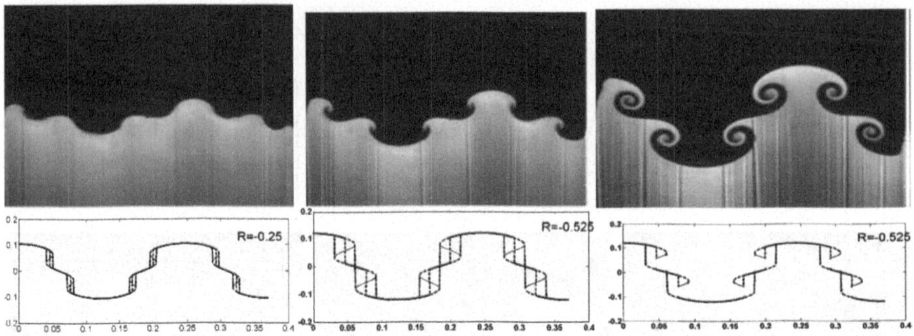

Fig. 4.25 An example of strongly nonlinear standing waves appearing on the interface of two liquids with different densities as a result of impulsive acceleration. The interface has the initial harmonic perturbation (upper). Results of analytic modeling (bottom). R is the transresonant parameter [7, 13, 22, 23] (see also Fig. 3.30)

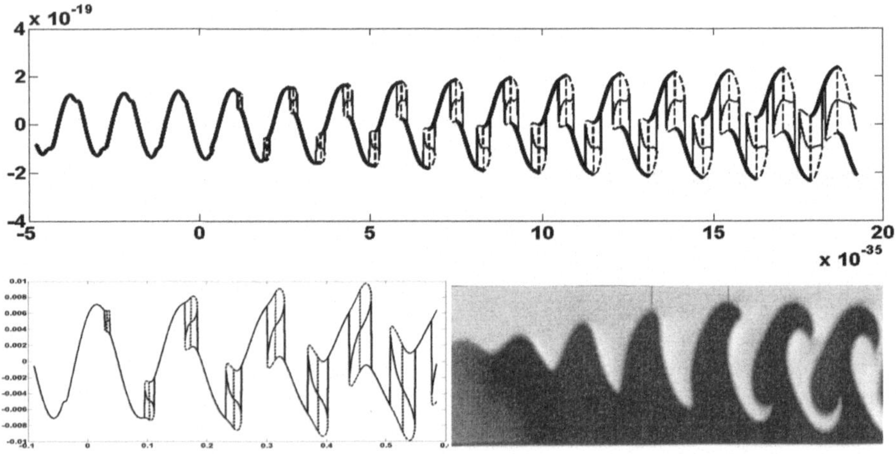

Fig. 4.26 Processes of the evolution of waves into vortices

with hair" (Figs. 3.37, 3.38, and 3.39). See also the evolution shown in Fig. 2.12. We presented in Fig. 4.27 a series of calculations (Figs. 2.37, 2.38, and 2.39). Calculations are illustrated by experimental data (bottom).

In Fig. 4.27, we presented a comparison of theoretical results, which can be interpreted as the appearance of the wave with hair (with hairstyle and experimental data [17–19, 26]. As in the case of wave-vortices, in this classification, we are trying to briefly define a very wide and at the same time rapidly variable (unstable) range of wave phenomena.

In Sect. 2.3.4, waves framed by the loops were considered ("ears"). On the whole, the curves from Sect. 2.3.4 agree with the data presented in Figs. 4.27 and 4.28.

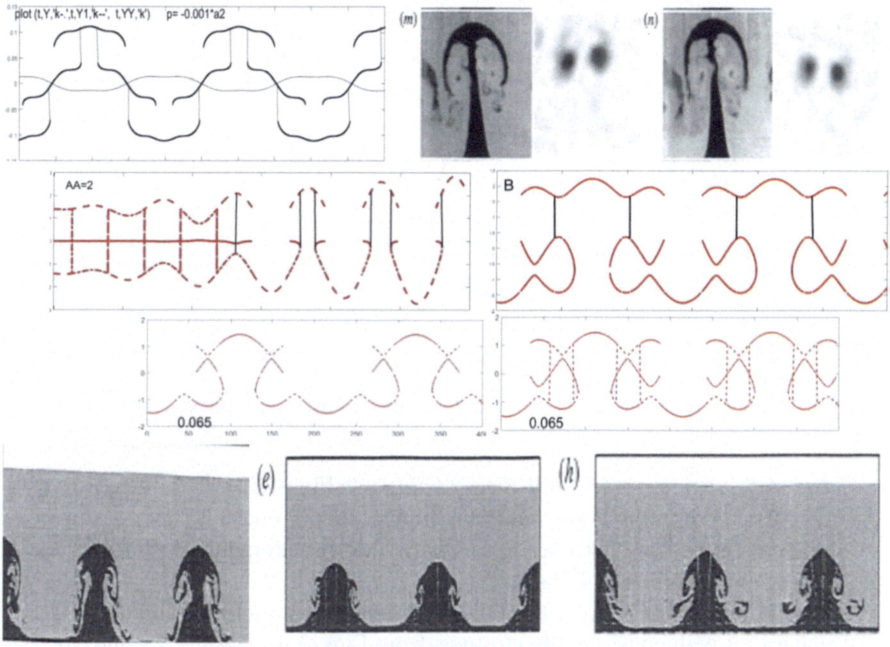

Fig. 4.27 Results of calculations and experiments. See also, the Sect. 2.5.2

Fig. 4.28 Two examples of the complex evolution of particle-waves and good agreement between the calculation results and experiment

4. *Wave-particles* (see Figs. 4.28 and 4.29). The fourth case is when, in the process of evolution, the wave profile goes beyond the scope of Euler theory and transforms into particle-waves (Fig. 4.28). In this case, various configurations may arise (Figs. 2.20, 2.21, 2.40, 2.41, 2.42, 2.43, 2.44, 2.45, 2.46, 2.47), and many others.

   Several more wave chains, additional to those indicated, are presented in Fig. 4.29.

   Subsequent chapters of the book will cover, to a large extent, the particle-wave. Let us single out isolated particle-waves among them.

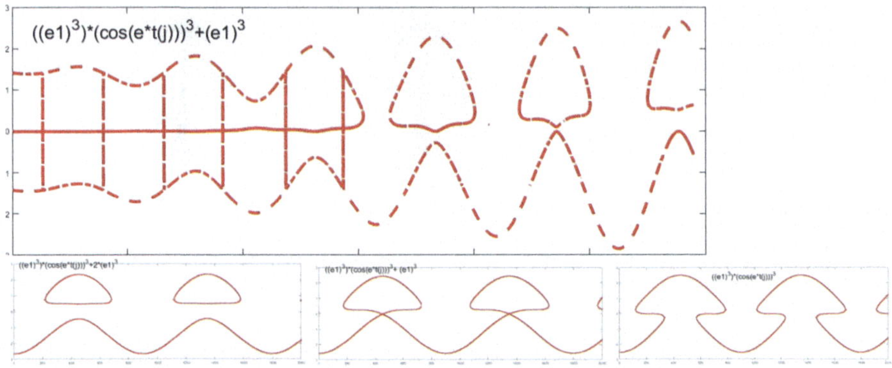

Fig. 4.29 Examples of the evolution of wave chains formed by wave particles

Isolated particle-waves and their chain (see Fig. 4.30). The idea is consistently carried out in the book that the nonlinear theory opens up the possibility of describing elementary particles not as harmonic functions but as particle-waves (see as examples, Figs. 1.4, 2.17, 2.18, 2.47, 2.49, 4.2, 4.3, 4.4, 4.5, 4.6, 4.7, 4.8, 4.9), and others. The possibility of the emergence of particles during the collision of scalar multivalued waves also follows from Figs. 3.43, 3.44, 3.45, and 4.1. It is also illustrated by Fig. 4.30 (left) taken from the book [22].

5. *Quantum particles* (see Fig. 4.31). It is known that within the framework of continuum mechanics, it is assumed that everything changes continuously. In particular, all differential equations are derived on this assumption, for example, Maxwell's theory of electromagnetism. At the same time, it is obvious that when breaking up the material environment, at some stage of the division, we will begin to deal with atoms, that is, with isolated particles that are not materially connected to each other. The same can be said for waves. Let us consider, for example, the vibrations of a string. Obviously, the length of the waves arising on it cannot be less than the distance between the atoms. To overcome the uncertainty that arose, Planck suggested that, starting from a certain energy level, oscillations exist not in the form of continuous waves, but in the form of isolated portions of energy. We emphasize that more than 100 years have passed since Planck introduced his postulate. Now no one doubts that matter consists of molecules and atoms. Therefore, Planck's postulate became obvious. At the same time, it seems that the process of decay of continuous oscillations into quanta (isolated particles) has not yet been studied mathematically. More accurately, we can say that the process of evolution of continuous waves into quanta has not been studied, although multivalued waves, in particular particle-waves, have been studied by the author for more than 20 years. Only in this book, we pay some attention to quanta (see Figs. 2.44, 2.46, 2.51, 2.53, 4.2, 4.3, 4.4, 4.5, 4.6, 4.7, 4.8, and 4.9).

The scalar field equations contain a constant C (see (1.13) or (4.2)), which is usually defined as the stationary part of the field energy. For example, for the case of

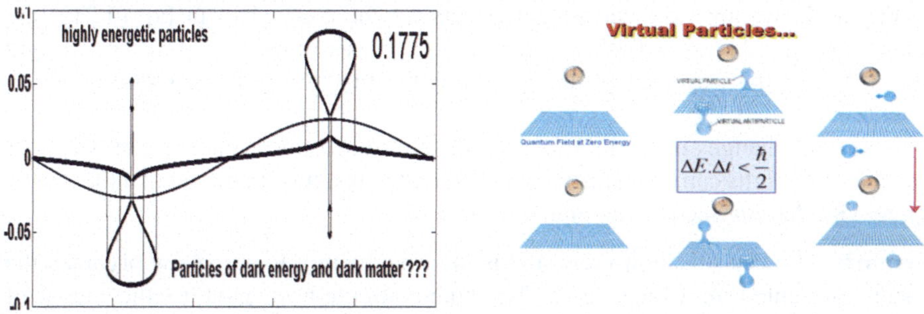

Fig. 4.30 Possible resonance mechanism for the emergence of particles, which corresponds to Fig. 4.1 (left). Possible mechanism for the emergence of particles [17] (right)

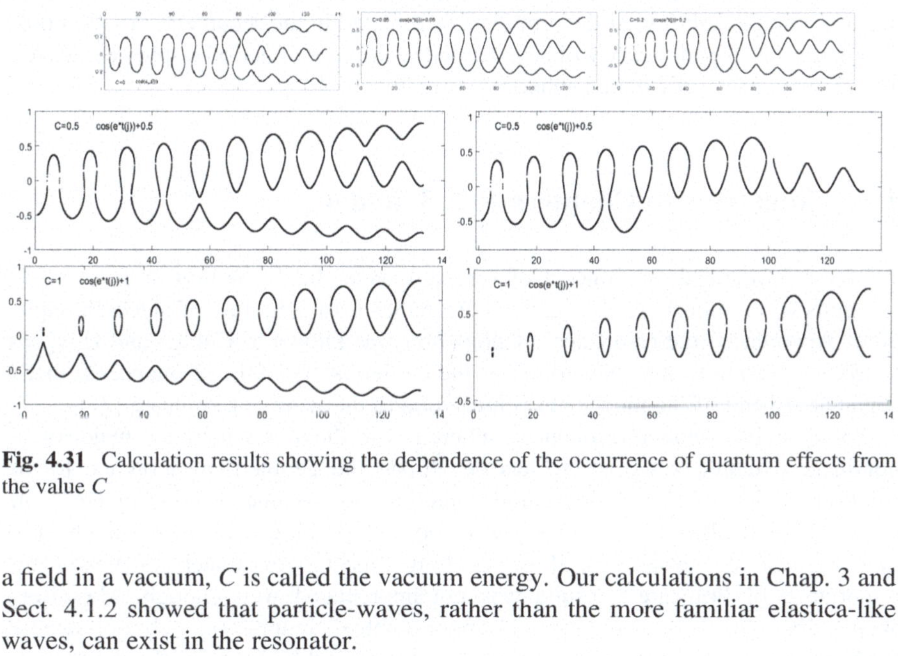

Fig. 4.31 Calculation results showing the dependence of the occurrence of quantum effects from the value C

a field in a vacuum, C is called the vacuum energy. Our calculations in Chap. 3 and Sect. 4.1.2 showed that particle-waves, rather than the more familiar elastica-like waves, can exist in the resonator.

In Sect. 4.1.3, we proceeded from the fact that the emergence of quanta is associated with the achievement of oscillation energy of a magnitude on the order of the stationary energy of the scalar field, that is, the value C. This energy can be achieved by increasing the frequency, but it is easier for us to change the oscillation energy by increasing the amplitude of the wave. Some results of the appearance of wave energy quanta are presented in Figs. 4.6, 4.7, 4.7a, 4.7b, and 4.8. Additional data is shown in Fig. 4.31.

We are interested in the influence of C on the occurrence of quanta.

We illustrate the transresonant evolution of forced waves from harmonic forms through the Euler elastica figures to particle-waves (left) or quana (right) to cnoidal waves and again harmonic forms (Fig. 4.31).

Figure 4.31 shows the calculation results for the case when in Eq. (4.21a) the value of C varied ($C = 0$, 0.05, 0.2, 0.5, and 1). We assumed $A = 0.05$, $c = 1$, and $\omega = 0.5$. The transresonant parameter R was the function of r and reduced slowly from 0.008 to -0.296.

In certain changes of C, the Euler figures are transformed into particle-waves or quanta. The results can be interpreted in different ways. It is important that the results agree with data in Sects. 4.1.2 and 4.1.3.

Remark The classification presented is the first attempt to somehow organize the results presented in Chaps. 2–4. The main attention is paid to the study of multivalued periodic waves arising at fixed resonant or near-resonant frequencies or multivalued waves arising when the frequency changes in the resonant regions of the corresponding wave systems. Of course, the listed cases do not exhaust the entire diversity of multivalued waves. Some additional cases will be considered in connection with the solution of the problems formulated in the second part of the book. There, the developed theory will be extended to the case of periodic spherical waves, as well as to cases of isolated spherical particle-waves.

4.7 Comments to Chapters 1, 2, 3, and 4

We have introduced the concept of resonant scalar fields. Within its framework, solutions of nonlinear wave equations are reduced to solutions of algebraic equations. Apparently, the resonance phenomenon was known to Galileo and Huygens nearly four centuries ago. Nevertheless, the modern story of the resonances actually began at the end of the nineteenth century due to the works of Poincaré [27].

Poincaré has shown resonances inherent for the overwhelming majority of nonlinear dynamic systems. He has developed the geometrical approach to the solution of the nonlinear differential equations, approximately, during the same time. All of it allowed him to suggest ways of probing a behavior of physical systems near the resonances. However, long time (approximately 50 years), the importance of Poincaré's results was not appreciated well enough. Max Born skeptically noticed: "It would indeed be remarkable if Nature fortified herself against further advances in knowledge behind the analytical difficulties of the many-body problems." [28].

Really, in some cases mathematicians and physicists could overcome technically this difficulty. For example, the well-known the "renormalization method", developed R. Feynman, J. Schwinger, and F. Dyson in 1944–1949, has allowed to eliminate from the theory the divergence arising at the solution of the equations by the perturbation method. We notice that this method, has laid down in the basis of the theory of quantum electrodynamics, which possesses the fantastic accuracy and the predictive force (Explanation from Wikipedia: Renormalization is a collection of techniques in quantum field theory, statistical field theory, and the theory of self-similar geometric structures, that are used to treat infinities arising in calculated quantities by altering values of these quantities to compensate for effects of their self-interactions).

However, when this method had been offered, some large physicists considered its as "dishonest". Indeed, the renormalization procedure does not work successfully during the solution of the similar problems arising during the development of "a theory of everything". The well-known physicist Richard Feynman was investigating till the end of his life the problem of the catastrophic infinity, which does not allow to connect the gravitation and the quantum mechanics: "On my sensations-though I can and be mistaken-exists not one mode of the decision of this problem. I don't think that there is only one mode which we can get rid of infinity»" [29, p. 144].

The development of science after Poincaré has shown, that his approach opened a way for new, statistical (chaotic) the formulation of the laws of dynamics. In particular, slightest changes of initial conditions can lead to chaotic development of many systems [27, 30, 31]. However, it became obvious [32] only almost sixty years after Poincaré due to discoveries of great mathematician A.N. Kolmogorov, developed by V.I. Arnold and J.K. Mozer.

Infinities arise in physical systems when certain resonance conditions are met. Despite these infinities, resonances, which at the beginning of the twentieth century became the basis of the development of means of communications, information systems and methods of scientific researches, and which play a fundamental role in the physicist, were successfully studied by means of physical methods. As a result, during the last century, six physicists were awarded the Nobel Prize for the study of different resonant phenomena (Isidor Isaac Rabi (1944), Felix Bloch and Edward Mills Purcell (1952), Rudolf Ludwig Mössbauer (1961), Alfred Kastler (1966), and Luis Walter Alvarez (1968)).

According to last researches in the field of theoretical physics, the resonances, in their widest understanding, lie in the base of the device of our Universe. According to the theory of "strings," the properties of elementary "particles"—their weights and constants of various interactions—in accuracy are defined by the resonant modes of oscillations of "strings" corresponding to particles [33, 34]. On the other side, it is possible to explain by the resonances the emergence of whirlwinds and spiral galaxies (see Chap. 7).

The resonant phenomena is adjoined by researches of coherent waves. Usually, the resonant phenomena are possible only for them.

Unlike the dispersive waves, the coherent waves can accumulate the nonlinear effects. Due to the coherent properties, the corresponding disturbance does not degenerate into the chaotic noise. It defines their wide use in technical applications. As a matter of fact, all modern information technology is based on the use of the coherent waves and the reduction of the influence of the dispersion on the transfer of wave energy.

This underlines the importance of the resonant phenomena. It was found that though the physical reasons for the formation and transformation of the resonant waves can sharply differ, the equations and the analytical solutions describing these processes are often similar (see Chap. 1). An evolution of the nonlinear waves excited in finite resonators possesses surprising beauty, symmetry, generality, and, at the same time, simplicity. The resonant waves in various resonators have many common features.

It is known that there are two basic methods of the solution of nonlinear problems of mathematical physics: a numerical method and a method of perturbations.

The realization of each of them demands deep and extensive knowledge. However, they do not guarantee the success of research. In particular, they do not eliminate completely the basic difficulty of the application of the method of perturbations, namely the occurrence of small divisors (or resonances) in the solutions. Despite huge efforts of outstanding scientists, this problem, according to the data of the author, is not solved completely. In [3, 14, 35, 36], we have offered new ways of the solution. As a result, we showed the internal unity of the various poorly studied waves and the nonlinear wave processes.

All of them have the resonant nature. It had appeared that within resonant bands, we must meet great difficulties connected with the simultaneous interaction of nonlinear, viscous, and dispersive effects. However, near exact resonances, an influence of the nonlinearity becomes the most important. It sharply simplifies the research up to the solution of the algebraic equation. Singular and secular terms (growing with time and/or with coordinates) do not disappear completely, but start to bear the new information. They define the width of resonant band and frequently allow to describe new physical effects.

Thus, the concept of the resonance is very important for the book. This concept is identified with the case when the most important linear terms of the nonlinear wave equation drop to zero. For example, it takes place in a finite resonator, when along its length, the whole number of waves is excited (Chaps. 3 and 4). Then linear terms in the corresponding wave equation can disappear near boundaries. In other cases, the catastrophic amplification takes place, if some perturbation begins to move with critical speed defined by the medium (Chaps. 1 and 2).

We do not know all possibilities of Nature, and we cannot imagine all cases of resonant strengthening or suppression in it and in biological communities of perturbations. Many of them lie out of possibilities of our researches. In attempts of the analysis, we resort to highly nonlinear equations which have in singular and resonant areas a large quantity of solutions. Here we have written a little from them in simplified forms. However, it was a complex problem to derive these solutions. All of it agrees with Richard Fejmana's words:

"The next great era of awakening of human intellect may well produce a method of understanding the qualitative content of equations. Today we cannot. Today we cannot see that the water flow equations contains such things as the barber pole structure of turbulence that one sees between rotating cylinders. Today we cannot see whether Schrodinger's equation contains frogs, musical composers, or morality—or whether it does not. We cannot say whether something beyond it like God is needed, or not. And so we can all hold strong opinions either way." Richard Feynman in Volume II, Section 41, page 12 of *The Feynman Lectures on Physics*, 1964.

References

1. Amin MA, Easther R, Finkel H, Flauger R, Hertzberg MP (2012) Oscillons after inflation. Phys Rev Lett 108(24):241302
2. Antusch S, Cefala F, Orani S (2017) Gravitational waves from oscillons after inflation. Phys Rev Lett 118(1):011303
3. Galiev SU, Galiyev TS (2001) Nonlinear trans-resonant waves, vortices and patterns: from microresonators to the early universe. Chaos: an interdisciplinary. J Nonlinear Sci 11(3): 686–704
4. Walker J (1978) Drops of liquid can be made to float on the liquid. Sci Am 238:151–158
5. Couder Y, Protière S, Fort E, Boudaoud A (2005) Walking and orbiting droplets. Nature 437: 208
6. Couder Y, Fort E (2006) Single-particle diffraction and interference at a macroscopic scale. Phys Rev Lett 97:154101
7. Galiev SH (2020) Evolution of extreme waves and resonances. CRC Press, p 508
8. Bush JWM (2015) Pilot-wave hydrodynamics. Annu Rev Fluid Mech 47:245–268
9. Bohr T (2018) Quantum physics dropwise. Nat Phys 14:209–210
10. Tambasco LD, Pilgram JJ, Bush JWM (2018) Bouncing droplet dynamics above the Faraday threshold. Chaos 28:096107. https://doi.org/10.1063/1.5031426
11. Pucci G et al (2018) Walking droplets interacting with single and double slits. J Fluid Mech 835: 1136–1156
12. Jiang L, Perlin M, Schultz WW (1998) Period tripling and energy dissipation of breaking standing waves. J Fluid Mech 369:273–299
13. Galiev SU (2011) Charles Darwin's geophysical reports as models of the theory of catastrophic waves. Moscow, p 655 (in Russian)
14. Galiev SU (2003) The theory of non-linear transresonant wave phenomena and an examination of Charles Darwin's earthquake reports. Geophys J Int 154:300–354
15. Proukakis NP, Parker NG, Frantzeskakis DJ, Adam CS (2004) Analogies between dark solitons in atomic Bose-Einstein condensates and optical systems. J Opt B Quantum Semiclass Opt 6: S380–S391
16. Marts B, Hagberg A, Meron E, Lin AL (2004) Bloch-front turbulence in a periodically forced Belousov-Zhabotinsky reaction. Phys Rev Lett 93(10):108305
17. Ilgamov MA (2023) Annotation book Galiev S.U. "Extreme multivalued waves in scalar fields" planned by Springer for publication in early 2024. Herald of Ufa Scientific Center, Russian Academy of Sciences. No 3, p 115–120
18. Liu D, Lin P (2022) Two-layer liquids Interface instabilities in Faraday waves of two-layer liquids with free surface. J Fluid Mech 941:A33. https://doi.org/10.1017/jfm.2022.259
19. Galiev SU (2024) Multivalued waves and the Richtmyer-Meshkov instability as the causes of the formation of galaxies, vol 1. Herald of Ufa Scientific Center Russian Academy of Sciences, p 5–17. https://doi.org/10.31040/2222-8349-2024-0-1-5-17
20. Li P-X et al (2022) 3D DNS of laminar Rayleigh-Bénard convection in a cylinder for incompressible fluid flow. Chin J Phys 79:374–394
21. Rabinovich MI, Trubetskov DI (2000) Introduction to the theory of oscillations and waves. Moscow (in Russian)
22. Galiev SU (2015) Darwin, geodynamics and extreme waves. Springer
23. Niederhaus CE, Jacobs JW (2003) Experimental study of the Richtmyer-Meshkov instability of incompressible fluids. J Fluid Mech 485:243
24. Corcos GM (1984) The mixing layer: deterministic models of a turbulent flow. J Fluid Mech 141:139–178
25. Julien S, Lasheras J, Chomaz J-M (2003) Three-dimensional instability and vorticity patterns in the wake of a flat plate. J Fluif Mech 479:155–189
26. Morgan RV et al (2012) On the late-time of the two-dimensional Richtmyer-Meshkov instability in shock tube experiments. JFM 712:354–383

27. Poincaré H (1899) Les méthodes nouvelles de la mécanique céleste, vol 3. Gauthier-Villars et fils
28. Born M (1960) The classical mechanics of the atom. Ungar, New York
29. Greene B (1999) The elegant universe. Vintage Books, New York
30. Пригожин И (2001) Конец определенности. РХД, Москва (Prigogine I (1997) The end of certainty. The Free Press)
31. Пригожин И, Николис Г (2008) Познание сложного. Издательство ЛКИ, Москва (Nicolis G (1989) Prigogine I Exploring complexity. W. H. Freeman, New York)
32. Tabor M (1989) Chaos and Integrability in nonlinear dynamics. Wiley
33. Witten E (1985) Cosmic superstrings. Phys Lett B 153(4–5):243–246. https://doi.org/10.1016/0370-2693(85)90540-4
34. Vilenkin A, Shellard EPS (1994) Cosmic strings and other topological defects. Cambridge University Press
35. Galiev SU, Ilgamov MA, Sadikov AV (1970) Periodic shock waves in a gas. Izv. Akad. Nauk USSR, Mech. Fluid and Gas 2:57–66
36. Galiev SU (1988) Nonlinear Waves in Bounded Continua. Naukova Dumka: Kiev (in Russian); Galiev SU (1972) Forced oscillations of a nonlinear elastic layer. Izv. Acad. Nauk USSR, Mech. Solid Body 1:58–63; Galiev SU (1972) Forced longitudinal oscillations of a nonlinear elastic body. Izv. Acad. Nauk USSR, Mech. Solid Body 4:80–87; Galiev SU (1973) Nonlinear gas oscillations maintained in a closed pipe by a heat source. Soviet Aeronautics 16(2):7–12; Galiev SU, Galiyev TS (1994) Linear and discontinuous forced oscillations of flow of bubbly liquid in a deformable pipe (survey). Strength Mater 9:633–65

Part II
From Resonant Fields to Origins of Particle-Waves, Gravity and the Universe

In the first part of the book, we introduced multivalued waves into consideration. From the perspective of these waves, many problems and mysteries of modern physics can be reconsidered.

The more high-energy processes we study, the more we must take into account their nonlinear nature. An example of this is acoustics. When studying weak (sound) waves, a linear relationship between density and pressure disturbances can be used. But with an increase in the energy transferred by waves, we must take into account the fact that the relationship between pressure, density, and temperature in a wave is, generally speaking, nonlinear. The most high-energy processes take place in space, and also during the interaction and collisions of elementary particles. Therefore, the idea arises that the main area of application of the nonlinear wave theory outlined in the book is space and quantum-level processes.

Awareness of the importance of taking nonlinearities into account in the above areas of physics came in different ways. The most general equations that claim to describe the Universe are Einstein's equations, and they were originally formulated as nonlinear.

In contrast, the founders of quantum (wave) mechanics thought more in terms of particles, quanta. It was only after de Broglie that the concept of a wave entered quantum mechanics and Schrödinger derived his famous equation. However, this equation was linear!

Thus, in my opinion, about 100 years ago, the two main physical sciences of modern times were created according to different principles. Cosmology from the very beginning used the concepts of field and nonlinearities, and quantum mechanics was more based on the concepts of particles and the linear Schrödinger equation. Of course, this initial situation began to change quickly. The Dirac and Klein-Gordon equations appeared. However, the linear Schrödinger equation is still widely used.

Thus, we lead the reader to the idea that after applying the developed theory to surface waves, waves in gas, and waves in scalar fields; one can try to use the multivalued waves when considering problems of cosmology and quantum mechanics.

The importance of nonlinear and resonant processes in the emergence of quantum particles and our Universe is emphasized. On this basis, definitions of virtual and real quantum particle-waves (quantum particles) are introduced. Resonance solutions describing particle-waves of a scalar field are associated with certain parameters of the origin and evolution of the Universe. A very simplified (qualitative) scheme of such a combination and some illustrations for it are proposed.

The idea that elementary quantum particles are not point objects is used. They are local (site) oscillations of fields that are described by resonant solutions of the nonlinear Klein-Gordon equation (NKGE). These approximate solutions qualitatively describe the appearance and movement, as well as some properties of particle-waves. On this basis, an attempt was made to describe (study) various quantum problems and effects. In particular, in Chap. 6, we discuss the results of double-slit experiments. An attempt is made to explain the results of experiments based on the solutions of the NKGE.

The last chapter is devoted to the problems of cosmogony.

Modern cosmology believes that the Universe arose from a certain singularity, very small in size and shaped like a perfect sphere (Big Bang (BB) theory). Along with this model, other models that are well known to specialists have been developed. In particular, models have emerged related to the possibility of our Universe "erupting" from the pre-Universe. The fact is that in 2022 the James Webb Space Telescope (JWST) was launched into space. Subsequently, he began to send out information that contradicted the standard model of cosmology. According to new information, developed and large galaxies existed already 300 million years after the BB. Thus, the existing theory of BB has come into conflict with observations and perhaps another theory is needed. The last chapter attempts to review the approximate results obtained in the book. It is emphasized that the book, based on a single mathematical model, in a single flow, describes the wave, highly nonlinear, basic fundamental processes and properties of the Universe, both on the microscale and on the macroscale, including its emergence and evolution.

Chapter 5
Spherical Particle-Waves and Modelling of Gravity

This chapter discusses spherical particle-waves. This allows to approach a more accurate description of elementary particles. Thus, the range of problems is expanded from the previously developed theory of plane waves that combines nonlinear, resonant and quantum effects. In particular, a wide multi-valued spectrum of spherical scalar waves is considered as they propagate in quantum time space. Against this background, a new interpretation of general relativity is introduced that takes into account quantum effects. In particular, a new explanation is given for the emergence and existence of gravity, gravitational waves and inertia.

5.1 Radial Scalar Field Waves

As a model of many wave phenomena, we consider forced oscillations of a scalar field in a spherical resonator.

5.1.1 NKGE and Its Approximate Solution

The nonlinear Klein-Gordon equation (NKGE) is written as follows:

$$(r\varphi)_{tt} - c_*^2(r\varphi)_{rr} + \ r\left(m^2\varphi + m_2\varphi^2 - \lambda\varphi^3\right) = 0. \tag{5.1}$$

At the surface $r = R$, we have

$$\varphi_r = -l\cos\omega t \quad (r = R) \tag{5.2}$$

S. U. Galiev, *Extreme Multivalued Waves as a Model of Everything*,
https://doi.org/10.1007/978-3-031-67027-5_5

Here l and ω are the amplitude and the frequency. Then, it is assumed that the speed equals zero infinitely closely to the center of the sphere [1]

$$4\pi r^2 \varphi_r = 0 (r \to 0). \tag{5.3}$$

Let us

$$\xi = ct - r \text{ and } \eta = ct + r. \tag{5.4}$$

Then

$$4c^2 r^{-1} (r\varphi)_{\xi\eta} + m^2 \varphi + m_2 \varphi^2 - \lambda \varphi^3 = 0. \tag{5.5}$$

Based on our experience in solving nonlinear wave equations, we will focus on the fact that the wave field in a sphere is determined by two oppositely traveling waves.

Method of solution. The solution of (5.1) is sought as a sum

$$\varphi = \varphi_1 + \varphi_2, \tag{5.6}$$

where φ_1 is first-order and φ_2 is second-order values ($\varphi_1 >> \varphi_2$). The sum (5.6) is substituted into (5.5). Then we write equations approximately corresponding to (5.5):

$$(r\varphi_1)_{\xi\eta} = 0, \tag{5.7}$$

$$4c^2 r^{-1} (r\varphi_2)_{\xi\eta} + m^2 \varphi_1 + m_2 \varphi_1^2 - \lambda \varphi_1^3 = 0. \tag{5.8}$$

A solution of (5.7) is a sum of two traveling waves:

$$\varphi_1 = r^{-1} (J + j) . \tag{5.9}$$

Then (5.8) takes the form

$$4c^2 (r\varphi_2)_{\xi\eta} + m^2 (J + j) + m_2 r^{-1} (J + j)^2 - \lambda r^{-2} (J + j)^3 = 0. \tag{5.10}$$

Let the oppositely propagating waves not interact. In addition, we will assume that the change r relatively to changes in functions J and j is very weak. This allows us to consider r as a constant value when integrating (5.10). As a result of integration, we have

$$4c^2\varphi_{2\xi} + m^2r^{-1}\left(\eta J + \int jd\eta\right) + m_2r^{-2}\left(\eta J^2 + \int j^2d\eta\right)$$
$$-\lambda r^{-3}\left(\eta J^3 + \int j^3d\eta\right) = J_1(\xi),$$
$$4c^2\varphi_{2\eta} + m^2r^{-1}\left(\int Jd\xi + \xi j\right) + m_2r^{-2}\left(\int J^2d\xi + \xi j^2\right)$$
$$-\lambda r^{-3}\left(\int J^3d\xi + \xi j^3\right) = j_1(\eta). \tag{5.11}$$

Next we find

$$\begin{aligned} &4c^2(-\varphi_{2\ \xi} + \varphi_{2\ \eta}) + m^2r^{-1}\left(\int Jd\xi + \xi j - \eta J - \int jd\eta\right) \\ &+m_2r^{-2}\left(\int J^2d\xi + \xi j^2 - \eta J^2 - \int j^2d\eta\right) \\ &-\lambda r^{-3}\left(-\eta J^3 - \int j^3d\eta + \int J^3d\xi + \xi j^3\right) = J_1(\xi) + j_1(\eta). \end{aligned} \tag{5.12}$$

Selecting accordingly $J_1(\xi)$ and $j_1(\eta)$, (see (4.43) and (4.60)) the last expression is transformed into the following:

$$\begin{aligned} &4c^2(-\varphi_{2\xi} + \varphi_{2\eta}) + m^2r^{-1}(-\eta J + \xi j) \\ &+ m_2r^{-2}(-\eta J^2 + \xi j^2) - \lambda r^{-3}(-\eta J^3 + \xi j^3) = J_*(\xi) + j_*(\eta). \end{aligned} \tag{5.13}$$

Next, selecting accordingly $J_*(\xi)$ and $j_*(\eta)$, we get

$$\begin{aligned} &4c^2(-\varphi_{2\xi} + \varphi_{2\eta}) - 2m^2(J + j) \\ &- 2m_2r^{-1}(J^2 + j^2) + 2\lambda r^{-2}(J^3 + j^3) = J^*(\xi) + j^*(\eta). \end{aligned} \tag{5.14}$$

Now we can consider the boundary conditions (5.3) at the center of the sphere

$$\begin{aligned} &4\pi r^2(\varphi_{1r} + \varphi_{2r}) = 4\pi r^2\Big[-r^{-2}(J + j) - r^{-1}J_\xi + r^{-1}\ j_\eta + \frac{1}{2}m^2c^{-2}(J + j) \\ &+ \frac{1}{2}m_2r^{-1}c^{-2}(J^2 + j^2) - \frac{1}{2}\lambda r^{-2}c^{-2}(J^3 + j^3) + \frac{1}{4}c^{-2}J^*(\xi) \\ &+ \frac{1}{4}c^{-2}j^*(\eta)\Big]. \end{aligned} \tag{5.15}$$

At $r \to 0$, we have $J = -j$ and $J^* = j - j^*$. Then the condition for $r \to 0$ is satisfied.

Let us consider the second condition (5.2). It gives

$$\begin{aligned}&-R^{-2}[J-J(\eta)]-R^{-1}J_{\xi}-R^{-1}J(\eta)_{\eta}+\frac{1}{2}m^{2}c^{-2}[J-J(\eta)]\\&+\frac{1}{2}m_{2}r^{-1}c^{-2}\left[J^{2}+J^{2}(\eta)\right]-\frac{1}{2}\lambda R^{-2}c^{-2}\left[J^{3}-J^{3}(\eta)\right]\\&+\frac{1}{4}c^{-2}J^{*}(\xi)-\frac{1}{4}c^{-2}J^{*}(\eta)=-l\cos\omega t.\end{aligned} \tag{5.16}$$

Linear analysis. Here we will consider the case of resonant oscillations when the frequencies are such that

$$-R^{-2}[J-J(\eta)]-R^{-1}J_{\xi}-R^{-1}J(\eta)_{\eta}=0. \tag{5.17}$$

For an example, if $J = A\sin\omega c^{-1}(ct - R)$ and $J(\eta) = A\sin\omega c^{-1}(ct + R)$, $J_{\xi} = \omega c^{-1}A\cos\omega c^{-1}(ct - R)$, $J_{\eta} = \omega c^{-1}A\cos\omega c^{-1}(ct + R)$, then (5.17) gives

$$\tan\omega c^{-1}R = R\omega c^{-1}. \tag{5.18}$$

From Eq. (5.18), we obtain the resonant frequencies:

$$\omega = \theta\pi c/R. \tag{5.19}$$

According to [1, paragraph 293], we can write six values of θ:

$$\theta = 1.4303, 2.4590, 3.4709, 4.4774, 5.4818, 6.4844. \tag{5.20}$$

These values are described approximately by the formula $\omega = (N + 1/2)$. Here $N = 0, 1, 2, 3, \ldots$ The linear solutions are not valid near the frequencies

$$\omega_{N} = (N + 1/2)\pi c/R. \tag{5.21}$$

since Eq. (5.21) approximates the resonant frequencies well (5.19).

5.1.2 Quadratic Nonlinear Analysis

The wave is strongly amplified, and its amplitude becomes much larger in resonance than the amplitude of the boundary oscillations. We will not take into account the cubic nonlinear term in (5.16) here. In this case, we name (5.22)

$$\begin{aligned}&\frac{1}{2}m^{2}c^{-2}[J(\xi)-J(\eta)]+\frac{1}{2}m_{2}r^{-1}c^{-2}\left[J^{2}(\xi)+J^{2}(\eta)\right]\\&+\frac{1}{4}c^{-2}J^{*}(\xi)-\frac{1}{4}c^{-2}J^{*}(\eta)=-l\cos\omega t.\end{aligned} \tag{5.22}$$

Next we will consider the cases of resonant frequencies (5.21). We accept that at these frequencies in Eq. (5.22)

$$J(\xi) = -J(\eta), \quad J^*(\xi) = -J^*(\eta). \tag{5.23}$$

As a result of all the above considerations, we obtain the quadratic algebraic equation

$$J^2 - m^2 m_2^{-1} RJ - \frac{1}{2} m_2^{-1} RJ^*(\eta) = -lm_2^{-1} c^2 R \cos \omega t. \tag{5.24}$$

We emphasize that $J^*(\eta)$ is an arbitrary function. Changing it, we can obtain various modifications of Eq. (5.24). Consider several wave solutions of Eq. (5.24) obtained for different $J^*(\eta)$.

Let

$$\frac{1}{2} m_2^{-1} RJ^*(\eta) = -\left(\frac{1}{2} m^2 m_2^{-1} R\right)^2 - lm_2^{-1} c^2 R - \mu J_\eta - kJ_{\eta\eta} - b \int J d\eta. \tag{5.25}$$

Then Eq. (5.24) can be rewritten as follows:

$$\left(J - \frac{1}{2} m^2 m_2^{-1} R\right)^2 + \mu J_\eta + kJ_{\eta\eta} + b \int J d\eta = \\ -lm_2^{-1} c^2 R \cos \omega c^{-1} (\eta - R). \tag{5.26}$$

Effect of viscosity and shock waves. We then assume that $b - 0$ and introduce the following notation

$$\varepsilon = -lm_2^{-1} c^2 R\,, \quad \overline{R} = \frac{1}{2} m^2 m_2^{-1} R \varepsilon^{-0.5}, \\ J = \varepsilon^{0.5} (\overline{R} + F(\eta) \cos \omega c^{-1} (\eta - R). \tag{5.27}$$

As a result, (5.26) takes the form

$$\mu \left[F_\eta - F \tan \omega c^{-1} (\eta - R)\right] + k\left[F_{\eta\eta} - 2F_\xi \tan \omega c^{-1} (\eta - R) - F\right] \\ = \varepsilon^{0.5} \left(1 - F^2\right) \cos \omega c^{-1} (\eta - R). \tag{5.28}$$

In a similar way, you can obtain the equations defining $J(\xi)$ and $F(\xi)$ after which it is approximately determined φ according to (5.35) and (5.38) taking into account $\varphi_1 >> \varphi_2$.

Now let us consider special cases of Eq. (5.28) and the corresponding wave solutions.

We consider the case $k = 0$, when the influence of the viscosity is much greater than the effect of the dispersion.

Then, Eq. (5.28) is approximated by the equation presented in [1]:

$$F_\eta - F \tan \omega c^{-1}(\eta - R) = \mu^{-1}\varepsilon^{0.5}\left(1 - F^2\right)\cos \omega c^{-1}(\eta - R). \tag{5.29}$$

Let us assume that in Eq. (5.29)

$$F_\eta >> F \tan \omega c^{-1}(\eta - R)\ . \tag{5.30}$$

In this case, an approximate solution of Eq. (5.29) is

$$F = \tanh\left[c\omega^{-1}\mu^{-1}\varepsilon^{0.5}\sin \omega c^{-1}(\eta - R)\right]. \tag{5.31}$$

If $\mu \to 0$, the solution describes the shock wave profile. Shock waves of this type often name as Taylor [2].

Now using (5.27), we find

$$J(\eta) = \varepsilon^{0.5}\left\{\overline{R} + \tanh\left[c\omega^{-1}\mu^{-1}\varepsilon^{0.5}\sin \omega c^{-1}(\eta - R)\right]\cos \omega c^{-1}(\eta - R)\right\}. \tag{5.32}$$

Expression (5.38) will be written as follows:

$$\begin{aligned}\varphi \approx \varphi_1 = {} & r^{-1}\varepsilon^{0.5}\left\{\overline{R} + \tanh\left[c\omega^{-1}\mu^{-1}\varepsilon^{0.5}\sin \omega c^{-1}(\eta - R)\right]\cos \omega c^{-1}(\eta - R)\right\} \\ & - r^{-1}\varepsilon^{0.5}\left\{\overline{R} + \tanh\left[c\omega^{-1}\mu^{-1}\varepsilon^{0.5}\sin \omega c^{-1}(\xi - R)\right]\cos \omega c^{-1}(\xi - R)\right\}.\end{aligned} \tag{5.33}$$

The effect of viscosity on the wave process is illustrated in Figs. 3.37 and 3.40.

Dispersion effect and solitary waves. Let us now consider the case when the influence of dispersion increases. Here, we assume that nonlinear and dispersive effects have the same order, and they are much more than the viscosity. In this case, Eq. (5.26) gives

$$k\left[F_{\eta\eta}\cos \omega c^{-1}(\eta - R) - 2F_\xi \tan \omega c^{-1}(\eta - R) - F\right] = \varepsilon^{0.5}\left(1 - F^2\right)\cos \omega c^{-1}(\eta - R). \tag{5.34}$$

We will consider cases when the function F changes so quickly that

$$F_{\eta\eta}\cos \omega c^{-1}(\eta - R) > > -2F_\xi \tan \omega c^{-1}(\eta - R) - F\ . \tag{5.35}$$

In this case, (5.63) gives

$$F_{\eta\eta} = \varepsilon^{0.5}k^{-1}\left(1-F^2\right)\cos\omega c^{-1}(\eta-R). \tag{5.36}$$

Let

$$F = \left\{A\sec h^2\left[\chi\sin\left(\omega c^{-1}(\eta-R)+\phi\right)\right]+C\right\}\cos\omega c^{-1}(\eta-R). \tag{5.37}$$

The last expression is substituted into (5.36). From the resulting equation, we determine approximately three equations:

$$1)\ 8\chi^2 \approx q_0A;\ 2)\ 8\chi^2-2\approx -3q_0C;\ 3)\ 4q_0^{-1}C\approx 3C^2-4. \tag{5.38}$$

Here $q_0 = \varepsilon^{0.5}k^{-1}c^2\omega^2$. Let us consider the special case, when $4q_0^{-1}C << 3C^2-4$. Then approximately, $A\approx 3.5$, $C\approx -\sqrt{4/3}$, $\chi\approx\sqrt{\frac{1}{2}q_0+\frac{1}{4}}$, and

$$F = \left\{3.5\sec h^2\left[\sqrt{\frac{1}{2}q_0+\frac{1}{4}}\sin\left(\omega c^{-1}(\eta-R)\right)\right]-\sqrt{4/3}\right\}\cos\omega c^{-1}(\eta-R). \tag{5.39}$$

The influence of dispersion on wave formation, as is well known, can be very significant. When k changing from significant values to zero, the wave profile (5.39) changes from periodic solitons to harmonic waves, as shown in Fig. 5.1.

Calculations were performed for six values k. Note that the expressions were used in the analysis of surface waves and described experiments well in a wide range of parameter changes.

$$J(\eta) = \varepsilon^{0.5}\overline{R} + \varepsilon^{0.5}\left\{3.5\sec h^2\left[\sqrt{\frac{1}{2}\,q_0\ +\frac{1}{4}}\sin\omega c^{-1}(\eta-R)\right]-\sqrt{4/3}\right\}\cos^2\omega c^{-1}(\eta-R). \tag{5.40}$$

Expression (5.9) will be written in a form similar to (5.33).

The effect of dispersion on the wave process is also illustrated in Fig. 3.37.

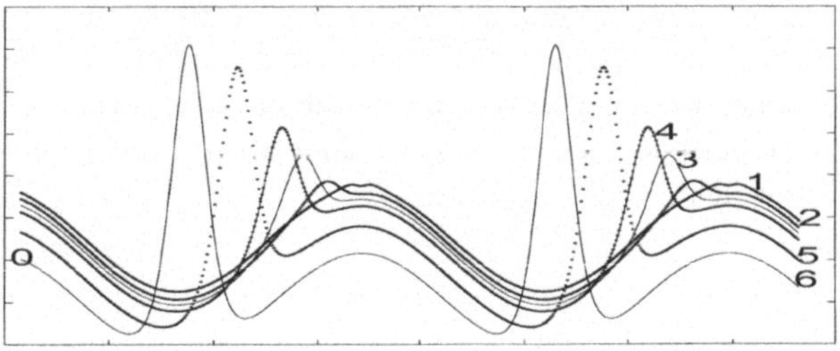

Fig. 5.1 Effect of the dispersion

5.1.3 Multiplevalued Waves (Effects of Vortex Formation, Singularities, and Motion Trajectories)

Let us substitute $\mu = 0$ and $k = 0$ into expression (5.26)

$$\left(J - \frac{1}{2}m^2 m_2^{-1}R\right)^2 + b\int J d\eta = -lm_2^{-1}c^2 R\cos\alpha c^{-1}(\eta - R). \tag{5.41}$$

Then we differentiate (5.41). As a result, we get the following equation:

$$\left(J - \frac{1}{2}m^2 m_2^{-1}R\right)J_\eta + b_1 J = lm_2^{-1}c\omega R\sin\omega c^{-1}(\eta - R). \tag{5.42}$$

We introduce the following notation:

$$\frac{1}{2}m^2\ /\ lc\ \omega = -2\omega_1, b_1\ /\ lm_2^{-1}c\omega R = b,\ lm_2^{-1}c\ \omega R = -(\gamma + 1). \tag{5.43}$$

As a result, Eq. (5.32) gives

$$(2\omega_1 - (\gamma + 1)J)J_\eta + b_1 J = \sin\omega c^{-1}(\eta - R). \tag{5.44}$$

We considered the above solution (5.32), which describes the appearance of a discontinuity in the wave if $\mu \to 0$. Let us consider the neighborhood of the indicated discontinuity using Eq. (5.44). Singularity occurs when

$$J = 2\omega_1/(\gamma + 1)\ . \tag{5.45}$$

On the other hand, when $b_1 = 0$ and $\omega_1 = 0$, we have a discontinuous solution. To consider these two features together, we must examine the neighborhood of relation (5.45). The nature of the special points can be classified in terms of ω_1 γ, b_1, and $(\eta - R)$ (Fig. 5.2).

Jimenez [3] introduced the following quantities to study the waves arising at various singular points,

$$J = 2\omega_1/(\gamma + 1) + g\quad ,\quad (\eta - R) = \phi + z \tag{5.46}$$

where ϕ is some general value of $(\eta - R)$ in the neighborhood of which we want to study Eq. (5.44). In (5.46), g and z are disturbances. Now we present the calculations obtained from the analytical solution of Eq. (5.44) in order to show the possibility of the appearance of discontinuous multivalued waves in the considered volume of the scalar field.

$\sin\phi \neq 2\dfrac{b_1\omega_1}{\gamma+1}$			I	
$b_1\omega_1 \leqslant \frac{1}{2}(\gamma+1)$	$\phi^+ = \sin^{-1}\dfrac{2b_1\omega_1}{\gamma+1}$	$\omega_1 < \dfrac{\gamma+1}{2b_1}\left\{1-\dfrac{b_1^4}{16\pi^2(\gamma+1)^2}\right\}^{1/2}$	II	
		$\omega_1 > \dfrac{\gamma+1}{2b_1}\left\{1-\dfrac{b_1^4}{16\pi^2(\gamma+1)^2}\right\}^{1/2}$	III	
	$\phi^- = \pi - \phi^+$		IV	

Fig. 5.2 Possible trajectories of the behavior of the solution near the discontinuities according to (5.43)

Figure 5.3 presents the possible trajectories (curves) of the behavior of the solution near the specified discontinuities. In particular, according to these curves, multivalued solutions are possible and not just discontinuous solutions.

Figures 5.1, 5.2, and 5.3 are drawn for $\gamma = 1.4$, although a simple change in scale will adapt them for other values of γ.

Let us now consider the waves

$$\varphi_{1r} = -r^{-2}[J(\xi) + j(\eta)] + r^{-1}[J(\xi) + j(\eta)]_r. \tag{5.47}$$

which to a certain extent can be put in correspondence with the pressure and density of the scalar field ((1.14) and (1.15)). An idea of their changes and movement trajectories is given by Figs. 5.4 and 5.5.

Figure 5.4 shows the results calculated for shock waves (5.32).

Results calculated for periodic solitons (5.40) are given in Fig. 5.5.

Thus, near and at resonance, localized oscillating spherical waves are excited, which differ significantly from the known sawtooth spherical waves [4].

From the above, the qualitative conclusion can be drawn: resonant nonlinear waves excited in a spherical resonator can resemble waves excited in long resonators.

We considered the case of the propagation and interaction of radial waves taking into account quadratic nonlinearity. The purpose of the example is to show how strongly the results of a nonlinear theory can diverge from linear analysis. Of course, when higher nonlinearities are taken into account, these results will diverge even more.

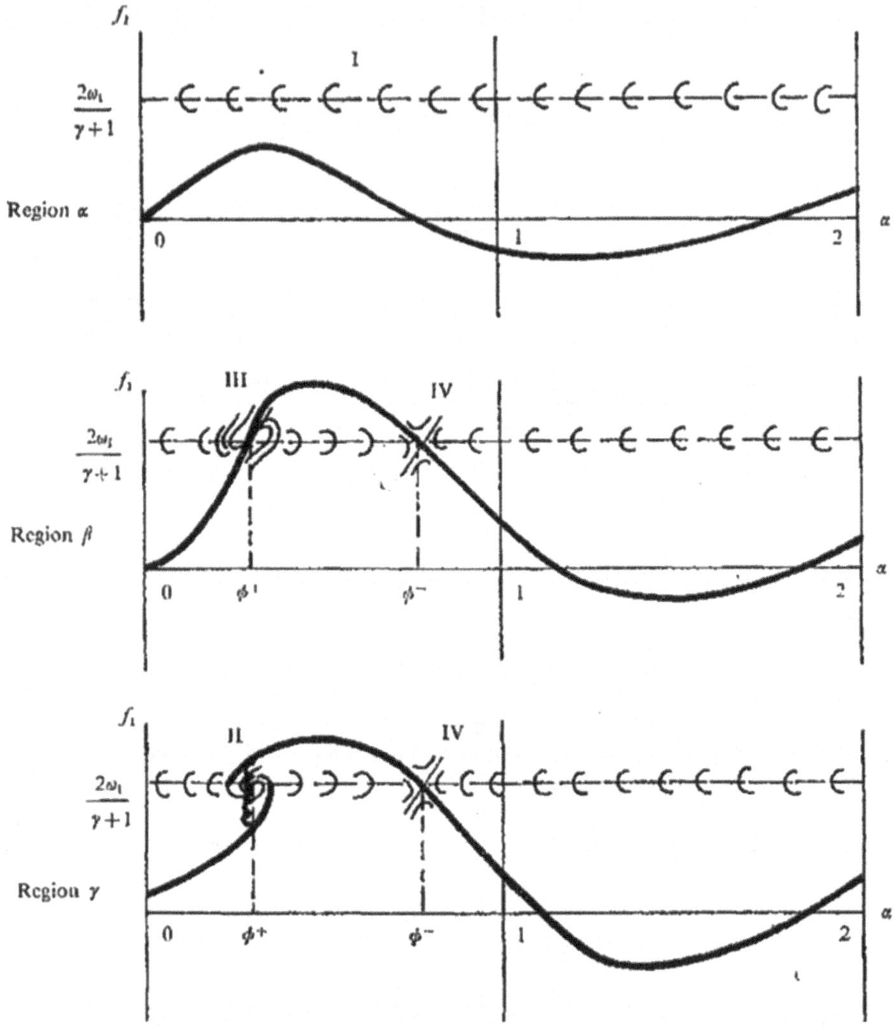

Fig. 5.3 An example of the evolution of a smooth solution (wave) into a two-valued solution (breaking wave) [1, 3]

5.1.4 Radial Cubic Nonlinear Waves of a Scalar Field

Let us return to the cubic nonlinear Klein-Gordon equation. Thus, following the calculations of Sect. 5.1.2, we will use their final Eq. (5.16) corresponding to the boundary condition (5.2). We accept as before $J(\xi) = -J(\eta)$, $J^*(\xi) = -J^*(\eta)$ (see (5.23)). As a result, (5.16) takes the form

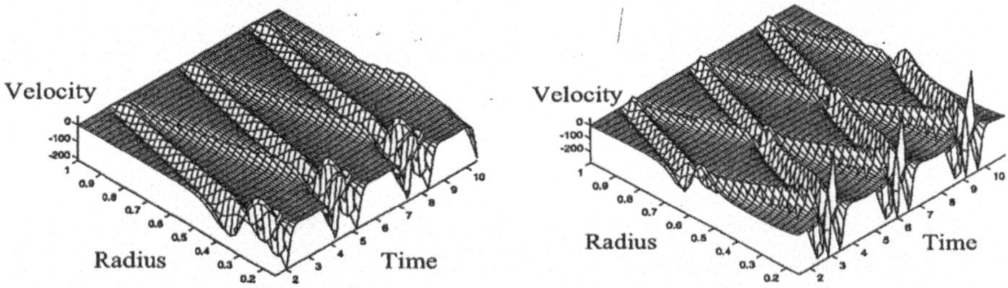

Fig. 5.4 Result calculations for the first mode (left) and the second mode (right)

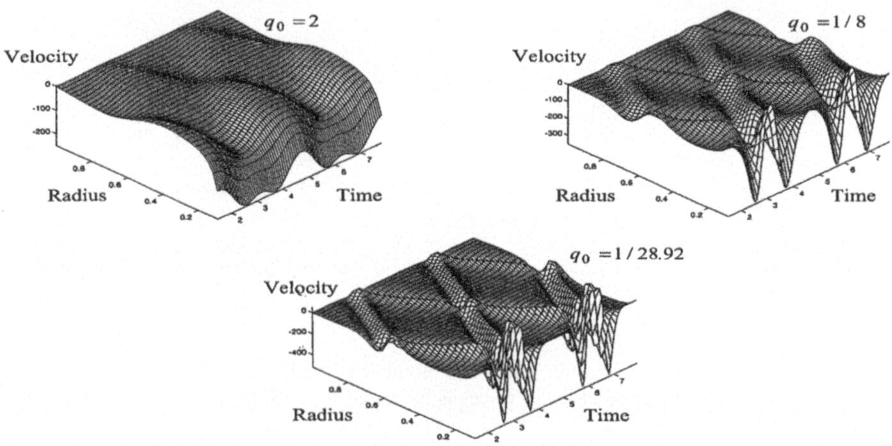

Fig. 5.5 Localization of waves due to dissipative effects

$$\begin{aligned} &- m^2 c^{-2} J(\eta) + m_2 R^{-1} c^{-2} J^2(\eta) + \lambda R^{-2} c^{-2} J^3(\eta) \\ &+ \frac{1}{4} c^{-2} J^*(\xi) - \frac{1}{4} c^{-2} J^*(\eta) = -l \cos \omega c^{-1} (\eta - R). \end{aligned} \tag{5.48}$$

Here we took into account (5.17). Let $\frac{1}{4} c^{-2} J^*(\xi) - \frac{1}{4} c^{-2} J^*(\eta) = -C$ and $m_2 = 0$. We rewrite Eq. (5.48) taking into account what has been said:

$$J^3(\eta) - m^2 \lambda^{-1} R^2 J(\eta) = \lambda^{-1} R^2 c^2 \left[-l \cos \omega c^{-1} (\eta - R) + C \right]. \tag{5.49}$$

Calculation results. Forced waves. The purpose of the calculations was to assess the influence of cubic nonlinearity on spherical waves φ in the resonator. We assumed absolutely arbitrarily in Eq. (5.49) that $m^2 \lambda^{-1} R^2 = 0.001$, $\lambda^{-1} R^2 c^2 l = 0.0001$, $\omega = 23.65$, $\omega = 23.65$, $C = 1$, and C = 0. Calculations for this case are presented on the left of Fig. 5.6. The curves on the right are calculated for similar data, but it is assumed that $C = -l$.

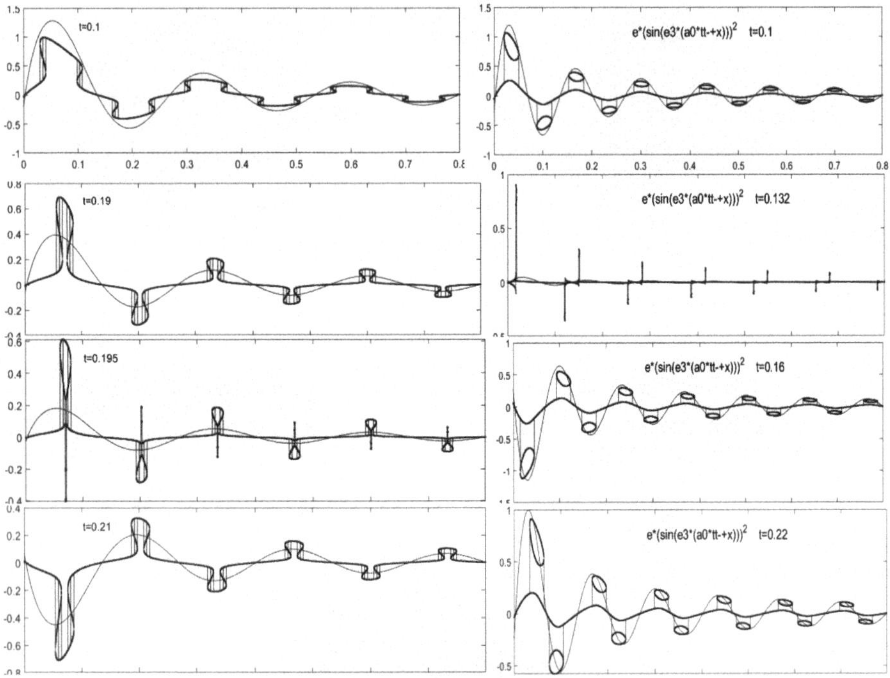

Fig. 5.6 Comparison of linear (thin lines) and nonlinear (thick lines) calculation of standing spherical waves. The amplitude of linear waves is increased by 1000 times

Calculations were performed for four moments of dimensionless time 0.1, 0.19, 0.195, and 0.21 (left), and 0.1 0.132, 0.16, and 0.22 (right). It can be seen that waves reminiscent of Euler figures (left) can appear inside a spherical particle. On the other hand (right), resonant particle-waves may appear inside a spherical particle.

When approaching the center of the sphere, the amplitude of the wave increases very strongly. At the same time, the nature of the wave change resembles the results obtained for one-dimensional, plane waves (Figs. 3.43, 3.44, and 3.45). If we do not consider that we are dealing with multivalued waves, then in general the change in multivalued waves resembles the change in a linear spherical wave, in particular spherical Bessel functions.

Let us study the influence of the initial energy of the scalar field C on the evolution of waves.

We carried out calculations based on Eq. (5.49) at constant time and $C = 0$, 0.5, 0.75, and 1. Moreover, the calculation at $C = 1$ was carried out again without printing continuous thick and thin curves. Thus, the evolution of the continuous curve $C = 0$ into a chain of isolated quanta in the case of $C = 1$ is shown.

Remark 1 From the calculation results, we can make certain assumptions about the shape of natural (natural) nonlinear spherical waves.

Let us now assume that in expression $\varphi = \varphi_1 + \varphi_2$ (5.6), there is a seed function φ_1. Then we can arrive at an equation of the form (5.8), namely

$$4c^2r^{-1}(r\varphi_2)_{\xi\eta} + m^2(\varphi_1 + \varphi_2) + m_2(\varphi_1 + \varphi_2)^2 - \lambda(\varphi_1 + \varphi_2)^3 = 0. \qquad (5.50)$$

If we choose here appropriately, then the boundary value problem can be reduced to (5.49). Consequently, some nonlinear spherical natural functions have modes reminiscent of those presented in Fig. 5.6.

As a result of this assumption, we can assert that the solutions to Eq. (5.49) describe, qualitatively, a strongly nonlinear version of the spherical Bessel functions. The above, in our opinion, explains the success of using spherical Bessel functions in analyzing the results of highly nonlinear processes. We mean the first moments of the existence of the Universe, when it increased from Planck sizes to the sizes of elementary particles. At the pressures and temperatures that existed at that time, all wave processes were extremely nonlinear. However, when analyzing the cosmic microwave background radiation, a linear version of the spherical Bessel functions is used, and apparently successfully. The success of their analysis is apparently due to the fact that linear theory usually predicts frequencies and wavelengths quite accurately. It can make a huge mistake in the amplitudes of the waves, as well as with all sorts of phase transitions in the original scalar field. The phase transitions can be determined by jumps and multivalued zones in the profiles of nonlinear waves similar to those shown in Fig. 5.6.

Remark 2 In the case of weak excitation, their further weakening leads to the fact that the amplitude of the waves excited inside the sphere increases even more strongly relative to the magnitude of the excitation amplitude. For example, during the excitation with an amplitude 10^{-21}, the waves appear with an amplitude of the order of 10^{-7}. That is, the difference in the indicated amplitudes is of the order of magnitude 10^{-14}. During the excitation with an amplitude 10^{-42}, the waves appear with an amplitude of the order of 10^{-14}. That is, the amplification of the wave amplitude is of the order of 10^{28} .

It must be emphasized that we are talking about averaged values. Near the center of the sphere, the wave amplification is maximum. We must also keep in mind that we are talking about a scalar field and the possibility of its amplification in the wave space of a spherical region. But we must also keep in mind that the pressure and field density are determined by the derivatives of the function (see (1.14) and (1.15)). From Figs. 5.6 and 5.7, it follows that this function may have jumps and ambiguity. That is, we have singularity points. At these points, the pressure and field density increase to infinite values. The above illustrates how closely the theories of infinite small and infinitely great come into contact in the problem under consideration. Indeed, solutions for scalar fields allow us, to some extent, to claim a description of wave phenomena on both cosmic and quantum length scales.

Remark 3 Folds in spherical waves have been observed recently in the plasma of solar wind. These folds were similar to those shown in Figs 5.6 and 5.7.

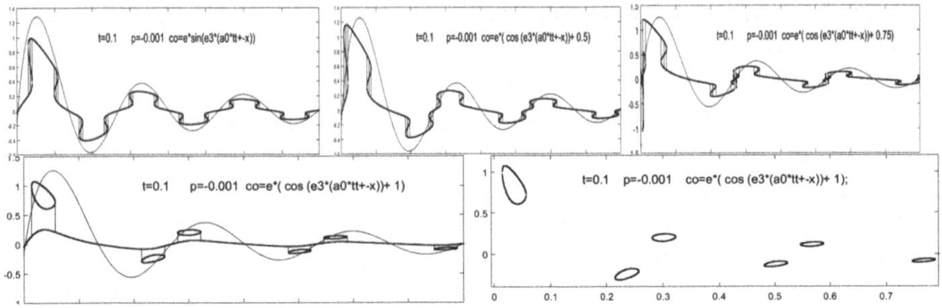

Fig. 5.7 Evolution of continuous radial nonlinear waves into a chain of quanta. The amplitude of linear waves (thin curves) is increased by 1000 times

5.2 The Casimir Effect and the Resonant Nature of the Particle-Wave Generation from Vacuum

Next, we will consider experiments in which the fact of the emergence of quantum particles from vacuum (from "nothing") is firmly established. We emphasized that during the experiments, the particles move almost in a vacuum (e.g., 10^{-8}mbar in [5]).

In 1948, Dutch theorist Hendric Casimir predicted that a generalized version of van der Waals forces would arise between two metal plates (or conducting materials) due to quantum fluctuations of the electromagnetic field (Fig. 5.8). When Casimir first calculated the effect, he used perfect "ideal" conductors. Later, more detailed calculations showed the effect for realistic conductors, and in 1997, the effect was confirmed experimentally. The most recent experiments get results to within 1% of the theoretical result. Strange as it is, the Casimir effect is very real. The Casimir effect is a great example of the strangeness of quantum theory, and how even some of its strangest predictions turn out to be true [1, 6–8].

Photons (energy) from nothing. More recently, Chris Wilson et al. [9] have tried to prove another eccentric prediction: that it is possible to use the effect to release latent energy. Instead of allowing the fluctuations to tug on the plates, you rapidly force the plates together to squeeze their wavelengths—and force out photons (see Fig. 5.8).

Generally speaking, in this experiment, vacuum fluctuation manifests itself only indirectly. However, possibly, taking into account nonlinear and resonant coupling of virtual particles allow to explane origination of the virtual particles [8].

The acoustic Casimir force. The term "acoustic Casimir force" (ACF) refers to the force between two parallel plates when they are placed in an acoustic random field [8]. This is a classical analog of the quantum Casimir force that results from quantum vacuum fluctuations (Fig. 5.9). Unlike the unbounded spectrum of the quantum case, the ACF has very interesting physical consequences. The most significant is that the ACF changes from attractive to repulsive depending on the plate separation and the frequency bandwidth.

Fig. 5.8 This scheme illustrates a prediction on how to pull energy from empty space and produce light [1, 6]

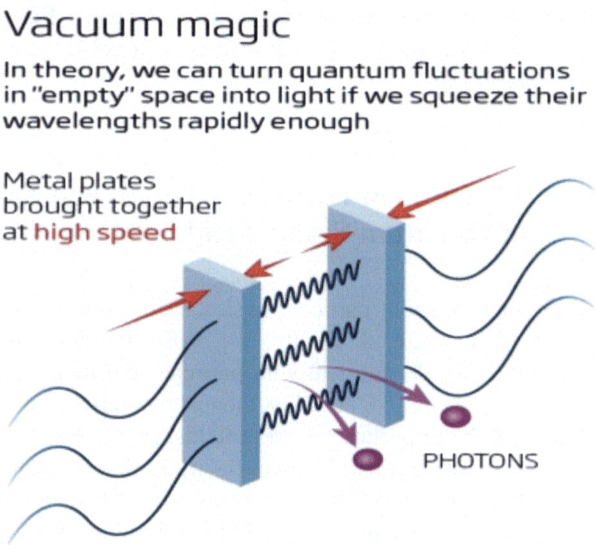

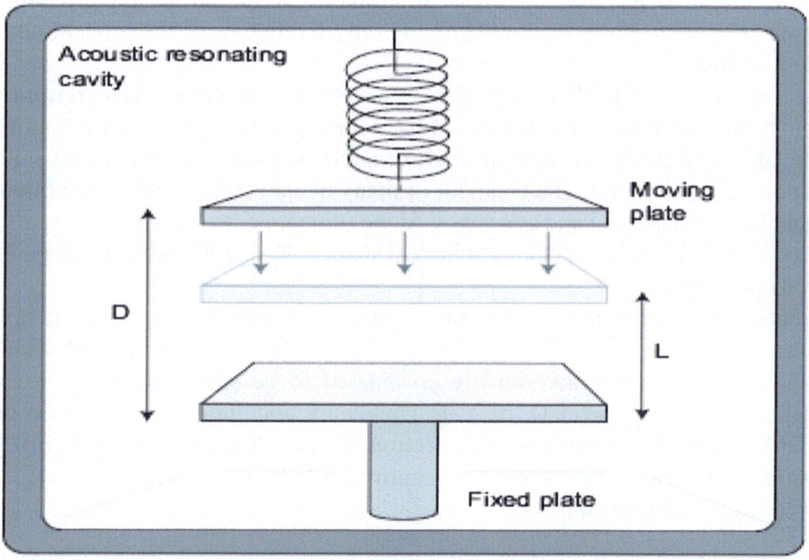

Fig. 5.9 Simple-lumped one-degree-of-freedom system considered in the calculation of the acoustic Casimir force [1, 10]

The schemes shown in Figs. 5.8 and 5.9 might be considered as some sources of the real particle-waves. If we excite one plate with certain resonant frequency of the system, perhaps, the real particles will be radiated in this case more effectively.

We emphasize that many of the results presented in this book are consistent with the indicated results of experimental studies of the Casimir effect. In particular, the appearance of resonant particle-waves is shown in Fig. 3.43–3.47; 4.1–4.8; and 5.6, 5.7. The appearance of similar effects is possible in space. Let us consider one of these possibilities.

5.3 Modeling of Particles—Waves and Some Properties of Quantum Scalar Fields

When we talk about particle physics, we do not usually emphasize that wave field physics is considered. Really, we deal with fields and corresponding equations [11]. Thus, we do not deal with particles and their trajectories. In reality, we have systems of waves and, possibly, the resonant effects of their amplification. Nonlinear effects can be important for these fields. In this case, nonlinear equations should be used to describe them. However, traditionally quantum mechanics uses linear equations. Sometimes their solutions are only some approximation to reality. The use of nonlinear equations is limited by the absence of a method for solving nonlinear problems. Therefore, in quantum mechanics, various approximate methods are used for solving nonlinear problems based on various perturbation methods. However, these methods provide only a refinement of the linear approximation results.

Let us consider NKGE (the φ^4 theory) as the fundament to the description of quantum particle-waves. We will build the solution of this equation without application of the perturbation method, and using, let it be a very rough, but initially nonlinear approximation. The behavior of many quantum particles was traditionally thought both as a particle and as a wave. At the same time, we can image the particles as very small wave packets having spherical shapes. Measuring instruments perceive them as particles.

Approximate equation. Until now, obtaining elementary particle images was notoriously difficult. It can be assumed that the particles are spherical. In particular, electrons are traditionally considered to be spherical. A group called ACME Lab, led by David D. of Yale University and John D. and Gerald G. of Harvard University, found that the electron looks spherical with an accuracy of 0.000000000000000000000000000001 centimeter [11].

We write NKGE as follows:

$$\varphi_{tt} - c_*^2 \sum_{n=1}^{N} \varphi_{x_n x_n} + \mu\varphi_t + m^2\varphi - \lambda\varphi^3 + C = 0. \tag{5.51}$$

Here the term $\mu\varphi_t$ can take into account the so-called vacuum friction (spatially separated bodies in relative motion in a vacuum experience a tiny frictional force known as quantum friction [12, 13]).

It can be assumed that this equation describes various scalar quantum fields and elementary quantum particles when coefficients a^{-2}, m^2, and λ have different values. Generally speaking, these coefficients are dependent on quantum fluctuations.

We introduce variable

$$r = R^2(t) - \sum_{n=1}^{N} (x_n - c_n t)^2. \quad (5.52)$$

Here c_n are constants and

$$R(t) = R_0 + l\cos\overline{\Omega}t \quad \text{and} \quad R_t = -l\overline{\Omega}\sin\overline{\Omega}t. \quad (5.53)$$

Here $R_0 >> l$. The variable (5.55) allows us to describe wave packets having spherical shapes. We gave

$$\varphi_t = \varphi_r r_t = \varphi_r \left[2RR_t - 2\sum_{n=1}^{N}(x_n - c_n t)(-c_n)\right],$$

$$\varphi_{tt} = \varphi_{rr}\left[2RR_t - 2\sum_{n=1}^{N}(x_n - c_n t)(-c_n)\right]^2 + \varphi_r\left[\left(R^2\right)_{tt} - 2\sum_{n=1}^{N} c_n^2\right],$$

$$\sum_{n=1}^{N} \varphi_{x_n x_n} = 4\varphi_{rr}\left\{\left[\sum_{n=1}^{N}(x_n - c_n t)\right]\right\}^2 - 2N\varphi_r. \quad (5.54)$$

Equation (5.51) yields

$$\begin{aligned} &4\varphi_{rr}\left\{\left[RR_t + \sum_{n=1}^{N}(x_n - c_n t)c_n\right]^2 - c_*^2\left[\sum_{n=1}^{N}(x_n - c_n t)\right]^2\right\} \\ &+ 2\varphi_r\left\{\left[\frac{1}{2}\left(R^2\right)_{tt} - \sum_{n=1}^{N} c_n^2\right] + c_*^2 N + \mu\left[RR_t + \sum_{n=1}^{N}(x_n - c_n t)c_n\right]\right\} \\ &+ m^2\varphi - \lambda\varphi^3 + C = 0. \end{aligned} \quad (5.55)$$

Let

$$C_1 = 4\left\{\left[RR_t + \sum_{n=1}^{N}(x_n - c_n t)c_n\right]^2 - c_*^2\left[\sum_{n=1}^{N}(x_n - c_n t)\right]^2\right\},$$

$$C_2 = 2\left\{\left[\frac{1}{2}\left(R^2\right)_{tt} - \sum_{n=1}^{N} c_n^2\right] + c_*^2 N + \mu\left[RR_t + \sum_{n=1}^{N}(x_n - c_n t)c_n\right]\right\}. \quad (5.56)$$

As a result, Eq. (5.55) is rewritten as

$$C_1\varphi_{rr} + C_2\varphi_r + m^2\varphi - \lambda\varphi^3 + C = 0. \tag{5.57}$$

Let coordinates, time, and R in (5.56) be weakly changing quantities relative to the change of φ. In this case, quantities C_1 and C_2 are considered as constants in Eq. (5.57).

We obtained an ordinary nonlinear equation containing a viscous term. The coefficients in (5.57) depend in a complex way on coordinates and time. We will go further in simplifications, assuming that coordinates and time can be considered as parameters of the equation that are in no way related to the main variable, which, of course, depends on x_n and t.

Using this assumption, we consider the linearized and nonlinear versions of Eq. (5.57).

5.3.1 Linear Analysis

In this case, we have a well-known equation describing the free vibrations of a point mass (see Sect. 2.4 of Chap. 2). The solution to this equation is well known. Let us write it in the following form:

$$\varphi = Ae^{\Omega r}\sin(\omega r + \phi). \tag{5.58}$$

Here A, Ω, and ω и ϕ are unknown constants. Expression (5.58) determines a solution in the form of a sphere-like wave packet moving in a certain direction. Following a well-known procedure, we find the unknowns:

$$\Omega = -\frac{1}{2}C_2C_1^{-1}, \quad \omega = \pm\sqrt{m^2C_1^{-1} - \frac{1}{4}C_2^2C_1^{-2}},$$

$$\phi = \arg\tan\left\{\left[C_1\left(\Omega^2 - \omega^2\right) + C_2\Omega + m^2\right]\left(2C_1\omega\Omega + C_2\omega\right)^{-1}\right\}. \tag{5.59}$$

Thus, accurate to amplitude A, we have found an expression for the scalar potential represented through previously determined quantities.

To give an idea of the shape of particle-waves described by expression (5.58), we calculate it for certain values of Ω, ω, and ϕ. ϕ. In our calculations, we assumed that C_1 and C_2 are constants and C = 0. The purpose of the calculations is to give the most approximate idea of the possibility of describing particle-waves by expressions of the form (5.58).

Figure 5.10 gives an example of particle-waves determined by the linear solution (5.61). There is a view of the particle from the side (on the left), as well as a view from above (on the right). The lower curves correspond to damped oscillations of the point mass. From Fig. 5.10, it is clear that mass fluctuations are some analog of a one-dimensional particle-wave. In general, as we believe, the nature of the oscillations of the scalar field inside the particle corresponds to the spherical Bessel functions.

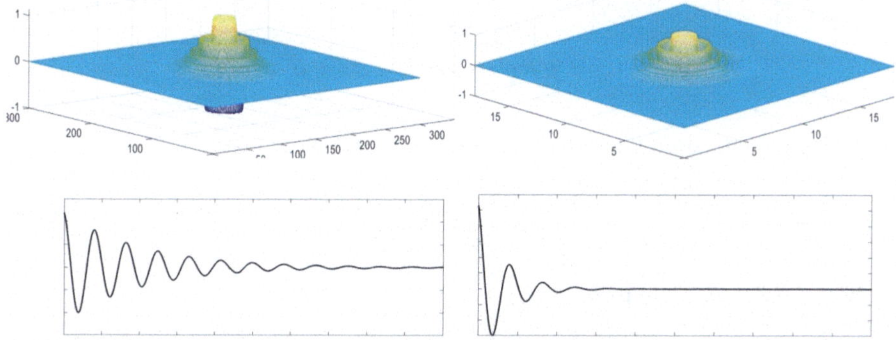

Fig. 5.10 The top two pictures define a particle-wave, and the bottom two curves define damped oscillations of a point mass

5.3.2 *Two-Valued or Twin Waves (Nonlinear Analysis)*

We will look for the nonlinear solution in a form reminding the linear solution (5.58):

$$\varphi = A\sec hB(r+\phi)\sin\overline{B}(r+\phi). \tag{5.60}$$

We wrote the solution in a form containing four arbitrary quantities. Expression (5.60) is substituted into the nonlinear Eq. (5.57). As a result, we have

$$\begin{aligned}
&C_1\left[2AB^2\sec h^3B(r+\phi)-AB^2\sec hB(r+\phi)\right.\\
&\quad-2AB\overline{B}\sec h^2B(r+\phi)\sinh B(r+\phi)ctg\overline{B}(r+\phi)\\
&\quad\left.-A\overline{B}^2\sec hB(r+\phi)\right]\sin\overline{B}(r+\phi)\\
&\quad+C_2\left[-AB\sec h^2B(r+\phi)\sinh B(r)\sin B(r+\phi)\right.\\
&\quad\left.+A\overline{B}\sec hB(r+\phi)\cos\overline{B}(r+\phi)\right]+m^2A\sec hB(r+\phi)\sin\overline{B}(r+\phi)\\
&\quad-\lambda\left[A\sec hB(r+\phi)\sin\overline{B}(r+\phi)\right]^3=0.
\end{aligned} \tag{5.61}$$

First, we consider the terms containing $\sin\overline{B}(r+\phi)$. They give the equation

$$\begin{aligned}
&C_1\left[2AB^2\sec h^3B(r+\phi)-AB^2\sec hB(r+\phi)\right.\\
&\left.-A\overline{B}^2\sec hB(r+\phi)\right]+m^2A\sec hB(r+\phi)\\
&-C_2AB\sec h^2B(r+\phi)\sinh B(r)\\
&\quad-\lambda\frac{3}{4}A^3\sec h^3B(r+\phi)=0.
\end{aligned} \tag{5.62}$$

From here, we get approximate expressions:

$$A = 0\,, \quad A_{\pm} = \pm B\sqrt{\frac{8}{3}C_1\lambda^{-1}}, \quad B^2 + \overline{B}^2 = m^2 C_1^{-1}. \tag{5.63}$$

The terms containing $\cos \overline{B}(r + \phi)$ give approximately that

$$B = \frac{1}{2}C_2 C_1^{-1}. \tag{5.64}$$

In this case, we have from (5.63)

$$\overline{B}^2 = -\frac{1}{4}C_2^2 C_1^{-2} + m^2 C_1^{-1}. \tag{5.64a}$$

Thus, to some rough approximation, we can now define three possible particle-wave amplitudes:

$$A = 0\,, \quad A_{\pm} = \pm C_2\sqrt{\frac{2}{3}C_1^{-1}\lambda^{-1}}. \tag{5.65}$$

Here the expression for ± A can be real or imaginary, depending on the coefficients C_1 and λ in (5.60).

$$A_{+} = C_2\sqrt{\frac{2}{3}C_1^{-1}\lambda^{-1}}, \quad A_{-} = -C_2\sqrt{\frac{2}{3}C_1^{-1}\lambda^{-1}} \tag{5.66}$$

Now let us determine the shift of the wave profiles determined by the amplitudes (5.66) relative to each other.

Let us first write down the expression

$$D = C_1\left(2AB^2 - AB^2 - A\overline{B}^2\right) + m^2 A - \frac{3}{4}\lambda A^3. \tag{5.67}$$

in the following form:

$$D_{\pm} = A_{\pm} C_1\left(2B^2 - B^2 - \overline{B}^2\right) + m^2 A - \lambda\frac{3}{4}A^3\,. \tag{5.68}$$

In (5.68), we took into account (5.67). Then from Eq. (5.61), it approximately follows that

$$\tan \overline{B}(r + \phi_{\pm}) = -C_2 D_{\pm}^{-1} A_{\pm}\overline{B}. \tag{5.69}$$

From here

$$\phi_{\pm} = \overline{B}^{-1} \arg\tan\left(-C_2 D^{-1} A_{\pm} \overline{B}\right) - r. \tag{5.70}$$

Obviously, two real quantities ± A determine two types of connected (similar) solutions (5.60). They describe two particle-waves existing at the same time. In other words, we can write for one particle-wave the following expression:

$$\begin{gathered}\varphi_{+} = A_{+} \sec hB(r + \phi_{+}) \sin \overline{B}(r + \phi_{+}), \quad \phi_{+} = \overline{B}^{-1} \arg\tan\left(-C_2 D^{-1} A_{+} \overline{B}\right) - r\,, \\ A_{+} = C_2 \sqrt{\frac{2}{3} C_1^{-1} \lambda^{-1}}\,, r = R^2(t) - \sum_{n=1}^{N} (x_n - c_n t)^2.\end{gathered} \tag{5.71}$$

For another particle-wave, we have

$$\begin{gathered}\varphi_{-} = A_{-} \sec hB(r + \phi_{-}) \sin \overline{B}(r + \phi_{-}), \phi_{-} = \overline{B}^{-1} \arg\tan\left(-C_2 D^{-1} A_{-} \overline{B}\right) - r\,, \\ A_{-} = -C_2 \sqrt{\frac{2}{3} C_1^{-1} \lambda^{-1}}\,, r = R^2(t) - \sum_{n=1}^{N} (x_n - c_n t)^2.\end{gathered} \tag{5.72}$$

Since $\phi_{-} = -\phi_{+,}$ we have

$$\varphi_{\pm} = \pm C_2 \sqrt{\frac{2}{3} C_1^{-1} \lambda^{-1}} \sec hB(r \pm \phi_{+}) \sin \overline{B}(r \pm \phi_{+}). \tag{5.72a}$$

In the general case, solutions (5.71) and (5.72) do not indicate any preferred localization of the place of the considered particle-waves. They can annihilate each other if they are located in the same place and $\phi_{+} = \phi_{-} = 0$. However, since ϕ_{+} and ϕ_{-} are arbitrary, they can be located as far from each other as desired.

It is also very important that the amplitude of the spherical waves concentrated in packets (5.71) and (5.72) becomes smaller and smaller with distance from the center of the packet, but nowhere does it vanish. Moreover, even if the wave packets are located very far from each other, they still lie in a single multidimensional surface, that is, they constitute a single connected system.

We hope that the solutions (5.71) and (5.72) give a qualitative description of the shape of wave packets (particle-waves) and the law of the motion of their in space. They contain several indeterminate constants and are limited, of course, by the ranges of application of using above assumptions and the initial Eq. (5.51).

If the scalar field and the scalar potential are known, one can calculate the energy pressure p and energy density ρ of this field according to expressions [14, 15]

$$p = \frac{1}{2}\varphi_t^2 - \frac{1}{6}(\nabla\varphi)^2 - V(\varphi), \tag{5.73}$$

$$\rho = \frac{1}{2}\varphi_t^2 + \frac{1}{2}(\nabla\varphi)^2 + V(\varphi). \tag{5.74}$$

We use the scalar potential $V(\Phi)$ presented as

$$V(\varphi) = \frac{1}{2}m_1\varphi^2 - \frac{1}{4}\lambda_1\varphi^4 + \overline{C}\varphi. \tag{5.75}$$

Here $\overline{C}$ is a constant.

Thus, we obtained two approximate solutions describing various wave packets, which we interpret as particle-waves. We emphasize that the resulting solution depends in a complex way on many parameters. In particular, the solution coefficients (5.60) can be imaginary. However, for the simplest understanding of the resulting solution, it can be represented as the planet Saturn. That is, like a ball with an equator extending to infinity. Below we consider only cases of real quantities.

Let us dwell on the solutions (5.71) and (5.72) additionally. It is known that the main equation of quantum wave mechanics is the Schrödinger equation. It describes a wave function that, when interpreted probabilistically, approximates a set of experimental data. One of the limitation on the use of the Schrödinger equation is well known. It is not applicable if the speed of the quantum particle is close to the speed of light.

Another model could be a round ball oscillating and moving on the surface of the ocean. No matter how small the movements and vibrations of the ball are, their wave traces, in principle, under ideal conditions, can be detected at any point in the surrounding space.

To describe relativistic particles, one can also use the Dirac equation. It turned out that the Dirac equation has two solutions with different signs: a solution with a plus sign corresponds to the positive energy, a solution with a minus sign – the negative energy. Dirac, in order to explain the existence of the second solution, suggested that in addition to particles, there are so-called antiparticles. Later it turned out that Dirac's solutions correspond to two real particles – an electron and a positron.

It also turned out that two solutions are inherent not only to the Dirac equation but also to other equations describing relativistic particles. Sometimes it turns out that the particle is the antiparticle of itself. It takes place if the two solutions of the equation coincide. For example, if $\phi_+ = 0$ and $\phi_- = 0$, then expressions (5.71) and (5.72) begin to describe two solutions that differ only in sign. They can annihilate each other. Galiev suggested that they may have some spatial or temporal shift relative to each other [16, 17]. For example, this shift can occur due to quantum uncertainty, both the localization and the lifetime of the specified wave particles (the Heisenberg uncertainty principle states that it is impossible to know the position and momentum of a particle with absolute accuracy. The more accurately you measure one quantity, the less you know about another quantity).

Comment 1 Quantum theory describes a Universe in which a particle can be in several places at the same time and instantly moves from one place to another. This concept puts a limit to our arrogance, because the world is much more complex and diverse than we can imagine based on our everyday experience [18]. As an example, consider equation (5.51) and its two solutions ((5.71) and (5.72)) describing two particle-waves. Their positions in space, in the general case, are not defined. We can imagine that these two solutions describe one particle moving along some complex, discontinuous trajectory. The particle can have a discontinuous amplitude. By analogy, in the case of a wave relativistic equation with nonlinearity to the power of 100, we can expect that it will have 100 solutions describing the simultaneous and separate existence of 100 particles. At the same time, we can imagine that in reality, these 100 solutions correspond to some multivalued wave moving along a complex trajectory. The shape and trajectory of the wave correspond to and take into account the individual properties of all 100 solutions. Let us emphasize that even in the case of a cubically nonlinear Eq. (5.51), there is a wide variety of waves (see (5.71) and (5.72)). In particular, if $\phi = N\pi/2$ ($N = 1, 3, 5,\ldots$), then φ_+ and φ_- will have the same sign. Therefore, the waves corresponding to this solution will not annihilate each other and they can be considered as particle-waves describing photons. On the other hand, for other values of ϕ_+, the waves φ_+ and φ_- have different signs and can partially annihilate each other. In the limiting case, when $\phi_+ = \phi_- = 0$, the sum $\varphi_+ + \varphi_-$ (see (5.71) and (5.72)) is zero. However, this cannot be true. Indeed, according to quantum mechanics, all quantities are defined with an accuracy of Planck's constant $\hbar = 1.054571817\ldots\times10^{-34}$ J·s. This leads to the fact that at any moment in time and at any point in space $\varphi_+ = \varphi_- = 0$.

We have considered the case of a cubically nonlinear field. Of course, this is the simplest case. Let us consider the motion of a charged particle in a vacuum, as a possible case of a highly nonlinear field. There, the particle is surrounded by a huge number of virtual particles. Their energy increases as they approach the real particle. Accordingly, the nonlinear properties of the field of virtual particles and the number of solutions describing the nonlinear, multivalued properties of this field increase.

Comment 2 We have written down solutions (5.71) and (5.72) describing the motions of elementary particles as particle-waves. The presented expressions, as one might think, determine the amplitude and location of elementary particles. However, this is not entirely true. In this book, we do not consider elementary particles as points; these are particle-waves. Therefore, as for other waves, we do not know their exact location, as well as their amplitude at any moment in time. We can only guarantee the existence of elementary particles in certain volumes determined by the solutions. In addition, the solutions include uncertain quantities. Generally speaking, they can be given some probabilistic interpretation. However, in what follows, we will consider these quantities as some precisely known constants.

Comment 3 Solutions (5.71) and (5.72) are the realization of an attempt to describe the birth and motion of elementary particles, considered as particle-waves. They describe, perhaps, the geometry of elementary particles and the wave formation existing in them. It is difficult for us to judge how successfully these solutions describe reality; however, we will use these solutions in an attempt to analyze two important aspects of fundamental physics. Thus, in Sects. 5.6, 5.7 and 5.8, we will consider the problem of gravity. In Chap. 7, the results of the double-slit experiments will be considered, which are reasonably considered "the heart of quantum mechanics."

5.3.3 Calculations Illustrating the Obtained Solutions and Describing Particle-Waves

We received the approximate solutions that can be physically interpreted in different ways. It was assumed that these solutions describe certain quantum waves which correspond to so-called quantum particles (particle-waves). Examples of such quantum particles constructed for five instants are shown in Fig. 5.11.

It is important that Fig. 5.11 models a particle-wave having a spherical shape and a rectilinear trajectory of motion. Let us consider the movement and interaction of these particles. Three moments of the movement of a particle (left) and interaction of particle-waves (right) are shown in Fig. 5.12.

It can be seen from Fig. 5.12 that the motion of quantum particles and their interaction may be presented as the motion and interaction of the particle-waves. Thus, it was shown that it might be possible to model certain aspects of quantum theory on the basis of self-consistent, scalar fields, where quantum particles are only concentrations of density of the fields. We remind that the words "quantum mechanics" mean "wave quantum mechanics."

We emphasize that the wave function (Figs. 5.11 and 5.12) does not correspond to the wavefunction well known in quantum mechanics (solutions of the

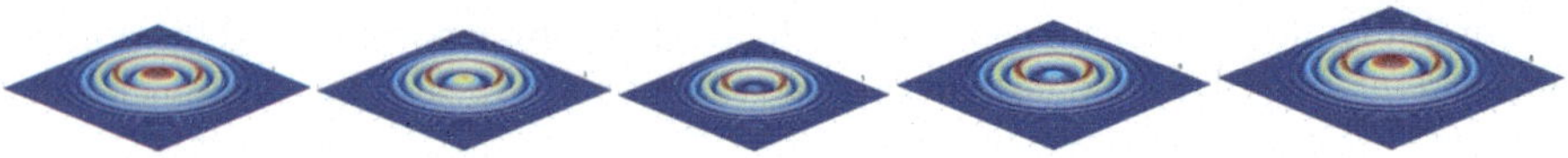

Fig. 5.11 Two-dimensional presentation of the wave packet calculated for five moments of the time

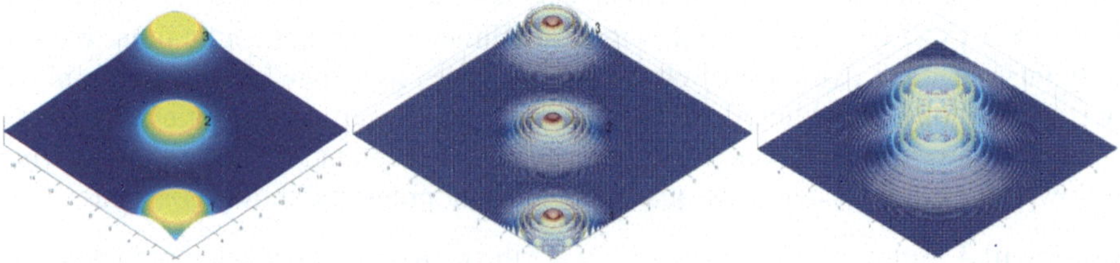

Fig. 5.12 2D picture of particle (its envelop) motion calculated for three instants of the time (left). 2D picture of wave packet motion calculated for three instants of the time (center). 2D picture of the interaction of two wave packets (right)

Schrödinger's equation). It is not also some kind of "pilot wave" (de Broglie-Bohm model). It is a nonlinear wave with a periodically oscillating center which is strongly localized near center. According to Fig. 5.11, during the half-cycle, the wave packet can be approximately considered as a particle, and during the next half-period, it is a typical wave packet. Recall that solutions (5.58), (5.71), and (5.72) were obtained for multidimensional waves. Therefore, according to our approach, any wave packet ("particle") influences the whole Universe and all "particles" of the Universe influence the wave packet.

Generally speaking, we can find the "mass" m of the wave packet according to de Broglie's theory ($m = \hbar\lambda^{-1}v^{-1}$, ħ is the Planck's constant, λ is the wave length, and v is the wave speed).

One can see that we do not agree with the Copenhagen interpretation of quantum theory, developed in the 1920s mainly by physicists Niels Bohr and Werner Heisenberg. They treat the wavefunction as nothing more than a tool for predicting the results of observations, and they caution physicists not to concern themselves with what the reality looks like underneath.

Figures 5.11 and 5.12 give only an approximate idea of the waves described by solution (5.60) since they give only a two-dimensional wave picture. In the general case, we have spherical volumes filled with ideally spherical, vibrating standing waves. The nature of vibrations and displacement of a spherical volume in space is determined by the decision arguments, which are characterized by quantities (5.52) and (5.53). To give at least some idea of the vibrations of the spherical volume (wave packet) moving in space, we give in Fig. 5.13 one-dimensional pictures of the movement of a spherical wave (packet). Now we fix our attention on the vibrations inside the sphere.

Two cases determined by the magnitude of the amplitude l in expression (5.53) are considered. For each case, three moments of movement of the wave packet in space are shown. In the case of $l = 0.005$, we have an almost stationary spherical particle; in the case of $l = 0.5$, the vibrations inside the particle are defined as an oscillating standing wave.

Thus, we described spherical wave packets spreading in space (see Figs. 5.10, 5.11, 5.12, and 5.13). These packages are very approximate solutions to the model Eq. (5.57). Of course, we are not talking here about any description of real quantum particles. Following the strategy of the book, we only assert that these wave packets can correspond to some solutions of the NKGE.

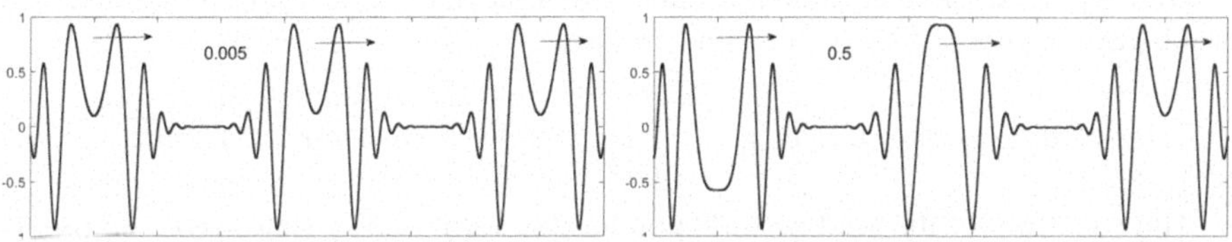

Fig. 5.13 Movements of wave packets in one-dimensional space which were calculated according to formula (5.60) for $l = 0.005$ (left) and $l = 0.5$ (right)

Comment. It is also important for us that the wave packets determined by the linear (5.58) and nonlinear (5.60) solutions do not differ significantly (compare Figs. 5.10, 5.11, 5.12, and 5.13). It is very important to emphasize that these solutions describe single-valued waves. It can be assumed that these waves determine the natural modes of vibration of the particles in question.

Thus, the linear solution (5.58) is well approximated by the nonlinear solution (5.60), which does not change its form in the linearized representation:

$$\varphi = A\,\mathrm{sec}\,hB(r+\phi)\cos\overline{B}(r+\phi). \tag{5.76}$$

In the last linearized case, we will assume that A is an undetermined quantity, and the quantities B, $\overline{B}$, and ϕ are determined according to Sect. 5.3.2.

Thus, we came to a description of elementary particles as a packet of spherical waves oscillating with a certain resonant frequency (mode). We described these packets using the linear solutions (5.58) and (5.76), as well as the nonlinear solution (5.60), as single-valued surfaces.

5.3.4 The Influence of Nonlinearity on the Formation of Discrete Spherical Multivalued Wave Packets (Particle-Waves)

We showed above that linear (5.58) and nonlinear (5.60) solutions determine qualitatively similar wave packets. This could be expected since the accepted analytical expressions for these decisions were similar. Strictly speaking, solution (5.60) is some refinement of solution (5.58).

However, this clarification does not take into account the possibility of the emergence of multivalued structures. Let us consider this possibility.

Inclusion in consideration of the possibility of multivalued solutions of Eq. (5.51) significantly complicates its analysis. We will use in our analysis the approaches developed, in particular, in Sects. 2.4 and 2.5 and Chaps. 3, 4, and 5. We will illustrate this approach using the example of Eq. (5.57). We assume that

$$\varphi = \varphi_1 + \varphi_2. \tag{5.77}$$

Here φ_1 is the initial seed perturbation. We assume $\varphi_1 << \varphi_2$. Expression (5.77) is substituted into (5.57). As a result, we have

$$C_1(\varphi_1+\varphi_2)_{rr} + C_2(\varphi_1+\varphi_2)_r + m^2(\varphi_1+\varphi_2) - \lambda(\varphi_1+\varphi_2)^3 + C = 0. \tag{5.78}$$

The equation for the seed perturbation has the form (5.57). Below we will use its linear approximate solution (5.76). For φ_2 we have

$$\varphi_2^3 + 3\varphi_1\varphi_2^2 + \left(3\varphi_1^2 - \lambda^{-1}m^2\right)\varphi_2 + \varphi_1^3 - \lambda^{-1}(C_1\varphi_{2rr} + C_2\varphi_{2r}) - \lambda^{-1}C = 0. \tag{5.79}$$

Thus, the posed research problem is reduced to solving Eq. (5.79). However, we do not have a mathematical tool to solve this equation. At the same time, our goal in this section is not to solve Eq. (5.79). Our task, first of all, is to demonstrate the possibility of the appearance of multivalued spherical wave packets in scalar fields. To realize this goal, we assume that

$$\varphi_2^3 + 3\varphi_1\ \varphi_2^2 + \left(3\varphi_1^2 - \lambda^{-1}m^2\right)\varphi_2 + \varphi_1^3 - \lambda^{-1}C > > \lambda^{-1}(C_1\varphi_{2rr} + C_2\varphi_{2r}). \tag{5.80}$$

In this case, Eq. (5.79) gives

$$\varphi_2^3 + 3\varphi_1\varphi_2^2 + \left(3\varphi_1^2 - \lambda^{-1}m^2\right)\varphi_2 + \varphi_1^3 - \lambda^{-1}C = 0. \tag{5.81}$$

We can now move on to describing these packets as multivalued wave surfaces. Expression (5.76) is used as the seed perturbation.

Equation (5.82) was written in the following highly simplified form:

$$\varphi_2^3 - \lambda^{-1}m^2\varphi_2 + A\sec hB(r + \phi_+)\cos\overline{B}(r + \phi_+) - \lambda^{-1}C = 0. \tag{5.82}$$

Thus, we took into account what $\varphi_1 << \varphi_2$ and accepted what $\left[\sec hB(r + \phi_+)\cos\overline{B}(r + \phi_+)\right]^3 \approx \sec hB(r + \phi_+)\cos\overline{B}(r + \phi_+)$. The last approximate equality is valid since A is an arbitrary quantity in (5.76) and $\varphi_1 << \varphi_2$. We have written down the equation for the particle-wave. The equation for antiparticle-waves is written similarly.

Numerous examples of the use of seed perturbations in solving nonlinear wave equations are given in the above chapters. Typically, a linear equation of the equation being studied was used as a seed (see, for example, Sects. 2.4 and 2.5 and Chaps. 3, 4, and 5). Examples of using a nonlinear solution as a seed solution are given in Sects. 3.3 and 3.4.

We will be particularly interested here in the wave dynamics within the wave. Let us emphasize once again that the following calculations are model ones and are intended to show the possibility of the evolution of the waves presented in Figs. 5.10, 5.11, 5.12, and 5.13 in multivalued particle-waves. It is almost impossible to do this since I do not have the ability to represent a multivalued spatial structure on a plane. In order to solve the problem to some extent, we will carry out calculations without taking into account the fact that the variable r depends on Cartesian coordinates (see (5.52)). We will assume that it varies linearly over a certain segment of length L.

The main constant parameters for which calculations will be carried out are as follows: $\lambda^{-1}m^2 = 10^{-14}$and $A = 10^{-21}$. By varying the values of B, D, and $\lambda^{-1}C$ in Eq. (5.82), we will obtain different wave profiles.

The wave profiles calculated for $B = 0.5^{*}10^{40}$ and $\lambda^{-1}C = 0$ are given in Eq. 5.82. In these cases, the value of D was equal to $0.45^{*}10^{41}$, $0.39^{*}10^{41}$, or $0.35^{*}10^{41}$. The results obtained do not lend themselves to simple interpretation. In the simplest sense, they should be considered as a multivalued one-dimensional analog of the two-dimensional wave patterns shown in Figs. 5.11 and 5.12. Thus, taking into account strong nonlinearity leads to the fact that the radial sections of the rings in Figs. 5.11 and 5.12 having the form of a harmonic function, as a result of taking into account strong nonlinearity, evolves into Euler figure no. 7 shown in Fig. 2.1.

On the other hand, Eq. (c) is quite similar to many of the equations discussed above. As an example, let us indicate Eq. (2.95). His solutions (see, for example, Fig. 2.35) qualitatively describe the results of many experiments with surface waves. Thus, we have no doubt about the accuracy of the solution to Eq. (5.82). We also have no doubt that it is possible to excite surface waves corresponding to those presented in Fig. 5.14.

The problems here are not in the accuracy of solutions to Eq. (5.82), but in the correctness of procedure for deriving this equation from Eq. (5.51). The problem is also how to visually represent the correspondence between one-dimensional multivalued curves 5.14 and the spherical multidimensional multivalued wave packets described by them, initially described in Cartesian coordinates x_n (5.52).

All wave patterns presented in Fig. 5.14 are determined by the Euler figures located at their centers and the curves 1, 2, and 3 bordering them. These curves correspond to three solutions of Eq. (5.82). From these three curves, let us choose the only one that will uniquely determine the wave profile outside the Euler figures. Three such options are shown in Fig. 5.15.

The wave profiles presented in Fig. 5.15 can be interpreted differently. We propose to consider them as multivalued waves propagating in various scalar fields. In particular, in a field characterized by values D equal to $0.45^{*}10^{41}$, $0.39^{*}10^{41}$, and $0.35^{*}10^{41}$. With this consideration, we can consider that the case $D = 0.45 * 10^{41}$ corresponds to the propagation of a wave in a positive scalar field, the case $D = 0.39 * 10^{41}$ corresponds to the propagation of a wave in a neutral field, and the case $D = 0.35 * 10^{41}$ corresponds to the propagation of a wave in a negative scalar field.

Qualitatively, the nature of the movement of the considered wave spherical ambiguous wave packet is shown in Fig. 5.16.

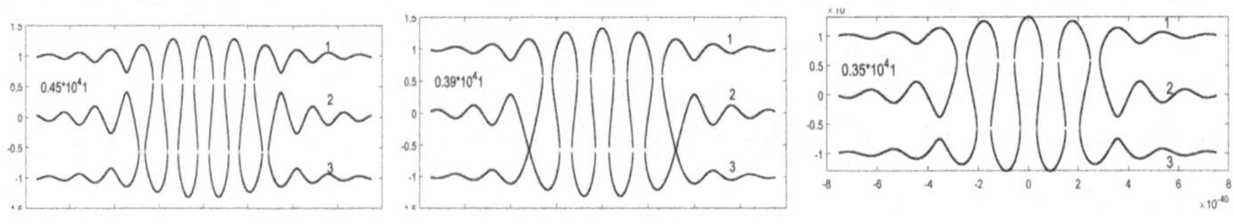

Fig. 5.14 Multivalued wave profiles calculated according to Eq. (5.82)

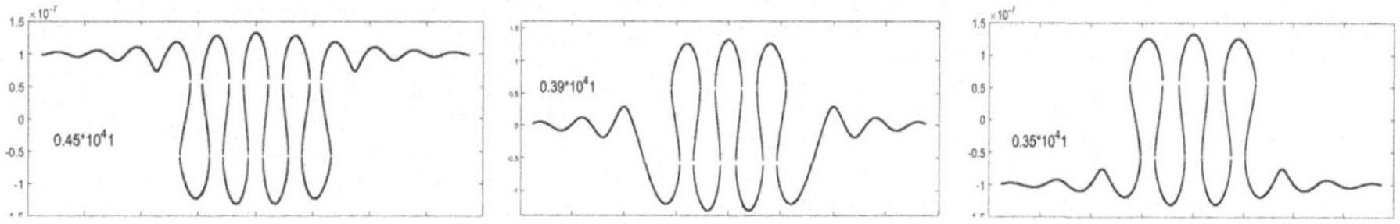

Fig. 5.15 Three variants of wave curves calculated for $D = 0.45 * 10^{41}$, $0.39^{*}10^{41}$, or $0.35^{*}10^{41}$ and following from the corresponding Fig. 5.14

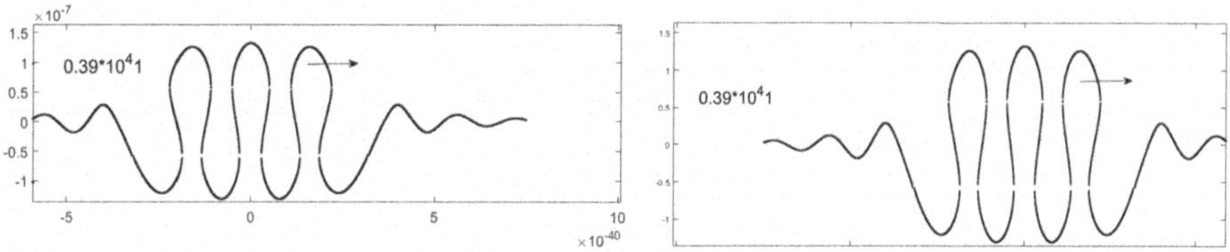

Fig. 5.16 An example of the movement of a wave packet

Figure 5.16 is some addition to Figs. 5.14 and 5.15. We have presented there the case when three multivalued figures are located in the central part of the wave. There may be more or less of them, which is determined by the quantities and D in (5.76). However, perhaps the number of multivalued figures in a wave cannot be arbitrary. Apparently, their number is determined by the own multivalued modes of each particle-wave.

We emphasize that the coefficients in (5.82) adopted in the last calculations were chosen completely arbitrarily.

This concludes our consideration of Eq. (5.82), when $\lambda^{-1}C = 0$.

5.3.5 "Quark" Model of Particle-Waves

In previous chapters, we examined many cases of transformation of Euler figures describing multivalued waves into particles-waves of a discrete nature. This transformation takes place by varying the influence of quadratic neutrality or the constant C in the governing equations of the form (5.82). Let us explore this possibility in the case of multivalued multidimensional spherical waves. The results of calculations based on Eq. (5.82) are presented in Fig. 5.17.

Let us look at the bottom three rows of paintings first. On the right, there are slightly modified copies of the waves presented on the left. Thus, the lines (imaginary solutions) connecting discrete wave particles have been removed.

The calculation performed for the case $C = 0$ and constant $D = 0.39 * 10^{41}$ is presented on the left of the top row of pictures. It fully corresponds to the picture in Fig. 5.14 calculated for $D = 0.39 * 10^{41}$. The only differences are in the details missed in Fig. 5.14. On the right of the top row is a calculation for a nonzero

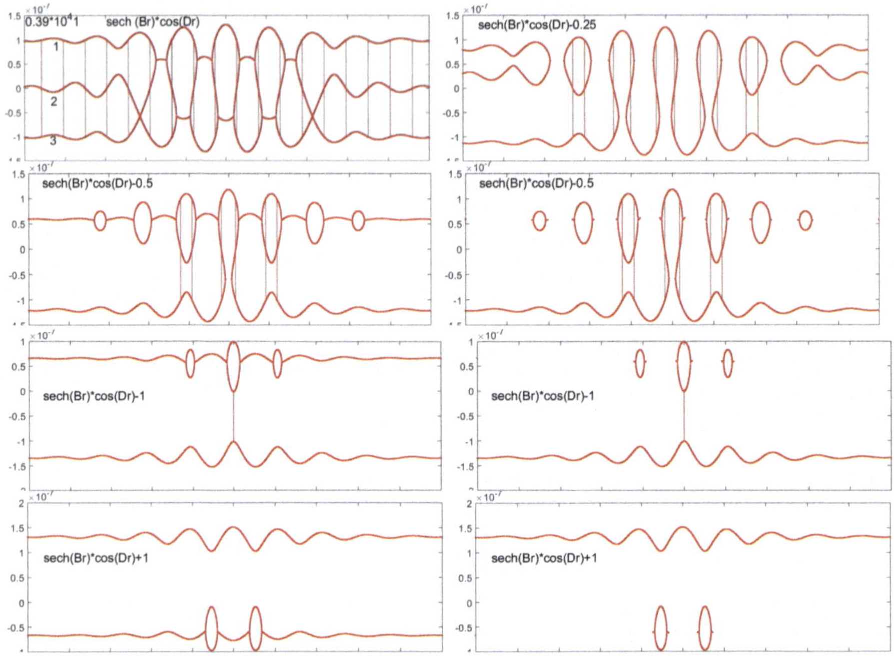

Fig. 5.17 Evolution of continuous multivalued waves into discrete multivalued waves

constant $C = 0.25$. It can be seen that the nature of the curves has changed significantly compared to the picture on the left. In particular, curves 1 and 2 moved. Thus, curve 1 was moved up. As a result, discrete particles were formed that framed the central group of Euler figures.

Next, let us look at the second row of paintings from the top. It can be seen that with an increase in C to $C = 0.5$, the situation has changed greatly. There is only one Euler figure left in the center. The number of isolated particles increased. With a further change in C to $C = 1$ and $C = -1$, the Euler figures disappear and the calculations begin to resemble those presented in Fig. 4.2 and other figures in Sect. 4.1.

It is very difficult to comment on the paintings in Fig. 5.17 from the standpoint of their correspondence to multidimensional multivalued strongly nonlinear packets of spherical waves, which the calculations are intended to illustrate. With the most daring interpretation of the calculations presented in the last two rows of the paintings in Fig. 5.17, one can imagine a particle-wave as 2 or 3 spherical "quarks" propagating in space, accompanied by some ripples of waves not associated with them.

It is much easier to consider the two-dimensional case and compare the calculations in Figs. 5.11, 5.12, and 5.17. From this comparison, it follows that there are possible cases of wave packets moving in space in which isolated particles exist above the wave crests. In particular, the presented results can be interpreted as waves propagating along the surface of a liquid (see Fig. 4.9). Interestingly, on the surface of a liquid, particles and ripples usually move in a circular motion, which also somewhat corresponds to the pictures discussed in this section. It is also interesting to compare the wave patterns in Fig. 5.17 with those presented in Figs. 2.14, 2.16 and 4.13, 4.30, 4.31. Finally, as before, the results of Fig. 5.17 can be interpreted as one-dimensional surface waves. In this case, it is interesting to compare them with the experimental data in Figs. 2.41 and 2.43. It can be seen that Euler figures and particles (drops) can appear simultaneously on the surface of the liquid, which corresponds to Fig. 5.17 (right). There, the continuous, wavy curves can be interpreted as the surface of a liquid, but it is not clear how to interpret these curves when considering the pictures as schemes of spherical particle-waves.

Comment 1 The latest calculations are very different from the linear and nonlinear particle-waves presented in Figs. 5.2, 5.3, and 5.5, although there is very important agreement between them. All of them indicate the possibility of the existence of multidimensional spherical particles having complex wave structures. The indicated spheres are framed by certain surfaces extending from the spheres to infinity. The indicated spherical particle-waves exist in the field where and when certain resonant conditions are met. Given the accepted Klein-Gordon equation, it can define three different unambiguous particle-waves. At the same time, extreme situations are apparently possible when three solutions are combined and determine a single multivalued spherical particle-wave.

In the last two sections, we looked at the multivalued structures that arise inside particle-waves. These structures appear to be characterized by the nonlinear eigenmodes of these multidimensional spherical wave packets. Studying them is a very difficult task, perhaps comparable to studying spheres inside the Calabi-Yau space [19, 20]. We have identified 2a types of these structures. In the first type, elastica-like waves (Euler figures) arise inside the particle-wave. In the second case, there are smaller particles inside the particle-wave. By analogy with chromodynamics, they can be called "quarks."

Comment 2 Just as Ptolemy's cosmological theory was very much limited by the hypothesis of uniform circular motions celestial bodies, we are also very limited by the fact that our research is based on the use of the Klein-Gordon equation. However, it opens up certain possibilities for explaining, qualitatively of course, some fundamental principles and effects of Nature. Let us consider some of them below.

5.4 Double Quantum Particles and Some of Its Consequences

Recall that expressions (5.71) and (5.72) for $A_{\pm}$ can be real or imaginary depending on the coefficients C_1, C_2, m_1, and λ_1 included in Eq. (5.57). Generally speaking, these coefficients are all the time under the influence of random quantum

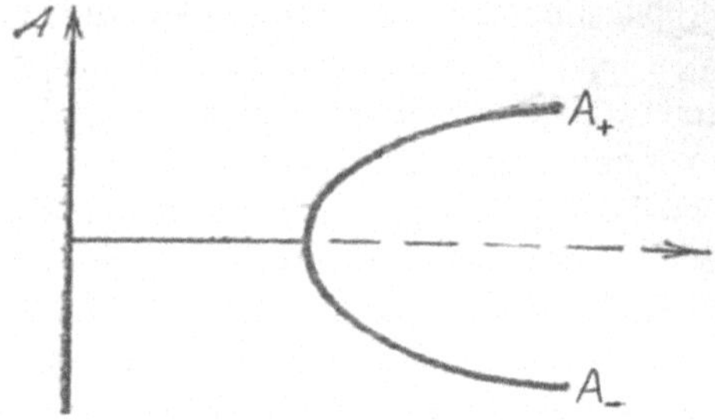

Fig. 5.18 Qualitative illustration of the results of Sect. 5.3.1

fluctuations, in the same way as an electron rotating in an atom is under the influence of random quantum fluctuations all the time. The creation of a pair of quantum particles is associated with such quantum fluctuations that instantly makes the quantity $A_{\pm}$ real (Fig. 5.18).

At some values of C_1, C_2, $m_{1,}$ and $\lambda_{1,}$ there is a change in the nature of solutions to Eq. (5.57). Together with the zero solution, two real solutions appear describing a quantum particle $\varphi_{+}$and a quantum antiparticle φ_{-}.

Of course, twin waves separated only by the appropriate space can also exist in the case of multivalued waves. As an example of this, we will indicate Fig. 4.30.

1. **Solutions (5.71) and (5.72), and the vacuum energy.** Generally speaking, the scalar fields described by the NKGE (5.51) can be different and describe different particles, which can have different energies. Particle-waves are born in pairs (positive and negative). Their sum can be nonzero. The possibility of the existence of a nonzero value of the sum can be called virtual. It is very important that the vacuum energy can be associated with particle-waves. On the one hand, they can annihilate each other. In this case, the vacuum energy is zero. However, some random, temporal, or spatial, very small displacement of the emerging particles relative to each other is possible. As a result, a certain finite, of course, very small (but not equal to zero) value of energy may appear. This value characterizes the result of the interaction of particles of the pair under consideration.

   On the other hand, the energy of quantum vacuum fluctuations in total can be unimaginably large, close to infinity [21–23]. It is very interesting that if a positive value is assigned to the total vacuum energy, then its action in space will manifest itself in the form of a force of universal repulsion.
2. **Repulsion.** Vacuum energy can be associated with Einstein's cosmological constant, which is included in Einstein's equations and explains the non-stationary (expansion) of the Universe. This value can be measured experimentally. It turned out that this constant is of the order of 10^{-120}. Thus, there is a colossal large discrepancy between the theoretically determined vacuum energy and the cosmological constant. This discrepancy is one of the greatest mysteries of modern cosmology [21–23].

   Thus, the location of the particle-waves (5.71) and (5.72) is known approximately to be within Planck's constant. Consequently, these two solutions constantly fluctuate chaotically relative to each other. The vacuum energy resulting from the partial annihilation of particles in these pairs can be very small. It can correspond to the cosmological constant. If we accept this, then in a certain sense we are building a bridge connecting the general theory of relativity with quantum wave mechanics.

Here we must keep in mind that the vacuum energy is the sum of the energies of all the fields included in it. Some fields can manifest themselves as repulsion, while other fields are responsible for attraction (gravity).

3. **Creation from "nothing."** We have considered some consequences of Eq. (5.51). The creation of particle-waves from "nothing" looks somewhat unusual from the point of view of classical physics. But for quantum mechanics, this result is not very new. We have already mentioned Dirac's solutions. Let us emphasize that there are developed theories describing the birth of the Universe from "nothing" [21–23].

   Note that in the context of the Universe, Fred Hoyle was apparently the first to point out the possibility of the formation of matter from "nothing." This idea underlies his theory of a steady-state Universe, which he successfully opposed to the Big Bang theory for almost 20 years, namely, from about 1948 to 1968 (Steady State Cosmology) [21, 24].

   In addition to what has been considered here, we will mention another option for eliminating the annihilation of pairs. Near a black hole, one particle of the pair will be "eaten" by the black hole, and the second will be able to escape according to Stephen Hawking's theory [21, 25].

   At the same time, we wonder whether the existence of expressions like solutions (5.71) and (5.72) is somehow related to quantum entanglement and quantum communication.

4. **Quantum entanglement** [21]. Thus, according to the Dirac [26] and NKGE equations, particles can appear in pairs. In this case, the mathematical expressions for them can differ only by one parameter, for example, the sign of the amplitude. Consequently, if we have determined the properties of one particle from the pair, then we can determine the parameters of the other particle from the pair without capturing and studying it specifically!

Let us explain this situation using a well-known interpretation (analogy). A pair of solutions to the Klein-Gordon equation can be considered as a pair of gloves. If we find one glove from the pair somewhere, then when we find another glove corresponding to it, we do not even need to consider it, since we know its shape in advance. Moreover, the accuracy of such a prediction is 100%. There is no uncertainty (probability) in this prediction.

In the general case, solutions (5.71) and (5.72) do not indicate any preferential localization of the place of the considered wave particles. These wave particles can be located at the same point in space. They can annihilate each other if they are in the same place and $\phi_+ = 0$. They can also weaken or strengthen each other depending on the values of ϕ_+ and ϕ_- in (5.71) and (5.72).

It is also very important that the amplitude of the spherical waves concentrated in the packets (5.71) and (5.72) becomes smaller and smaller with distance from the center of the packet, but never vanishes. Moreover, even if the wave packets are very far from each other, they still lie on the same multidimensional surface, that is, they represent a single connected system.

The obtained result may be related to quantum teleportation. Quantum teleportation is the ability to transmit the quantum state of one particle of a pair to another particle of the pair instantly over any distance. Thus, it is implicitly assumed that it is possible to transmit information faster than the speed of light. Many experiments confirm the possibility of quantum teleportation. However, its theory is based on the probabilistic interpretation of quantum mechanics, and the experimental results violate the locality principle. According to it, only its immediate environment can influence the state of an object. This "contradiction" is related to the Einstein-Podolsky-Rosen (EPR) paradox [21] and is one of the main conceptual difficulties of quantum mechanics (at least in its Copenhagen interpretation). Thus, if we proceed from the NKGE (5.51) and reject the probabilistic interpretation of the wave function in quantum mechanics, then the effect of quantum entanglement of particle-waves is explained by the results presented above. If we have a particle-wave described by solution (5.71), then we should not be surprised that we also know the parameters of the particle-wave corresponding to solution (5.72). Thus, our purely approximate study supports the EPR point of view, and not the Copenhagen school, on the issue of quantum entanglement.

Remark Thus, the latest calculations differ greatly from the linear and nonlinear particle-waves presented in Sects. 5.1.2–5.1.4, although there is a very important agreement between them all. They point to the possibility of the existence of particles with a complex wave structure if certain resonance conditions are met. Finally, it should be recalled that our results are based on an approximate analysis of Eqs. (5.51) and (5.57), the applicability of which to the problems under consideration is far from obvious. Moreover, the solutions of this equation were obtained under very strong assumptions and rough assumptions. The author understands this. His analysis pursues the main goal of this book, which is intended to show the possibility of describing, albeit very approximately, a very wide range of problems in the modern science of "physics."

5.5 Conclusions and Comments to Sects. 5.1–5.3.3

A. We emphasize that during a certain time, when there is no interaction, all particles move in a vacuum. Due to the uncertainty principle, the vacuum hums with particle-antiparticle pairs that appear and disappear. The pairs include, among others, electron-positron pairs and photons, which are their own antiparticles. Normally, these "virtual" particles cannot be captured directly. But they have a subtle effect on the "real" world. For example, virtual photons, appearing and disappearing from existence, create a randomly fluctuating electric field. In particular, physicists discovered in 1947 that the field

shifts the energy level of the electron inside the hydrogen atom and, therefore, the spectrum of radiation emitted by the atom.
According to our theory, cases where particles that are born as a pair only partially annihilate are considered cases of the birth of virtual quantum particles. The case when particles that are born as a pair do not interact in any way (no annihilation) corresponds to the birth of a particle and an antiparticle (the Dirac case).

B. It can also be assumed based on the experiments presented above that the birth of particles is associated with the resonance effect and resonators.
C. It can be assumed that when a real quantum particle appears in a resonator (e.g., between the plates), then somewhere in a neighboring galaxy its twin can appear. That is, it is located arbitrarily far away.
D. (1) Outside the resonators, virtual vacuum particles oscillate chaotically relative to each other, and virtual vacuum particles constantly appear and disappear chaotically. Chaotic quantum oscillations become more ordered in resonators. This is evident from the Casimir effect. Therefore, the vacuum field there acquires energy, mass, and pressure. The field acquires certain parameters that are practically zero outside the resonator. Perhaps the situation resembles the situation with electromagnetic waves. Their resonant amplification was shown in the Hertz experiment. Only then they began to be used. However, the vacuum energy resulting from full and partial annihilations of the particles can be very small. It can correspond to the cosmological constant included in Einstein's equations. If we accept this, then in a certain sense we are building a bridge connecting the general theory of relativity with quantum wave mechanics.

   Moreover, it is very important that the vacuum energy depends on the number of resonators. Resonators can be material (real) structures (e.g., plates in Casimir's experiments, a crystal lattice or atoms). Thus, implicitly, but the indicated vacuum energy (or Einstein's constant) can depend on the mass of matter that makes up our Universe.

   (2) On the other hand, the vacuum energy resulting from the partial annihilation of particles of these pairs can be negative. This case can correspond to Newton's law of universal gravitation.
E. At the same time, we ask ourselves maybe the existence of expressions like solutions (5.71) and (5.72) is somehow connected with quantum entanglement and quantum coupling.
F. Virtual particles and ordinary matter are interconnected environments.
G. Recall that the solutions of Sect. 5.3 describe particles and antiparticles, which in the cases indicated there can annihilate each other upon emergence. This result differs fundamentally from other results presented in the book. See, for example, the appearance of resonant particle-waves in Figs. 3.43–3.47; 4.1–4.8; and 5.6, 5.7. According to these calculations, particle-waves without an antipartner can appear in a scalar field! That is, there is an asymmetry of quantum particles and antiparticles.

This result can qualitatively explain the weak presence of antimatter in the Universe.

H. We usually meant and mean a particle-wave when we spoke about an elementary particle or even just a particle in this chapter.

5.6 Solar Wind and the General Theory of Relativity (GTR)

Recall that when we spoke about an elementary particle or even just a particle, in this chapter, we usually meant and mean a particle-wave.

Perhaps the developed theory and the results presented above allow a new, from the point of view of multivalued waves, look at very important areas of physics. Let us first recall some of these results.

From the very beginning, we built a theory that, on the one hand, did not pretend to be accurate, but, on the other hand, pretend the possibility of describing the Universe and its properties based on an extremely simple equation describing the relationship between resonances and multivalued waves. From this position, we considered many problems. Do these problems include gravity and how it related to particle-waves? Of course, these phenomena are related to each other, since in Nature everything depends on everything. But these dependencies are so subtle that they are at the very edge of the possibilities of studying modern science. In particular, wave particles do not generate in isolation (see Sect. 5.3), but only in pairs, with the components of the pair coupled to each other. Recall that the concept of electron-positron pairs arose from a paper by Paul Dirac [26] (1930). At the same time the origin of single photon is fixed in experiments (Fig. 5.8).

Thus, there are no absolutely individual particles. There is a kind of quantum dependence through scalar fields of everything with everything. Perhaps our devices register this dependence as gravity.

Nonlinearity and Quanta The Dirac equations describing these pairs are to some extent analogous to the Klein-Gordon equation and are not simply a combination of special relativity and quantum mechanics, but also imply that the number of particles may not be conserved if very intense electromagnetic fields act on the vacuum [27]. Indeed, in 1936, Heisenberg and Euler noted that, due to quantum mechanical effects and the presence of the Dirac sea, linearity ceases to hold in strong fields [28]. Pair coupling can clearly manifest itself in very strong nonlinear fields, such as those found in neutron stars or heavy ion colliders. Recently, it has become possible to study the influence of nonlinearities on quantum processes in Dirac materials, where nonlinearity manifests itself even at not very high field values [29–31].

Continuity, Multivalues, and Quantumness It is no secret that the foundations of modern physics were laid in the nineteenth and early twentieth centuries, when nonlinear science as such did not exist and all physical processes were interpreted as continuous and smooth. In particular, Riemannian geometry, which Einstein used when constructing the general theory of relativity, was exactly this. Planck's revolutionary idea did not have a great influence on the construction of the mathematical foundations of quantum mechanics. In particular, Schrödinger derived his fundamental equation for the wave function, which was assumed to be linear and continuous. Perhaps this is why so many additional assumptions were required to bring the solutions of this equation into agreement with observations.

But even with them, this equation did not advance much beyond describing hydrogen and helium atoms! Of course, despite the above, quantum electrodynamics has made enormous strides in describing and predicting the results of experiments and observations. But this does not change the fact that these successes are built on significant assumptions that we are accustomed to, but which no one understands to this day. In particular, R. P. Feynman stated: "The theory of quantum electrodynamics describes Nature in a way that is absurd from the point of view of common sense. And this is in complete agreement with experiments. So I hope that you will be able to accept Nature as it is – absurd" (see Sect. 6.1).

The approach proposed in the book is to abandon the fundamental assumption of physics and mathematics about the uniqueness of processes. To do this, we assumed that multivalued waves arise in resonance regions. Outside these regions, the waves can be described by linear equations and their linear solutions. As the waves approach the resonance regions, we begin to take into account nonlinear terms in the wave equations. In this case, the abovementioned linear solution is used as a seed perturbation when solving the nonlinear equation in the resonance region. The book mainly considers cubic nonlinear problems. For them, the multivalued solution of the nonlinear equations is constructed from solutions, each of which satisfies the d'Alembert equation. As a result of all these superpositions, the solution of the nonlinear problem and the problem of constructing a multivalued wave profile are reduced to solving a cubic nonlinear equation. All this is valid inside the resonance region (see Sect. 2.2.1).

In addition to the above, our study, in addition to many studies, has once again emphasized that nonlinear equations have many solutions! In reality, it may not be a quantum superposition, but a manifestation of a multiplicity of solutions for the same processes.

Resonances and Singularities Finally, in our studies, we paid special attention to resonance and singular effects. The fact that we abandoned the idea of continuity of processes allowed us to explore these areas. We found that it is precisely for these areas that multivalued solutions are characteristic. Thus, it is possible, as we have already noted in publications [32–40] that a huge number of processes in nature are associated with resonances and multivalued wave processes. The author manages to study them only by simplifying their description to the extreme, namely, by reducing them to algebraic equations.

The fact that wave profiles are constructed under certain resonant conditions is so important that the fields used can be called resonant scalar fields. Namely, particle-waves are origined, possibly, in resonant fields (see Fig. 5.8).

Theory of Relativity (GTR) Currently, there are two generally accepted theories of gravity that describe a huge range of physical phenomena [41–51]. These are Newton's theory and the immeasurably more complex general theory of relativity

(GTR) created by Albert Einstein. The comparison of the results of these theories shows that their results almost always agree, except in very special cases. First, let us consider the case of two bodies placed at an ultra-small distance. In these cases, Newton's theory describes the experimental data perfectly. Of course, when the distance between the masses tends to zero, Newton's theory stops working, but today, at ultra-small distances, GTR apparently works no better than Newton's theory. Let us consider the problem posed in relation to cosmology. Surprisingly, in the vast majority of cases, Newton's theory describes reality no worse than GTR. How can this be explained? To do this, let us consider when GTR turns into Newton's theory. This takes place if the outer space is homogeneous and isotropic. In addition, the smallness of the velocities of cosmic bodies relative to the speed of light is required. Then, to describe cosmic bodies, we can use the equation of state of "dust" matter. If the specified conditions are met and after linearization of the equations of general relativity, they can be reduced to the Newtonian model. It is easy to see that all of the specified conditions are met, on average, in cosmology. It is especially important that, according to geometric (geodetic) studies of the Universe, it is homogeneous and isotropic on scales of tens of millions of light years. Of course, what has been said is not enough to talk about the fulfillment of the specified conditions on all other scales of the Universe, for example, on the scale of galaxies, or many billions of light years. In the last two cases, Newton's theory obviously does not work. Apparently, it is in the description of such cases that the value of general relativity lies. However, in the last two cases, general relativity can also give results that are not consistent with our expectations. When we try to understand gravity on cosmic scales, at the scale of galaxy clusters and beyond, we encounter obvious inconsistencies with the predictions of general relativity (see Symmetry 2022, 14(7), 1331; https://doi.org/10.3390/sym14071331; Robin Y. Wen et al JCAP03(2024)045; Physics Letters B, Volume 856, id.138882 https://doi.org/10.1016/j.physletb.2024.138882). For example, it cannot always describe the gravitational fields of galaxies. To explain the peculiarities of gravity of some galaxies, the so-called "dark" matter was introduced. The equations of general relativity contain components of the space-time tensor, which are equated to components of the energy tensor. Thus, space-time ceased to be a fixed, unchanging background. It acquired the ability to bend and stretch in response to the presence and changes of matter and energy. The latter acquires additional motion during the evolution of the background. Thus, energy dictates to space-time how to evolve, and space-time dictates to energy (matter) how to move [20–22]. The space-time changes when mass and/or energy change. This change is perceived by us as gravity. For example, planets with mass and energy bend the space around them, forming an axisymmetric depression in it. A similar depression, but much deeper, can be formed by a star adjacent to a planet. If the planet has sufficient speed, it begins to rotate around the star, similar to the way planets rotate in the solar system. We perceive this result of the interaction of the two depressions and speed as the result of the gravitational attraction of the planet to the star.

Effect of the Velocity of Elementary Particles It is important for us that Newton's theory and GTR do not take into account the influence of the motion of elementary particles on gravity, which constitute the gravitating bodies considered in these models. The contribution of the radiation existing in space is also not taken into account [5, 6]. In the following sections, we try to give a new understanding of the effect of gravity, which is based on taking into account the velocity of motion of elementary particles described by solutions (5.71) and (5.72). Thus, we try, to some extent, to unite gravity with relativistic quantum mechanics presented by Equation (5.51).

5.6.1 Particle-Wave Traces in Quantum Space-Time (QST) and Gravity

Description of the gravitation is the problem much more complex than the problems considered above. Therefore, in order to be able to perform a very rough analysis, we will make some assumptions. The main assumption is related to the so-called fundamental background. In this section, it is something like the space-time. Many leading physicists of the past and present suspect that the continuous space-time, despite their omnipresence, are not truly fundamental concepts [18–23, 41–46]. The smooth form of space-time perceived by our senses arises from averaging out all the fluctuations that are inaccessible to our perception. We assume that space-time is quantized. This means that by dividing space-time into ever smaller parts, at a certain level of fragmentation, we will arrive at quanta of space-time. After that, further fragmentation of the medium under consideration is impossible. Quantization implies that the maximum degree of discreteness, granularity of space-time, has been achieved.

The sizes of the granules (quanta) of this medium can correspond to the Planck length 10^{-33}cm and the Planck time 10^{-43} s. When grouped, these quanta can create continuous spaces in which structures much larger than 10^{-33}cm can be placed.

Thus, on large scales, quanta of space-time form a certain medium against the background of which everything exists. We emphasize that against the background of this medium (jointly, in parallel), all physical fields and elementary particles exist. As some analogs of the specified medium, we will indicate the ether, introduced in ancient times, and, much more modern, the scalar Higgs field.

Strictly speaking, this is all that can be said, about the described quantum space-time (QST). The rest that we will say about this medium are our assumptions on the basis of which we can describe the considered model of gravity.

In particular, we propose to consider QST as a two-phase medium.

It can be assumed that the QST consists of spherical granules (quanta). They are separated by voids, the volume of which is much smaller than the volume of the granules. If the volume of voids, for some reason, increases to the volume of the granules, then this volume turns into a granule. However, difficulties arose in the mathematical description of the specified model, so we use an even simpler model of the QST, namely, we model it with an inviscid bubble liquid (a liquid containing gas bubbles).

The movement of a particle-wave in the specified medium with a constant speed is not associated with any difficulties, but does not remain traceless for the medium. In an attempt to give an idea of these consequences (traces), we turn to experiments with spherical bodies flown around by liquid [52–54].

As examples of such experiments, we consider cavitation traces from a moving solid sphere and from a shot underwater from a Kalashnikov assault rifle (Fig. 5.19).

The cavitation cavities (we can tell "emptinesses") are formed in the liquid by a moving solid sphere (Fig. 5.19, upper). At the bottom, a cavitation cavity is a trace from a bullet moving in water. The traces are formed by gases and vapors of the liquid.

The shapes of these traces can be different and depend, first of all, on the speed of the liquid. A stationary liquid does not leave a trace behind the sphere, and a very fast flow can leave a trace similar to a tunnel, the length of which is much greater than the initiating solid body (sphere or bullet). Against the background of such a trace, the contours of the body itself may not even be distinguishable.

Something similar happens, in our opinion, when particle-waves move in quantum space-time (QST). Behind the particle-wave, there is a rarefaction zone (i.e., a trace zone) in which space-time is significantly distorted or can even disintegrate into space-time quanta.

Thus, a solid body (sphere or bullet) corresponds to a particle-wave in experiments, and the flowing liquid corresponds to the QST.

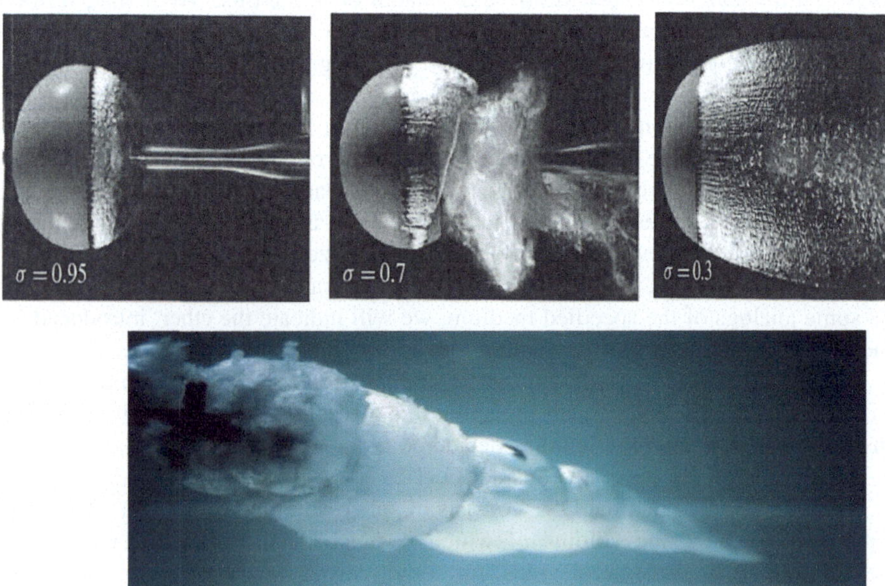

Fig. 5.19 Cavitation characteristics around a sphere obtained for different cavitation numbers [53, 54] (upper). The cavitation numbers are 0.95, 0.5, and 0.3. Cavitation traces from a shot underwater from a Kalashnikov assault rifle (bottom)

The number of elementary particles depends on the amount of matter and energy filling the considered unit of material volume of macro-substance. All elementary particles propagating in this volume (space-time) are followed by traces of varying duration. There, a restructuring and even destruction of the QST are possible.

Consequently, the total volume of traces, i.e., the volume of a distorted or destroyed QST, depends on the amount of matter and energy contained in the original considered macro-volume.

According to our hypothesis, the volume of the specified traces determines the structure of quantum space-time. Many researchers speak in such cases about the landscape of multidimensional space-time. Thus, we come to the conclusion that the structure of space-time depends on the amount of matter and energy contained in a unit volume. Einstein came to this conclusion in his time. Unlike him, we made an attempt to link the structure of space-time with the quanta that compose them.

The looseness of the QST depends on the matter and energy wave particles moving in space-time. The movements of these particle-waves are described by the corresponding wave equations. In particular, we have considered above the cases of particle-waves of scalar fields. That is, the specified looseness is determined by the traces of matter and energy wave particles. As a consequence of this chain of connections, the QST depends on motion and the location of matter and energy there.

Moving along the landscape of quantum space-time, a person experiences sensations that he interprets as gravity.

Note According to the proposed hypothesis, all the processes of the microworld known to us occur on the stage determined by quantum space-time. However, the participants in the play are not soulless performers; they move around the stage and, thanks to their footprints, to some extent edit Einstein's general theory of relativity (GTR).

5.6.2 Solar Wind and Gravity

Disturbances formed by traces exist wherever various particle waves arise and exist. Of course, the Sun is the largest source of such corpuscular waves in the vicinity of the Earth. A stream of matter called the solar wind rushes from the Sun into interplanetary space. It fills the solar planetary system. This wind consists of protons, alpha particles, electrons, photons, and many other various particles that arise inside the Sun as a result of a thermonuclear reaction. We emphasize that in March 1997, the Pioneer 10 spacecraft reported the presence of solar wind at a distance of 67 AU from the Earth. In addition to the wind, the Sun creates a magnetic field that covers a vast area (Fig. 5.20).

What is very interesting to us is that the fixed zone of influence of solar gravity is also limited to a distance of ~60 AU.

Thus, the zone of influence of solar gravity roughly corresponds to the zone of influence of the solar wind and magnetic field. Therefore, it is possible that solar

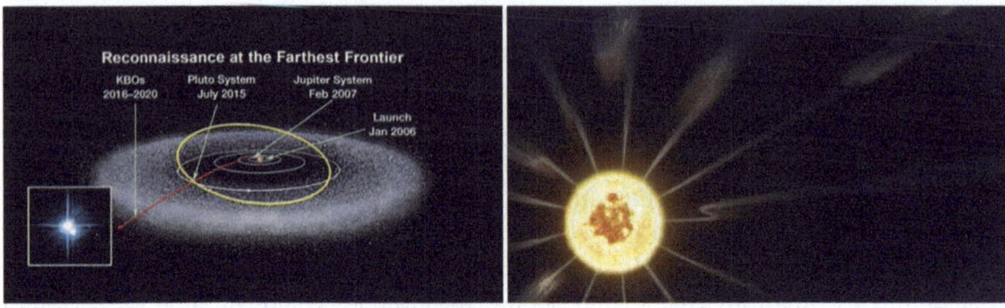

Fig. 5.20 The Kuiper belt is a disk-shaped region beyond Neptune, extending approximately 60–80 astronomical units (compared to the Earth, which is one astronomical unit, or AU, from the Sun) (left). The Sun ejects a highly variable stream of plasma from its corona, forming the solar wind and multivalued Alfvén waves (right). http://www.science.org/doi/10.1126/science.adk6953; https://www.cfa.harvard.edu/news/smithsonian-scientists-help-uncover-how-solar-wind-gets-its-energy

gravity is a consequence of the existence of the solar wind and magnetic fields, or, in other words, a consequence of the existence of traces in QST that arose due to the corpuscular particles emitted by the Sun.

Let us repeat what has been said and develop it a little. Each material body or clot of energy consists of wave fields and corresponding wave-particles.

These particle-waves move in a scalar field, which, at the same time, exists against the background of the fundamental quantum space-time, where the particle-waves leave some traces.

Their traces lead to a kind of local loosening of space-time. As a result, a certain dynamic, curved landscape of QST appears. As a consequence of all that has been said, the dynamics of the curvature of this landscape is determined by the initial location and dynamics of the generating material body or clot of energy.

As an example, let us consider two material bodies consisting of moving elementary particles that create curved landscapes around the bodies with the said particles. The interaction of these landscapes can be defined as the gravitational interaction of these bodies.

The moving elementary particles of the material body increase the looseness of the QST. Let us assume that as a result, a depression appears under the body. Moreover, the more massive the body and the higher the speed of the elementary particles that make it up, the greater the depth of the depression. In this case, the less massive body and the bodies whose particles move more slowly (e.g., less heated bodies) will be attracted to the more massive body. It is possible to select such masses and velocities of rectilinear motion of these bodies, at which the less massive body (e.g., the Earth) will rotate around the more massive body (e.g., the Sun).

This picture qualitatively corresponds to the process calculated for the solar system according to Einstein's theory [46].

Thus, according to the proposed hypothesis, the evolution of gravity depended on the dynamics of the interaction of the Universe with the QST. This interaction led to a local loosening of space-time. These loosenings formed and form a unique space-time landscape of the Universe, which we perceive as gravity.

5.7 Conclusions and Comments to Sect. 5.6

According to the proposed model, all gravitational phenomena have a secondary effect in relation to the fields that describe and determine the motion of elementary particles.

5.7.1 Changes in the Structure of Space-Time and the Gravitation

As already noted, the gravitational field of the Sun is generated by a flow of various particles emitted by it all the time. These particles leave traces that seem to loosen space-time and to a certain extent determine its landscape. Our senses perceive the change in landscape as gravity. Since the Earth moves around the Sun and the change of the distance from it and the change in radiation on the surface of the Sun is small, we do not feel a change in solar gravity as the Earth moves. In other words, there is a small change of the solar wind and the solar gravity along the trajectory of the Earth.

All gravitational phenomena have a secondary effect in relation to elementary particles. Therefore, a huge flow of the corresponding elementary particles is necessary to create such a significant and intense region of space-time loosening that they could be interpreted as gravity and that they could be recorded by the researcher's instruments. If there are no such flows, then space-time remains practically unperturbed. Such flows are created in space during super-extreme events, such as the mergers of black holes and quasers [47–51]. During such events, peaks of radiation of various elementary particles are created. Each emitted particle is followed by its trace (a zone of destruction or distortion of space-time). In the zone of merging of these cosmic bodies, space-time is fragmented into quanta of space-time. This can be interpreted as a local occurrence of a certain depression (hole, dip, and trough) at the places of mergers. The resulting situation can be considered as analogous to the case of an explosion in the ocean [55, 56]. As a result, a rapidly expanding region of water destruction appears, filled with liquid vapor and explosion products. The formation of this bubble is usually a very fast process, accompanied by oscillations of the bubble as it rises. Moreover, the period of oscillation of the bubble during its ascent is much longer than the time that occurred during the explosion.

Something similar happens during the merger of cosmic bodies. The resulting peak of radiation of elementary particles leads to the destruction of space-time at the place of merger. Then the volume of destruction is greatly reduced, since the elementary particles leave the place of merger and, as it were, carry away part of the destruction with them. Since a significant part of the radiation consists of high-energy photons, this part of the disturbance of space-time reaches the experimenters together with the said photons (electromagnetic waves).

The experimenters register these disturbances as disturbances in the gravitational field. This wave arrives to the observer practically simultaneously with the electromagnetic wave. This fact was recently confirmed [50, 51]. In these publications, gravitational and electromagnetic waves coming from the same source were described for the first time. The ~100 s gravitational wave signal (GW170817) was followed by electromagnetic waves in the ultraviolet, optical, infrared, and radio ranges—for hours, days, and weeks. The results demonstrate the importance of combined gravitational wave, electromagnetic and neutrino observations and mark a new era in multichannel, time-based astronomy.

Slower moving elementary particles arrive at the experimental setup later. They bring with them their traces, the sum of which has a smaller amplitude and a much longer duration.

Finally, we can expect the arrival of a third group of space-time waves. These are the compression-expansion waves that occur when the confluence vibrates. These waves apparently correspond to the waves that occur in the ocean when the products of the explosion float up.

We do not know what the speed of the expansion-compression waves of space-time is. If we attribute some properties of the continuous space-time model introduced by Einstein to the quantum model of space-time used here, then we can assume that the speed of expansion-compression waves in the quantum model is also equal to the speed of light.

Let us repeat that the simultaneous detection of gravitational waves and a peak in electromagnetic radiation arising from the merger of two neutron stars was recorded in 2017.

5.7.2 *Modification of the General Relativity*

The general theory of relativity has great advantages. For example, it provides a clear geometric interpretation of gravity. However, the legitimacy of linking geometric abstractions with physics is not obvious to the author of the book. In addition, GTR is not connected with quantum wave mechanics. Our hypothesis is an attempt to provide some physical basis for geometry and modify GTR in such a way that, while retaining its main previous advantages, it would receive a fundamental explanation from the point of view of quantum mechanics. Our model introduces quantum space-time into consideration. It does not postulate the possibility of changing the curvature of space-time, but directly links, of course, implicitly, the dynamics of change in this curvature with the emergence and existence of particle-waves in the Universe. As they move, they leave certain disturbances ("emptiness") in space-time, which determine the change in curvature in the equations of Einstein's general theory of relativity. In this regard, in our opinion, the proposed modification of GTR does not affect the predictive power of this theory. Moreover, in addition to the known successful predictions of this theory, after modification, it predicts the connection between quantum and gravitational effects. In particular, electromagnetic extreme impulses are followed by gravitational waves. Here it is necessary to understand that, according to our modification, the amplitude of gravitational waves is secondary in relation to the elementary particles that caused them. Remember these moving particles are the clots of corresponding material or energetic fields. The amplitudes of the gravitational and field events reaching the experimenter can differ, literally, by a dozen or more orders of magnitude!

The ancient Greeks believed that nature does not like "emptiness." If "emptiness" arises, then the surrounding bodies strive to fill this "emptiness," that is, to destroy the disturbance that has arisen. I do not think that this old idea is so bad if we take into account the motion. Taking it into account allows us to qualitatively understand the possible trajectory of a body falling into "emptiness." In fact, this idea underlies the modification of GTR proposed in Sect. 5.6.

5.7.3 Approximate Theory of Particle-Waves and the "Malice" of Nature

The results of Sect. 5.6 are obtained under the assumption of sphericity of particle-waves, that is, on the basis of essentially approximate solutions (5.71) and (5.72). The approximate theory of particle-waves has not been tested by specially designed and conducted experiments. Conducting such experiments is a difficult task in many ways. In particular, the dual nature of particle-waves is of great importance. Indeed, when planning an experiment, the researcher sets up the equipment for some expected result. Usually, in an experiment, one expects to detect either a wave or a particle. But the elementary particles discussed in this book have two faces.

Let us consider the first face. This is the case when the experimenter expects to detect a wave and sets up the equipment accordingly. In accordance with expectations, the particle-wave will be registered as a wave, since the equipment is set up for this!

However, if the equipment is to some extent set up to capture a point particle, then in such experiments the particle-wave will manifest itself as a particle, for example, when the particle-wave hits the surface of a screen.

Even greater difficulties may arise when attempting to experimentally demonstrate the multiplicity of elementary particles!

Trying to explain and describe the described diversity of elementary particles, discussed in many books, the creators of quantum mechanics introduced a probabilistic interpretation of the results of experiments and observations with micro-objects.

The success of the probabilistic interpretation of events in the microworld, in my opinion, is due to the fact that many events there occur in strictly defined resonance bands, in other words, in singularity zones, in particular, where the influence of nonlinearity increases very strongly. This leads to the fact that quantum systems always live in the risk zone of "bifurcations." That is, where disturbances that are not felt by the experimenter can practically throw the system into states that were not expected. In this case, we can talk about quantum uncertainty and hidden parameters. The experimenter can accept such arbitrariness as a consequence of probability or even the "evil will" of Nature. It is noteworthy that Einstein, who in his youth discussed the "goodness" of Nature, before his death began to note the possibility of "malice" in Nature.

The basic idea of the proposed in Sect. 5.6 approach is reduced to the assumption that space-time depends on quantum processes. Of course, this assumption places restrictions on the area of applicability of GTR, which is valid only for smooth, continuous processes. Moving elementary particles-waves loosen and change space-time up to its local destruction. The degree of change can be different. For example, we can distinguish the case when continuous jumps arise in space-time, similar to weak shock waves in gas dynamics, or weak folds of space-time. Let us call this case a weakly nonlinear change in space-time. time. It can be assumed that it can be described, qualitatively, by the above-proposed modification of GTR. The second case corresponds to the possibilities of strong loosening and even destruction of space-time by a moving particle. This variant can be called strongly nonlinear, in which one can expect the appearance of multivalued waves. The question of how gravity could be described in this case is considered in the next section. Here, We remind the reader once again that we are considering complex problems whose models are very crude. These are very crude attempts to create at least some, but simple, understanding of the fundamental problems of Nature.

5.8 "Nature abhords a Vacuum" and a New Theory of Quantum Gravity

Even 100 years ago, knowledge about the structure of the Universe and the processes occurring in it was rather vague and obscure. Assumptions about the smoothness and homogeneity of processes seemed no worse than other assumptions and were accepted not in accordance with deep knowledge, but rather for reasons of constructing the simplest and most integrable model of a specific process or observation. Then, sometimes, the accepted assumptions and models became so familiar that they were considered obvious.

If the above assumptions are changed, this will not be the only example of a revolution in cosmology. At the very beginning of their development, the ideas of the geocentric and heliocentric pictures of the world were equally promising. But then, for almost two millennia, the heliocentric system of the world was almost not considered. All the experience of mankind, which knew only one Universe—our planet Earth, convinced it of the geocentricity of the world. Having accepted this point of view, scientists have created theories of astonishing beauty and predictability, which have not become more correct because of this. And only with the expansion of human experience beyond the Earth, after the invention of the telescope, the theory of the heliocentric world began to develop, although the merits of this theory were obvious 2000 years before the birth of Galileo Galilei.

The same experience exists with theories about the structure of matter and the causes of gravity. For more than 2000 years, the generally accepted explanations were those dating back to Aristotle. In particular, the atomic theory completely triumphed only 100 years ago.

The ancient Greeks believed that "Nature abhords a vacuum" and associated gravity with this.

In this section, we give a sketch of a new theory of gravity. It is based on an old idea of the Greeks and a well-known assumption of Paul Dirac [57–59]. I repeat, a scheme of the theory is proposed. The content of the components of this scheme is not entirely clear even to the author himself. But perhaps the chain of assumptions and some results presented below may be of interest to the reader.

5.8.1 The Fundamental Background (FB) Model

If a "void" has arisen, then the surrounding bodies tend to fill this void, that is, to destroy the disturbance that has arisen. It seems to me that this old idea is not so bad if we take into account the motion. Taking it into account allows us to qualitatively understand the possible trajectory of a body falling into a void. In fact, this idea underlies the modification of GTR proposed above.

We can say that from the standpoint of detached philosophy, the models of the Greeks, Newton and Einstein, do not differ much. The Greeks deified the void. Newton introduced a god instead, who occasionally corrects the Universe. Einstein combined the geometry of space with gravity, taking on the functions of a god, so to speak. However, in terms of simplicity and generality, the latter approaches, it seems to me, are philosophically inferior to the Greek models.

In my modification of GTR, I am trying to return, to some extent, to the ancient Greeks. The essence of the above, new understanding of GTR, is that all bodies emitting energy or consisting of moving particles create peculiar "emptiness" (areas of loosening or "holes") in quantum space-time. The "massive" body, the larger "emptiness." All bodies roll to one degree or another into these voids of QST. If the speed of movement of bodies, as a whole, is low, then they fall and connect with some local "emptiness" of quantum space-time. If the initial speed of the body is high enough, then it begins to rotate around the local "emptiness." For example, this is the case of planets in the solar system or electrons in an atom. Finally, the velocity of a body can be so great that in the vicinity of "emptiness," the body only slightly changes its trajectory. This is the case, apparently, for photons moving near stars or galaxies.

Below we make the next step compared to the model considered in Sect. 5.6. A model of the gravitational effect is constructed without resorting to a parallel with the GTR, but based on the nonlinear theory of a quantum scalar field and a new model of the fundamental background (FB) on which everything exists. Unlike the background used above, defined as the quantum space-time, the new background is not connected with space-time in any way. We consider this background as a kind of stage (medium) on which all other events in the Universe unfold after the Universe has emerged. This background can be considered as an analog of the ether or the Higgs field. The aim of the study is to demonstrate the possibility of mathematical description of gravity as an effect of FB loosening associated with quantum effects. Namely, we believe that the level of loosening determines the effect of ravitation. The complexity of the problem, in particular, is that both real particles and virtual particles affect loosening (gravity). A scheme is proposed in which virtual particles determine the existence, albeit very short-term, of local voids in FB. The duration of existence of the corresponding loosening in FB depends on the characteristics of real particles moving in FB. In theory, it is implied that real particles are described by solution (5.71) or (5.72). Virtual particles are described by the sums of these solutions, that is, they are determined by the process of annihilation of oppositely charged real particles.

The introduced new background corresponds to some extent to the concept of luminiferous ether as a medium transmitting light waves. The latter was discarded after the advent of the special theory of relativity. However, back in 1951, P. A. M. Dirac published two articles devoted to the rehabilitation of the ether, in which he indicated that the new model should take into account quantum fluctuations in the ether flow [58, 59]. This idea was developed in a number of works [60, 61]. However, the results obtained there are difficult for us to use in relation to the approach to explaining the effect of gravity developed here.

Here, it is proposed to model the fundamental background (FB) as a two-phase ideal medium. Here, we propose an FB that develops Dirac's idea, assuming that the quantum fluctuations indicated by him are described by solutions (5.71) and (5.72). There, in the general case, the quantities ϕ_+ and ϕ_- may be nonzero. In particular, the following three cases can be considered.

(1) The first when $\phi_+ = -\phi_- \approx \hbar$. In this case, oppositely charged similar particles arise in the ether. Of course, they destroy each other, but it takes time. As a result, the particles do not completely destroy each other in the entire volume of space, at each moment of time. Therefore, FB acquires a small but finite value of energy if $\phi_+ = -\phi_- \approx \hbar$.

Let me remind you that solutions (5.71) and (5.72) describe spherical particle-waves. Therefore, in the simplest consideration, each pair of these particles will be approximately modeled as the smallest bubble arising from a very small embryo. Then it expands and contracts during the Planck time (Fig. 5.21).

Thus, it is proposed to model one of the phases of the FB, corresponding to quantum fluctuations, as bubbles. To describe the other phase of the FB, we propose to use the simplest model of an ideal condensed medium. Thus, to describe the FB, it is proposed to use the model of an ideal liquid containing oscillating gas bubbles.

Now it is proposed to slightly deviate from the given abstract description and to concretize it a little with more familiar images. Let us imagine the FB as some strictly two-dimensional flat surface like a perfectly smooth sea surface. It is raining. As a result, the surface is covered with many craters, bubbles and drops. In this state, the surface resembles the FB described above. A three-dimensional particle moves along it. At a low speed of its movement, the particle does not affect the formation

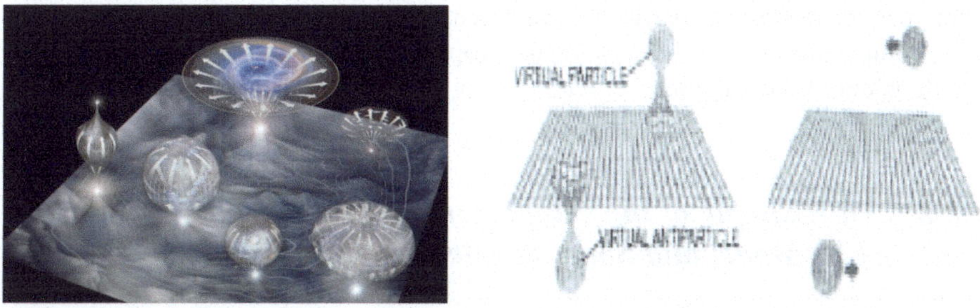

Fig. 5.21 The cases are presented, (1) when $\phi_+ = -\phi_- \approx \hbar$ (four bubbles at the bottom left), (2) when $\phi_+ > \hbar$ (two bursting bubbles at the top, left), and (3) when $\phi_+ >> \hbar$, and the transformation of a pair of virtual particles into real particles occurs (right)

of surface craters. The influence of the particle on the occurrence of craters, bubbles and drops, however, is possible with an increase in its speed. A moving three-dimensional particle leaves a trace behind itself. From the standpoint of the flat FB, this trace is a zone of destruction of a flat smooth surface. As another useful analogy to the problem under consideration, consider the deep ocean in which a submarine is moving. The underwater animals surrounding it can be considered as small zones of water destruction. If the submarine is stationary or moving at low speed, the influence of the animals on the wake that appears behind the vessel is very small. However, as the submarine's speed increases, the influence of these animals on the destruction of the liquid in the wake behind the vessel can become significant.

Of course, other cases are also possible, when the values of ϕ_+ and ϕ_- may be nonzero.

(2) In the case $\phi_+ > \hbar$, the solutions (5.71) and (5.72) approximately describe the possible formation of larger bubbles than in the first case. However, they do not yet have enough energy, so they collapse before they have time to completely complete the formation of a large bubble. This case is important because in it, the FB can receive more energy than in the variant (1).

(3) Finally, there is a case when the values of $\phi_+ >> \hbar$ differ very strongly. In this case, we assume the possibility of forming two "bubbles" corresponding to a pair of "twin" particles. Thus, the case n corresponds to the emergence of a pair of real elementary particles (Fig. 5.21, right). For their mathematical description, the solutions and figures presented in Sects. 5.3.4 and 5.3.5 may also be useful.

Of course, we interpret all three cases considered in relation to the case of the elementary particle motion in the FB under consideration. We believe that if its speed is not very high, then the case $\phi_+ = -\phi_- \approx \hbar$ takes place. As the particle speed increases, it begins to disturb the adjacent region of the FB. As a result, cases can be realized there when $\phi_+ > \hbar$. Finally, when the particle speed approaches the speed of light, then it can generate particles similar to itself (third case). There when ϕ_+ and ϕ_- may be very larger. Finally, when the particle speed approaches the speed of light, then it can generate particles similar to itself. Of course, we have touched upon very complex processes here, and the proposed chain of hypotheses is only a development of a certain theory that does not pretend to describe reality.

5.8.2 An Example of the Fundamental Background (FB) Model and Some Results of Its Application

We assume that $\varphi_+ \approx \hbar$ liquid.

In this section, the first case is considered. Thus, we assume that the speed of the elementary particle moving in FB is such that $\varphi_+ = -\varphi_- \approx \hbar$, and the FB can be considered as a bubbly liquid. Bubbles whose volume varies to the emergence, evolution, and disappearance of virtual particles—twins. With the help of the proposed model, we want to describe the emergence of compression zones and rarefaction in FB zones correspond to the "emptiness" ("voids") described in Sect. 5.6. The "voids" open the way to describe the effects of gravity. The above cases can be mathematically described in the same way, if we assume that ϕ_+ depend on the velocity of elementary particles in FB. In this case, the physical picture begins to resemble, to a certain extent, the problem of the influence of nonstationary cavitation on the motion of deformable bodies in a bubbly liquid, in particular submarines [52, 55, 56]. Of course, the problem considered here is much more complicated than those considered in [52, 55, 56]. In particular, this is due to the fact that a huge number of elementary particles move in FB. Take into account their influence on bubbles, which take into account the presence of quantum disturbances in FB, impossible. In order to come to at least some foreseeable result, we are considering here option 1) noted above. Thus, we assume that the speed of the elementary particle moving in FB is such that the FB can be considered a bubbly liquid. We will simplify it as much as possible. Thus, we will take into account the influence of bubble oscillations as the dynamics of a certain gas phase entering FB. We can call this gas as quantum perturbation gas (QPG). In the steady state, the gas phase and also the ideal liquid are uniformly distributed over FB. We initially assume that quantum fluctuations always arise and disappear in FB. We do not know where and when they arise. However, if we consider a sufficiently large volume of FB in some unperturbed state, we can assume that quantum perturbations are distributed there almost uniformly over time and coordinates. On average, the quantum perturbations form some uniformly distributed phase. We assume that its density is much less than the density of the other phase (ideal liquid). In relation to the density of the liquid, the density of the gas of quantum perturbations is very small, and it will be associated further with "emptiness." In the described two-phase medium, elementary particles move. The sizes of which larger are on tens of orders the sizes of quantum fluctuations (bubbles described above). Of course, elementary particles influence quantum fluctuations (bubbles or gas phase) very strongly. We assume that in the traces left by elementary particles, the volume of the gas phase increases very strongly. These rarefaction zones correspond to the "emptiness" (voids) described in Sect. 5.6. They determine the effects of gravity with a very strong increase in their volume.

Basic Equations As in Sects. 5.6 and 5.7, we will limit ourselves to a brief presentation of the main conceptual issues of the proposed FB model. The description of the background as a two-phase medium allows us to write down a system of equations for averaged quantities. We assume the following classical equations of motion and of continuity for average parameters of FB:

$$\frac{d\overline{v}}{dt} = -\frac{1}{\rho}\nabla p, \tag{5.83}$$

$$\frac{d\rho}{dt} = -\rho \mathrm{div}\overline{v}, \tag{5.84}$$

where $\overline{v}$ is a velocity vector. The equation of state may be written approximately as

$$\rho = \rho_0 \left[(1-\varphi)\left(\frac{p_0 + B}{p + B}\right)^{1/k} + \varphi\left(\frac{p_0}{p}\right)^{1/\gamma} \right]^{-1}, \tag{5.85}$$

Here, the first term of the sum corresponds to the contribution to the average density of the condensed component, and the second term corresponds to the contribution of QPG (the gas-like component contribution or "emptiness"-like component contribution). The value determines the initial volume fraction of QPG.

To simplify the analysis, the state equation (5.85) was presented in the form of the Tait-type equation. Following [56] and neglecting small (cubic nonlinear) terms, we can write Eq. (5.85) in the form

$$\frac{p + y}{p_0 + y} = \left(\frac{\rho}{\rho_0}\right)^x, \tag{5.86}$$

where

$$y = \frac{\lambda(1-\phi_0)\gamma^2 p_0^2 + \phi_0 \gamma p_0}{(k+1)(1-\phi_0)\lambda^2\gamma^2 p_0^2 + \phi_0(\gamma+1) - [\lambda\gamma p_0(1-\phi_0) + \phi_0]^2} - p_0,$$
$$X = [\lambda(1-\phi_0) + \phi_0/\gamma p_0]^{-1}(y + p_0)^{-1}.$$

Then the velocity potential is introduced $\overline{V} = \nabla\varphi$. The assumption of an irrotational motion has allowed the determination of the velocity, the pressure, and the density with the help of a single arbitrary function φ [52, 55, 56]

$$c^2\nabla^2\varphi = \varphi_{tt} + \left[(\nabla\varphi)^2\right]_t + (\nabla\varphi)^2\nabla^2\varphi, \tag{5.87}$$

where ∇^2 is the Laplacian operator and $c^2 = X(y + p_0)\rho_0^{-1} - (X - 1)[\varphi_t + ½(\nabla\varphi)^2]$. The density is determined as

$$\rho = \rho_0\left\{1 - (X - 1)\rho_0 X^{-1}(p_0 + y)^{-1}\left[\varphi_t + \frac{1}{2}(\nabla\varphi)^2\right]\right\}^{1/(X-1)} \tag{5.88}$$

Here we use the notation from [52, 55, 56]. Of course, they need in some clarification. We will not dwell on this, since these clarifications are not essential for a general understanding of the proposed approach.

Thus, the problem has been reduced to solving equation (5.87). It must be solved taking into account the motion of numerous different elementary particles in FB. That is, we have obtained a wave problem of studying the potential motion of a medium in which there are uniformly distributed, rectilinearly moving spherical particles. Apparently, this problem is reduced to solving a complex boundary value problem. Or we must move from discrete (that is, from considering each elementary particle in FB separately) to describing the problem using some averaging methods. Only after determining the function can we calculate the density and distribution of rarefactions (voids) in FB. This distribution determines variation of the gravity in space and time.

Let us emphasize that here we do not consider specific parameters of FB. However, let us dwell on the important properties of the two phase mediums, which are largely described by the equations given.

It is known that the velocity of the propagation of disturbances in mixtures can be much less than in their components.

As example we consider water/gas mixture. Figure 5.22 (left) shows the curves corresponding to the equation of state (5.85) for the discussed model of the mixture. The curves 3, 4, and 5 are calculated for $\phi_0 = 10^{-3}$, 10^{-4}, and 10^{-5}, respectively. It is seen that the nonlinearity of the mixture depends on the gas concentration. It is seen that the pressure in the mixture is always positive even when $\rho \to 0$. For example, let $\rho_0\rho^{-1} >> (1 - \phi_0)(1 + \lambda p_0)$. Then $p \approx -\phi_0 b^{-1} p_0 \approx \phi_0 \rho \rho_0^{-1} p_0$. If $\rho_0\rho^{-1} \approx \infty$, then $p \approx 0$.

The nonlinearity increases, when the gas concentration decreases. The last case, the main property of the bubbly liquid, is possible to be described by two segments (so-called two-line model). One segment is almost perpendicular to density axis, and other segment is almost parallel to this axis.

The curves on the right (Fig. 5.22) determine the dependence of the disturbance velocity on the gas concentration. Equation (5.87) (dashed lines) and some of its variants (solid lines) were used. When calculating curves 1–3, different values p and p_0 were used: $p = 0.06$ MPa, $p_0 = 0.1$ MPa (curve 1); $p = 0.16$ MPa, $p_0 = 0.2$ MPa (curve 2); and $p = 0.3$ MPa, $p_0 = 0.5$ MPa (curve 3).

We do note that $c = 1450$ m/sec at $\phi_0 = 0$ for all of the curves of Fig. 5.22 (right). At the same time, the curves in Fig. 5.22 reach different values of c (sound velocity) at $\phi_0 = 1$, because these values are different for gases with different values of p and p_0.

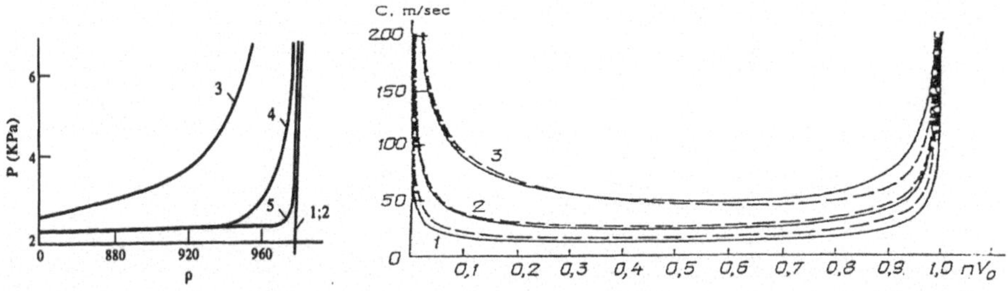

Fig. 5.22 Curves, which illustrate different laws $\rho = \rho(p)$ (left). Curves show the change of the sound speed with increasing of the gas component (right) [55, 56]

Here it is important to emphasize once again that according to the proposed theory the speed of sound in a mixture can be much less than in the components of the mixture. Let us assume that the above is also true for FB under consideration. The speed of propagation of disturbances in it is much less than in environments that do not contain virtual particles. On the other hand, this speed is much less than in a medium consisting only of virtual particles. Thus, we can assume that elementary particles propagate in FB without causing any disturbances if $\phi_+ = -\phi_- \approx \hbar$. However, this is probably true if the velocity of the elementary particle is such that it does not affect the rate of emergence and disappearance of virtual particles.

We will return to the discussion of this issue later.

Here we note that the decrease in the speed of sound in various mixtures relative to the speed of sound in the components of these mixtures is a well-known fact. It is illustrated by the curves presented in Fig. 5.23.

Curves of Fig. 5.23 (left) illustrate the gas influence on properties of water-based mud located on depths up to 5 km [56]. The sound velocity of the mud can reduce strongly, if the gas phase volume is slightly larger than 0 or slightly smaller than 1. It is interesting that the curves of Fig. 5.23 (left) qualitatively describe also the dependence of the sound velocity of the bubbly liquid from the gas component (Figs. 5.23, right and 5.22, right).

Therefore, we assume that many results and models of the theory of bubbly liquids are valid for modeling of the propagation of elementary particles in FB. In particular, the proposed approach may open the possibility of greatly simplifying of the understanding of such a complex phenomenon as the effect of gravity.

The above equations, figures, and texts are intended to provide a certain understanding of what we mean by FB, how this medium is related to the particle-waves described above in this chapter, and, finally, what are the main properties of FB for us. This understanding will simplify the presentation of the subsequent material of the chapter. We emphasize once again that the equations and figures are presented for a better, qualitative, understanding of the material. Since we will not perform any calculations based on the given equations, we only note that numerous results of studying the influence of nonstationary cavitation on the dynamic behavior of various deformable bodies can be found in my books and articles [55, 56].

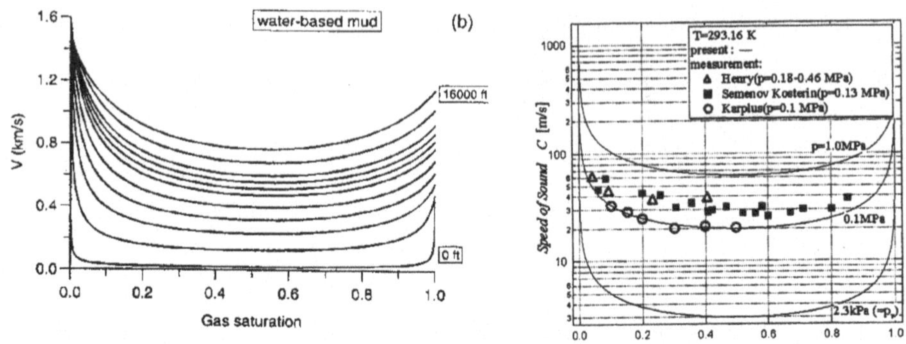

Fig. 5.23 Sound velocity of water-based drilling mud versus gas saturation. The depth varies, and the frequency is 25 Hz (left). Sound speed of bubbly liquid under isothermal condition (right) [41, 310, 329, 330]

Discussion Remark 1 Let us consider Figs. 5.22 and 5.23. It follows from the curves presented there that there are mixtures for which the speed of sound is minimal. It is much less than that of the mixture components. Let us assume that this is also true for the FB.

Let the state of the FB does not change and the speed of propagation of disturbances in the FB is minimal. Let us assume that it is zero. Thus, if the speed of motion of an elementary particle is not very high, it does not disturb the FB, does not affect quantum fluctuations, and remains $\phi_+ = -\phi_- \approx \hbar$. A stationary cavity appears behind the particle, which can be called a trace or the void ("emptiness").

As the particle speed increases, it begins to affect quantum fluctuations and situations arise when $\phi_+ > \hbar$. As a result, disturbances begin to propagate in the FB. This occurs directly at the front of the moving particle. A kind of compression shock or "boundary layer" or "attached" mass of the particle appears in front of it.

With a further increase in the particle speed, the thickness of the "boundary layer" increases. The influence of the particle speed on ϕ_+ and ϕ_- becomes stronger. More and more energy accumulates in front of the moving particle. As a result, solutions (5.71) and (5.72) can be realized, describing the appearance of a new pair of quantum particles. As the particle's velocity approaches to the speed of light, its "added" ("attached") mass grows without limit. The above may be true for particles having the rest mass. Massless particles have no mass by definition. But they have the impulse. We believe they leave a trace behind them.

Remark 2 Now let us recall that the FB under consideration contains a huge number of different moving elementary particles. We are interested in the interaction of these particles with the FB. In this interaction, as we believe, rarefaction zones (emptiness) arise, which ultimately determine the effect of gravity. To a certain extent, the problem of studying the described problem resembles the study of the motion of spherical dispersive particles immersed in two-phase media. Some idea of the complexity of the similar problems and the methods for solving it can be found in the specialized literature (see, for example, [55, 56] and Journal of Fluid Mechanics, 77(1), 27–44 (1976)).

Remark 3 The magnitude of gravity of a body is determined by its speed and the speed of the particles that form the body, since the volume of voids (traces) following the body in the FB depends on the specified speeds. The higher the speed of the body and its constituent particles, the larger and deeper the gravity well that arises in the FB and the stronger the gravitational effect of this body on other surrounding bodies. In order to give some idea of the results that can follow from the latest model, we will consider below some of the simplest, and almost obvious, consequences of it.

5.8.3 Nonlinearity and Cavitation "Emptiness"

Let us emphasize again that it is assumed that to a certain extent the properties of the FB are similar to the properties of a liquid containing gas bubble nuclei. That is, it is a strongly nonlinear medium.

1. Let me first dwell on the effect of nonlinearity. In contrast to what Newton would expect, identical experiments with identical boundary and initial conditions do not necessarily lead to identical results in the case of nonlinear problems. In our opinion, these are the ones that are characteristic of quantum mechanics. Therefore, the possibility of different results appearing when conducting identical experiments is not necessarily associated with the so-called probability wave. It is more natural to accept that Nature has several solutions for nonlinear problems.

In particular, elementary particles have a very definite number of variants of their behavior, determined by the governing equations, and one of them is realized in the experiment. Thus, if we have three solutions of the equations corresponding to the experiment, then the results of these three solutions will be repeated no matter how many times we conduct the same experiment!

Here we need to make a clarification. Just as it is impossible to step into the same river twice, it is also impossible to repeat the experiment. At best, we can say that the difference in the experiments is negligibly small.

2. Another circumstance is especially important for the problem of quantum gravity under consideration. In it, an elementary particle is considered as a spherical clot of a nonlinear scalar field moving in a scalar field. At the same time, it also propagates in the FB. Apart from the motion, the particle-wave is in no way connected with the FB. The situation is similar to the case of a rigid ball moving in an ideal fluid. It is assumed that after the particle acquires a certain speed, a trace appears behind the particle. This possibility is illustrated above (Fig. 5.19) using the example of a rigid sphere in a liquid flow. A more detailed picture of the emergence of cavitation bubbles (traces) is shown in Fig. 5.24 (on the left). Such traces arising behind spheres or moving particles will further

be called "emptiness." The volume of these voids can be calculated using Eqs. (5.83)–(5.87).

Figure 5.24 shows on the left a cloud of bubbles (trace) arising in liquid during flow around a rigid sphere. In the center is shown a cavitation zone generated by rapid deformation of an elastic ring in water. On the right is a picture of cavitation flow around a rigid plate by liquid.

Of course, Fig. 5.24 (left) is a very approximate illustration of the processes that are assumed to occur behind an elementary particle moving in FB. There, this trace is so weak that we do not have the equipment to record it and study it experimentally. The influence of these traces can be felt only through other effects. In particular, we associate the effect of gravity with the existence of the traces ("voids") considered here.

Experiments show that in many cases (Fig. 5.19 and 5.24), the dimensions of the cavitation zone are quite comparable with the dimensions of the streamlined sphere. In the case of a moving bullet, these dimensions can greatly exceed the volume of the bullet. Something similar may also occur when elementary particles move in FB, if their speed is high enough. When the size of an elementary particle, modeled as a sphere, is much smaller than the length of the corresponding trace, this size can be neglected and the sphere can be considered as a very thin plate (Fig. 5.24, right). In this case, the Kirchhoff scheme of cavitation flow around a plate can be used to qualitatively estimate the length of the trace from an elementary particle. As a rough approximation, it can be assumed that the particle leaves behind a tunnel of infinite length! Of course, there can be different options for calculating the volume of "emptiness" behind a moving particle. These variants can be based on Eqs. (5.83)–(5.87), solved in a region with variable geometry.

For example, from the very beginning of the calculation, the FB can be considered as an ideal liquid. In the process of calculation, some conditions for the occurrence or disappearance of cavitation behind a moving particle are constantly checked (some of them are considered in [55, 56]). In this case, a two-segment cavitation model can be used (see Fig. 5.22, curve 5). Points of the medium at which the cavitation conditions are met are considered at the next step as filled with gas (vapor of the liquid). If, during the calculation, the condition for the occurrence of

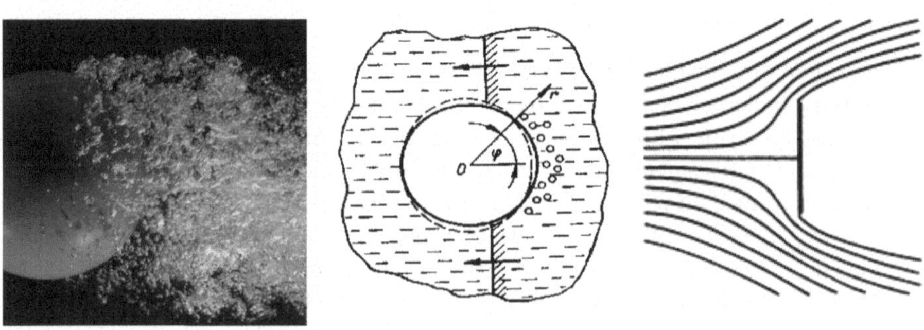

Fig. 5.24 Examples of cases of cavitation and gas bubbles in liquid

cavitation at a point in the medium is not met, then such a point corresponds to a liquid.

The proposed algorithm allows us to take into account changes in the volume of voids of the FB during the motion of an elementary particle. The described approach is qualitatively similar to those we developed for calculating nonstationary cavitation arising from the interaction of underwater shock waves with deformable structures [55, 56].

5.8.4 Limiting Cases of the Gravity Theory: Planetary Model of the Atom and the Dark Matter in Galaxies

Model of the atom. In 1911, Rosenford proposed his model of the atom. According to the model, the structure of the atom resembles the structure of the planetary system of the Sun. According to it, in the center of the atom is a nucleus, the size of which is several orders of magnitude smaller than the atom itself. Electrons move around the nucleus in circular orbits. All this occurs in a vacuum.

Note that according to modern views, the localization of electrons in an atom is known very, very approximately. In particular, de Broglie assumed that an electron should be considered not only as a particle but also as a wave oscillating with its own frequency in space-time (Fig. 5.25).

Let us look at this model assuming that the atom is immersed in the FB. Electrons will be modeled as ideally spherical solids.

Many quantum particles, while propagating, do not fix themselves in space-time in any way. They declare themselves only by interacting with the researcher's device. In particular, it is believed that an electron moving in orbit around the nucleus of an atom is not located at a specific point in space. We often learn about its existence only by its destruction. However, if we proceed from the theory being developed, the electron moves leaving behind a trace of a destroyed or distorted FB.

The proposed model, to some approximation, corresponds to the case of a solid sphere moving in an ideal (nonviscous) fluid. As we have already noted (see Figs. 5.19 and 5.24), a cavitation cavity arises behind the sphere at a certain speed of its movement.

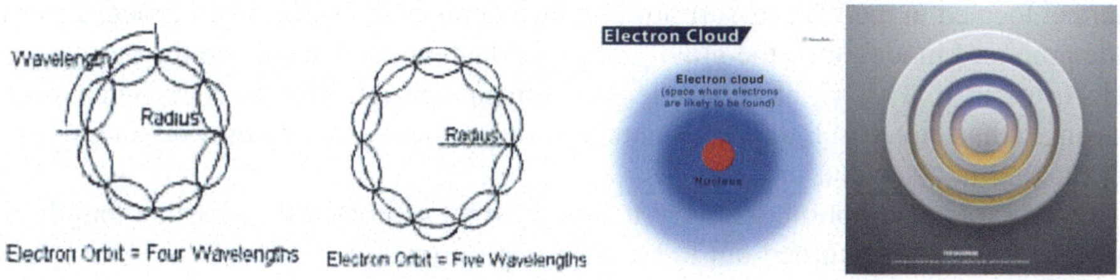

Fig. 5.25 Models of planetary models of the atom

It can be assumed that something similar can take place in the case under consideration. In particular, we will assume that if the electron speed is very high, then its trace in the FB can greatly exceed the size of the electron. For example, Fig. 5.19 shows that the trace of a bullet can be huge compared to the size of the bullet.

According to this, it is possible that some electrons move in a certain destroyed volume of the FB, which is not fully defined, both in radial and angular coordinates. We have attempted to reflect this situation in Fig. 5.25. The left part of the figure shows certain trajectories of an electron in the Rosenford-de Broglie model. In the center, multicolored rings nested inside each other are shown. Each of them corresponds to a certain spread of trajectories in the Rosenford-de Broglie model. Of course, the thicknesses and radii of the rings are in no way related to the exact position of the electrons.

The right figure shows, qualitatively, the traces of electrons in the FB. It is assumed that the electrons are located in the recesses of the circular rings shown there. They highlight, as it were, the contours of the circular tunnels in the FB along which the electrons move. Their locations in the tunnels remain completely unknown.

We have presented a diagram of the electron structure of an atom assuming that when an electron moves in a circle, the length of its trace can cover the entire trajectory of the movement (Fig. 5.25).

Thus, cases are considered when the electrons move with a sufficiently high speed. Then a trace ("emptiness") can arise in FB behind the electron which covers the entire circular trajectory of the electron in an atom.

Let us consider as an example a hydrogen atom. The size of the atom is 10^{-8} cm.

Let an electron fly around the nucleus of the atom in a time of about 10^{-16} s and the speed of the electron is about 10^8 cm/s. In this case, the electron makes about 10^{16} flybys per second around the nucleus of the atom. We assume that these parameters are sufficient to ensure the complete or partial destruction and disappearance of FB on the trajectory of the electron and in the vicinity of the trajectory.

From the above, the following conclusion can be drawn.

The more electrons an atom of a substance contains, the more emptiness it has. In other words, the more massive the body, the more emptiness it has and the deeper the gravitational well corresponding to this body. Thus, two opposites – mass and emptiness – coexist together. They determine the gravity. In this, in our opinion, the dialectic of nature manifests. Indeed, "opposite things are identical" according to Heraclitus.

According to this, the electron in an atom moves in a very long and very narrow tunnel located in the FB and surrounding the center of the atom. If we proceed from the proposed model, then the gravity of individual atoms depends on the number of electrons in it. That is, atoms of heavy elements are attracted to each other more strongly than atoms of lighter elements. Of course, we are talking only about gravity, not taking into account other forces.

Using this assumption, we have to some extent estimated the possible length of the electron trace. It turned out to be very, very large.

Above, as in the entire book, I discard details that are unimportant for understanding the main idea under discussion. Of course, they can be very important, but to understand the main idea, maybe false, we accept that these details are not essential.

Of course, in the general case, we are dealing with an almost infinite number of elementary particles and their traces. It is difficult to imagine that when it will be possible to calculate their influence on the FB and the dynamics of gravity with any precision. Apparently, a radical simplification of even the described, extremely simplified, approach will be required.

For example, it can be assumed that the contribution to the general loosening of the FB is determined by electrons moving in atoms, and they move (exist) in tunnels (inside the "emptiness" zones). The "volumes" of such voids in the FB in each type of atom can be calculated relatively simply, and this opens the way to assessing the influence of moving electrons (particle-waves) in atoms on the gravity surrounding us!

Variation of Gravity in Galaxies and the Dark Matter Model A publication from 1930 points to Swedish Knut Lundmark being the first to realize that the Universe must contain much more mass than can be observed. In the 1930s, American astrophysicist Fritz Zwicky observed the rotation of galaxies in a spiral shape. According to the laws of mechanics, stars located at the edges of the galaxy should move more slowly than stars in the center of the galaxy; otherwise they would fly apart. But Zwicky found that all the stars in the center of spiral galaxies move approximately the same, and the stars on the periphery move as if the mass of the galaxy were much greater than that predicted by existing theories of gravity.

There is a discrepancy between the observed rotation curves of spiral galaxies and the theoretical prediction of the existence of a central dominant mass in them (Fig. 5.26). In particular, the rotational/orbital velocities of stars on the outskirts of

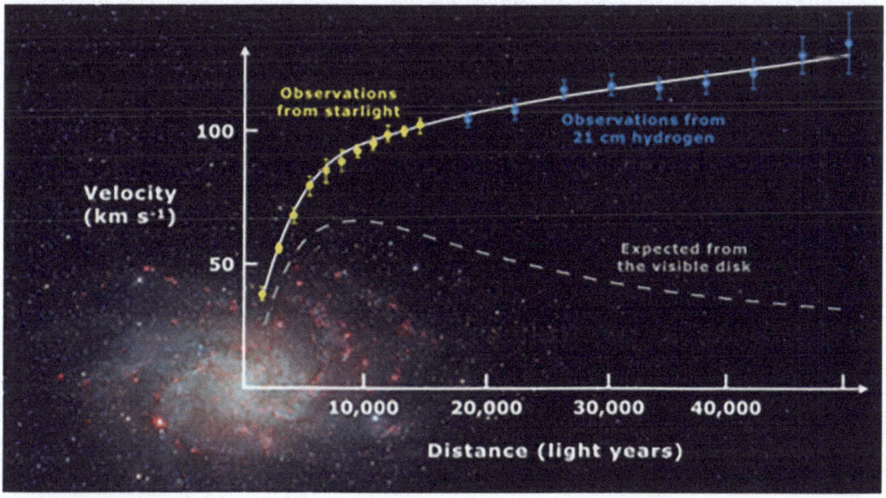

Fig. 5.26 Rotation curve of spiral galaxy Messier 33 (yellow and blue points with error bars), and a predicted one from distribution of the visible matter (gray line). https://en.wikipedia.org/wiki/Galaxy_rotation_curve

spiral galaxies do not follow the rules found in other orbital systems, such as stars/planets and planets/moons. To solve this puzzle, a mysterious substance was introduced that can only be detected through its gravitational interaction with other objects, which was called "dark matter." Let us consider this puzzle from the point of view of the proposed model of gravity. It is easy to see that for the problem under consideration, the results of the proposed theory are determined by the velocities of the stars (and, in particular, the elementary particle-waves which the galaxy consists). According to Fig. 5.26, when a star is near the center of the galaxy, its velocity is practically independent of its orbital velocity. Apparently, therefore, the gravitational properties of the stars located there and the center of the galaxy are quite consistent with the expected ones. However, as we move away from the center of the galaxy, its gravity begins to depend more and more on the orbital motion of the galaxy. Let us consider the gravitational situation of stars located in the center and on the periphery of the galaxy additionally. In the first case, the gravitational properties of the star are completely determined by the depth of the gravitational well in the FB located in the center of the galaxy. The influence of the gravitational well of the star itself is negligible. However, as the star moves away from the center, the influence of its orbital motion increases, and the depth of its own gravitational well in the FB grows. It is also important that the star is increasingly approaching the edge of the central gravitational well. As a result, the gravitational influence of the latter becomes equal to the local influence of the gravitational well of the star (see Fig. 5.26). As the star moves further away from the center and the orbital velocity of the star increases, the depth of the gravitational well of the star begins to exceed the magnitude of the gravitational influence of the center of the galaxy. Of course, this does not mean that the star begins to move independently of the galaxy. Indeed, the well of its gravitational influence is very local relative to the influence of the center of the galaxy. It should be borne in mind that the volume of the peripheral part of the galaxy can be an order of magnitude greater than the volume of the central part; in addition, the orbital velocity of the particle-waves can be very large. This can lead to the fact that the influence of the peripheral stars on the total gravity of the galaxy can significantly exceed the influence of the central stars. The described result does not correspond in any way to the Newtonian theory of gravity. At the same time, the proposed model of gravity, in our opinion, qualitatively explains the results of observations (Fig. 5.26) and does not require additional assumptions such as "dark matter." At the same time, we understand that the outline of a new theory of gravity discussed here may be at odds with the data of many other experiments (see https://www.nature.com/articles/nature25767; https://iopscience.iop.org/article/10.3847/1538-4357/ace101 and https://doi.org/10.1103/PhysRevLett.127.161302). The larger the galaxy, the greater the contribution of the orbital motion of its stars (particles) to gravity! Here we must keep in mind that the orbital velocity can be so large that it affects the mass of the particle-waves. Whether this added mass affects gravity is an open question!

5.8.5 "Emptiness," Motion, Gravity, Mass, and Inertia

Gravitation Thus, the proposed understanding of the causes of gravity is primarily associated with the assumption of the existence of the FB, where there are the quantum fluctuations, as well as with the movement of elementary particles. As soon as their movement arose, voids began to form in the FB. The more elementary particles in the body, the more movement in it, the more voids in the FB arise in place of this body.

Let us repeat what has been said for the sake of clarity. Every material body or energy clot consists of wave fields and corresponding wave-particles. The book considers cases when these wave-particles move in a scalar field. They correspond to certain pressures and densities (see (5.73) and (5.74)). At the same time, these particles move in a fundamental background space, where the wave-particles leave certain traces in the form of "emptinesses" ("voids").

These voids lead to a kind of local loosening of FB. As a result, a certain dynamic, curved landscape of FB appears. As a consequence of all that has been said, the dynamics of the curvature of this landscape is determined by the initial location and dynamics of the generating material body or energy clot. Moving or living in this space-time landscape, we perceive it as gravity.

Let us assume that as a result, in the FB, a depression arises under the body. Moreover, the more massive the body and the higher the speed of its constituent elementary particles, the greater the depth of the depression. In this case, the less massive body and the bodies whose particles move more slowly (e.g., less heated bodies) will be attracted to the more massive body. It is possible to select such masses and speeds of rectilinear motion of these bodies, at which the less massive body (e.g., the Earth) will rotate around the more massive body (e.g., the Sun). The mutual attraction of two bodies is stronger, the greater the speed of the particles that make them up. By increasing the speed of these particles, the gravity that binds them can be strengthened.

What Gives an Object the Property of Inertia? The answers to this question can be found in the works of many scientists. The most famous answers are those of Newton, Mach, and Einstein [62]. Most modern physicists associate inertia and resistance to acceleration with the influence of the Higgs field on objects. In particular, it is assumed that atoms, electrons, and other elementary particles that make up our bodies are immersed in the ocean (field) of the Higgs. It interacts with these particles, and this manifests itself in the effects of inertia and resistance to change in speed.

We will consider the case when quantum fluctuations exist in the FB. Therefore, the speed of propagation of disturbances in the FB is extremely small (see Figs. 5.22 and 5.23). Therefore, the interaction of the FB and the moving particle in a stationary mode occurs almost without using the wave mechanism.

Thus, if an elementary particle begins to move, it begins to depend only on the FB particles adjacent to it. As a result, the elementary particle acquires a kind of boundary layer (additional mass). If its speed increases further, the thickness of this boundary layer increases. The speed of propagation of disturbances in this boundary layer is not now zero.

Moreover, the density of the boundary layer increases so much that the occurrence of low-intensity quantum fluctuations there becomes difficult.

Such a process of increasing, as it were, the mass of the particle, continues with an ever-increasing increase in the particle's speed. Moreover, this layer exists only on the front of the moving elementary particle. Of course, as soon as the particle stops, everything returns to its original state.

As a result, the greater the speed of the particle, the greater its "mass" and the more difficult it is for the particle to continue to increase its speed.

In the limit, when the particle's speed reaches a certain critical value, the thickness of the accompanying boundary layer becomes so significant that further increase in the particle's speed becomes impossible. The above allows us to think that inertia is a certain analogue of the dynamic pressure that arises in front of a body moving in a liquid or gas.

Thus, as the particle's speed increases, its "mass" and the tendency to conserve the acquired energy (inertia) increase. When the particle stops, the boundary layer disappears. It increases only with accelerated motion of the particle. Only then does the thickness of the "boundary layer" increase.

The particle velocity, when it grows strongly, begins to influence the intensity of the appearance of quantum fluctuations. Thus, the density of the boundary layer increases so much that the appearance of low-intensity quantum fluctuations there becomes difficult. The speed of propagation of disturbances there begins to change, as it should be in a two-phase liquid.

Of course, all this is already far beyond the description of our crude theory. However, let us see what can follow from it.

With a further increase in the particle velocity, only very intense quantum fluctuations arise inside the "super-compressed" boundary layer. As the particle velocity tends to the speed of light, the density of the boundary layer tends to infinity, and the growth of the speed of propagation of disturbances there stops and can fall to zero.

On the other hand, when increasing the acceleration of the elementary particle, it is possible to obtain very different values ϕ_+ and ϕ_-. I believe that these values can take on the values $\phi_+ = -\phi_- \approx \hbar$, $\phi_+ > \hbar$, or $\phi_+ >> \hbar$ depending on the particle speed (see Fig. 5.21). In the last case, they arise so far from each other that their pair does not constitute a virtual quantity, but forms a pair of separate real quantities quite comparable with the particle that generated them as a result of acceleration.

Final Remark In this chapter, it was assumed that the particle-waves move along well-defined trajectories. Many questions may arise about this assumption. They are discussed to some extent in the next chapter. This chapter is devoted to the probabilistic interpretation of quantum mechanics based, to a large extent, on the results of double-slit experiments.

References

1. Lamb H (1932) Hydrodynamics, 6th edn. Dover Publications, New York
2. Taylor GI (1910) The conditions necessary for discontinuous motion in gas. Proc Roy Soc Lond A 84:371–377
3. Jimenez J (1973) Nonlinear gas oscillations in pipes. Part 1. Theory J Fluid Mech 59:23–46
4. Galiev SU (1999) Passing through resonance of spherical waves. Phys Lett A 260:225 –233. Bhattacharyya D, Galiev SU, Panova OP (2002) Trans-resonant evolution of spherical waves governed by the perturbed wave equation. Strength Mater 14(4) : 359–368; Galiev SU, Panova OP (2002) Localization of resonant spherical waves. Strength Mater 34(1) : 73–79
5. Cowen R (2013) Single electrons make waves. Nature. https://doi.org/10.1038/nature.2013.14015. Dubois J et al (2013) Minimal-excitation states for electron quantum optics using levitons. Nature 502:659 –663; Controlling light propagation. Science 340:247 ; Rodríguez-Fortuño FJ et al (2013) Near-filled interference for the unidirectional excitation of electromagnetic guided modes. Science 340:328 –330
6. Adam J et al (2021) Measurement of e+e− momentum and angular distributions from linearly polarized photon collisions. Phys Rev Lett 127:052302
7. Wilson CM et al (2011) Observation of the dynamical Casimir effect in a superconducting circuit. Nature 479:376–379. https://doi.org/10.1038/nature10561. Esquivel-Sirvent R, Reyes L (2008) Pull-in control in microswitches using acoustic Casimir forces. Europhy Lett 84(4): 48002
8. Yoo D et al (2021) Ultrastrong plasmon–phonon coupling via epsilon-near-zero nanocavities. Nat Photonics 15:125 –130
9. Howgego J (2015) Can We Get Energy from Nothing? https://www.newscientist.com/article/mg22730370-800-can-we-get-energy-from-nothing
10. Martin F (2015) Quantum mechanics: thought experiments made real. Nat Photonics 9:76 –77
11. Close F (2008) Nothing: a very short introduction. Oxford University Press; Acme Collaboration, Baron J, Campbell WC et al (2013) Order of magnitude smaller limit on the electric dipole. Science 343(6168):269–272 https://doi.org/10.1126/science.1248213; Roussy TS et al (2023) An improved bound on the electron's electric dipole moment. Science 381(6653):46–50. https://doi.org/10.1126/science.adg408
12. Sonnleitner M, Trautmann N, Barnett SM (2017) Will a decaying atom feel a friction force? Phys Rev Lett 118(5) : 053601
13. Barnett SM, Sonnleitner M (2018) The vacuum friction paradox and related puzzles. Contemp Phys 59(2) : 145–154. https://doi.org/10.1080/00107514.2018.1439333
14. Mo H, van den Bosch F, White S (2011) Galaxy formation and evolution. Cambridge University Press
15. Ijjas A, Steinhardt PJ, Loeb A (2013) Inflationary paradigm in trouble after Planck2013. Phys Lett B 723:261 –266
16. Galiev SU, Galiyev TS (2021) Modeling of an appearance of particle-waves and the Universe as the extreme wave resonant events. Herald of Ufa Scientific Center Russian Academy of Sciences. № 2. pp 5–23
17. Galiev SU, Galiyev TS (2017) Nonlinear quantum waves in the light of recent slit experiments. The University of Auckland. http://hdl.handle.net/2292/3675
18. Cox B, Forshaw J (2013) The quantum universe: everything that can happen does happen. Hachette Books, London
19. Green MB et al (2012) Superstring theory. Cambridge University Press
20. Greene B (1999) The elegant universe. Vintage Books, New York
21. Liddle A, Ljveday J (2008) The Oxford companion to cosmology. Oxford University Press
22. Susskind L (2006) The cosmic landscape. Little, Brown and Conpany, New York
23. Halpern P (2024) The allure of the multiverse. Basic Books
24. Livio M (2006) From Darwin to Einstein. Simon & Schuster, New York

25. Hawking SW (1982) The development of irregularities in a single bubble inflationary universe. Phys Lett B 115:295
26. Dirac PA (1930) A theory of electrons and protons. Proc R Soc Lond Ser A 126:360 –365. https://doi.org/10.1098/rspa.1930.0013
27. Granz LF, Mathiak O, Villalba-Chávez S, Müller C (2019) Electron-positron pair production in oscillating electric fields with double-pulse structure. Phys Lett B 793:85 –89. https://doi.org/10.1016/j.physletb.2019.04.026
28. Heisenberg W, Euler H (1936) Consequences of Dirac theory of the positron. Z Phys 98:714
29. Sun Z, Basov DN, Fogler MM (2018) Universal linear and nonlinear electrodynamics of a Dirac fluid. Proc Natl Acad Sci 115(13) : 3285–3289
30. Keser AC, Lyanda-Geller Y, Sushkov OP (2022) Nonlinear quantum electrodynamics in Dirac materials. Phys Rev Lett 128:066402. https://doi.org/10.1103/PhysRevLett.128.066402
31. Jorge AR, Martínez AP, Querts ER (2024) Nonlinear electrodynamics for the vacuum of Dirac materials: photon magnetic properties and radiation pressures. Phys Rev A 110(1):012201; Sun Z, Basov DN, Fogler MM (2018) Universal linear and nonlinear electrodynamics of a Dirac fluid. PNAS, 115(13):3285–3289. https://doi.org/10.1073/pnas.1717010115; Yuli Lyanda-Geller K, Sushkov OP (2022) Nonlinear quantum electrodynamics in dirac materials. Phys Rev Lett 128:066402. https://doi.org/10.1103/PhysRevLett.128.066402; Romero Jorge AW, Pérez Martínez A, Querts R (2024) Nonlinear electrodynamics for the vacuum of Dirac materials. Photon magnetic properties and radiation pressures. arXiv:2402.01042v5 [hep-th] 14 Feb 2024
32. Galiev SU, Galiyev TS (2001) Nonlinear trans-resonant waves, vortices and patterns: from microresonators to the early universe. Chaos: an interdisciplinary. Journal of Nonlinear Science 11(3):686–704
33. Galiev SU (2008) Strongly-nonlinear wave phenomena in restricted liquids and their mathematical description. In: Perlidze T (ed) New nonlinear phenomena research. Nova Science Publishers, New York, pp 109–300
34. Galiev SU (2011) Charles Darwin's geophysical reports as models of the theory of catastrophic waves. Moscow, p 655 (in Russian)
35. Galiev SU (1999) Topographic effect in a Faraday experiment. J Phys A Math Gen 32:6963–7000
36. Galiev SU (2015) Darwin, geodynamics and extreme waves. Springer
37. Galiev SU (2003) The theory of non-linear transresonant wave phenomena and an examination of Charles Darwin's earthquake reports. Geophys J Int 154:300 –354
38. Galiev SU, Ilgamov MA, Sadikov AV (1970) Periodic shock waves in a gas. Izv. Akad. Nauk USSR, Mech. Fluid and Gas 2:57 –66
39. Galiev SU (1988) Nonlinear Waves in Bounded Continua. Naukova Dumka: Kiev (in Russian); Galiev SU (1972) Forced oscillations of a nonlinear elastic layer. Izv. Acad. Nauk USSR, Mech. Solid Body 1:58 –63; Galiev SU (1972) Forced longitudinal oscillations of a nonlinear elastic body. Izv. Acad. Nauk USSR, Mech. Solid Body 4:80 –87; Galiev SU (1973) Nonlinear gas oscillations maintained in a closed pipe by a heat source. Soviet Aeronautics 16(2) : 7–12; Galiev SU, Galiyev TS (1994) Linear and discontinuous forced oscillations of flow of bubbly liquid in a Deformable Pipe (survey). Strength Mater 9:633 –65
40. Galiev SU, Galiyev TS (2014) Nonlinear scalar field as a model describing the birth of the Universe. Известия Уфимского Научного Центра Российской Академии Наук (Herald of Ufa Scientific Center, Russian Academy of Sciences (RAS)) 2:7 ; Galiev SU, Galiyev TS (2016) Simple model of the origin of the Universe. Известия Уфимского Научного Центра Российской Академии Наук (Herald of Ufa Scientific Center, Russian Academy of Sciences (RAS)) 4:27
41. Serjeant S (2010) Observational cosmology. Cambridge University Press, Cambridge
42. Сибгатуллин НР (1984) Колебания и волны в сильных гравитационных и электромагнитных полях. Наука, Москва. (Sibgatullin NR (1991) Oscillations and waves in strong gravitational and electromagnetic fields. Springer)

43. Vilenkin A (2006) Many worlds in one. Hill and Wang, New York
44. Vilenkin A, Shellard EPS (1994) Cosmic strings and other topological defects. Cambridge University Press
45. Greene B (2004) The fabric of the cosmos. New York
46. Renn J, Gutfreund H (2023) The Einsteinian revolution: the historical roots of his breakthroughs. Princeton University Press
47. LIGO Scientific Collaboration et al (2017) Multi-messenger observation of a binary nentron star merger. arXiv:1710.05833 [astro-ph.HE] https://doi.org/10.3847/2041-8213/aa91c9
48. Abbott BP et al (2016) Observation of gravitational waves from a binary black hole merger. Phys Rev Lett 116:061102
49. Decarli R et al (2024) A quasar-galaxy merger at z~6.2: rapid host growth via the accretion of two massive satellite galaxies A&A, Forthcoming article Received: 15 January 2024. Accepted: 10 June 2024 https://doi.org/10.1051/0004-6361/202449239
50. Abbott BP et al (2017) Focus on the electromagnetic counterpart of the neutron star binary merger GW170817. Astrophys J Lett 848:L12. https://doi.org/10.3847/2041-8213/aa91c9; Li, Q. et al. Measurements of the Gravitational Constant Using Two Independent Methods. Nature, 560, 582-588 (2018) https://doi.org/10.1038/s41586-018-0431-5; Patrik Švančara et al. Rotating curved spacetime signatures from a giant quantum vortex. Nature, 628, pages 66–70 (2024); Burulo C., Liberati S., Visser V. Analogue gravity. Living Rev. Relativ. 14, 3 (2022)
51. Nicholl M, Berger E, Kasen D, Metzger BD, Elias J, Briceño C, Alexander KD et al (2017) The electromagnetic counterpart of the binary neutron star merger LIGO/Virgo GW170817. III. Optical and UV spectra of a blue kilonova from fast polar ejecta. Astrophys J Lett 848(2):L18
52. Galiev ShU (1981) Dynamics of Hydroelastoplastic Systems. Naukova Dumka, Kyev (in Russian); Galiev ShU (1988) Nonlinear waves in bounded continua. Naukova Dumka, Kyiv (in Russian)
53. de Graaf KL, Brandner PA, Pearce BW (2017) Spectral content of cloud cavitation about a sphere. J Fluid Mech 812:10. https://doi.org/10.1017/jfm.2016.819
54. Venning J, Pearce B, Brandner P. Scale effects on cavitation about a sphere. https://figshare.utas.edu.au/articles/conference_contribution/Scale_effects_on_cavitation_about_a_sphere/23100857
55. Galiev ShU (1997) The influence of cavitation upon anomalous behavior of a plate/liquid/underwater explosion system. Int J Impact Eng 19:345–359 ; Galiev ShU (1999) A method of solving nonstationary three-dimensional problems of hydroelasticity with allowance for fluid failure. Int J Impact Eng 22:469–483; Galiev ShU, Flay RGJ (2014) Interaction of breaking waves with plates: the effect of hull cavitation. Ocean Eng 88:27–33
56. Galiev ShU (2020) Extreme waves and shock-excited processes in structures and space objects. CRC Press
57. Majer U, Sauer T (2005) Hilbert's "World Equations" and His Vision of a Unified Science. Einstein Stud 11:259–276. arXiv:physics/0405110. https://doi.org/10.1007/0-8176-4454-7_14. ISBN 978-0-8176-4454-3. S2CID 985751; Weinberg S (1992) Dreams of a final theory: the search for the fundamental laws of nature. Random House; Stephen Hawking THE THEORY OF EVERYTHING Phoenix Books 9465 Wilshire Boulevard, Suite 315 Beverly Hills, CA 90212; Davies PCW, Brown J (eds) (1988) Superstrings: a theory of everything? Cambridge University Press
58. Dirac PAM (1951) Is there an Æther? Nature 168(4282):906–907. https://doi.org/10.1038/168906a0. S2CID 4288946
59. Dirac PAM (1952) Is there an Æther? Nature 169(4304):702. https://doi.org/10.1038/169702b0. S2CID 4279193
60. Sinha KP, Sivaram C, Sudarshan ECG (1976) Aether as a superfluid state of particle-antiparticle pairs. Found Phys 6(1):65–70
61. Dzhunushaliev V, Zloshchastiev KG (2013) Singularity-free model of electric charge in physical vacuum: non-zero spatial extent and mass generation. Cent Eur J Phys 11(3):325–335
62. Bartusiak M (2000) Einstein's unfinished symphony: listening to the sounds of space-time. National Academies Press, Washington DC

63. Galiev ShU (1981) Dynamics of hydroelastoplastic systems. Naukova Dumka, Kyev (1981) (in Russian); Galiev ShU (1988) Nonlinear waves in bounded continua. Naukova Dumka, Kyiv (in Russian); Galiev ShU (1997) The influence of cavitation upon anomalous behavior of a plate/liquid/underwater explosion system. Int J Impact Eng 19:345–359; Galiev ShU, Flay RGJ (2014) Interaction of breaking waves with plates: the effect of hull cavitation. Ocean Eng 88:27–33; Galiev ShU (2000) Extreme waves and shock-excited processes in structures and space objects. CRC Press
64. Michael KL et al (2021) Experimental measurement of the intrinsic excitonic wavefunction. Sci Adv. https://doi.org/10.1126/sciadv.abg0192, https://doi.org/10.48550/arXiv.2011.13104

Chapter 6
Nonlinear Quantum Waves in the Light of Recent Slit Experiments

The history of the double-slit experiment (DSE) is more than two centuries old. Its results played a significant role in the discovery of the wave nature of light and then formed the basis for important provisions of quantum mechanics. However, the results are still difficult to interpret unequivocally. In particular, they have been treated recently even outside the framework of the universally accepted, experimentally verifiable science. For example, they were associated with the influence on the experiment of the experimenter's consciousness or with the influence of worlds which exist parallel to our world.

At the same time, indeed, the interpretation of the results is extremely difficult and the situation with the understanding of these results, especially in the light of recent studies, is becoming more and more complicated. The purpose of our study is to analyze the results of several recent experiments and to introduce a new understanding by taking into account the possibility of synchronous interaction of all the components of the process of the experiment. We believe that it is necessary to take into account the nonlinear nature of the interacting fields.

First, we consider the most significant experiments and their analysis in the framework of traditional approaches. Then nontraditional approaches are considered. Some of them seem implausible, but they lie in line with the most high-profile ideas of modern physics. We draw readers' attention to the fact that practically no researcher pays attention to the influence of quantum fluctuations in slits and the interaction of the physical fields on the results of the experiment.

A systematic study of these issues has apparently not been carried out. Therefore, we propose a number of thought experiments. At the end of the study, we use the mathematical modeling of the physical processes responsible for the results of the experiments. Plasma waves and waves in vacuum are considered. The interaction of these waves is studied in the attempts to model some processes occurring in the DSE.

Thus, the main goal of our research is to introduce the reader to the new understanding of certain positions of quantum mechanics on the basis of the new understanding of the results of the DSE. We believe that in order to explain the

S. U. Galiev, *Extreme Multivalued Waves as a Model of Everything*,
https://doi.org/10.1007/978-3-031-67027-5_6

results of the DSE, it is first necessary to introduce greater clarity in quantum mechanics.

6.1 Introduction

Niels Bohr said to Werner Heisenberg "Your theory is crazy, but it is not crazy enough to be true." At the same time, Albert Einstein wrote "Most of the fundamental ideas of science are essentially simple, and may, as a rule, be expressed in a language comprehensible to everyone." "The great tragedy of science," Thomas Huxley observed, "the slaying of a beautiful hypothesis by an ugly fact." Science is the clash of opinions.

Short history of the puzzling experiment: Quantum mechanics is the best theory we have for describing the world at the nuts-and-bolts level of atoms and subatomic particles. According to this theory, particles behave both as a particle and as a wave. This point of view was supported by the results of the double-slit experiment (DSE), which is one of the most beautiful experiments in physics. It is frequently used in classic textbooks on quantum mechanics. The experiment demonstrates the principle of wave-particle duality. The double-slit experiment is schematically illustrated in Fig. 6.1. As we see from Fig. 6.1, particles illuminate a thin solid (metal) barrier in which two slits are cut. A photographic plate records the light that gets through the slits—brighter areas of the photograph indicate more incident light.

Particles pass through two narrow slits in a screen and then impinge on a second screen making small dots. After many particles have hit the second screen, an interference pattern develops in the form of light and dark stripes similar to light interference fringes. It looks like we are dealing with waves passing through the two slits. On the other hand, the particles make small dots on the second screen! It looks like the particles behave as moving point in space. Most amazingly, the pattern appears even if the particles arrive at the screen one at a time. It appears that a particle somehow interferes with itself, but how can this be possible?

Following Aatish Bhatia [1, 3], we put a few questions and give several answers to highlight the situation with this experiment.

Let us consider an electron that arrives at the screen. Which slit did it go through?

Fig. 6.1 The data obtained when electrons are fired and both slits are open [1, 2]

1. Did the electron go through the left slit?

No! Because when you cover up the right slit, the stripy pattern disappears and you get a boring single band instead.

2. Did the electron go through the right slit?

No! For the same reason as above. When you cover up the left slit, instead of the stripy pattern you get a single band.

3. Does the electron go through both slits?

No! Because if that were true, we'd expect to see the electron split into two, and one electron (or maybe half) would go through each slit. But if you place detectors at the slits, you find that this never happens. You always see only one electron at a time. It never, ever splits into two.

4. Did the electron go through neither slit?

No! Of course not, that's just silly. If you cover both slits, nothing happens.

However, to this day, physicists do not agree on the best way to interpret the results of these quantum experiments. It is hard to avoid the implication that we can describe quantum effects but we do not have a clear understanding of them. Richard Feynman famously said that "the double-slit experiment has in it the heart of quantum mechanics. In reality, it contains the only mystery" [1, 4, 5]. Below, there is an extract from *QED—The Strange Theory of Light and Matter* of R.P. Feynman [6].

> "Before I go into the main part of this lecture, I'd like to show you another example of how light behaves. What I would like to talk about is very weak light of one color—one photon at a time going from a source, at S, to a detector, at D (see Fig. 6.2a).
>
> Two tiny holes (at A^* and B^*) in a screen that is between a source S and a detector D let nearly the same amount of light through (in this case 1%) when one or the other hole is open. When both holes are open, "interference" occurs: the detector clicks from 0 to 4% of the time, depending on the separation of A^* and B^*—shown in Fig. 6.3a.
>
> When we put in detectors at A^* and B^* (Fig. 6.2), we changed the problem. ....The complete story of this situation is very interesting: if the detectors at A^* and B^* are not perfect and detect photons only *some* of the time, then there are three distinguishable final conditions." Here we finished the citation of the text from R.P. Feynman [1, 5] (see also (a), (c), and (d) of Fig. 6.3).

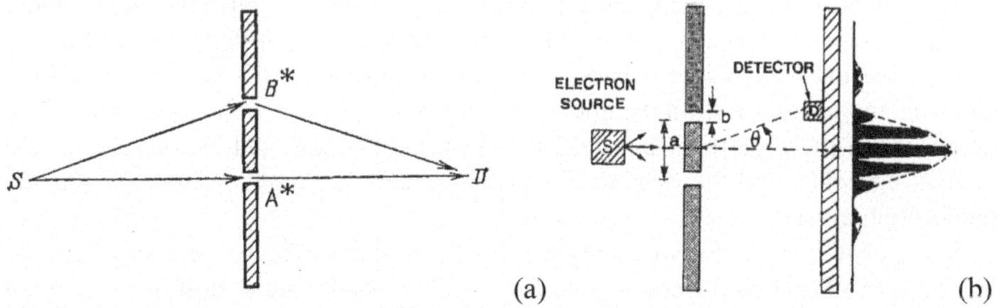

Fig. 6.2 A rough scheme of the double-slit experiment presented by R.P. Feynman (**a**) [1, 5]. An electron propagates like a wave from source S through the two slits to the detector D but always registers as a discrete particle. The particle distribution shows the oscillatory behavior of wave interference. The broken line which envelops the interference fringes is the one-slit electron diffraction pattern (**b**) [1, 4]

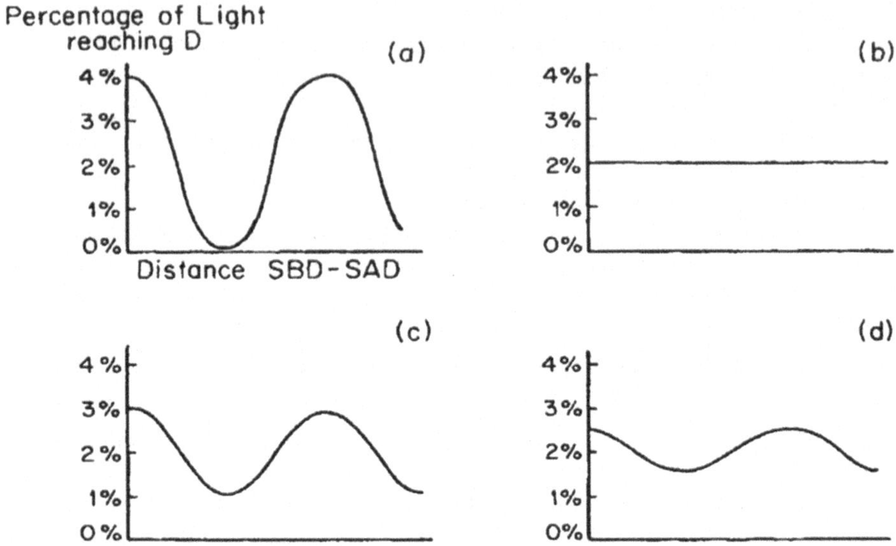

Fig. 6.3 Effects of the distance SBD-SAD and of the quality of detectors on the interference. Here the value SAD is similar SA*D, and the value SBD is similar SB*D in Fig. 6.2 (see [6]). (**a**) no detectors. In this case, the detector clicks from 0% to 4% of the time, depending on the separation of A and B; (**b**) perfect detectors. The amount of light is constant—2%; (**c**) slightly reliable detectors. The amount of light varies from 3% to 1%; (**d**) detectors close to perfect. The amount of light varies near 2%

It is important for us that experimental results depend strongly on the distance SBD-SAD. In other words, the results are determined by the location of the experimental equipment. In considering the case, the dependence on the distance is described by the harmonic law. According to Fig. 6.3, there is a periodical amplification of the light up to maximal value when the distance SBD-SAD increases. Is it possible that this amplification is connected with its own modes of oscillations of the system and certain resonances?

We consider the experimental equipment as a system of physical fields. Their strong interaction might be awakened by the first particles. A part of them hits the barrier (the interslit material) and excites its oscillations. The following particles move in the space of strongly interacting wave fields. The trajectories and wave properties of these particles are determined by the "nodes" and "antinodes" of the interacting fields, and these "nodes" and "antinodes" are fixed on the detector (the photographic plate).

In contrast, considering his thought experiment and its results, Feynman declared "The theory of quantum electrodynamics describes Nature are absurd from the point of view of common sense. And it agrees fully with experiments. So I hope you can accept Nature as She is – absurd."

At this point, you are probably thinking that this is getting a bit ridiculous. *Why can't we just look at the damn electron and see which path it took?* The problem with this idea is that looking at something means shining light on it, and shining light on it means bumping it with a photon. If you're a tiny electron, this bump disturbs your original path. Perhaps the most renowned of its mysteries is the fact that the outcome of quantum experiment can change depending on whether or not we choose to measure some property of the particles involved.

The double-slit experiment is often used to highlight the differences and similarities between the various interpretations of quantum mechanics. It must be remembered that the initial explanations of the double-slit experiments was obtained when concepts of quantum fields and vacuum energy were absent. However, these explanations were very interesting. Furthermore, these explanations formed the basis of the quantum theory (quantum mechanics).

However, even from the point of view of quantum mechanics, the results of the double-slit experiments are difficult to explain. To clear these difficulties, theorists resort to ever-increasing levels of mathematical sophistication and abstraction. As a result of many attempts to explain, researchers again and again encounter difficulties that can be explained by a lack of understanding at the fundamental level.

Taking this into account, we suggest a new understanding of the fundamental experiment of quantum mechanics. As we see from Fig. 6.1, the vertical length of the slits is much greater than their width. We can consider the strip of the metal between the slits as a one-dimensional resonator. The slits can also be considered resonators for vacuum fluctuations. So we have three "pianos" which can play "resonantly." We think that the results of the double-slit experiments are determined by the resonant interaction of vacuum and plasmonic oscillations (waves) with particles.

According to modern quantum field theory, absolutely everything is made of a field or a combination of fields. What we call "particles" are tiny local vibrations in these fields. The results of the DSE are merely the results of resonant interaction of quantum fields. We put these ideas in a basis our analysis of "the heart of quantum mechanics."

A systematic study of this issue has apparently not been carried out. Therefore, we propose a number of thought experiments that can help a reader formulate the answers we need most. At the end of the study, we use the mathematical modeling of the physical processes responsible for the results of the experiments. Plasma waves and waves in vacuum are considered. In particular, solutions (5.63) of the NKGE that describe highly localized wave packets (particles) are used. The interaction of these waves is studied in the attempts to model some processes occurring in the DSE.

Thus, the main goal of our research is to introduce the reader to the new understanding of certain positions of quantum mechanics on the basis of the new understanding of the results of the DSE.

6.2 Experiments Using Different Kinds of "Slits" and the Beginning of the Discussion

Quantum mechanics is built on experiments and solutions of mathematical equations. It is important that the solutions describe the experimental data very well. But these solutions in themselves do not bring the understanding of the experiments. As Richard Feyman wrote some 50 years ago, as soon as we try to build the understanding, we come to conclusions that contradict common sense. Therefore, as he writes, "nobody understands quantum mechanics." Apparently, the situation in quantum mechanics has not improved over the past 50 years. Perhaps it even worsened. This is well demonstrated by the data of the DSE that were obtained in the last few decades. There is a significant variety of experiments in both the particles used and schemes and experimental equipment. Below, we describe three types of experiments that follow the DSE and briefly discuss their results.

Experiment

1. Diffraction of electrons. Until 1961, the double-slit experiments were performed with light beams. In 1961, this experiment was performed with electron beams (Claus Jönsson of the University of Tübingen) [7].

There is a source that emits a stream of electrons onto a photosensitive screen (Fig. 6.4). And there is an obstruction in the way of these electrons, a copper plate with two slits. What kind of picture can be expected on the screen if the electrons are imagined as small charged balls? Two bands illuminated opposite to the slits.

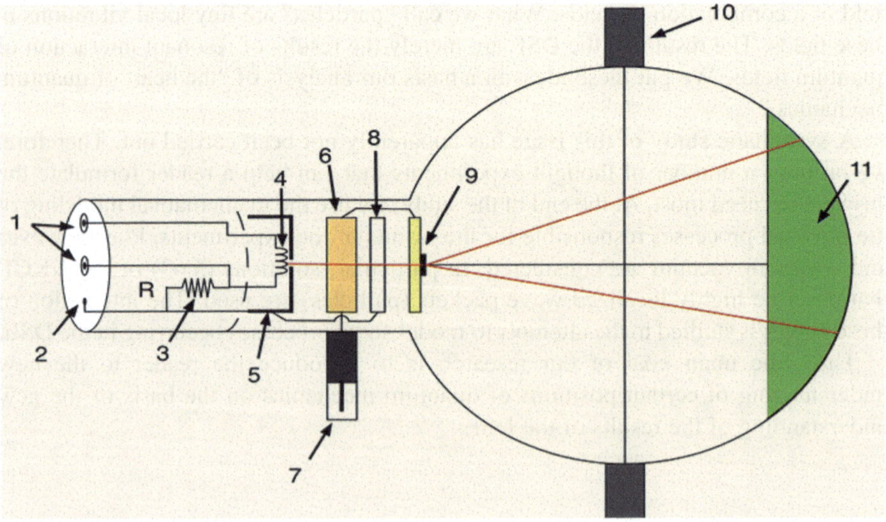

Fig. 6.4 There is a source that emits a stream of electrons onto a photosensitive screen. And there is an obstruction in the way of these electrons, a copper plate with two slits

In fact, the screen displays a much more complex pattern of alternating black and white bands. This is due to the fact that, when passing through the slit, electrons begin to behave not as particles, but as waves. These waves interact in space, and as a result, a complex pattern of alternating light and dark bands appears on the screen. In 1974, the Italian physicists [8] repeated the experiment using single electrons and biprism (instead of slits, see Fig. 6.5), showing that each electron interferes with itself as predicted by quantum theory. In 2002, the single-electron version of the experiment was voted "the most beautiful experiment" by readers of Physics World [10]. What was its nature?

2. In 1989, the wave-particle duality of electrons was demonstrated in a kind of double-slit interference experiment using an electron microscope equipped with an electron biprism and a position-sensitive electron-counting system [9] (see Fig. 6.5).

Electrons are emitted one by one from the source in the electron microscope. They pass through a device called the "electron biprism," which consists of two

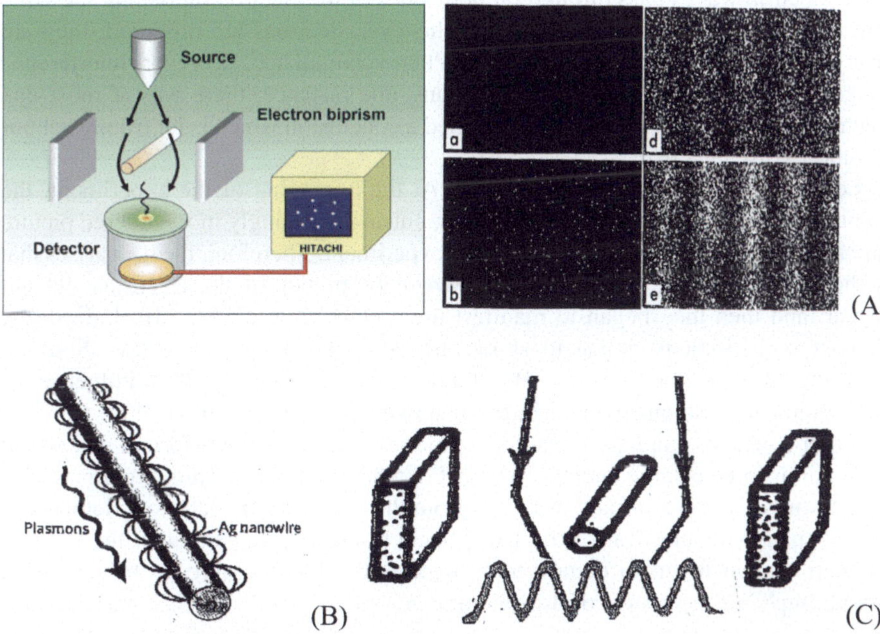

Fig. 6.5 One electron double-slit experiment by Akira Tonomura (see http://www.hitachi.com/rd/portal /highlight/quantum/doubleslit/index.html) (left (**A**)). Single electron events build up to from an interference pattern in the DSE (right (**A**)). Number electrons 10(*a*), number electrons 100(*b*), number electrons 20,000(*c*), number electrons 70,000(*d*)]. Each time a bright spot is seen, as a result, an electron detected as a "particle." However, over time, an unmistakable interference pattern which is undoubtedly a signature of waves [1, 9] (**A**). Plasmons are excited by electrons in a fine filament (nanowire) (**B**) (see also Figs. 6.17 and 6.18). Vacuum fluctuations (standing waves of the Casimir effect) between two parallel plates which interact with the electrons and the plasmons (**C**)

parallel plates and a fine filament (nanowire) at the center. The filament is thinner than 1 micron (1/1000 mm) in diameter. Electrons having passed through on both sides of the filament are detected one by one as particles at the detector. This detector was specially modified for electrons from the photon detector produced by Hamamatsu Photonics (PIAS). It could detect even a single electron with almost 100% detection efficiency.

Interference fringes are produced only when two electrons pass through both sides of the electron biprism simultaneously. If there were two electrons in the microscope at the same time, such interference might happen. But this cannot occur, because there is no more than one electron in the microscope at one time, since only 10 electrons are emitted per second. When a large number of electrons is accumulated, something resembling regular fringes begins to appear in the perpendicular direction as Fig. 6.5A(c) shows. Clear interference fringes can be seen in the last scene of the experiment (Fig. 6.5A(d)). It should also be noted that the fringes are made up of bright spots, each of which records the detection of an electron.

We have reached a mysterious conclusion. Although electrons were sent one by one, interference fringes could be observed. These interference fringes can form only when electron waves pass through on both sides of the electron biprism at the same time, but we do not have any of that. Whenever electrons are observed, they are always detected as individual particles. When accumulated, however, interference fringes are formed. Please recall that at any one instance, there was at most one electron in the microscope. We have reached a conclusion which is far from what our common sense tells us.

Let us consider Fig. 6.5A attentively. At the beginning of the experiment, the points on the screen are scattered randomly, but an increasingly more ordered picture appears and bands appear at the end of the experiment. Therefore, the idea arises that at the beginning (Fig. 6.5A(a, b)), the wave properties of the electrons did not appear, and then they began to manifest themselves (Fig. 6.5A(c, d)). Indeed, the element of periodicity began to show only after the passing of a few thousand electrons through the biprism. We think it is the result of the influence of the periodical plasmons excited in the nanowire by these electrons (Fig. 6.5). On the other hand, we emphasize that the cylindrical wire and plates form a pair where the Casimir force is large enough (see Sect. 6.6 and Fig. 6.17). Thus, we think, after the passing of a few thousand electrons from the source to the detector, elements of system (nanowire, electrons, and plates) begin to interact. In particular, the vacuum fluctuations can begin to demonstrate themselves. The interference fringes are a "photograph" of the results of this interaction.

3. Interference patterns similar to those presented in Figs. 6.1, 6.2, and 6.5 have been observed in experiments using beams of electrons, neutrons, and even heavier particles like fullerenes moving through carefully designed slits. The challenge in these experiments is that the slits should be sufficiently close to each other—the distance between slits must be comparable to the wavelength associated with the particle beam. In this case, it is possible to determine through which slit the particle passes. However, according to quantum mechanics,

position and momentum of particle cannot be measured accurately at the same time. One of the most famous discussions of this hypothesis involved the two great physicists Albert Einstein and Niels Bohr. Einstein suggested a gedanken, or thought, experiment—the so-called Einstein-Bohr recoiling double-slit gedanken experiment—in which one of the two slits in the experiment is allowed to move.

Xiao-Jing Liu and colleagues report the first experimental realization of the Einstein-Bohr gedanken experiment using a molecular double slit [11]. Elegant experiments performed with X-rays and a double slit formed from molecular oxygen have finally made it possible to realize and test a long-standing and famous gedanken experiment in quantum mechanics (Fig. 6.6).

According to quantum mechanics, the determination of which slit the particle passes through inevitably destroys the wave aspects and implies the disappearance of the interference. In [1, 11, 12], it is noted that the results of this experiment are in full agreement with Bohr's complementarity principle.

Here we should emphasize that it is difficult to name the last two experiments fully as double slits. Indeed, Figs. 6.5 and 6.6 resemble the one-slit experiment more when the interference was not observed. However, these figures show clearly the interference patterns. The noted discrepancy is very surprising. May the interference pattern be some kind of "photograph" of the interaction of the electrons with the vacuum fluctuations which exist between the two atomic centers? We emphasize that the Casimir effect is largest between metal balls (see Fig. 6.6).

4. One of the earliest and strangest predictions of quantum physics is the idea of wave-particle duality, that everything in the Universe has both particle and wave nature. Generally speaking, it is true only for elementary particles. However, the experiment can be done with entities much larger than electrons and photons, although it becomes more difficult as size increases. The largest entities for which the double-slit experiment has been performed were molecules (Fig. 6.7), each of which comprised 810 atoms (whose total mass was over 10,000 atomic mass units) [13–15].

These experiments demonstrated interference between matter waves passing through two or more slits cut in a barrier. Back in the day, Richard Feynman famously said that the interference of particles captures the essential mystery of quantum physics. Indeed, it is a mystery since the matter waves of large molecules are negligible and small relative to the dimensions of the molecules!

The interference of waves is determined in part by the wavelength. According to quantum physics, the wavelength of a massive particle is inversely proportional to its momentum: the mass multiplied by the particle's speed. In other words, the heavier the object, the shorter its wavelength at a given speed. A kicked football, for example, has a very tiny wavelength compared to the size of the ball because it has a relatively large mass and speed measured in meters per second (rather than nanometers or such). In contrast, an electron has a relatively large wavelength (though still microscopic) because it has a small mass. Longer wavelengths make

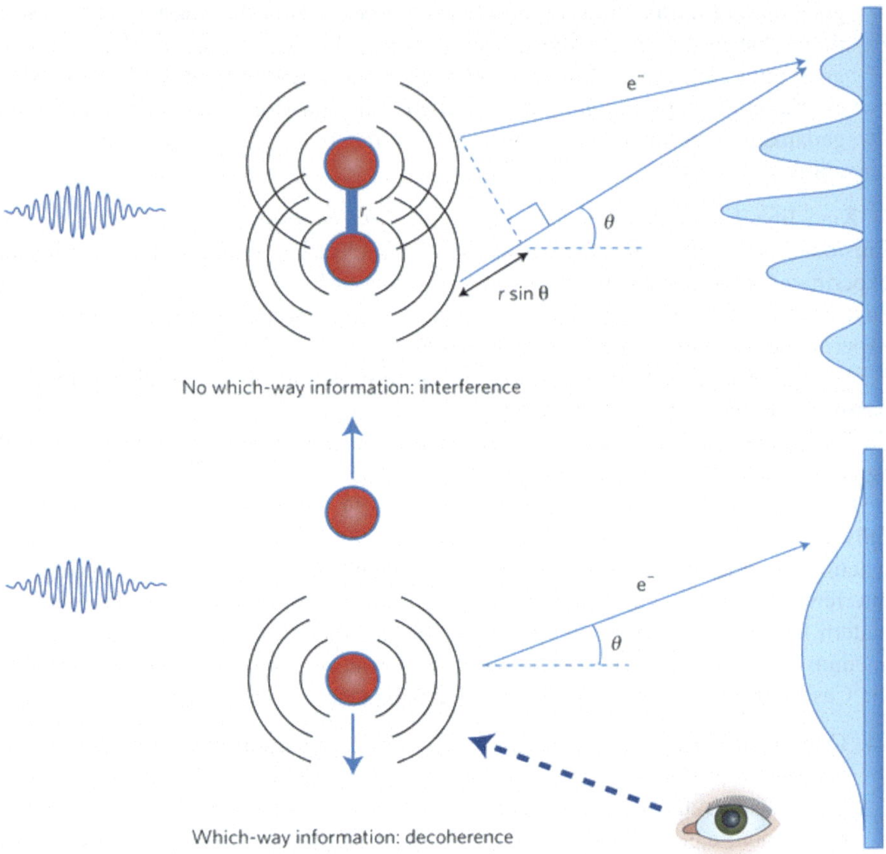

Fig. 6.6 Upper: A pulse of light (left) impinges on a diatomic molecule, leading to coherent electron ejection (e^-, blue arrows) from the two atomic centers (the "two slits," red circles) in the form of waves (black lines). These waves reach a screen (right) where the intensity of the electron signal is recorded. Constructive interference between electron waves generated by the two atomic centers is obtained at observation angles θ satisfying $n\lambda = r\sin\,\theta$, where r is the internuclear distance, λ is the electron wavelength, and n is an integer. No information about the path followed by the electron can be obtained from this experiment. Bottom: molecular version of the Einstein-Bohr gedanken experiment. Same as above except that the two atomic centers separate in opposite directions (thick blue arrows). An observer measures the momentum of the atomic fragment from which the electron wave is emitted. As a result of this measurement, which provides information about the path followed by the electron, interference patterns are no longer observed on the screen [1, 12]

it easier to generate interference, so, while it is not going to be possible to make two footballs interfere with each other (in the quantum sense!), it is comparatively straightforward to produce electron interference.

The researchers observed the particle nature of the molecules in the form of individual light spots appearing singly in the fluorescent detector as they arrived.

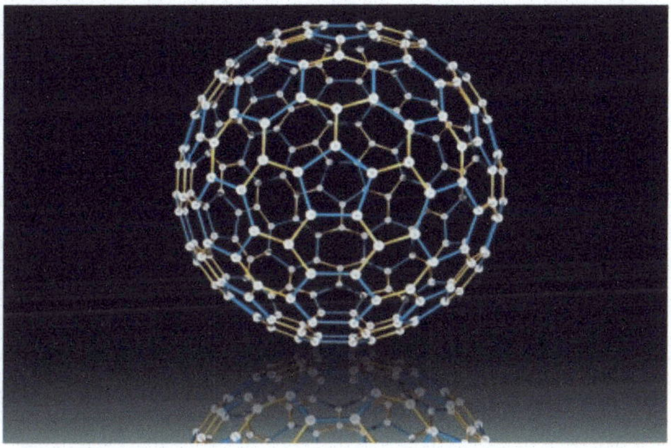

Fig. 6.7 Experiments were also held with large molecules in a form very similar to a soccer ball: a hollow sphere made of pentagons and hexagons

But, over time, these spots formed an interference pattern due to the molecules' wave-like character.

As Juffmann et al. [15] point out, no other explanation but quantum interference can account for the pattern that appears in the fluorescent detector. Since the phthalocyanine- and phthalocyanine-derived molecules are relatively large and massive, their behavior approaches the limits at which macroscopic properties begin to exhibit themselves. It should be noted that the described experiment is very complex and is carried out at extremely low temperatures. Perhaps the Van der Waals forces and the Casimir effect manifest themselves especially strongly in this case. Future experiments with even larger molecules may be able to examine the transition between everyday physics, in which quantum interference does not play a role, and the underlying quantum world.

A beam of phthalocyanine molecules passes through a nanofabricated diffraction grating. The fluorescence light detects single molecules striking a glass plate on the far side. Each dot in Fig. 6.8 represents a single molecule, and their arrival times and locations are essentially random. The compilation of all the detections, though, reveals a pattern of bright and dark bands that is the result of interference. You can even spot the relationship between momentum and wavelength in the angle of the bands—the spacing is wider toward the bottom of the figure because those molecules are moving slower, and took more time to cover the distance between the grating and the detector. Slower speeds mean lower momentum, which means a longer wavelength, and thus a larger spacing between the bands. During the experiments, moving molecules were irradiated by a laser ray. After being heated by an external source, molecules began glowing; thus, they became visible to the observer. After this change was added and used in multiple experiments, molecules altered their behavior.

In this case, it was possible to heat the molecules by means of a laser beam, which changed their internal temperature. Now let us recall that any heated body, including a molecule, emits thermal photons. In this case, it is possible to determine the trajectory of the emitting molecule with accuracy of the wavelength of the emitted quantum. In the experiment, it was found that in the absence of laser heating, one

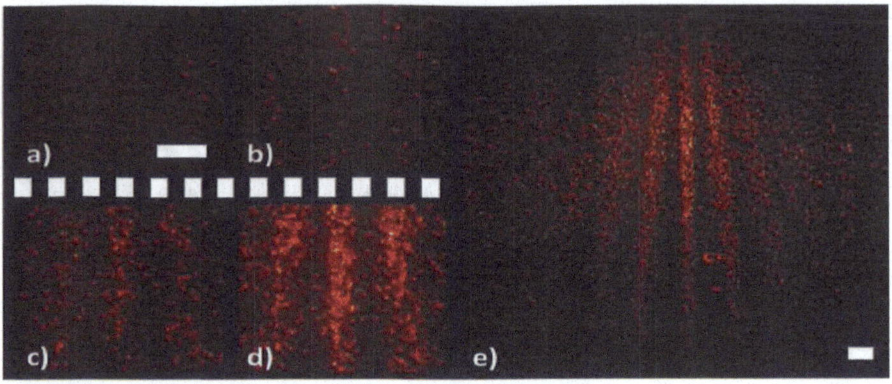

Fig. 6.8 Image of interference of phthalocyanine molecules; each dot is a single molecule detected by fluorescence, and the pattern of bands is indicative of wave behavior. The images were recorded (**a**) before deposition and after PcH2 deposition for (**b**) 2 min, (**c**) 20 min, (**d**) 40 min, and (**e**) 90 min [1, 15]

observes an interference pattern, completely analogous to the picture from two slits in the experiment with electrons. If there is heating, then, of course, there will be no Casimir effect. The addition of laser heating leads first to a weakening of the interference contrast, and then, as the heating power increases, to the complete disappearance of interference effects.

The Beginning of the Discussion The possibility of controlled heating of the molecules made it possible in this experiment to study the transition from the quantum to the classical regime. In other words, proceeding from this experiment, it can be concluded that the observed reality is based on a nonlocalized and "invisible" quantum reality that becomes localized and "visible" during the experiment, namely, when the effect of irradiation is reduced.

In other words, *we* connect the results with the fact that at low temperatures, some "coherent invisible" quantum reality exhibits some resonant properties.

If we place a detector inside or just behind one slit, we can find out whether any given particle goes through it or not. In that case, however, the interference vanishes. Simply by observing a particle's path—even if that observation should not disturb the particle's motion—we change the outcome. Until a particle is observed, an act that causes the wave function to "collapse," we can say nothing about its location. Albert Einstein, among others, objected to this idea. He asked, "does the moon exist only when I look at it?"

Can the observation disturb the results of the experiments? Taking into account similar results of the experiments, certain researchers *declared that objectivity is an illusion*. This point of view was supported by a paper [16] where the authors consider closely some results of the DSE.

Then, in 2011, Aephraim Steinberg and his colleagues from the University of Toronto [17] succeeded in showing that the measurement does not necessarily lead to the collapse of the wave function and that the trajectory of photons before the collapse of the wave function exists! We emphasize that after much debate among the *Physics World* editorial team, year's honor (2011) went to Aephraim Steinberg and colleagues for their experimental work on the fundamentals of quantum mechanics. Using an emerging technique called "weak measurement," the team is the first to track the average paths of single photons passing through Young's double-slit experiment—something that Steinberg says physicists had been "brainwashed" into thinking is impossible.

Steinberg's work stood out because it challenges the widely held notion that quantum mechanics forbids us any knowledge of the paths taken by individual photons as they travel through two closely spaced slits to create an interference pattern.

In these new experiments, the technique of "weak measurements" was developed [18–20]. As a result, physicists managed to measure the average momentum of photons passing through a specific slit, i.e., determine their averaged trajectories and at the same time preserve the interference pattern on the screen.

We emphasize that this result corresponds to the experiments with molecules.

This contradicts the Copenhagen interpretation of quantum mechanics and is more consistent with the so-called hidden-parameter theory that Einstein defended in his time—"God does not play dice".

A group of scientists tried a variation of the DSE, called the delayed choice experiment [21, 22]. This experiment was first proposed as a thought experiment (gedanken experiment) by John Wheeler as a way of exploring the counterintuitive aspects of particle-wave duality. According to Wheeler, "It is hard to avoid the implication that consciousness and quantum mechanics are somehow linked" [21–24].

That possibility was admitted in the 1930s by the Hungarian physicist Eugene Wigner. "It follows that the quantum description of objects is influenced by impressions entering my consciousness," he wrote. "Solipsism may be logically consistent with present quantum mechanics."

Wheeler even entertained the thought that the presence of living beings, which are capable of "noticing," has transformed what was previously a multitude of possible quantum pasts into one concrete history. In this sense, Wheeler said, we become participants in the evolution of the Universe since its very beginning. In his words, we live in a "participatory Universe."

Wheeler's idea was to imagine a "cosmic interferometer" (see also [21] and Fig. 6.22). Suppose light from a distant quasar were to be gravitationally lensed by closer galaxy. As a result, light from a single quasar would appear as coming from two slightly different locations. Wheeler then noted that this light could be observed in two different ways. The first would be to have a detector aimed at each lensed image, thus making a particle measurement. The second would be to combine light from these two images in an interferometer, thus making a wave measurement. According to quantum theory, the results of these two types of experiments (particle or wave) would be exactly as we have observed in their standard form. But the light began its journey billions of years ago, long before we decided on which experiment to perform. Through this "delayed choice," it would seem as if the quasar light "knew" whether it would be seen as a particle or wave billions of years before the experiment was devised (see also Fig. 6.22 (Schematic diagram of a hybrid Hanbury Brown-Twiss and Aharonov-Bohm experiment)).

Although the quasar experiment Wheeler proposed is not practical, modern experimental equipment allows us to perform a similar experiment in the lab, where the decision to measure a particle or wave is made at random after the quantum system is "committed." For example, in 2007 [21], a delayed-choice double-slit experiment was made using laser light. In [22], the team used an ultracold helium atom to do a similar delayed-choice interference experiment. With both experiments, the results were exactly as predicted by quantum theory. So both matter and light exhibit this strange quantum effect.

While that might seem strange, it is not magical or mystical. The Moon would not vanish from existence if everyone closed their eyes, and reality is not dependent upon us observing it. Although if the second particle is detected after the first particle hits the screen, it still ruins the interference pattern. This means that observing a particle can change events that have *already happened*.

Scientists are still unsure how exactly this whole thing works. It is one of the greatest mysteries of quantum mechanics. Perhaps someday someone will finally be able to solve it.

Conclusion Quantum theory is strange, but very real. Through countless experiments, we have found that quantum objects have both particle-like and wave-like properties. In some experiments, the particle nature dominates, while in others the wave nature dominates. Some experiments can even show the effects of both properties. This duality between particles and waves in quantum theory is deeply counterintuitive, which means often the results of quantum experiments are interpreted incorrectly.

From a philosophical point of view, this may be the case. Science has repeatedly refuted itself and recognized truths that were well consistent with experiment as erroneous.

6.3 Explanations and Discussion of the Experimental Results

We should remember that the greatest puzzles of the DSE must have explanations. Let us consider the most well-known explanations.

Classical explanations

1. Particles can be thought of as a kind of wave, and when waves emerge from two slits like this, they can interfere with each other. If their peaks coincide, they reinforce each other, whereas if a peak and a trough coincide, they cancel out.

This diffraction produces a series of alternating bright and dark bands on the back screen, where the waves are either reinforced or canceled out.

However, in this case, we do not take into account the particle as an element of physical reality. This is just a wave in which there is nothing from the particle. However, for example, when experimenting with electrons, each electron forms on the screen a tiny dot, as if it was a particle! At the same time, the mathematical description of the experiment as the wave process is in good agreement with its results.

2. On the other hand, the basis of quantum mechanics lies in the mathematical formalism, in which there is no absolute predictions, characteristic of classical physics. If you made identical experiments with two particles, then there is a chance you will get different results. In particular, during the passage through the slits, the particle can be fixed at different points on the screen. However, if through the slits consistently pass a lot of particles, they in the process form an increasingly clear interference pattern! And this picture does not depend on the type of particles or the features of the experiment. As in the case of a coin toss, you cannot guarantee the result of tossing. However, when the number of tosses is increased, the total result will tend to a probability equal to 0.5. And this result, in an ideal experiment, does not depend on a size of the coin. Is it possible that during the experiments the particles also exhibit their corpuscular properties?

The standard interpretation by which a wave function expresses probability helps to understand the results of the experiment, since it can be argued that we are dealing with probabilities rather than with actual events. It was this strange phenomenon and the inability to explain it that motivated Niels Bohr to develop his idea of complementary. After having struggled with this riddle and discussing it with leading scientists over a long time, he came to believe that it was impossible to explain the wave-particle duality, and we just have to accept that nature is strange. Somehow matter is both particle and wave.

It is well known that quantum objects have both particle-like and wave-like properties. However, Bohr suggested that particles can demonstrate corpuscle-like or wave-like properties during the same experiment. For example, near slits, the wave nature dominates, while on the detector the particle-like nature is demonstrated. According to the classical quantum superposition, the physical system simultaneously exists in all theoretically possible states, but when measured or observed gives a result corresponding to only one of the possible states. But in this case, we can ask "what is really real?" [23, 24].

When physicists, during similar experiments, tried to determine with the help of instruments which slit the electron actually passes through, the image on the screen had changed dramatically and became a "classic" pattern with two illuminated sections opposite to the slits and no alternating bands displayed. Electrons did not seem to show their wave nature under the watchful eye of observers. At the same time, the particle demonstrates its corpuscle-like properties in the detector! Is this

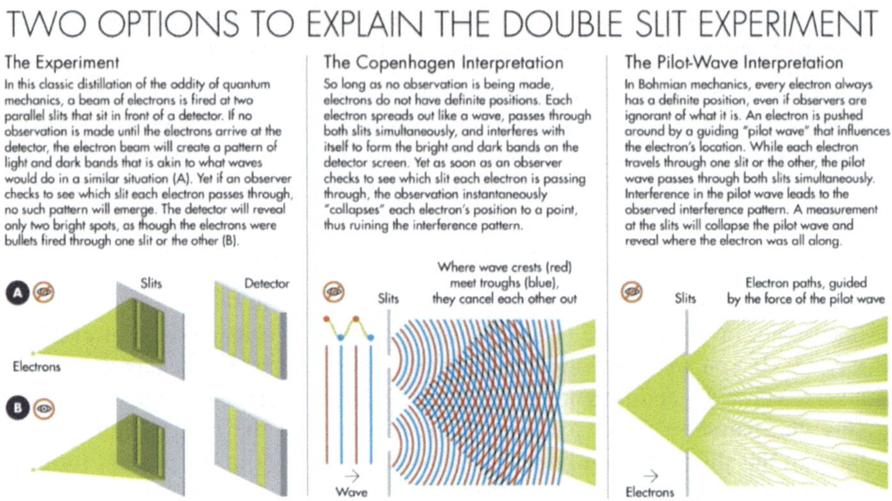

Fig. 6.9 Scheme of the DSE and basic explanations of its results [1, 20]

some kind of mystery? There is a simple explanation if we accept it. No observation of a system can be carried out without physically impacting it.

However, we can imagine a very weak impact which practically does not perturb the particles and at the same time determines the path.

3. The latest experiments [17–20] perturb the role of the classical quantum superposition. There is an alternative view, known as pilot-wave theory. According to de Broglie-Bohm theory, every particle has an actual, definite location, even when it is not being observed. Changes in the positions of the particles are given by the "pilot wave" equation. The theory is fully deterministic; if you know the initial state of a system, and you have got the wave function, you can calculate where each particle will end up (Fig. 6.9).

According to the de Broglie-Bohm model, particles have definite locations and properties, but are guided by some kind of "pilot wave" that is often identified with the wave function. The de Broglie-Bohm interpretation imagined that the wave function of a particle is additional part of reality which exists in addition to the particle itself. In the DSE, for example, each particle goes through one slit or the other, while its wave function goes through both and suffers interference.

On a historical note, Einstein lived just long enough to hear about Bohm's revival of de Broglie's proposal—and he was not impressed, dismissing it as too simplistic to be correct. In a letter to physicist Max Born, in the spring of 1952, Einstein weighed in on Bohm's work:

> Have you noticed that Bohm believes (as de Broglie did, by the way, 25 years ago) that he is able to interpret the quantum theory in deterministic terms? That way seems too cheap to me. But you, of course, can judge this better than I.

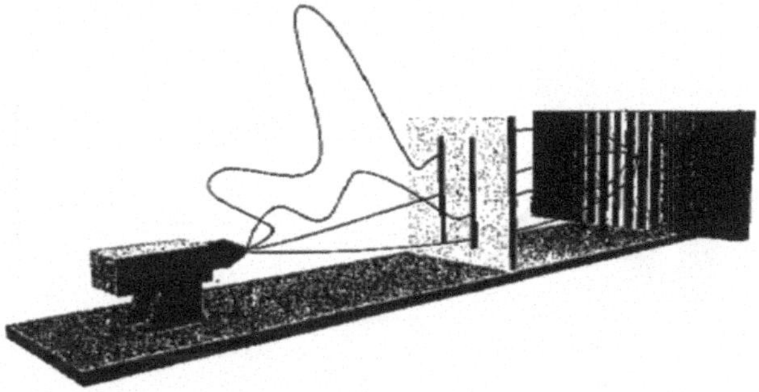

Fig. 6.10 According to Feynman's formulation of quantum mechanics, particles must be viewed as traveling from one location to another along every "possible" path

Nonclassical Explanations Thus, there are a number of approaches which tried to explain the DSE results. Further, we describe three more interpretations.

1. Feyman proclaimed that each electron that makes it through to the phosphorescent screen actually goes through both slits. Feyman argued even more that in traveling from the source to a given point on the phosphorescent screen, each individual electron actually traverses every possible trajectory simultaneously. As an example, some of the infinity of trajectories for a single electron traveling from the source to the phosphorescent screen is shown in Fig. 6.10. We stress that this one electron actually goes through both slits.

The effect of one slit on another in the quantum language is easier to explain using the alternative descriptions of quantum physics developed by Feynman. According to his approach, known as "path integrals," when a particle moves from one point to another, it passes right through all trajectories connecting these points, but each trajectory has its own "weight." The greatest contribution is made by trajectories close to those predicted by classical physics. Therefore, the quantum laws reduce to classical ones in the limit. But other trajectories are also important.

Among these trajectories, there may be those that are completely impossible in the classical sense [27, 28]. They can contain areas on which the particle moves in opposite directions (Fig. 6.11). In the case of an experiment with slots, for example, these are trajectories that first enter one slit, then pass through the other, and then exit through a third one. These strange trajectories explain the effect of one slit on another, because only they are missing when one of the slits is closed.

2. The probabilistic interpretation of quantum laws within the framework of the Copenhagen interpretation was not very popular with many researchers, in particular, Einstein. American scientist Hugh Everett proposed a different look at the process of collapse (reduction) of the wave function. It seems that he came under the influence of Richard Feynman. Everett generalized Feynman's idea and suggested that each trajectory corresponds to its own Universe.

The many-worlds picture was used to roughly describe some results of the DSE. Wiseman and colleagues [26] showed that with just 41 worlds, it is possible to produce the same pattern as in some two-slit experiment. Their publication does not

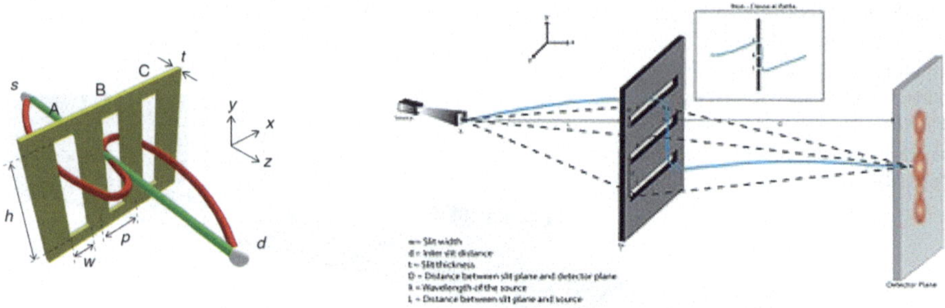

Fig. 6.11 Schemes of the three-slit interference [1, 26]. Path integrals in a laboratory. The green line demonstrates a representative classical path. The purple line demonstrates a representative nonclassical path (left). The dashed lines demonstrates a representative classical paths, and the blue line demonstrates a nonclassical path (right)

contain a wave function: particles obey classical rules such as Newton's laws of motion. The weird effects seen in quantum experiments arise because there is a repulsive force between particles and their clones in parallel Universes.

However, the theory of the Multiverse is not the most incredible, which can be told with reference to quantum paradoxes and riddles. Is quantum teleportation or quantum coupling applicable for explanation of the results of the DSE?

3. Can an existence of the observer influence the experimental results? Can quantum mechanics deal with the intervention of conscious thought in material reality. Or is it mystic?

We are only one step away from admitting that the world around us is just an illusory product of our mind [16, 23, 29]. Scary, isn't it? Let us then again try to appeal to physicists. Especially when in recent years, they favor less the Copenhagen interpretation of quantum mechanics, with its mysterious collapse of the wave function, giving place to another quite down to earth and reliable terms like the coherence (decoherence) and resonances.

We emphasize that in all these experiments with the observations, the experimenters inevitably impacted the system. This is a common and very important principle: you cannot observe the system or measure its properties without interacting with it. And where there is an interaction, there will be a modification of properties, especially when a tiny quantum system interacts with colossal quantum objects. In this case, it is difficult to build a good theory describing experiments.

At the same time, quantum mechanics is a good theory which often gives very good predictions of the results of experiments. However, physics is not just about making predictions. There's a difference between *making* predictions and *understanding* them. The ability to predict behavior is a big part of physics' power, but the heart of physics would be lost if it did not give us a deep understanding of the hidden reality underlying what we observe.

Certain Comments on the Experiments and Their Results We are not going to discuss the subtleties of the experiments and their interpretations. However, there are some features that are common to most of them.

First, coherent particles are used; in particular even in the case of experiments with a single particle, the same particles are used (e.g., only photons or electrons). The same is true for particle beams.

Second, In a typical DSE, the slits are located close enough to each other (in some one-slit experiments, the slit is formed by plates, or plate and cylinder (Fig. 6.5), or balls (Fig. 6.6)). The ratio of the width of the slit to the distance between them varies in some cases over a wide range, but for the case of photons and electrons, this ratio is often approximately 1/5. The slit width was often of the order of 100 nm. We remember that the de Broglie wavelength λ of an electron depends on its energy. For example, an electron with an energy of 1.5 eV has $\lambda = 1$ nm, an electron with an energy of 15 keV has $\lambda = 0.01$ nm, and an energy of 40 keV has $\lambda = 0.006$ nm. The wavelength of photons also reduces when the energy increases. In particular, the photon wavelength was typically near 800 nm [17, 27]. Thus, typically, the electron wavelength was much smaller than the width of the slits, and the photon wavelength could be larger than the width of the slits.

The wavelength of particles is usually much smaller than the width of the slits (e.g., 50 pm relatively 62 nm in [30]). There are very counterintuitive results when the wavelength of the photon was 800 nm while the slit width was 200 nm [27]. Generally speaking, the slits should be close enough to each other so that the distance between them is comparable with particle wavelength. If the distance between the slits increases greatly, then the interference patent disappears (see Figs. 6.2 and 6.3).

In the DSE, the part of the screen between the slits is either metallic or covered by a metal that conducts electricity well. The wavelength of the particles is significantly smaller than the distance between the slits.

Third. Conditions are created under which the particles move almost in a vacuum.

Fourth circumstance is due to the fact that, in interpreting the results, the Schrödinger equation is mainly used and its linear version (classical version).

Fifth, it was found recently that if we use very weak measurements, then we manage to overcome the classical prohibitions of quantum mechanics.

Sixth, the interference patterns are also formed by very heavy molecules (drops) that do not have wave properties that can affect the intensity picture. But the patterns form as shown in Fig. 6.8.

All the DSE show that when one slit is closed, the interference patent disappears. In particular, this effect was thoroughly studied in [30]. However, there are experiments in which there is one "slit," but the interference is strong (see Fig. 6.6). We think that the question of the existence of interference in the case of one slit requires additional careful study. Probably, in this case, the interference is very weakly expressed and therefore practically not visible. We will discuss this issue below during the theoretical modeling of the problem.

Thus, the results of the latest experiments emphasized in paragraphs 4–6 contradict Feynman's ideas, the Multiverse hypothesis, and the basis of quantum mechanics. This does not mean that quantum mechanics does not describe reality, but it means that certain results of it require new understanding.

We think that the understanding of the results of the experiments depends on the idea underlying the method of its analysis. For example, the method used by Everett is a new understanding of the solutions offered by the Schrödinger equation. Another way is to reject the use of the Schrödinger linear equation. In general, we can use the nonlinear Schrödinger equation that takes into account the interaction of the probability peaks of the wave function. We will use nonlinear wave equations in the future to try to somehow represent the motion of particles and their interactions, but now we will focus on using a more radically new approach. We try to give a physically new understanding which is not completely a mathematical description of the results of the experiments.

6.4 Casimir Effect and Standing Waves in the Slits

We emphasized that during the experiments, the particles move almost in a vacuum (e.g., 10^{-8} mbar in [15]). Thanks to the uncertainty principle, the vacuum buzzes with particle-antiparticle pairs popping in and out of existence. They include, among many others, electron-positron pairs and pairs of photons, which are their own antiparticles. Ordinarily, those "virtual" particles cannot be directly captured. But like some spooky chorus, they exert subtle influences on the "real" world. For example, the virtual photons fitting in and out of existence produce a randomly fluctuating electric field. In 1947, physicists found that the field shifts the energy level of an electron inside a hydrogen atom and hence the spectrum of radiation the atom emits.

Two Plates A year later, Dutch theorist Hendric Casimir predicted that a generalized version of van der Waals forces would arise between two metal plates (or conducting materials) due to quantum fluctuations of the electromagnetic field (see Sect. 5.2 and Fig. 5.8). The Casimir effect is a great example of the strangeness of quantum theory, and how even some of its strangest predictions turn out to be right [25, 31–36] (Fig. 6.12).

Two parallel plates can be considered as a resonator, in which exist only those waves for which the resonance condition is met: at an interval L between plates, an integer n of half-waves is stacked. The force of attraction is inversely proportional to the fourth power of the distance between the plates. As the distance decreases, the force increases sharply. But even at submicron distances, the Casimir force remains so small that it was observed only 10 years after the prediction. It was measured for plates only in 1996.

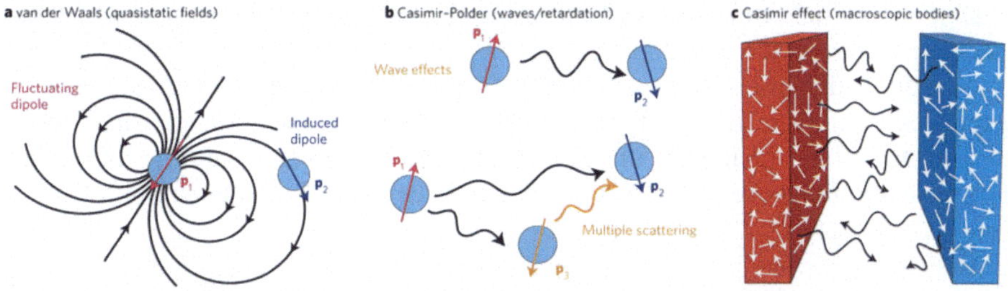

Fig. 6.12 Relationship between van der Waals, Casimir-Polder, and Casimir forces, whose origins lie in the quantum fluctuations of dipoles. (**a**) A fluctuating dipole p_1 induces a fluctuating electromagnetic dipole field, which in turn induces a fluctuating dipole p_2 on a nearby particle, leading to van der Waals forces between the particles. (**b**) When the particle spacing is large, retardation/wave effects modify the interaction, leading to Casimir-Polder forces. When more than two particles interact, the nonadditive field interactions lead to a breakdown of the pairwise force laws. (**c**) In situations consisting of macroscopic bodies, the interaction between the many fluctuating dipoles present within the bodies leads to Casimir forces [35]

Fig. 6.13 Quasi-one-dimensional structures such as nanowires or carbon nanotube and a plate [1, 25]

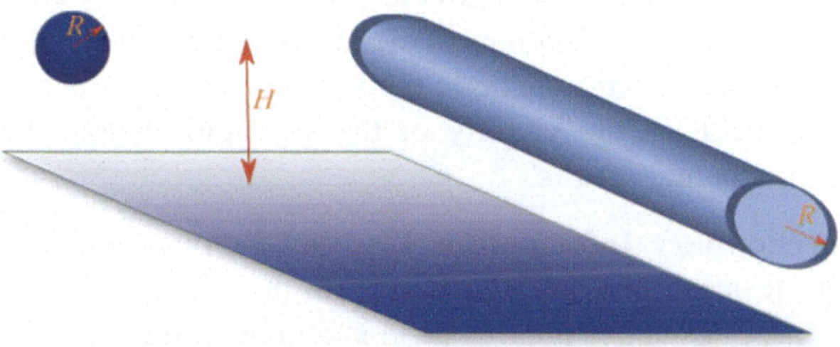

Curvature Effect on the Casimir Force Ensuring the parallelism of the plates with a submicron slit is extremely difficult, so most experiments with the Casimir effect were carried out by replacing one of the plates with a sphere (ball). In this case, the force of attraction is inversely proportional to the cube of the distance between the sphere and the plate. The same is true for the interaction of plate and cylinder (Fig. 6.13) [25].

This explains why in the DSE, cylinder and spheres are often used as certain element of the system (Figs. 5.8, 5.9, 6.5, and 6.6).

Another important circumstance related to the Casimir effect (or the van der Waals force) and the double-slit experiments is determined by the resonance. Apparently, only when the slits, plate, cylinder, or balls form a resonant system, only then the interference picture appears on the screen.

Thus, the vacuum is not "empty," but filled with virtual particles that are very difficult to register, but under certain conditions become real—for example, when an external field of high energy is applied. In addition, they can have an effect on the actual particles and fields introduced into the vacuum. According to our theory (see Chap. 5), the virtual particles (the standing waves of them) form in the slits. They change the particle trajectories. As a result, the particles fix the structure of these standing waves on the detector.

We think that the results of certain experiments considered above might be explained by the interaction of these standing waves and the particles which move through the slits. Apparently, this is not valid for experiments with the molecules. For them, the influence of the van der Waals force might be the most important.

6.5 Thin Metal Layer and Plasmons as Synchronizators of Waves in the Slits

In order to prove the existence of "nonclassical" trajectories, Robert W. Boyd [27] and his colleagues proposed to excite the near-surface plasmons in the interslit material. The existence of plasmons increases the influence of one slit on another. In the experiment, the slits were in a layer of gold deposited on a transparent glass. Since gold is a good conductor, the plasmons are easily excited in it.

Collective oscillations of electrons in the thin metal, that is, surface plasmons, can radiate the electromagnetic field [37]. Can this radiation disturb the electromagnetic near-fields in the vicinity of the slits and change the right understanding of the experimental results?

Preliminary Information Plasma of typical metals can be viewed as a kind an electron Fermi liquid. Plasma is on the average electrically neutral. Due to fluctuations, plasma oscillations arise in a good conductor. Surface plasmons are quanta of vibrations of the density of free electrons of a metal, propagating only along its boundary with a dielectric or vacuum. Surface plasmons can be excited by means of a laser beam directed toward the metal surface. Scientists established that under certain conditions, plasmon waves can oscillate at the same frequency as external electromagnetic waves. So plasmons are a bound state of a photon and an electron in a metal which can form the standing plasmon wave between the edges. Similar to what happens with the standing waves on springs, only waves with appropriate characteristics can appear for a certain width.

Surface plasmon polaritons (SPP) can confine electromagnetic fields in subwavelength spaces. In analogy to photons, they exhibit wave–particle duality [37] (Fig. 6.14).

We believe that plasmons could be excited in interslit material of the typical of the DSE. coated with a thin layer of well conductive metal (silver or gold) [7, 30, 36, 38]. When both slots are open, each of them influences the other, and the particle, generally speaking, now passes the slits differently than if the other slit was closed. The difference is not great. However, as it turns out, the influence of the slits on each other can be amplified.

Of course, the question of the possible excitation of plasmons by successive hits of particles remains unclear [39–43]. However, recently similar questions were discussed.

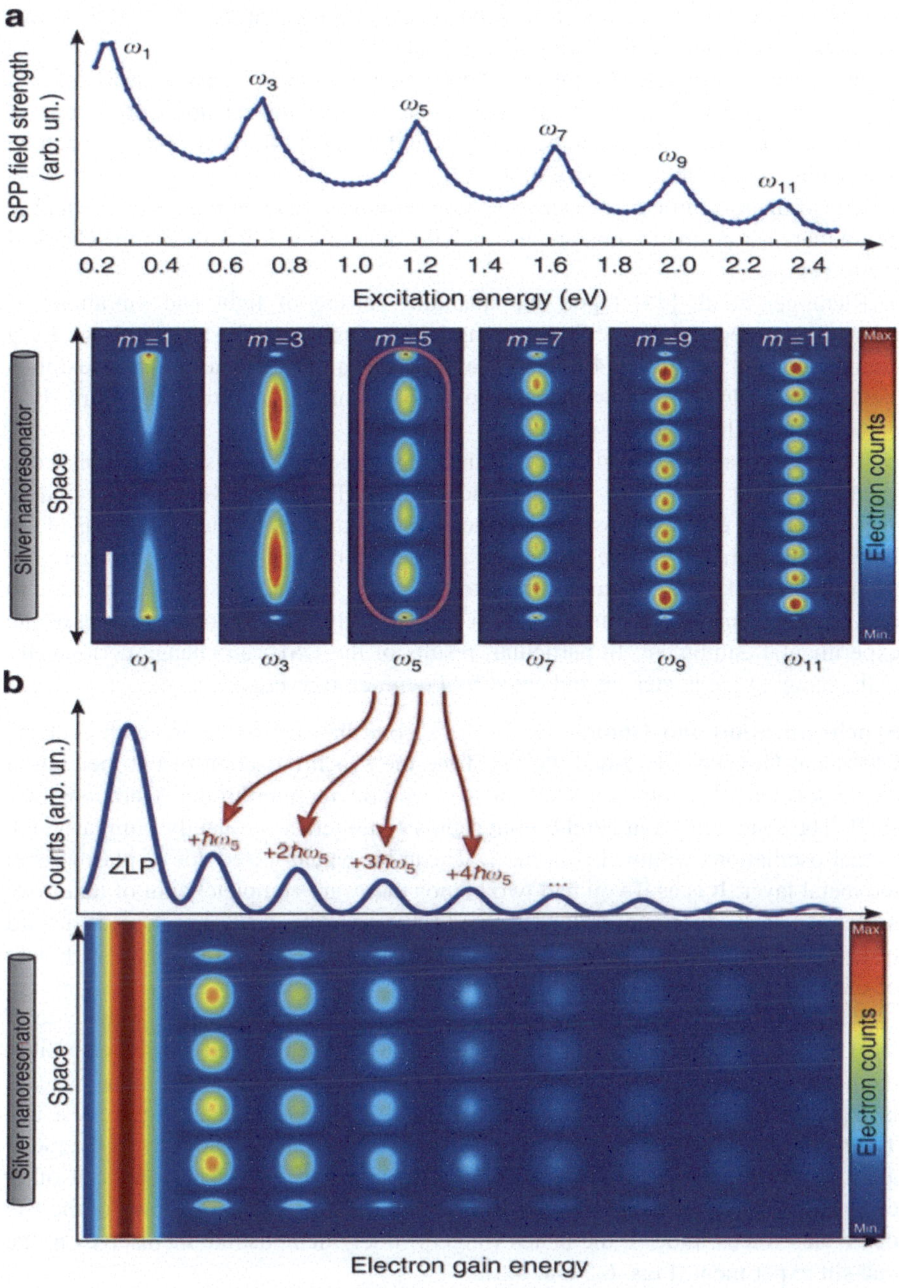

Fig. 6.14 Excitation energy-dependent imaging versus energy-space imaging. (**a**) Finite-element simulation of the excitation energy dependence of the photoinduced SPP (surface plasmon polartions) field strength. The vertical scale bar in the image of the $m = 1$ SPP mode corresponds to 500 nm and holds for all images. (**b**) Selectively photoexciting only one of the SPP modes (here $m = 5$) [1, 37]

Can a sequence of electrons (photons) awake plasmons in the DSE? The answer depends on the interslit dimension and its material.

Everyday experience, of course, indicates that big objects behave classically. In special labs and with a lot of effort, we can observe the quantum properties of photons or electrons. But even the best labs and greatest efforts are yet to find them in anything approaching the size of a cat.

One of the most important experimental questions in quantum physics is whether or not there is a point or boundary at which the quantum world ends and the classical world begins.

Riedinger et al. [44] report the quantum pairing of light and vibrations of microscopic mechanical oscillators comprising more than 10^{12} atoms—large for a quantum object (see also [45–49]). We should remind that the plasmon exhibits wave-particle duality and wavelength of photon may be of order 1000 nm. It is comparable to the length of plasmons.

Thus, there are testimonies that quantum particles (electrons and photons) can excite waves in sufficiently large atomic systems. Therefore, we do not exclude a very significant effect of plasmons on the results of the DSE. Thus, the results of the DSE are determined by the resonance interaction of fields. In the slits, there are waves of virtual particles, and in the interslit space, excited plasmons exist. Due to this resonance nature, the interferention may be only for certain parameters of the experimental equipment. In particular, results of the DSE can change periodically with changing parameters as indicated by Feynman (see Fig. 6.3).

Synchronization and Comments In the second half of the seventeenth century, Christiaan Guygens described the first time the synchronization of two pendulum clocks located on a common shelf due to weak interaction through vibrations of a shelf. The same pianos in neighboring rooms can interact resonantly. Similarly, the virtual oscillations within slits in the DSE can be synchronized due to plasmons in the metal layer. It is as if you had two pianos that were completely out of tune with respect to each other and ordinarily would not resonate at all, but there is a third instrument in the room, like a violin, that has enough flexibility to resonate with both of them.

Plasmons synchronize the oscillations in the slits. We assume that in the case of the DSE, the plasmons appear in the surface layer of the metal at the very initial moments of the experiment, as a result of the impact of particles on the metal, when it is impossible to speak of any occurrence of bands on the screen (see Fig. 6.5A(a, b)). The bands arise when synchronization of resonance oscillations occurs in the metal and the slits. The latter is valid only for the DSE (Fig. 6.1). The results of other experiments are explained by Casimir effect (i.e., by vacuum oscillations) or the van der Waals forces. Indeed, the bands (fingers) finely demonstrate themselves in the one-slit experiment (Figs. 6.5 and 6.6).

Thus, there is a strong interaction of the physical fields of the equipment elements with moving particles. The foregoing determines that the study and description of these fields are of the utmost importance for understanding the experiment.

6.6 Rutherford's Gold Foil Experiment as a Key to Understanding the Results the Slit Experiments

Proceeding from what has been said, we assume that in the series of the DSE, the wave packet (wave-particle) passes through the regions of standing waves formed by vacuum fluctuations. Thus, the wave packet interacts with standing waves (vacuum fluctuations) during passing through the slit. For example, it can be assumed that the particles cannot pass through the compression zones, but they can pass the rarefied zones of the standing waves. With this understanding, the interference on the fixing screen does not arise as a result of the interaction of the particles emitted in the experiment. The bands correspond to the zones of condensation and rarefaction existing on the standing wave in the slits, like in the famous Rutherford's gold foil experiment, the flying particles show the condensation zone (Fig. 6.15).

In our case, the particles show the zones of condensation of the vacuum fluctuations on the detector (screen) (see also Figs. 6.16, 6.17, 6.18, 6.19, 6.20, 6.21, 6.22, 6.23, 6.24, and 6.25).

When we talk about particle physics, we do not usually emphasize that we are actually talking about field physics. But we are. Our aim is to reorient our intuition, in order to appreciate how quantum fields are the ultimate building block of reality as we currently understand it.

Traditionally, during the analysis of the experimental results, it is not taken into account that we are dealing with describing the all-embracing interaction of field waves. This introduces confusion and misunderstanding of the results of experiments. We are not dealing with points and their trajectories, but with wave systems and effects of the resonant amplification.

In particular, the thin metal (gold) layers are resonators for the photon (electron) gas. On the other hand, the ripple of the field in the slits can change the trajectory of the motion of the moving waves (particle) so that they are fixed in certain zones of the screen!

Effects that manifest themselves in experiments are very weak and capricious. The attempts at tracking the experiment can disturb the synchronization of oscillations in the elements of the system and can completely destroy the interference.

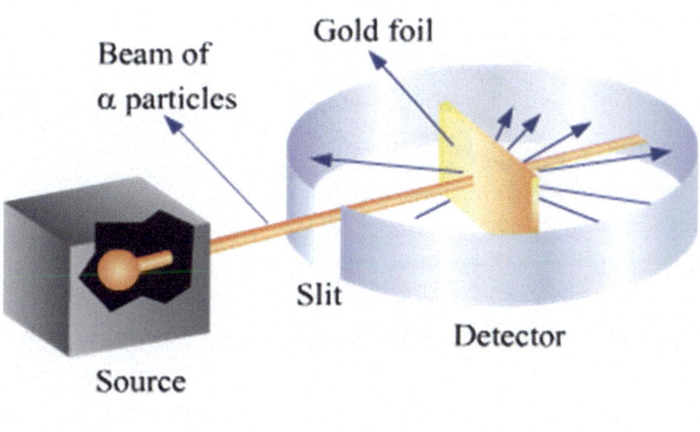

Fig. 6.15 Scheme of Rutherford's gold foil experiment

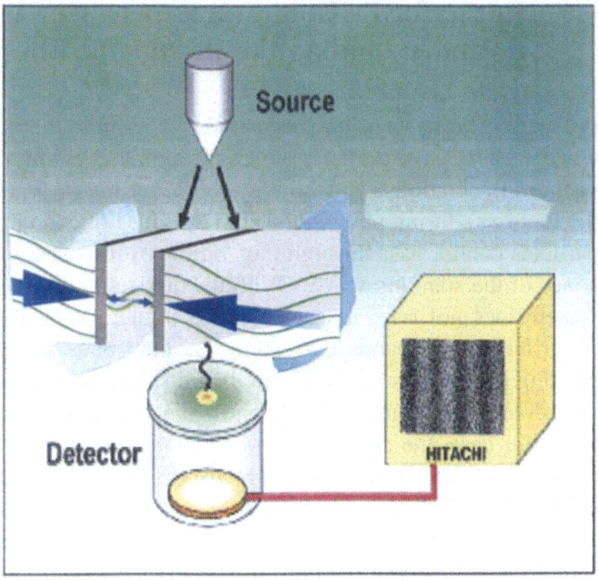

Fig. 6.16 The thought experiment where the biprism effect is replaced by the Casimir effect (the two plates)

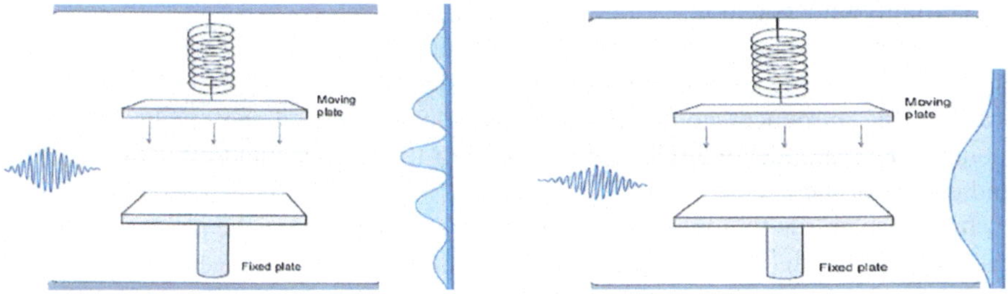

Fig. 6.17 The thought experiments with the plates (the Casimir effect) and the wave packets (showed as short harmonics). Resonant version of the experiment (left) and nonresonant version of the experiment (right)

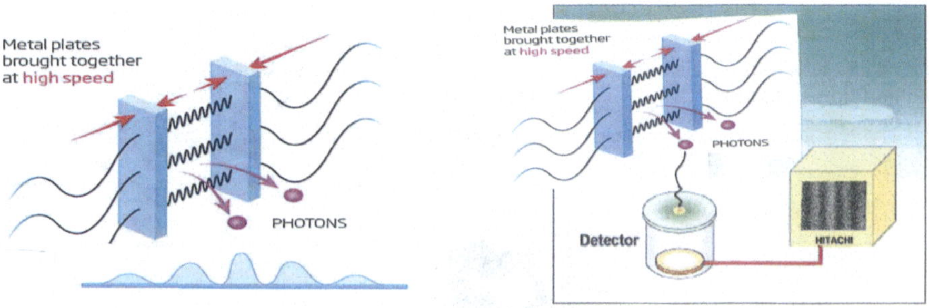

Fig. 6.18 Two versions of the thought one-slit experiment with radiating vacuum and the Casimir plates

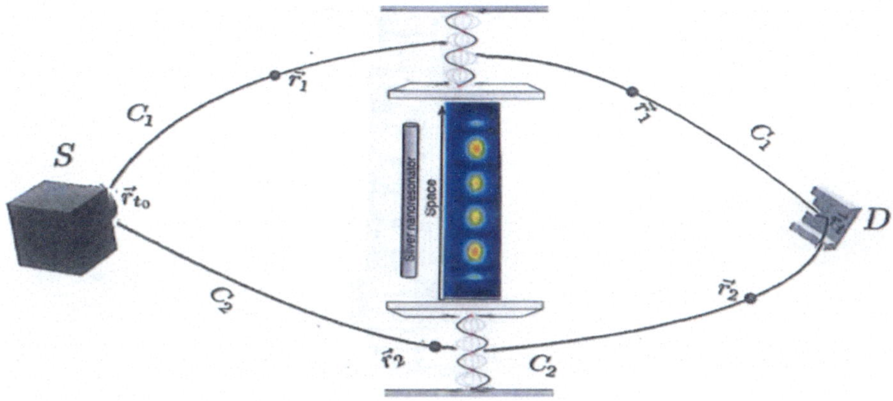

Fig. 6.19 Certain model of the thought experiments of Feynman. The springs model the influence of van de Waals or Casimir forces

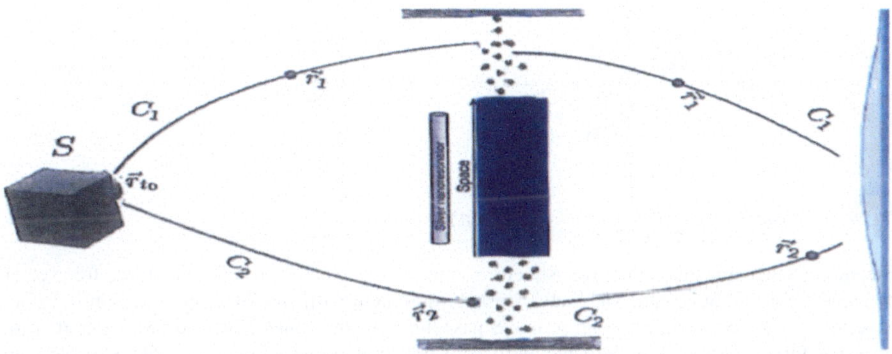

Fig. 6.20 Certain model of the thought experiments of Feynman. The thin layer of metal is excluded from the experiment. As a result, the plasmons, the synchronization, and the fingers disappear from the experiment. The vacuum particles are shown as the points in the slits

However, if very weak perturbations are introduced, interference appears, which fundamentally contradicts the basic principles of quantum mechanics. Thus, the carefully conducted very weak control does not fundamentally change the results as it would be expected. Apparently, in single-slit experiments with elementary particles, it is possible to obtain the bands (fingers) if we manage to strengthen the Casimir effect. If there is an interaction of fields, the interference also appears in single-slit experiments (Figs. 6.5 and 6.6).

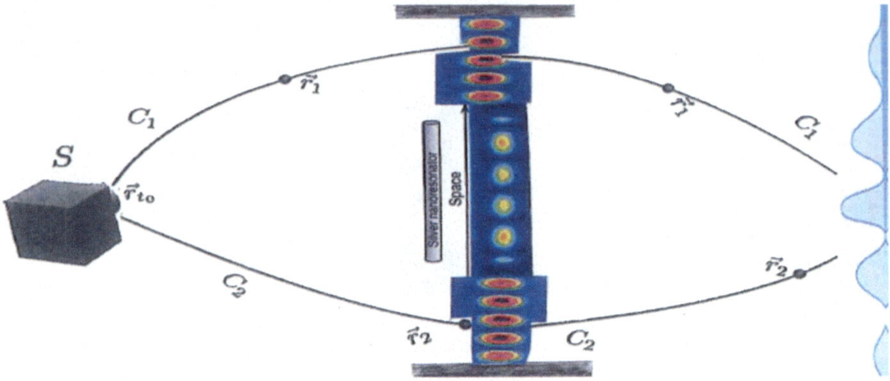

Fig. 6.21 Certain model of the thought experiments of Feynman

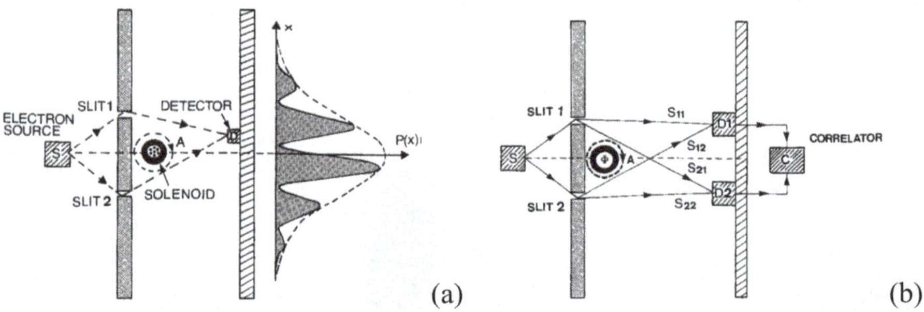

Fig. 6.22 Schematic diagram of two-slit electron interference in the presence of a force-free vector potential field (**a**). Schematic diagram of a hybrid Hanbury Brown-Twiss and Aharonov-Bohm experiment. Electrons emitted from source S pass through slits 1 and 2, around the solenoid (with magnetic flux F directed into the page and vector potential field A circulating clockwise) and are received at detectors D1 and D2 (**b**) [1, 25]

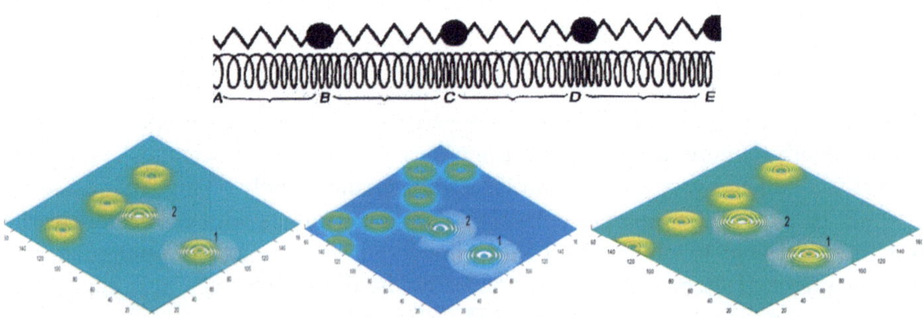

Fig. 6.23 Models of vacuum standing waves in slits. In the up chain, the compression zones are modeling as atomic nuclei in the Rutherford's experiments. In the down chain, the compression zones are modeling as the condensations of the spring (upper row). Three rough schemes of the beginning of the penetration of a wave packet (instances 1 and 2) into a chain of the wave packets simulating the vacuum standing wave in the slit (down row)

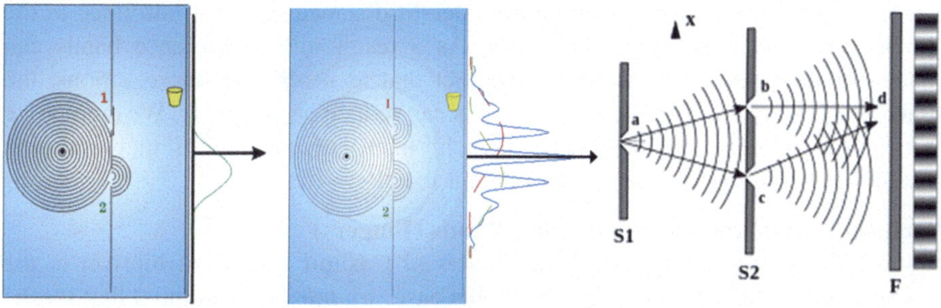

Fig. 6.24 Passing of the wave packet through one slit (on the left). Passing of the wave packet through two slits (center). The scheme and results of the Young's experiment (right)

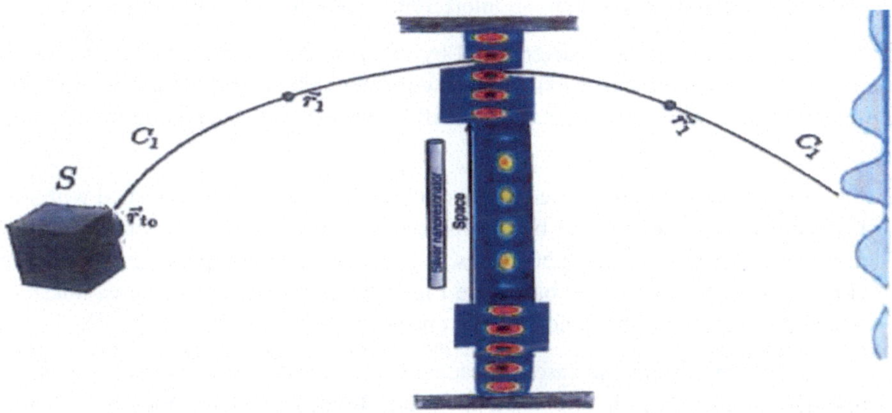

Fig. 6.25 The appearance of bright (light) bands and shadows in thought DSE experiment where only one slit is illuminated

6.7 Testing Thought Experiments

We can change (in this case mentally) the components of the experiment so as to explicitly test our conclusions. For example, one can remove the thin layer of metal to exclude the appearance of plasmons and the synchronization of vacuum oscillations in the slits. As a result, the Casimir effect disappears and the standing waves in the slit are not excited. As a result, the fingers can disappear!

We formulated several provisions which were not directly confirmed by experiment up to this moment. However, following great physicists, we can suggest the thought experiments so that to make our assumptions more real. This suggestion is no more than ideas. But there are several ways in which we can test their usefulness.

Following Feynman (Fig. 6.3), we have no doubt that changing the geometric and other characteristics of the experiments can affect the final interference pattern. In thought experiments, instead of slits (or in slits), one can place the Casimir plates. Or it is possible to significantly change the interslit distance. We can introduce in the system disturbing electromagnetic field. As a result, the interference bands can change or disappear. This is due to the fact that as a result of these actions, the coherence of vacuum oscillations in the slits can reduce or disappear. We emphasize that, in our opinion, this coherence (synchronism) of oscillations in the slits is ensured by waves (plasmons) in the interslit metal.

Vacuum Waves and Corresponding Bands (Fingers)

1. It would be interesting to use two plates of Casimir instead of biprism in the experiment illustrated by Fig. 6.5. In this case, we must bear in mind that for the biprism (for the plate and cylinder), the Casimir effect is an order of magnitude greater than for the plates. But if we take this into account (i.e., pick up the right material for the plates and the correct distance between them), then it will probably be possible to obtain an interference pattern, even for a single slit.

Since the Casimir force is determined by the virtual harmonics arising between the plates, we assume that the experiment will reflect this (Fig. 6.16). The effect of these harmonics on the moving particles will be shown on the screen in the form of bands (fingers).

2. Another important circumstance of the double-slit experiment associated with the Casimir effect is determined by the resonance. Apparently, only when the slits and the material between them (Figs. 6.2 and 6.3) or the plates and cylinder (Fig. 6.5) or closely located balls (see Fig. 6.6) form resonant systems, interference takes place and the bands appear on the screen.

On the left (Fig. 6.17), the case is presented when the distance between the plates is optimal for the Casimir effect. At the same time, one plate vibrates with the resonant frequency of the vacuum oscillations in the slit. As a result, the bands appear on the screen. If the oscillations do not occur with the resonance frequency, then the bands do not appear.

3. Generally speaking, we do not need the passing particles for the formation of the bands. This possibility is illustrated in Fig. 6.18.

In Fig. 6.18, the plates oscillate with the resonant frequency and an amplitude such that the vacuum begins to radiate particles. Naturally, these particles arise at the top of harmonics excited between the slits. The particles are fixed in the detectors in the form of interference fringes corresponding to the vacuum harmonics excited in the slit.

We have shown that the presence of the Casimir force in experiments ensures the appearance of the bands even if there is only one slit. Consequently, the appearance of the bands in no way, in the general case, is connected with the passage of particles through two slits. Reducing the power of the Casimir effect can achieve the disappearance of the bands. For example, this can be achieved by placing sensors near the slits or irradiating them with a laser beam.

Interaction and Synchronization of the Fields A certain model of the thought double-slit experiment of Feynman (Fig. 6.2a) is schematically shown in Fig. 6.19. We present the slits as springs (the vacuum oscillations are the springs). The particles light a thin solid (metal) barrier and the springs. The detector D records the particles that get through the slits. The vertical length of the springs is much greater than their thickness. We can consider the strip of the metal and the springs as certain one-dimensional resonators (indeed, e.g., let the length of the strips be L and its ends be fixed as in a typical one-dimensional resonator).

However, the frequencies of the longitudinal vibrations of the metal layer and the springs can be different. In this case, the waves radiated by them can be incoherent. Therefore, no interference bands can form in the detector.

If there is no synchronization of the vacuum oscillations in the slots, then the virtual particles can appear independently (even chaotically). Therefore, one cannot speak of the possibility of the appearance of interference fringes. Even if Casimir forces (or van de Waals forces) appear there, the vacillations will still occur with different frequencies. As a result, in this case, the appearance of interference fringes is unlikely (Fig. 6.20). The synchronizing role of the metal coating (plasmons) is illustrated by the mental experiments shown in Fig. 6.21.

Of course, we can imagine that the bands disappear if the particle energy is so small that the plasmons do not appear or if the passing particles are separated by very long time intervals (e.g., several minutes) sufficient to completely attenuate of plasmon oscillations. In these cases, the experiment with slits will not show the appearance of the bands.

The above experiments (Figs. 6.16, 6.17, 6.18, 6.19, 6.20, and 6.21) complement the results of Feynman's thought experiment. We believe that the synchronization was lost and restored (Fig. 6.3) when the geometric dimensions of the elements of the experiment were changed.

Plasmons and Corresponding Bands (Fingers) We have considered several thought experiments designed primarily to show the possibility of the influence of quantum fluctuations on the interference. Another important element of the proposed understanding is plasmons. According this understanding, the interference fingers will be changed if the plasmons are disturbed.

In [50], we described thought experiments designed to demonstrate the effect of an external electric field on the final results of experiments.

In Fig. 6.22a, the magnetic field inside the solenoid is directed perpendicularly into the page; the external vector potential field has a clockwise sense about the solenoid axis. Although the diffraction envelope remains undeviated from the forward direction, the interference fringes are displaced by a relative phase shift between the components of the electron wave issuing from slits 1 and 2. This phase shift is proportional to the magnetic field within the solenoid, the region from which the electrons are excluded. The schematic diagram (Fig. 6.22b) illustrates the effect of strong amplification of the solenoid field.

We think that the confined magnetic flux changes the picture of plasmon oscillations in the interslit metal; as a result, the interference pattern is changed.

6.8 Main Thought Experiment

Let us summarize the results of the conducted thought experiments [1, 51]. Imagine that in the slit, there is a spring with weights (Fig. 6.23 (upper row)), and we emit particles one by one. We used the expression (5.63) to calculate the wave packets. These packages have the form shown previously in Figs. 5.13 and 5.14.

Of course, many particles will pass through the slit, but some will be reflected from the weights (as in Rutherford's experiments, some of the particles were reflected from the atomic nucleus, Fig. 6.15) or will significantly change their trajectories. Flying particles will be recorded on the screen, but, of course, at the beginning of the experiment, no stripes appear on the screen.

Indeed, at the beginning of the experiment, particles that have slightly changed their trajectory will manifest themselves more strongly than particles reflected from the chain.

However, over time, the stripes on the screen will appear more and more clearly, as shown in Fig. 6.5.

In this way, we can simulate the results of DSE. It can be argued that in real experiments, there are no springs with weights in the slit. However, standing vacuum waves can exist, having antinodes and nodes, frequencies, and amplitudes. Qualitatively, they correspond to the Casimir forces presented in Figs. 6.16, 6.17, and 6.18.

But why are the stripes not visible in the case of one slit? This is not entirely true. We can consider the cases presented in Figs. 6.5 and 6.6, as diagrams corresponding to one slit. For them, the Casimir effect is very strong, and therefore the stripes appear. On the other hand, in the case of a single slit, the virtual vibrations of the vacuum in the slit are very weak and do not affect the flying particles at all. However, in the case of two slits separated by a metal layer, in which plasma oscillations are possible, synchronization of plasma and vacuum oscillations is possible. As a result, a resonant amplification of vacuum waves in the slit occurs. Waves appear there, the shadow of which appears more and more clearly on the screen over time.

Of course, plasma waves do not appear in the metal immediately, but only after many intense impacts by particles on the metal at the beginning of the experiment. Time is needed to synchronize the oscillations of the plasma and vacuum waves. In general, the picture described resembles a tuning fork excited by waves in the air. If the frequency of the waves coincides with the natural frequency of the tuning fork, then there is a strong amplification of its oscillations.

Thus, we come to the conclusion that in the slit experiments with sole particles, the picture of processes is completely different from Young's interference experiment (Fig. 6.24). In the last case, the second slit provides the two sources of coherent waves. The presence of these waves and their superposition, as in the case of water surface waves, creates the interference pattern. In the case of experiments with a sequence of particles (Fig. 6.1), there is nothing like shown in Fig. 6.24 (right). The particles passing through the slits meet their vacuum oscillations forming standing waves like those existing in the experiments of Casimir! These waves, like, for example, longitudinal waves in a spring, have compression and rarefied zones, and also nodes where the wave parameters practically do not change (see Fig. 6.23). We showed two instances of the wave packet motion but did not show how the chain changed during this time. In accordance with the structure of these standing waves, the particles (wave packet) change their trajectories. Thus, a part of them easily passes through the rarefied zones, while the others meet the compression zones. Some of these particles may not pass through compression zones, similar to the way it was in Rutherford's experiments (see Fig. 6.15).

We emphasize that in Rutherford's experiments, the atomic nucleus zone was fixed on the screen only as a shadow. Similarly, in slit experiments, the compression zones of vacuum particles are fixed by detectors as shadows (Fig. 6.25). However, the detectors fix also the light bands. But these light bands and shadows are not interference fringes (Figs. 6.21 and 6.25)!

Thus, the appearance of these bands and shadows, in principle, is not related to the presence of the second slit! Of course, the presence of the second slit can enhance the effect, since it leads to the appearance of the interslit material. This material, under some additional conditions, forms a resonator. In particular, there resonant plasma oscillations (plasmons) arise. Under their influence, as experiments show [52], vacuum standing waves in the slits are amplified and synchronized.

As a result of the appearance of resonant oscillations of the plasmons, the vacuum fluctuations in the slits become coherent (this resembles the appearance of coherent light waves due to the slits in Young's experiment (Fig. 6.24), and their amplitude is intensified strongly. As a result, the interaction of the passing particles with the vacuum standing waves increases and the resulting difference of light bands and shadows becomes so bright that they are fixed by the detectors.

Conclusion In particular, a chaotic appearance and disappearance of virtual particles take place in the slit space. However, because of this randomness, virtual particles do not manifest themselves. Indeed, their total effect is zero. In particular, they do not in any way affect the particles passing through a slit. The situation fundamentally changes if there are two slits separated by a certain resonator (e.g., a finite layer of a good conducting metal). Plasma of solids consists of ions, which are usually inactive, and electrons, which are moving. Plasma is on average electrically neutral. However, due to fluctuations, plasma oscillations arise in it. For their description, a quantum of plasma oscillations called the plasmon is introduced. Surface plasmons, which, in particular, take place in the DSE, are quanta of vibrations of free electrons of the metal, propagating only along its boundary with a dielectric or vacuum.

Free electrons in the metal can be excited by hits of photons or electrons. As a result, collective oscillations (surface plasmons) may be excited in the thin metal layer surface during the first moments of the experiment. If in this layer there are ordered vibrations of charged particles (e.g., photons or electrons—a photon or electron gas), then these oscillations order the vibrations of virtual particles in the slits. There standing waves of virtual particles are formed. Each set of waves has its own characteristic set of nodes and crests. These waves begin to affect the passing particles in such a way that their trajectories vary depending on whether the nodes or the crests of these virtual particles meet during their passage through the slit. As a result, the passing particles in the detector "draw" the interference pattern, which is a certain kind of "photograph" of the standing wave in the slit (in the slits). Thus, in our opinion, along with the Casimir effect and the Lamb shift, the results of the DSE are evidence of the existence of virtual particles in a vacuum (the Lamb shift, named after the American physicist, Willis Lamb. His work was carried out in the late 1940s and showed the effect of zero-point fluctuations of the electromagnetic field on electron orbits). Considering the above-mentioned explanations, the results of DSE could not be formulated by the founders of quantum theory who were not aware of the existence of the fluctuating background (vacuum) field.

Remark In presenting the material presented above, we illustrated it mainly with data for two slit experiments. However, our thoughts may be more valid for the cases shown in Figs. 6.5, 6.6, 6.13, 6.16, 6.18, and 6.25, since for them the Casimir effect is stronger than for narrow slits. In general, our understanding boils down to the fact that in the experiments considered, not so much of the properties of the world of quantum particles are revealed, as the features of the interaction of particle-waves with experimental equipment. It is very important to begin to understand the features of the interaction of physical fields that describe both the elements of equipment and the emitted particle-waves. This interaction may have a resonant nature. The latter seems to be indicated by Feyman's results (Fig. 6.3). The Casimir effect is no less important.

Thus, in the author's opinion, the results of double-slit experiments are still being misinterpreted. This book attempts to develop a new view of the problem, which lies in line with the idea of describing everything using the simplest equations.

References

1. Galiev SH (2020) Evolution of extreme waves and resonances. CRC Press, p 508; Tirole R et al (2023) Double-slit time diffraction at optical frequencies. Nature Physics 19, 999–1002; Devrari V, Singh M (2024) Quantum double slit experiment with reversible detection of photons. Sci Rep 14, 20438. https://doi.org/10.1038/s41598-024-71091-1; Hui Peng (2023) Mystery of double slit experiments—non-interference patterns and interference patterns showing on screen simultaneously. TechRxiv. December 24, 2022. https://doi.org/10.36227/techrxiv.21754220.v1
2. Ball P (2018) Two slits and one hell of a quantum conundrum. Nature 560(7717):165. https://doi.org/10.1038/d41586-018-05892-6. Ananthaswamy A (2020) Through two doors at once: the elegant experiment that captures the enigma of our quantum reality. Duckworth; Greene B (2007) The fabric of the cosmos: space, time, and the texture of realit. Random House, New York
3. Bhatiab A (2014) Hey there little electron, why won't you tell me where you came from? Date of Publication. https://www.wired.com/2014/09/%20double-slit-empzeal
4. Feynman RP, Leighton RB, Sands M (2005) Feynman lectures on physics volume 3: quantum mechanics. Addison-Wesley, Boston. Silverman MP (2002) A universe of atoms, an atom in the universe. Springer; Greenstein G, Zajonc AG (2006) The quantum challenge: modern research on the foundation of quantum mechanics. Jones and Bartlett Publishers

5. Feynman RP (1985) QED-the strange theory of light and matter. Princeton University Press, New York
6. Bullough RK, Jack PM, Kitchenside PW, Saunders R (1979) Soliton in laser physics. Phys Scr 20:364
7. Jönsson C (1961) Elektroneninterferenzen an mehreren künstlich hergestellten Feinspalten. Z Phys 161(4):454–474
8. Merli PG, Missiroli GF, Pozzi G (1976) On the statistical aspect of electron interference phenomena. Am J Phys 44:306–307
9. Tonomura A, Endo J, Matsuda T, Kawasaki T, Ezawa H (1989) Demonstration of single-electron buildup of an interference pattern. Am J Phys 57:117–120. Gähler R, Zeilinger A (1991) Wave-optical experiments with very cold neutrons. Am J Phys 59:316–324
10. Taylor GI (1909) Interference fringers with feeble light. Proc Camb Philos Soc 15:114–115. Ananthaswamy A (2018) Through two doors at once: the elegant experiment that captures the enigma of our quantum reality. Dutton; Crease RP (2002) The most beautiful experiment. Phys World 15(9):19
11. Liu X-J et al (2015) Einstein–Bohr recoiling double-slit gedanken experiment performed at the molecular level. Nat Photon 9:120–125
12. Martin F (2015) Quantum mechanics: thought experiments made real. Nat Photonics 9:76–77
13. Hackermueller L et al (2003) Wave nature of biomolecules and fluorofullerenes. Phys Rev Lett 91:090408
14. Hackermueller L et al (2004) Decoherence of matter waves by thermal emission of radiation. Nature 427:711
15. Juffmann T et al (2012) Real-time single-molecule imaging of quantum interference. Nat Nanotechnol 7:297–300
16. Englert BG, Scully MO, Sussman G, Walther H (1992) Surrealistic Bohm trajectories. Z Naturforsch 47a:1175–1186
17. Kocsis S et al (2011) Observing the average trajectories of single photons in a two-slit interferometer. Science 332:1170–1173
18. Menzel R, Puhlmann D, Heuer A, Schleich WP (2012) Wave-particle dualism and complementarity unraveled by a different mode. PNAS 109(24):9314–9319
19. Mahler DH et al (2016) Experimental nonlocal and surreal Bohmian trajectories. Sci Adv 2(2); e1501466
20. Falk D (2017) New support for alternative quantum view. https://gravityphysics.com/2017/05/16/new-support-for-alternative-quantum-view/
21. Jacques V et al (2007) Experimental realization of Wheeler's delayed-choice gedanken experiment. Science 315:966–968
22. Manning AG et al (2015) Wheeler's delayed-choice gedanken experiment with a single atom. Nat Phys. https://doi.org/10.1038/nphys3343
23. Merali Z (2015) What is really real? Nature 521:278–280
24. Ball P (2017) The strange link between the human mind and quantum physics. http://www.bbc.com/earth/story/20170215-the-strange-link-between-the-human-mind-andquantum-physics
25. Emig T, Jaffe RL, Kardar M, Scardicchio A (2006) Casimir interaction between a plate and a cylinder. Phys Rev Lett 96(8):080403
26. Hall MJW, Deckert D-A, Wiseman HM (2014) Quantum phenomena modeled by interactions between many classical worlds. Phys Rev X 4:041013
27. Magaña-Loaiza OS et al (2016) Exotic looped trajectories of photons in three-slit interference. Nat Commun 7:13987
28. de Oliveira Jr JGG, de Souza G, Cabral LA, da Paz IG, Sampaio M (2017) Exotic looped trajectories via quantum marking. Ann Phys 387:222–238
29. Hiley BJ, Callaghan RE, Maroney OJE (2000) Quantum trajectories, real, surreal or an approximation to a deeper process? arXiv:quant-ph/0010020v2 5
30. Bach R, Pope D, Liou S-H, Batelaan H (2013) Controlled double-slit electron diffraction. New J Phys 15:033018

31. Wilson CM et al (2011) Observation of the dynamical Casimir effect in a superconducting circuit. Nature 479:376–379. https://doi.org/10.1038/nature10561. Esquivel-Sirvent R, Reyes L (2008) Pull-in control in microswitches using acoustic Casimir forces. Europhy Lett 84(4): 48002
32. Adam J et al (2021) Measurement of e+e− momentum and angular distributions from linearly polarized photon collisions. Phys Rev Lett 127:052302
33. Yoo D et al (2021) Ultrastrong plasmon–phonon coupling via epsilon-near-zero nanocavities. Nat Photonics 15:125–130
34. Lamoreaux SK (2005) Casimir force: background, experiments, and applications. Rep Prog Phys 68:201–236. https://doi.org/10.1088/0034-4885/68/1/R04
35. Rodriguez AW, Capasso F, Johnson SG (2011) The Casimir effect in microstructured geometries. Nat Photonics 5:211–221. https://doi.org/10.1038/nphoton.2011.39
36. Lambrecht A et al (2007) The Casimir effect for silicon and gold slabs. EPL 77:44006. https://doi.org/10.1209/0295-5075/77/44006. Zhao R et al (2019) Stable Casimir equilibria and quantum trapping. Science 364:984–987
37. Piazza L et al (2015) Simultaneous observation of the quantization and the interference pattern of a plasmonic near-field. Nat Commun 6:6407. https://doi.org/10.1038/ncomms7407
38. Frabbori S (2012) The young-Feynman two-slits experiment with single electrons: build-up of the interference pattern and arrival-time distribution using a fastreadout pixel detector. Ultramicroscopy 116:73–77. https://doi.org/10.1016/j.ultramic.2012.03.017
39. Gerchikov LG et al (1998) Excitation of multipole plasmon resonances in clusters by fast electron impact. J Phys B Atomic Mol Phys 31:3065–3077
40. Hathiramani D et al (2000) Electron-impact induced fragmentation of fullerene ions. Phys Rev Lett 85:3604
41. Verkhovtsev AV, Korol AV, Solov'yov AV. (2013) Plasmon excitations in photo- and electron impact ionization of fullerenes. J Phys Conf Ser 438:012011. https://doi.org/10.1088/1742-6596/438/1/012011
42. Gildenburg VB, Kostin VA, Pavlichenko IA (2016) Excitation of surface and volume plasmons in a metal nanosphere by fast electrons. Phys Plasmas 23(3)
43. Ringermacher HI, Mead LR (2015) Observation of discrete oscillations in a modelindependent plot of cosmological scale factor versus lookback time and scalar field mose. Astron J 149(4):137
44. Riedinger R et al (2016) Non-classical correlations between single photons and phonons from a mechanical oscillator. Nature 530:313–316
45. Blencowe M (2016) Quantum physics: photons paired with phonons. Nature 530:284–285. https://doi.org/10.1038/530284a
46. (2015) Quantum leap. Nature 528:435–436. https://doi.org/10.1038/528435b
47. McConnell R et al (2015) Entanglement with negative Wigner function of almost 3,000 atoms heralded by one photon. Nature 519:439–442
48. Dubois J et al (2013) Minimal-excitation states for electron quantum optics using levitons. Nature 502:659–663. https://doi.org/10.1038/nature12713
49. Flindt C (2013) Quantum physics: single electrons pop out of the Fermi Sea. Nature 502(630–632). https://doi.org/10.1038/nature12699. Eddi A, Fort E, Moisy F, Couder Y (2009) Unpredictable tunneling of a classical wave-particle association. PRL 102:240401
50. Филиппов AT (1986) Многоликий солитон. M., Наука, 1986 (in Russian) (Filippov AT (2010) The versatile soliton, Birkh user. Boston)
51. Galiev SU, Galiyev TS (2017) Nonlinear quantum waves in the light of recent slit experiments. http://hdl.handle.net/2292/3675; García-Ñustes MA, Marín JF, González JA (2017) Bubblelike structures generated by activation of internal shape modes in two-dimensional sine-Gordon line solitons. Phys Rev E 95:032222
52. Galiev SU (2003) Transresonant evolution of wave singularities and vortices. Phys Lett A 311: 192–199

Chapter 7
Model of Eruption of the Universe Out of a Pre-Universe and Its Early Evolution

In the previous chapters of the book, we examined macrowave processes and to some extent touched upon the problems of the mikoworld. Of course, when considering problems, we were greatly limited by the mathematical apparatus used. This apparatus, as a whole, boiled down to solving the simplest nonlinear wave equations considered for a very special case, namely cases of resonance. The book uses the following mathematical understanding of resonance. Resonance occurs when the influence of the nonlinear terms of the wave equation becomes much stronger than the influence of the linear terms and the differential equation is reduced to an algebraic one.

Now we turn to building a theory that explains the origin of our Universe. At the same time, we are going to remain within the mathematical apparatus described above. Moreover, we will limit ourselves to using the nonlinear Klein-Gordon equation (NKGE) written in its simplest form (the φ^4 theory).

We are fully aware of the complexity of the task at hand. Of course, we are not going to answer all the questions of cosmology. Together, we want to build a theory that answers several fundamental questions about the origin of the Universe.

Here it must be emphasized that when constructing any theory, one usually focuses on some initial assumptions, explicitly formulated hypotheses and the very goals of the study. After constructing a theory, their validity is verified by an experiment or experience of practical use of the results of the theory. With the development of science, when it is increasingly immersed in the study of the depths of the microworld and space, the possibilities of experimental testing of a theory or the possibility of comparing its results with observations have sharply decreased for many reasons. Mathematical research, which usually follows the construction of a theoretical model of the process under study, is becoming increasingly important in fundamental research. Experimentation and observation are practically excluded from the research process.

The matter is also complicated by the fact that experimental data now cannot be interpreted in any way without an initial theoretical understanding of it, which, in principle, may not be correct. Generally speaking, in modern science,

S. U. Galiev, *Extreme Multivalued Waves as a Model of Everything*,
https://doi.org/10.1007/978-3-031-67027-5_7

experimentation and theorizing are so closely intertwined that the further development of science will depend on how successful their interaction will be.

As an example, consider the evolution of our ideas about the Universe.

In 1917, Albert Einstein formulated his equations of the general theory of relativity, which when applied to the Universe as a whole predict that the Universe is unstable as a consequence of gravitation and energy. But the concept of an evolving Universe, contradicted accepted the ideas of that time. Therefore, Einstein rejected the idea about instability of the Universe. Later, after Aleksander Friedmann's (1922) calculations and Edwin Hubble's observations (1929), the expansion of the Universe was accepted by everyone, including Einstein. Now that once unacceptable theory of the expansion of the Universe has become a well-documented fact. It formed the basis of the standard cosmological model. In 2022, the James Webb Space Telescope (JWST) was launched into space [1, 2]. In the last year, he began sending information that contradicted the standard model. According to it, well-developed and large galaxies already existed 300 million years ago. Therefore, the existing theory has come into conflict with observations and another theory is needed.

The given example shows how difficult ideas are developing about very complex processes in which the applicability of a previously developed theory is not clear, and experimental facts are not entirely reliable. Thus, this situation will arise more and more often as we delve deeper and deeper into the study of problems of cosmology and the microworld.

Modern science is no more than 500 years old. Ahead, I hope, are many more millennia of its development. It is difficult to assume that we have comprehended the basic things that characterize the Universe and Nature, without leaving something fundamental to explore for the remaining millennia. I am inclined to assume the opposite! We know very little of what is truly fundamental. Perhaps in 100 years, our understanding of the Universe will be very different from the standard cosmological model.

7.1 Introduction

Existence of the Universe and its origin are some of the greatest puzzles. There were many tales, myths, and ideas that attempted to provide an answer. The scientists, however, came up with only a few fundamental theories. We start by briefly introducing some of those concepts.

7.1.1 The Problem of the Origin and Initial Development of the Universe

Up to the middle of the twentieth century, many researchers thought that the Universe always existed and did not change with time. This opinion was supported by the general theory of relativity of Einstein. He considered that the Universe is stationary. In particular, it has no beginning and no end. This point of view held in cosmology for a long time after Einstein. It still has its supporters up to the present time. In particular, models of slowly varying (nearly stationary) Universe without the beginning and the end were suggested [3].

However, more modern cosmological theories have been developed since then.

The classical cosmology is based on Albert Einstein's general theory of relativity (GTR). It describes the evolution of the Universe very well, as long as you do not approach to birth of the Universe (Big Bang). It is the initial singularity where relativity theory fails to provide what must be demanded of a final theory of space and time. The modern theoretical cosmology was based by Alexander Friedmann [4]. He started from the fundamental results of Einstein but rejected the static paradigm. As a result, he formulated a mathematical theory of the evolutionary Universe, which has a beginning and an end. Friedmann's results were the starting point for many works. In most of them, the researchers assumed the beginning of the Universe. Lemaître assumed that at the begining, "the energy of the Universe packed in a few or even in a unique quantum" and "the beginning of the world happened a little before the beginning of space and time" [5]. In 20 years, George Gamov has started to develop and advocate actively the theory of the hot extending early Universe in which the radiation dominated [6]. Then fundamental articles by Sakharov [7], Zel'dovich, and Sunyaev [8] were published. New ideas were formulated in [4–8]. For example, the Universe did not necessarily begin from a singularity. Maybe the Universe passed smoothly through the beginning [4–11]?

But how did this beginning come about? Perhaps the most original opinion was formulated in [12]. The Universe can be born instantaneously, as a result of a quantum vacuum fluctuation, from "nothing" along with space and time as a clot of energy and matter. This suggestion opened a way to use physical laws to study the beginning of the Universe's evolution [13–17].

The model [12] was criticized since a vacuum may only exist in some preexisting space-time (existing or existed before our Universe). Now it is well known that a vacuum is very different from "nothing" [18, 19]. The vacuum can have energy. The vacuum cannot be considered an empty place. It is possible to assume that the vacuum of [12] belongs to some preexisting Universe. Over the last few decades, several scientific models have been developed taking into account a possibility of existence of the pre-Universe and some space-time before the beginning.

The idea of existence of pre-Universe or some global multiverse opened up opportunities for constructing theories which are far from models of a single stationary or quasi-stationary Universe. One of them is the theory of the eternal inflation. According to it, our Universe came into existence along with an infinite number of other Universes (multiverse) due to a phase transition in some scalar field [20–23]. The origin resembles the birth of bubbles in boiling water. The emergence of the bubble Universe is determined by the quantum fluctuations when the conditions of the scalar field are close to the conditions of "boiling" [24–27]. In this model, a bubble Universe can collide with another bubble Universe [24, 27].

The parallel deep idea in scientific cosmology is that the Universe was born not just once, but multiple times in an endless cycle of fiery deaths and rebirths. These cyclic Universe models were popular in the years 1920–1930. But they were replaced by the model of the Big Bang. Now they reappear [28–36]. Perhaps, the most well-developed version of this model is presented in [30]. Roughly speaking, it suggests the existence of two Universes separated by an extra spatial dimension (parallel Universes). The Universes exist independently, but periodically collide when the extra dimension disappears. The moment of this collision can be regarded as an analogy of the Big Bang. Thus, the model contains some elements and many results of the Big Bang model and the inflationary scenario. After the collision the extra dimension increases and these Universes diverge. Some researchers speak about a single cyclic Universe. Our Universe may have begun as the Big Bounce instead of the Big Bang [37].

Above we described two fundamental approaches to the origin of the Universe. Many new models may be constructed within these approaches. Some of them have been developed in recent years [38–43].

For example, the Universe could form as a result of a collapse of a star located in the pre-Universe into a black hole [38–40]. The collapsing black hole causes the emergence of a new Universe, whose fundamental properties may differ from the pre-Universe where the black hole collapsed. Each Universe gives rise to as many new Universes as it has black holes. Or our 3D Universe can emerge from a 2D reality as some holographic picture [43].

Thus, there are a sufficient number of well-developed theoretical models of the origin of the Universe. Quantum cosmology has been developing rapidly in recent years. Quantum cosmology is the attempt in theoretical physics to develop a quantum theory of the Universe. This approach attempts to answer open questions of classical physical cosmology [29]. Such an approach is tried to develop with the string theory too [40]. However, even the most popular standard cosmological model (the Big Bang model modified by the inflationary scenario) cannot claim that its predictions are fully consistent with the observations [44–47]. In principle, the results of the collisions of the Universes or the Big Bounce results can manifest themselves in the cosmic microwave background radiation (CMB) [48–50]. We can assume that the precision of modern experiments and cosmological observations is sufficient to allow us to start testing existing theories. But, for example, up to now, traces of such collisions have not been detected [51]. The latest experiments did not support the holographic Universe idea [52].

Over the past decades, a standard cosmological model has been formulated. It is based on Einstein's equations of general relativity. There are two big assumptions. First, that the Universe is homogeneous, looking roughly the same in all locations. Second, it is isotropic, looking roughly the same in all directions from any standpoint. According to the model, the Universe has existed for about 13.8 billion years. Stars began to form due to gravity approximately 100 million years ago. The very first galaxies arose about a billion years ago, and the development of their forms to modern forms took up to 3 billion years. Almost all cosmologists agree with these results [53].

However, it is found today that galaxies exist in clusters and filaments of matter distributed around the boundaries of huge, bubble-shaped voids. These voids have roughly one-tenth of the clusters' matter density, but account for more than 60% of the Universe's volume. New cosmic portraits (the cosmic microwave backgrounds) contain new strange anomalies. For example, a huge swathe of sky at a slightly cooler temperature than the rest (the Axis of Evil) which was observed in the 2013 Planck map. There is no theory which explains the appearance of some another anomalies.

As a result, the Big Bang theory and a standard cosmological model become slippery concepts. "The great tragedy of science," as Thomas Huxley observed, is "the slaying of a beautiful hypothesis by an ugly fact." He was talking about the origin of life, but it is true for all theories of the origin of Universe.

To put an end to these and other important cosmological problems in 2022, the James Webb Space Telescope (JWST) was launched into space [54–56]. In the last year, it began sending information that contradicted the standard model. According to it, well-developed and large galaxies already existed 300 million years ago. Therefore, the existing theory has come into conflict with observations and another theory is needed.

Some researchers think that we need now a simple, radical idea that will point toward new approaches to the puzzles of the origin of the Universe. We emphasize that in modern science, there are no generally accepted methods suitable for studying the Universe. In this regard, we do not have many more tools and global ideas than the ancient Greeks. Like them, we are faced with the problem of choosing a model to describe a completely uncertain process.

A model of the origin presented in [57–60] is considered below. A scalar field described by a nonlinear Klein-Gordon equation is considered as a fundamental field in this model. The origin of the Universe is being studied as an evolution of an initial scalar field into other fields (particles), energies, waves, and vortices.

The model that is based on simple ideas and equations is chosen. Of course, this is here only a rough model. Modeling the birth of the Universe is a bit like playing whack-a-mole. All time, the new results knock out the old theories.

7.1.2 Nonlinear Klein-Gordon Equation and the Galiev-Galiyev Model

Due to the above, when constructing a theory, it is necessary, at a minimum, to clearly formulate the questions that the theory is going to answer.

What kind of result are we going to get in line with the approach developed in the book? Based on nonlinear Klein-Gordon equation, we will answer two fundamental questions:

1. Where was the energy found for the origin of the Universe and how was it transformed?
2. Where did the so-called fine-tuning of the Universe come from?

Let us clarify the last question. The fact is that modern physical theories that describe reality at a fundamental level include many different "free" parameters, the numerical values of which are not yet explained by currently existing theories and are simply taken from experiment. Since the numerical values of these parameters are not determined by existing theories, they appear arbitrary, capable in principle of taking on any values. Moreover, these parameters look as if they are finely tuned for the possibility of the existence of life.

In general, the answers to these questions given in the book are not entirely original. But in connection with the latest JWST observational data, they began to be discussed again. The fact is that the latest observations do not fully correspond to the theory of the Big Bang and the known theories of galaxy formation. Here it must be emphasized that well-known theories, in particular, the standard cosmological theory, have been tested for decades and, in general, have responded satisfactorily to more and more new data.

However, this does not exclude the fact that it is incorrect. As you know, in order for a theory to be correct, it must correctly describe all the phenomena related to it, but it becomes incorrect if it does not describe the only new phenomenon!

This part of the book answers the questions posed above. Usually, the result of a chapter is formulated at the end of it, but the questions discussed below are so diverse in topic that it makes sense to first present the outline of the research and intermediate results.

Below we will describe the model of the birth of the Universe advanced by model of Galiev and Galiyev (MGG) [57–60]. It proceeds from the assumption that there is a pre-Universe. It is filled with a high-energy, multidimensional, highly structured field. This field, due to some resonant processes, constantly radiates the seeds of the Universe, just as extreme Faraday waves can radiate isolated drops and bubbles from a liquid surface, that is, to radiate particles from a two-dimensional surface into a new dimension. It is accepted that there are not many stable multidimensional spaces, so the emitted particles die en masse, but some, upon entering a new space, losing stability, disintegrate into fundamentally one-dimensional objects, which can be considered as extended strings or thin pipes. From these objects arise structures that vibrate with unimaginably high energy and radiate this energy. As a result, as is accepted in the model under consideration, a new three-dimensional space and time is formed.

After which the extreme dense formation that has arisen begins to rapidly expand, from some starting, final in size, values and evolving into the Universe familiar to us.

Based on the above, the following characteristics of the Universe are formulated at various stages of its origin and initial development:

1. There is a pro-Universe filled with a scalar energy field. It can be assumed that it is filled with bosons that have an unimaginably high energy. The field is stirred up by tiny quantum fluctuations. The pro-Universe is located in such a multidimensional space that ensures its stability. At any given time, the pro-Universe gives birth to billions of "seeds" of rapidly evolving Universes. Most of them are "flickering" Planck-sized Universes that instantly appear and disappear.

But there are some Universes that evolve to larger sizes. One of them accidentally formed our Universe with the fundamental parameters that we have. Thus, our Universe appeared as a particle (bubble) of a multidimensional scalar field, emitted from the pro-Universe into a new space (see Sects. 7.2 and 7.3).

2. It is assumed that the emitted particle (bubble) disintegrated into an unimaginably dense unity of independent one-dimensional oscillating structures. This unity can be imagined as a conglomerate of oscillating strings. Each one-dimensional formation has practically infinite energy and vibrates with its own frequency (see Sect. 7.4).
3. Strong nonlinearity leads to the fact that each oscillation (each one-dimensional wave) radiates particle-waves beyond the one-dimensional formations. The emitted particles correspond to the natural frequencies of the one-dimensional formations. As a result, the areas beyond the formations are filled with, literally, an infinite number of diverse particles. A huge number of identical particle-waves begin to interact and create a new space (certain sphere) (see Sects. 7.4 and 7.5).
4. At a certain critical density, bosons begin to transform into fermions (due to the Pauli exclusion principle, only one fermion can occupy a given quantum state at a time. Fermions are usually associated with matter, while bosons are usually force-carrying particles. However, in the current state of particle physics, the distinction between the two is unclear. Weakly interacting fermions can also exhibit bosonic behavior under extreme conditions). More and more fermions appear. They push the material of the original sphere apart. A colossal repulsive force arises, and the Big Bang begins. Thus, the expansion does not begin with a singularity. At the beginning, due to the increasing density of fermions, the expansion rate is very high and even increases. However, as the size of the sphere increases and the density of matter decreases, the expansion rate begins to decrease (see Sect. 7.6).
5. During the expansion, some dynamic equilibrium of the numbers of baryons and fermions arises. At the same time, a huge number of baryons that arose in the first moments of the Universe's existence are still traveling through it, forming the cosmic microwave background (relic radiation).
6. The surface of the expanding sphere loses stability. Therefore, waves and vortices, the embryos of future galaxies and stars, begin to appear on it (see Sects. 7.6.1 and 7.6.2).
7. The particles that filled the expanding sphere have traces ("voids" behind them). These traces generate gravity. It depends on the speed of the particles, their number, and changes as the sphere expands (see Sects. 5.6, 5.7, 5.8 and 7.7).

The described elements of the theory have been presented in various publications. Here it is slightly modified to more closely match the latest observational data from the James Webb Space Telescope (JWST) [54–56].

7.2 Scalar Field Dynamics of a Pre-Universe

We begin our study with a description of the scalar field of the pre-Universe. The fields themselves are not "made of" anything—fields are what the world is made of. At the same time, apparently, fields are often the easiest way to describe different natural phenomena. In particular, field theories are often used for purposes of introduction of novel concepts and techniques.

7.2.1 *Basic Equation and Relations*

The pre-Universe model is described here using the nonlinear Klein-Gordon equation (NKGE). It is known that this equation describes a wide variety of physical phenomena, such as wave propagation in different artificial and natural systems and the "birth" of spinless particles in relativistic quantum mechanics. Above we stressed that this equation plays a fundamental role in certain cosmological models.

1. Versions of NKGE. We assume that there is nothing in the pre-Universe except the scalar field. Let it be described by NKGE in the following form:

$$\widehat{\Phi}_{tt} - c_*^2 \sum_{i=1}^{I} \widehat{\Phi}_{ii} = -\partial V\left(\widehat{\Phi}\right)/\partial\widehat{\Phi}. \tag{7.1}$$

   Here c_* denotes a constant, I is an integer value ($I \geq 1$), and V is the scalar field potential. The indexes denote the differentiation; $\widehat{\Phi}_{tt} = \partial^2\widehat{\Phi}/\partial t^2$, $\widehat{\Phi}_{ii} = \partial^2\widehat{\Phi}/\partial x_i^2$, where x_i are rectangular Cartesian coordinates.

   We stress that rectangular Cartesian coordinates are used. These coordinates are used widely for the description of complex processes in different areas of physics. In particular, turbulent processes are often described using these coordinates. Recently, the collisions of Universe was modeled with the help of Cartesian coordinates [50, 61]. We note that the Friedmann-Robertson-Walker metric is used traditionally for the study of the homogeneous and isotropic Universe at the earliest epoch of its evolution. But this metric is not applied to massless scalar fields.

   Different expressions for the function V can be found in many books and papers. We assumed that

$$V\left(\widehat{\Phi}\right) = \frac{1}{2}m^2\widehat{\Phi}^2 - \frac{1}{4}\lambda\widehat{\Phi}^4. \tag{7.2}$$

   Here m^2 and λ are constants. A solution of (7.1) is represented as a sum,

$$\widehat{\Phi} = \overline{\Phi} + \Phi + \bar{f}. \tag{7.3}$$

Here $\overline{\Phi}$ is a stationary part and Φ is a dynamic part of the scalar field, and $\overline{f}$ takes into account the quantum fluctuations. Using (7.3), we obtain from (7.1) equations for $\overline{\Phi}$, Φ, and f:

$$-c_*^2\sum_{i=1}^{I}\overline{\Phi}_{ii}=-m^2\overline{\Phi}+\lambda\overline{\Phi}^3, \tag{7.4}$$

$$\Phi_{tt}-c_*^2\sum_{i=1}^{I}\Phi_{ii}=-m^2\Phi+\lambda\left[3\overline{\Phi}^2\Phi+3\overline{\Phi}\Phi^2+\Phi^3+3(\overline{\Phi}+\Phi)\overline{f}^2+3\left(\overline{\Phi}^2+2\overline{\Phi}\Phi\right)\overline{f}\right], \tag{7.5}$$

$$\overline{f}_{tt}-c_*^2\sum_{i=1}^{I}\overline{f}_{ii}=-m^2\overline{f}+\lambda\overline{f}^3+3\lambda\Phi^2\overline{f}. \tag{7.5.1}$$

Equations (7.5) and (7.5.1) are focused on the interaction of the fields. We will link the origin of the particles with interaction of the scalar fields. Equations (7.4), (7.5), and (7.5.1) are a simple example of similar interaction. It is important that the right-hand side of (7.5) contain the quantum perturbation. The simplest version of this equation may be presented in the form of

$$\left(-\Box+\mu^2\right)G(x,y)=\delta(x,y)-\left(3/8\pi^2\right)H^4. \tag{7.5.2}$$

Here $\delta(\eta-\xi)$ is the Dirac delta function (the impulse function), the square in (7.5.2) is the d'Alembert operator. The last equation was used by Hawking [24, 62]. In general, the equation does not contradict modern ideas. For example, an equation similar to the Hawking equation and resembling Eq. (7.5) is discussed in [3, 63]. The equations of scalar fields containing quantum perturbations are discussed in books [64, 65].

2. New variables and a simplified model of NKGE. Equations (7.4), (7.5), and (7.5.1) are not integrable in the general case. Therefore, we sought approximate solutions. On the other side, we cannot tell what of the pre-Universe scalar field looked like. However, we assume that it contains multidimensional spherical structures similar considered in the Chaps. 1 and 5. The following new independent variables are used to describe these structures and different scenarios of the evolution of the scalar fields:

$$\eta=-\overline{K}\sum_{i}^{I}\sin^2\vartheta x_i\ ,\xi=R^*+B\sin^2\omega\ t-K\sum_{i}^{I}\sin^2\vartheta x_i. \tag{7.6}$$

Here $\overline{K}=\overline{K}(x_i,t)$, $\vartheta=\vartheta(x_i,t)$, $R^*=R^*(x_i,t)>0$, $B=B(x_\iota,t)$, $\omega=\omega(x_i,t)$, and $K=K(x_i,t)$. Let us assume that the dependence of these coefficients upon coordinates and time is very weak. It is accepted that the coefficients $\overline{K}$, ϑ, R^*, B, and ω are constants when considering any local structures of the scalar field.

Using (7.6), we can find expressions Φ_{tt} and Φ_{ii}. For example,

$$\Phi_i = \Phi_\xi \omega B \sin 2\omega t, \quad \Phi_i = -2\Phi_\xi \vartheta K(a_i + \sin \vartheta x_i) \cos \vartheta x_i,$$

$$\Phi_{tt} = \frac{1}{2}\Phi_{\xi\xi}\omega^2 B^2(1 - \cos 4\omega t) + 2\Phi_\xi \omega^2 B \cos 2\omega t \approx \frac{1}{2}\Phi_{\xi\xi}\omega^2 B^2,$$

$$\begin{aligned}\Phi_{ii} &= 4\Phi_{\xi\xi}\vartheta^2 K^2\left(a_i \cos \vartheta x_i + \frac{1}{2}\sin 2\vartheta x_i\right)^2 - 2\Phi_\xi \vartheta^2 K(\cos 2\vartheta x_i - a_i \sin \vartheta x_i) \\ &\approx \frac{1}{2}\Phi_{\xi\xi}\vartheta^2 K^2\left(4a_i^2 + \frac{1}{4}\right).\end{aligned} \tag{7.7}$$

Then, we ignored in (7.7) the terms explicitly dependent on the harmonics. In this case, we rewrite Eqs. (7.4) and (7.5) using (7.6):

$$-2c_*^2\vartheta^2\overline{K}^2\left(\sum_{i=1}^{I} a_i^2 + \frac{1}{4}I\right)\overline{\Phi}_{\eta\eta} + m^2\overline{\Phi} - \lambda\overline{\Phi}^3 = 0, \tag{7.8}$$

$$\left[\frac{1}{2}\vartheta^2 B^2 - 2c_*^2\vartheta^2 K^2\left(\sum_{i=1}^{I} a_i^2 + \frac{1}{4}I\right)\right]\Phi_{\xi\xi} + \left(m^2 - 3\lambda\overline{f}^2\right)\Phi - \lambda\Phi^3 = 0. \tag{7.9}$$

We neglected the influence of boundaries in (7.9). Equation (7.9) shows that the coefficient of the linear term clearly depends on the quantum fluctuations. Similarly, the remaining coefficients in (7.8) and (7.9) can depend on these fluctuations.

We simplify the problem to the solution of ordinary differential equations. In general, this approach is widely used in physics and cosmology for the modeling of dynamics of scalar fields [15–19, 21–24]. This approximation enables us to describe observations and to identify main trends in the development of fundamental processes.

We used the variables t, x_i and η, ξ (7.6). We will assume, for the most part, that the scalar field and its corresponding derivatives are finite and continuous functions of these variables. At the same time, we will consider also cases of exceptions, when Eq. (7.9) has discontinuous solutions.

3. The influence of boundaries (static field structure). Let us consider the influence of boundaries. We stress that the values η and ξ are independent. At the same time, these values may be equal

$$\eta = \xi, \tag{7.9.1}$$

if

$$t = \omega^{-1}\operatorname{argsin}\sqrt{\left[\left(K - \overline{K}\right)\sum_{i}^{I}\sin^2\vartheta x_i - R^*\right]B^{-1}}. \tag{7.9.2}$$

In particular, if $x_i = 0$, we have $t = \omega^{-1} \operatorname{argsin} \sqrt{-R^*/B}$. Cases may occur when the fields $\overline{\Phi}$ and Φ interact. In this case,

$$\Phi_{tt} - c_*^2 \sum_{i=1}^{I} \Phi_{ii} = \begin{array}{l} -m^2\Phi + \lambda\Big[3\overline{\Phi}^2\Phi + 3\overline{\Phi}\Phi^2 + \Phi^3 + 3\left(\overline{\Phi} + \Phi\right)\overline{f}^2 \\ +3\left(\overline{\Phi}^2 + 2\overline{\Phi}\Phi\right)\overline{f}\Big]\delta(\eta - \xi). \end{array} \tag{7.9.3}$$

It can be easily shown that during collisions with a boundary, the field can be significantly enhanced. Let us accept here that the scalar field under consideration describes the periodic evolution of the Universe as a whole. Then these collisions correspond to the model of a cyclic Universe, which periodically expands spherically from the center. It then hits and reflects off the bounding spherical surface to focus back at the center.

Remark 1 The expression ξ (7.6) can be compared with the Minkowski formula linking time and space. Indeed, if $R^* = 0$ and $\omega\, t$, ϑx_i are very small, the expression ξ describes the distance from an arbitrary point of space-time till the point $t = 0, x_i = 0$.

Remark 2 In (7.9), the value $\frac{1}{2}\omega^2 B^2 - 2c_*^2 \vartheta^2 K^2 \left(\sum\limits_{i=1}^{I} a_i^2 + \frac{1}{4} I \right)$ may be considered as the small divider. At the same time, this value determines the width of the resonant band.

Remark 3 Further, for simplicity, we will assume that $a_i = 0$ in (7.8) and (7.9).

7.2.2 Basic Solutions and Bubbles of Energy

We stress that in spite of the observational triumphs, the standard cosmological model remains an unfinished work of art. Many of its recent successes can be traced to the initial conditions postulated for very early stages of the Universe. The model assumes that the Universe had to start with certain properties that have never been successfully explained by the fundamental physics. So to explain new observational data, researchers are introducing more and more suggestions which are not fully supported by observations, experiments, and the results of fundamental physics. As a result, the initially simple Big Bang theory became quite complex.

In contrast with this, we hope [57, 59, 60] that some solutions of NKGE can describe the strict sequence of stages of an eruption of the Universe out of some pre-Universe. In particular, the solutions should explain an appearance of the initial energy of the Universe. Possibly it is a result of the interaction of scalar fields. But how can they interact?

On the first stage of the study we will not take into account the influence of тхэ quantum perturbations. We will postpone their consideration until Sect. 7.3. Here we will study the structure of the scalar field formed by the superposition of its static and dynamic parts. Periodic, localized solutions of Eqs. (7.8) and (7.9) under the above conditions are sought as expressions

$$\overline{\Phi} = \overline{A}\,\mathrm{sec\,h}\eta, \Phi = A\,\mathrm{sec\,h}\xi. \tag{7.10}$$

Using (7.7), we approximately found from (7.8) and (7.9) that

$$\overline{A} = A = 0, \quad \overline{A}_{\pm} = A_{\pm} = \pm\sqrt{2}\lambda^{-0.5}m, \tag{7.11}$$

$$m^2 = 2c_*^2\vartheta^2K^2\left(\sum_{i=1}^{I}\left(a_i^2 + \frac{1}{4}I\right), \quad \omega^2B^2 = c_*^2\vartheta^2K^2\left(\sum_{i=1}^{I}\left(4a_i^2 + I\right) - 2m^2. \tag{7.12}$$

The three amplitudes (7.11) are determined by the values m and λ (see also (1.93) and (1.107)). We do not know what scalar fields are represented by these formulas. It is, however, possible that these three fields are certain analogs of true and false vacuums. Values ϑ, $\overline{K}$, B, ω, and K are not completely determined by (7.12). We will consider these values as arbitrary.

How can we interpret the solutions and expressions written using the variables η and ξ? Let us return to the Cartesian coordinates.

At first, we consider the landscape of the scalar field so as to appreciate its dynamics. Generally speaking, using the solutions (7.10) and (7.11), we can construct smooth as well as discontinuous landscapes. Here smooth landscapes are considered. We find assuming (1.11) and using $\widehat{\Phi} = \overline{\Phi} + \Phi$ (7.3),

$$\widehat{\Phi} = -m\sqrt{2/\lambda}(\mathrm{sech}\eta + \mathrm{sech}\xi). \tag{7.13}$$

The expressions (7.13) and (7.2) allow us to calculate the scalar function and the scalar potential. In particular, the expression $\widehat{\Phi}$ (7.13) determines a landscape of the scalar field which is described by static and dynamic parts. We stress that if η and ξ are very large, then $\widehat{\Phi} \approx 0$.

For simplicity, the two-dimensional landscapes will be calculated. In Figs. 7.1 and 7.2, the results of the model calculations are presented. It is assumed that $R^* = 5$, $B = 25$, $\omega = 1$, $K = 300$, $\vartheta = 0.15$, and $\overline{K} = 0.85$. We stress that $K = 300$ is used. In this case, the dynamical part describes a structure having a very thin wall. We assumed above that $a_i = 0$.

The landscape of the scalar field calculated according to (7.13) is shown in Fig. 7.1 (left). For simplicity, we assume that $m\sqrt{2/\lambda} = 1$. The stationary part of the scalar field describes the landscape, which consists of ridges (hills) and valleys (Fig. 7.1, left). The highest value of the function is reached at the top of the hills. The dynamical part of the field disturbs the stationary picture in the valleys where a thin ring radially oscillates (Fig. 7.1, right). This ring may be named as the oscillon [66–72].

This oscillon corresponds to a multidimensional oscillating bubble (sphere and clot) which can have a very thin wall. The lowest value of the dynamic field

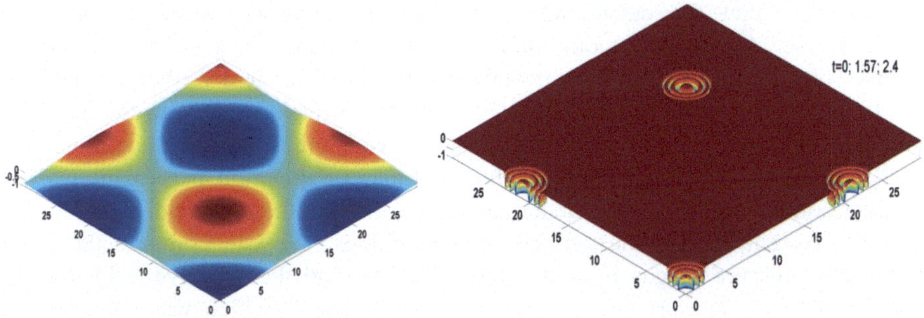

Fig. 7.1 The static (left) and dynamic (right) two-dimension maps of the landscape of the scalar field. The dynamic landscapes were calculated at three dimensionless moments of time: 0, 1.57, and 2.4

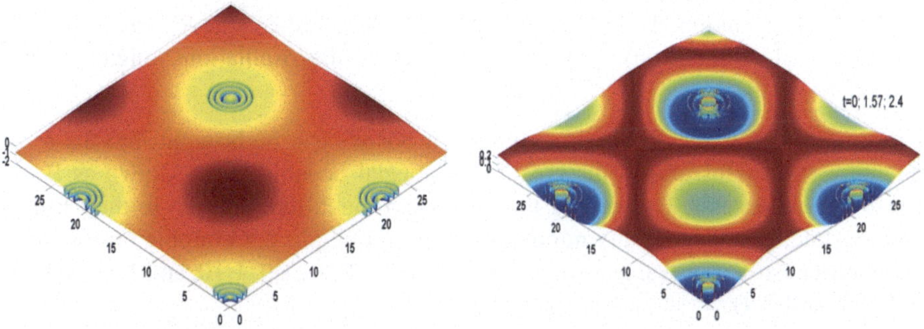

Fig. 7.2 Two-dimensional maps of the combined landscapes. The map of the scalar field (left) and the map of the scalar potential (right). The calculations were made at three dimensionless moments of time: 0, 1.57, and 2.4

(negative field) is reached in this wall. Within the bubble $\Phi \approx 0$, if the wall is very thin, $(K \to \infty)$. This case is some analog of the well-known thin-wall approximation.

Substituting (7.13) into (7.2), we approximately find that

$$V\left(\widehat{\Phi}\right) = \lambda^{-1} m^4 \left[(\mathrm{sech}\eta + \mathrm{sech}\xi)^2 - (\mathrm{sech}\eta + \mathrm{sech}\xi)^4\right]. \tag{7.14}$$

The expression (7.14) determines a landscape of the scalar potential (Fig. 7.2 bottom) which takes into account static and dynamic parts of the field. For the calculation, we assume that $\lambda^{-1} m^4 = 1$ in (7.14).

The dynamic part (bubble, sphere, and clot) of the scalar field oscillates inside the scalar potential well (Fig. 7.2, right). This part cannot escape from the well unless it is given a large enough energy influx. We assume that similar landscapes describe the scalar potential of some pre-Universe.

We think that the calculations describe qualitatively a structure of some pre-Universe. According to our model, the pre-Universe is a system of infinite number of bubbles oscillating in potential wells. The hills and wells are described by the static part of the scalar field. The bubbles correspond more to the dynamic part. The bubbles may be slightly different from each other, since the coefficients in (7.8) and (7.9) may be different slightly because of quantum perturbations. The bubbles contain a scalar field which may be very close to the zero field. The negative field is concentrated in the surface of the bubbles. We can compare the inner field with pressure of gas in some elastic bubble, and the negative field may be some analog of the surface tension. The negative field (which was described as $-\sqrt{2}\lambda^{-0.5}m\text{sech}\xi$ ((7.8) and (7.9))) does not allow us to increase strongly the volume of the bubble. We can use another parallel for the oscillating bubbles. Let us suggest that the "potential" energy of bubbles transforms partly into "kinetic" energy and vice versa. In other words, the zero field (pressure) transforms partly into the negative field (tension) and back.

Generally speaking, the bubbles might be very small (the value R^* may be like the Planck's scale dimension), and the amplitude of the oscillations might be also extremely small.

Comment Let us recall that any number of bosons can be brought into the same quantum state. This allows the creation of special bosonic states known as Bose-Einstein condensates. By cooling the bosons so much that they go to their lowest energy state, you can put any number of them in one place. Grouping bosons form symmetrical structures such as rings or spheres (see Figs. 1.9, 1.10, and 1.11 and 7.1, 7.2, 7.3, and 7.4).

7.2.3 *Oscillons and Bose-Einstein Condensate as Models of the Scalar Pre-Universe*

We discussed oscillons in Sect. 1.5.2. Examples of such waves arising on a layer of small balls are shown in Fig. 7.3 (left). These excitations may be considered as strongly nonlinear waves, where the vertical motion of the particles is connected with their horizontal motion. Similar spatiotemporally oscillating localized nonlinear

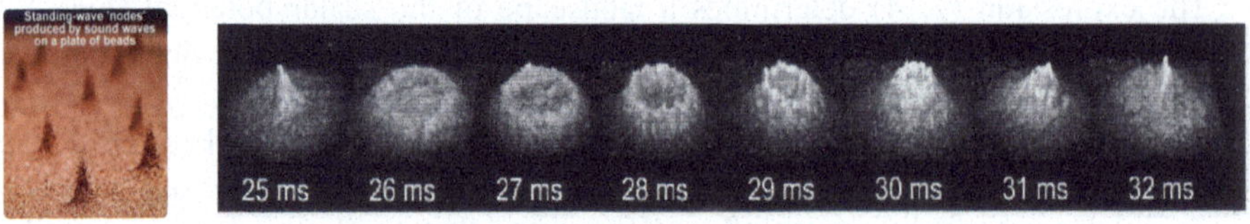

Fig. 7.3 Periodical granular columns and craters on the surface of a vertically excited layer. Each granular column has two craters near it [57, 73] (left). Dynamic wave process observed for Bose-Einstein condensate (right)

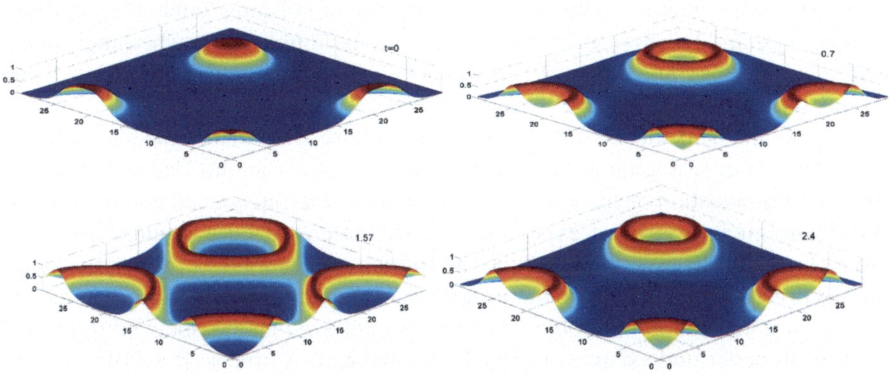

Fig. 7.4 Radially oscillating rings are shown which form the peak during the convergence. The data simulate the experiments with the Bose-Einstein condensate [57, 73] (see also http://www.lkb.ens.fr/recherche/atfroids/anglais/vortex_an.html)

waves having properties of both standing and travelling waves were described in [66–82]. In particular, the analytical theory of these waves was developed.

On the whole, this process also describes qualitatively oscillons in the Bose-Einstein condensate (BEC) (Fig. 7.3 (right)). It is possible to see periodic formation of peaks; they transform into the crater which generates new peaks. BEC was obtained for the first time in the year 1995, and then the study of its properties began. Some of them were explained by strongly nonlinear properties of BEC [57, 73]. In particular, localized regions of periodic oscillations were found in the condensate (Figs. 1.10 and 7.3).

A possibility of the existence of space oscillons in the very early Universe immediately after the end of the inflationary stage of its expansion was found [78–80]. It is very important for us that those oscillons were described by the solutions of NKGE.

In Fig. 7.4, we present data of calculations of the landscape of the scalar field which simulate the Bose-Einstein condensate oscillations.

It is assumed that $R^* = 0.001$, $B = 3.5$, $\omega = 1$, $K = 5$, $a_i = 0$, and $\vartheta = 0.15$. We stress that in contrast with Fig. 7.1, we have assumed $K = 5$ in Fig. 7.4. As a result, unlike Fig. 7.1, the thickness of the rings in Fig. 7.4 increased. Figure 7.4 begins to resemble data of the experiments with the Bose-Einstein condensate (Figs. 1.10 and 7.3). Oscillons are usually modeled with the help of numerical calculations [69, 70, 72, 75, 79, 80]. However, we have shown that they may also be studied analytically in the context of the theory of extreme waves [2, 57, 58]. It was shown in 2001 that local oscillons similar to those found in BEC can exist in cosmic space [83].

Conclusion We explored in this section the properties of the vacuum that possibly fills the pre-Universe. The possibility of structuring this vacuum, which is filled with potential wells containing pulsating clots of energy, has been discovered. These clumps can be considered as oscillons.

We emphasize that the generation and existence of pulsating bubbles is observed in many systems [82, 84], in particular, in various liquid media when their phase state changes. For some time now, this process has also begun to be associated with various theories of the origin of Universes [85].

In quantum field theory, the change in the phase state of the medium and the emergence of a real ground state is known as "false vacuum decay." It occurs through the nucleation of spatially localized bubbles. Despite great theoretical efforts to estimate nucleation rates, experimental observations of false vacuum decay are still lacking. In [86, 87], the first time the formation of bubbles in supercooled gas was observed. The transition to the ground state (liquid phase) is carried out due to the resonant nucleation of bubbles. The authors believe that the obtained result will allow us to better understand some models of the early Universe [88–90].

Based on the above results and the latest publication [91], it can be assumed that the appearance of bubbles and oscillons in pre-Universes can be studied experimentally!

In the future, we will consider the vacuum of the pre-Universe as a false vacuum in which there are potential wells containing bubbles of scalar energy. Under the influence of quantum disturbances, they can acquire sufficient energy to tunnel out of the well.

In quantum field theory, the decay of an extended metastable state into the real ground state is known as "false vacuum decay," and it takes place via the nucleation of spatially localized bubbles. Despite the large theoretical effort to estimate the nucleation rate, experimental observations were still missing. Here, we observe bubble nucleation in isolated and highly controllable superfluid atomic systems, and we find good agreement between our results, numerical simulations, and instanton theory opening the way to the emulation of out-of-equilibrium quantum field phenomena in atomic systems.

7.3 Pop-up from the Pre-Universe as the Transresonant Process

Here we ask why and how a scalar field particle could be emitted from the underlying material of the pre-Universe.

We have the solutions describing many dynamic scalar fields. Our interpretation of the above results is that the space-time of the pre-Universe is pervaded by a number of scalar fields. It is possible to assume that they correspond to some boson-like fields. The fields are localized in some space, which is determined by the coordinate ξ. All fields focus at the same place where they could interact. As a result, the external sources may appear in the field equations. The simplest version of the similar equation has a form (7.5.2). This equation was used by Hawking [24]. Similar equations take into account the quantum effects.

We assumed that the quantum perturbations exist in the potential well of a pre-Universe. We also found that these perturbations, developing within the scalar field, act on certain bubbles located in the bottom of the well. In this case, the equation describing the periodical oscillations of the bubbles should be supplemented with a term which takes into account these perturbations (7.9).

We noted earlier that there are, at least in theory, different types of vacua. In particular, solutions to the cubic nonlinear Kdein-Gordon equation can be interpreted as describing various types of vacua. Obviously, the cubic equation has three solutions. In the case of real solutions, they can be interpreted as a false vacuum, a true vacuum, and a low vacuum.

We assumed that the pre-Universe is filled with the false vacuum [86, 87]. Our solutions describe the oscillations of energy bubbles (Figs. 7.1, 7.2, and 7.4) in potential wells. Let us consider the effect of a quantum perturbation on such a bubble. Generally speaking, as a result of such an impact, it can tunnel out of the potential hole up to the point of leaving the landscape of the false vacuum.

Comment An experiment conducted in Italy has produced the first experimental evidence of vacuum decay. The experiment uses a supercooled gas at a temperature of less than a microKelvin from absolute zero. At this temperature, bubbles are seen to emerge as the vacuum decays [91].

7.3.1 Theoretical Models

Modeling of instant quantum actions. Using the results of Sect. 7.2.1, we will consider Eq. (7.9) subject to an instant external quantum action. In this case, we have

$$\frac{1}{2}\left(\omega^2 B^2 - c_*^2 \vartheta^2 K^2 I\right)\Phi_{\xi\xi} + m^2\Phi - \lambda\Phi^3 = \varphi(\Phi)\widetilde{A}\delta\left(\overline{\xi}\right). \tag{7.15}$$

Here $\overline{\xi}$ is a point subject to the quantum action, $\delta\left(\overline{\xi}\right)$ is the Dirac delta function, and $\widetilde{A}$ is the amplitude of the quantum fluctuation. The function $\varphi(\Phi)$ in (7.15) corresponds to $3\lambda\widetilde{A}^{-1}\Phi f^2$ in (7.9). Let

$$\Phi = A\mathrm{sech}\xi \text{ and } \varphi(\Phi) = \mathrm{sech}^2\overline{\xi}\sinh\overline{\xi}. \tag{7.16}$$

We will consider the field in the vicinity of $\overline{\xi}$. It is assumed that near $\xi \approx \overline{\xi}$, the expression $3\lambda\widetilde{A}^{-1}\Phi\overline{f}^2$ in (7.9) may be approximated as $\sec h^2\overline{\xi}\sinh\overline{\xi}$. There the value of the field can change discontinuously as a result of the quantum kick (fluctuation). This discontinuous change in the field is computed by integrating (7.15) from $\xi_j = \overline{\xi} - \varepsilon$ to $\xi_{j+1} = \overline{\xi} + \varepsilon$:

$$\frac{1}{2}\left(\omega^2B^2 - c_*^2\vartheta^2K^2I\right)\left[\Phi_\xi(\overline{\xi}+\varepsilon) - \Phi_\xi(\overline{\xi}-\varepsilon)\right] + \int_{\overline{\xi}-\varepsilon}^{\overline{\xi}+\varepsilon}\left(m^2A\mathrm{sech}\xi - \lambda A^3\mathrm{sech}^3\xi\right)d\xi = \widetilde{A}\int_{\overline{\xi}-\varepsilon}^{\overline{\xi}+\varepsilon}\varphi(\Phi)\delta(\overline{\xi})d\xi. \tag{7.17}$$

This equation is rewritten using (7.16). As a result, we have

$$\begin{aligned}
&C_1\left[-\left(A\mathrm{sech}^2\xi\sinh\xi\right)_{\overline{\xi}+\varepsilon} + \left(A\mathrm{sech}^2\xi\sinh\xi\right)_{\overline{\xi}-\varepsilon}\right]\\
&+C_2\left[-\left(A\mathrm{sech}\xi\right)_{\overline{\xi}+\varepsilon} + \left(A\mathrm{sech}\xi\right)_{\overline{\xi}-\varepsilon}\right]\\
&+2m^2\left[\left(A\mathrm{argtan}(\exp\xi)\right)_{\overline{\xi}+\varepsilon} - \left(A\mathrm{argtan}(\exp\xi)\right)_{\overline{\xi}-\varepsilon}\right]\\
&-\frac{1}{2}\lambda\left[\left(A^3\mathrm{sech}^2\xi\sinh\xi\right)_{\overline{\xi}+\varepsilon} - \left(A^3\mathrm{sech}^2\xi\sinh\xi\right)_{\overline{\xi}-\varepsilon}\right]\\
&-\lambda\left[\left(A^3\mathrm{argtan}(\exp\xi)\right)_{\overline{\xi}+\varepsilon} - \left(A^3\mathrm{argtan}(\exp\xi)\right)_{\overline{\xi}-\varepsilon}\right] = \widetilde{A}\int_{\overline{\xi}-\varepsilon}^{\overline{\xi}+\varepsilon}\mathrm{sech}^2\overline{\xi}\sinh\overline{\xi}\delta(\overline{\xi})d\xi.
\end{aligned} \tag{7.18}$$

Let $\varepsilon \to 0$ and $(A)_{\overline{\xi}+\varepsilon} \gg (A)_{\overline{\xi}-\varepsilon}$. First we collect the terms with

$$\left(\mathrm{sech}^2\xi\sinh\xi\right)_{\overline{\xi}+\varepsilon}. \tag{7.19}$$

As a result, we have an equation linking $A_{\overline{\xi}}$ and $\widetilde{A}$:

$$A_{\overline{\xi}}^3 + \lambda^{-1}\left(\omega^2B^2 - c_*^2\vartheta^2K^2I\right)A_\xi + 2\widetilde{A}\lambda^{-1} = 0. \tag{7.20}$$

Then we collect the terms with

$$\mathrm{argtan}(\exp\xi)_{\overline{\xi}+\varepsilon}. \tag{7.21}$$

In this case, Eq. (7.18) approximately yields that

$$m^2 = \frac{1}{2}\lambda\left(A_{\overline{\xi}}\right)^2. \tag{7.22}$$

Thus, we have the algebraic equation (7.20) coupling the nonlinear properties of the field, the amplitude of the quantum action, and coefficient $\omega^2B^2 - c_*^2\vartheta^2K^2I$. If $\widetilde{A} = 0$, then (7.20) yields the former solutions (7.11) for the dynamic part of the field. However, for some values of $\widetilde{A}$, a solution of (7.20) can be larger $A_+ = m\sqrt{2/\lambda}$ (7.11). In this case, the vicinity of the point $\overline{\xi}$ can erupt out of the potential well.

Finite Time Quantum Actions Equation (7.20) shows the possibility of a very strong change in the amplitude of the dynamic part. It opens a possibility for the energy bubble to jump instantly out of the potential well. However, generally speaking, this phenomenon may not be instantaneous. It is determined by the time of the quantum fluctuations. For example, if the quantum fluctuation is determined as $f(\xi)=\widetilde{A}$ sech $\Omega\xi$, Eq. (7.15) yields

$$\frac{1}{2}\left(\omega^2B^2 - c_*^2\vartheta^2K^2I\right)\Phi_{\xi\xi} + m^2\Phi - \lambda\Phi^3 = \widetilde{A}\,\sinh\Omega\xi. \tag{7.23}$$

Let $\Phi = A$ sech $\Omega\xi$. In this case, Eq. (7.23) yields the equation for the amplitude

$$\lambda A^3 + \left(\omega^2B^2 - c_*^2\vartheta^2K^2I\right)A + \widetilde{A} = 0. \tag{7.24}$$

It is easily seen that this equation practically coincides with Eq. (7.20) for the instant quantum action. Therefore, all results obtained for the last case are valid for the quantum fluctuation $\widetilde{A}\mathrm{sech}\Omega\xi$. But this approach will not give us any new information about the evolution of the scalar field in a potential well. To get the new information, let us radically simplify Eq. (7.15). Thus, we will consider the behavior of the scalar field near the bifurcation point, where $\omega^2B^2 \approx c_*^2\vartheta^2K^2I$.

For this case, the scalar field and the quantum action will be described by the equation

$$\Phi^3 + \overline{R}\Phi = \lambda^{-1}f(\xi). \tag{7.25}$$

Here $\lambda^{-1}f(\xi)$ describes the quantum action and $\lambda^{-1}f(\xi) = \lambda^{-1}\varphi(\Phi)\widetilde{A}\delta(\overline{\xi})$ in (7.15). The quantity $\overline{R} = -m^2/\lambda$ is transresonant parameter.

Thus, we obtained approximate Eqs. (7.20) and (7.25). We hope that they qualitatively describe some options for tunneling of the energy bubble from its potential well.

Below several examples of the quantum action are presented.

7.3.2 *Discontinuous Jump of the Scalar Field*

Our aim in this section is to study the influence of $\omega^2B^2 - c_*^2\vartheta^2K^2I$ and $\widetilde{A}$ on solutions of Eq. (7.20). Effects of λ^{-1} are also analyzed, but for most calculations, we use $\lambda = 10^{-15}$ [57].

We recall that ϑ, B, ω, and K in (7.20) are arbitrary values which determine the coefficient $\omega^2B^2 - c_*^2\vartheta^2K^2I$. Let us examine the dependence of the solution to the equation on this coefficient. This coefficient will be considered as a very slow variable function of ξ.

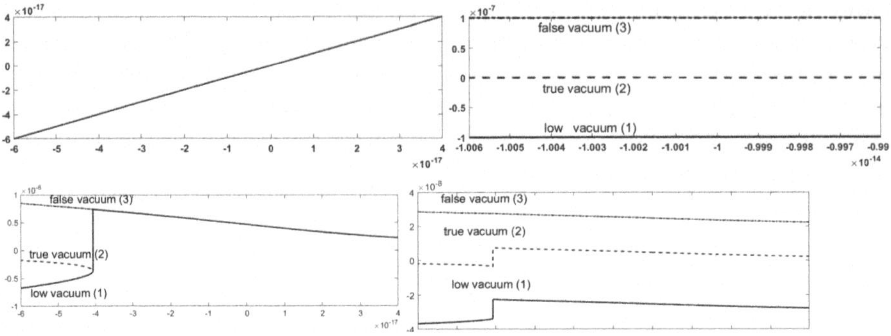

Fig. 7.5 Typical pictures calculated for $\lambda = 10^{-15}$ and different values $c_*^2\vartheta^2K^2I - \omega^2B^2$ and $2\widetilde{A}\lambda^{-1} = 10^{-25}$

Discontinuous Jump Figure 7.5 shows calculations of the scalar field for the case of a value $2\widetilde{A}\lambda^{-1} = 10^{-25}$. In this case, the coefficient $\omega^2B^2 - c_*^2\vartheta^2K^2I$ was changed according to a linear law. It is shown in the top row of Fig. 7.5 (left). For each coefficient value, three roots (solutions) were calculated. The values of the roots calculated for various intervals of change in the coefficient $\omega^2B^2 - c_*^2\vartheta^2K^2I$ were used to construct the curves presented in Fig. 7.5. The calculations showed that almost always the solutions to Eq. (7.20) change very slowly with a change in the coefficient $\omega^2B^2 - c_*^2\vartheta^2K^2I$, and the solutions do not interact in any way (see the top, right picture).

However, a resonant region of change of ξ has been discovered when these fields begin to interact and combine into one field (see figure below, left). To make the nature of the merging of fields clearer, we separated the fields and presented their changes without interaction in the picture placed below, on the right.

Recall that each of the three solutions corresponds to a specific scalar field. We, with a great deal of convention, called these fields false vacuum, true vacuum, and low vacuum. Thus, according to calculations, situations are possible in which low-energy fields can abruptly evolve into a high-energy field under the influence of the quantum pulse.

It is important that low-energy fields (curves 1 and 2) can experience a jump to positive energy values (curve 3). It is seen that values of these fields may be very different. However, during the jump, these fields coalesce and begin to interact in such a way that they transform into a high-energy field.

Additional calculations performed with modified problem parameters are shown in Fig. 7.6.

We have examined the effects of the instant quantum actions on the scalar field and obtained the results linking directly nonlinearity of the scalar field and the emergence of the Universe. In particular, Eq. (7.20) has a bifurcation at the point where $\omega^2B^2 \approx c_*^2\vartheta^2K^2I$.

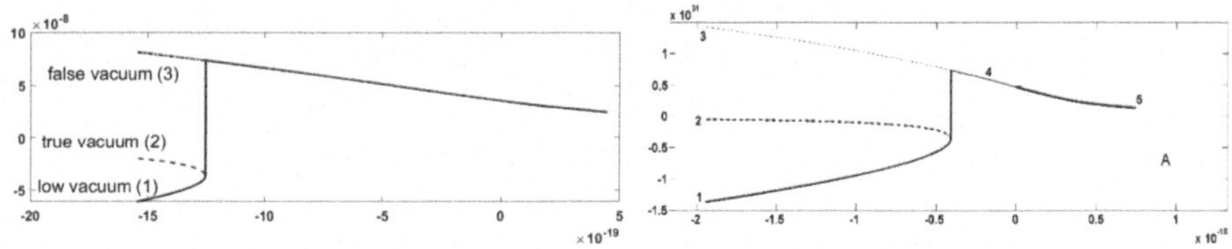

Fig. 7.6 Another example of extreme amplification of the scalar field calculated for different values of $c_*^2\vartheta^2K^2I - \omega^2B^2$

These results are limited by the action of the instant quantum fluctuation. Let us consider more general cases of finite time quantum fluctuations. Researches had shown [92, 93] that quantum jumps are not instantaneous. Let us take this possibility into account.

7.3.3 Finite Time Quantum Actions

Now we consider the bifurcation point more attentively. For this case, the scalar field and the quantum action will be described by Eq. (7.25).

1. At first, we consider a soliton-like quantum fluctuation when $\lambda^{-1}f(\xi) = 0.3^*$ sec $h^3(10^{29}\xi)$ in (7.25). For this case, the solutions of Eq. (7.25) are shown in Fig. 7.7 for different $\overline{R}$.

   Figure 7.7 (top, left) shows the impulse function $\lambda^{-1}f(\xi)$. To the right of the picture, curves are calculated for the case $\overline{R} = -10^{-20}$. There are curves characterizing three different fields that begin to distort in a certain area. The distortion leads to the interaction of all three scalar fields, which is calculated at $\overline{R} = -10^{-20.0725}$. It is obvious that different interpretations of the interaction of these fields are possible. We have adopted an interpretation in which false vacuum instantly drops to the low vacuum level, and then, bouncing off the bottom, it turns into true vacuum. The value of the latter begins to decrease rapidly (bottom, right picture).

   The opposite interpretation of the lower, left picture is also possible. Let us imagine it schematically like this: true vacuum → low vacuum → false vacuum.

   Many calculations have been carried out, which, in general, correspond to Fig. 7.7. Let us give two of them.
2. In Figs. 7.8 and 7.9, the change in the false vacuum is shown by curve 3, the true vacuum is curve 2, and the low vacuum is curve 1. Let the parameters of the quantum fluctuation be $\widetilde{A} = -E \cdot 10^{-30}$, $\Omega = 5 \cdot 10^{29}$. For this case, the solutions of Eq. (7.25) are shown in Fig. 7.8 for different E.

   According to Fig. 7.8, the three independent fields (curves 1, 2, and 3) are described by (7.25), if the amplitude of the quantum action is small enough ($E < 0.385$). The fields begin to interact and form a new composite field when the quantum action increases. In particular, the composite discontinuous solution

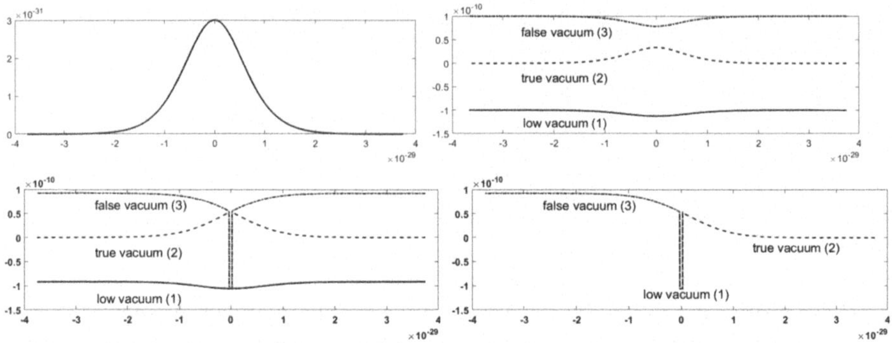

Fig. 7.7 Typical pictures calculated for finite time quantum actions

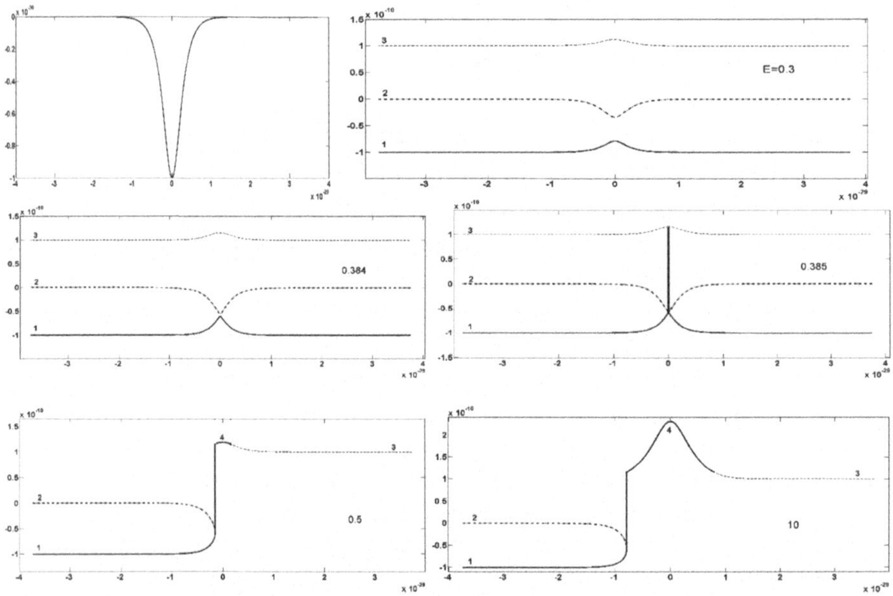

Fig. 7.8 Continuous and discontinuous curves described by the solutions of (7.25), where $\overline{R} = -10^{-20}$. Curves 1, 2, 3, and 4 are determined by different roots of the equation. Pictures $E = 0.3$ and $E = 0.384$ are determined by three continuous solutions. These solutions begin to interact when $E = 0.385$. The pictures $E = 0.385$, 0.5, and 10 are determined by discontinuous solutions. The quantum fluctuation is the upper picture, left.

appears when $E = 0.385$. In this case, curves 1 and 2 coincide near the point 0 and jump to curve 3. As a result, the complex composite field is formed if $E \geq 0.385$. The composite solutions (pictures $E = 0.5$, 10) describe the discontinuous slope of the hill-like scalar fields and the strong amplification of fields 1 and 2. They can reach the hilltop of the composite scalar field. Then the composite scalar field reduces.

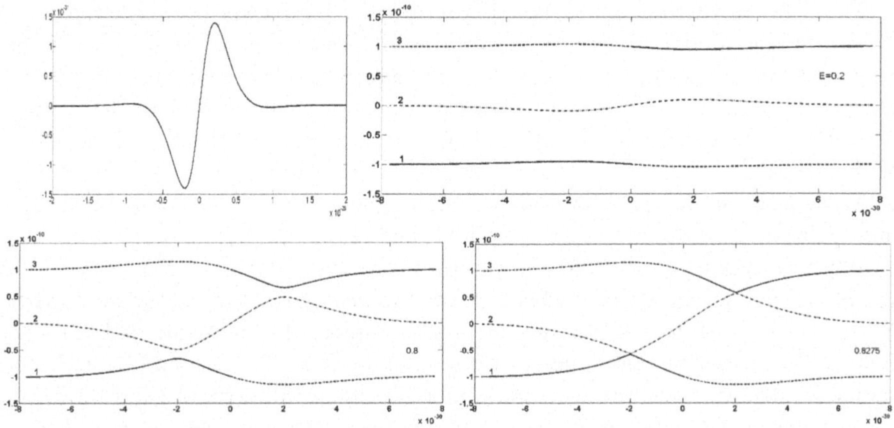

Fig. 7.9 Continuous curves described by the solutions of (7.25), where $\overline{R} = -10^{-20}$. Curves 1, 2, and 3 correspond to different roots of the equation

We stress that some theories describe the initial evolution of the Universe as rolling down of the scalar field from some local maximum [94–96]. Figure 7.8 shows how the scalar field can jump to this maximum. At the same time, we do not touch here the process after the jump. We only consider here the possibility of the jump. The study of the jump and the following process are special complex problems.

3. Now we consider a case when $\lambda^{-1} f(\xi) = \widetilde{A}\,\mathrm{sech}\,\Omega\xi \sin \overline{\Omega}\xi$ in (7.25) (see Figs. 7.9). In this case, the quantum fluctuation changes its sign, if $\xi = 0$. Let the parameters of the quantum fluctuation be: $\widetilde{A} = -E \cdot 10^{-30}$, $\Omega = 5 \cdot 10^{29}$, $\overline{\Omega} = 4 \cdot 10^{29}$ (Fig. 7.9 the upper picture, left). The solutions of Eq. (7.25) are shown for different coefficient E.

   The pictures of Fig. 7.9 are determined by three continuous solutions. If the amplitude of the action is small, then the fields are independent and curves 1, 2, and 3 do not interact. These solutions begin to interact when $E = 0.8275$.

7.3.4 Modulated Quantum Actions

We assume in Eq. (7.25) that

$$f(\xi) = \widetilde{A}\lambda \ \mathrm{sech}\ \Omega\xi \cos \overline{\Omega}\xi, \tag{7.26}$$

where Ω and $\overline{\Omega}$ are constants. Equation (7.25) yields

$$\Phi^3 + \overline{R}\Phi = \widetilde{A}\ \mathrm{sech}\ \Omega\xi \cos \overline{\Omega}\xi. \tag{7.27}$$

Thus, the quantum action is being described by the packet of oscillations. We expect that the amplitude and the form of oscillations of the dynamic part can be amplified strongly inside the potential well as a result of the quantum action (7.26). On the other hand, we found that the amplification is possible only for certain transresonant values $\overline{R}= -m^2/\lambda$ [57, 59, 60]. Below several examples of the action are presented.

Modulated "short" quantum actions. Let the parameters of the quantum fluctuation (7.26) be $\widetilde{A}=0.3^*10^{-38}$, $\Omega = 5 \cdot 10^{39}$, and $\overline{\Omega}=5 \cdot 10^{40}$ (Fig. 7.10).

When calculating, we change $\overline{R}= -10^{-25.25}$; $\overline{R}= -10^{-25.5}$; $\overline{R}= -10^{-25.75}$; and $\overline{R}= -10^{-27}$ in (7.27). The top row of paintings was calculated when $\overline{R}= -10^{-25.25}$. Moreover, the first picture determines the right side of Eq. (7.27) (that is, the profile and amplitude of the quantum pulse). The second picture in the same row defines three curves corresponding to three solutions of Eq. (7.27). We assume that in the case under consideration, these curves do not interact. Let us emphasize that we have accepted that each of the three solutions of Eq. (7.27) corresponds to a scalar field, which we have, with a great deal of convention, called false vacuum, true vacuum, and low vacuum. From the picture in the top row on the right, it follows that these three fields, when accepted, do not interact. However, with a slight increase to $\overline{R}= -10^{-25.5}$ and $\overline{R}= -10^{-25.75}$, these fields begin to interact strongly and form multivalued profiles.

These profiles are in the second row of Fig. 7.10. Moreover, the picture on the left corresponds to $\overline{R}= -10^{-25.75}$, and the picture on the right corresponds to $\overline{R}= -10^{-25.75}$. The interaction of fields is interesting to us from the point of view of the possibility of the emergence of a field that could potentially become a source of energy that will subsequently evolve into our Universe. The curves are very intricacy. It is possible to create various continuous, but jumped curves using of

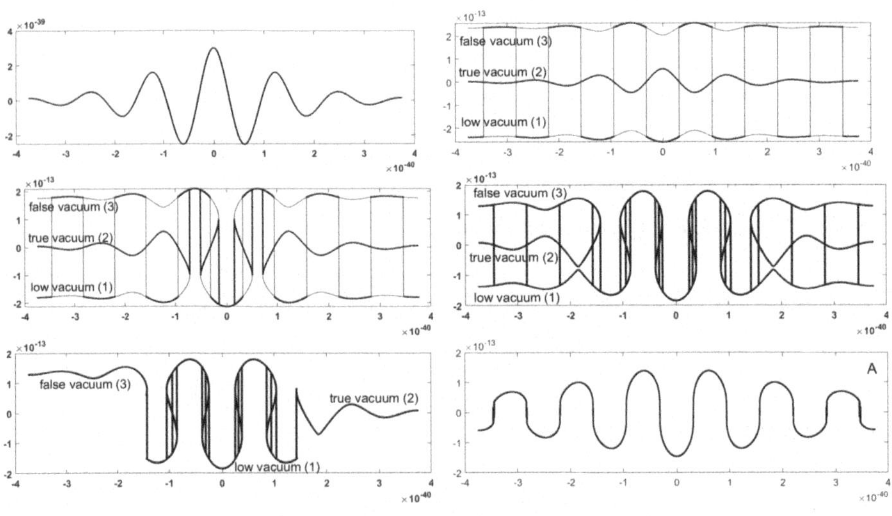

Fig. 7.10 Effects of resonance and alternating quantum action on the energy of a scalar field bubble

them. Those curves can determine the various evolutions of the fields into each other. We illustrated the possibilities of such evolutions in the figure placed in the bottom row, on the left. It is shown there that as a result of complex transitions and jumps, false vacuum energy can evolve into true vacuum energy.

The last picture (a) shows the curves calculated at $\overline{R} = -10^{-27}$. From this figure, it follows that the three fields have practically merged, there are practically no multivalued areas in them, and the amplitude of field oscillations has begun to decrease.

Due to the noted changes of $\overline{R}$, the transresonant transition from three almost harmonic, noninteracting, scalar fields to one, almost harmonic field, in which the three indicated fields merged took place. This transition is accompanied by a very complex interaction of fields within the "resonance band." We believe that something similar could have happened during the eruption of a very heavy field from the pre-Universe. This field, as a result of complex evolution, was transformed into a lump (bubble) of energy, which, in the process of subsequent expansion, formed our Universe.

An important result of the calculations is that a very weak quantum action can form the composite scalar field within the resonant band. According to the calculations, the scalar field can jump between the upper and lower branches (between positive and negative values of the energy).

"Long" quantum actions. Let us study a case of the quantum fluctuations having longer duration (Fig. 7.11). It is also assumed that the quantum action function can have fast-varying parts (Fig. 7.11 center and right). These parts are much shorter than the action duration.

The solutions of Eq. (7.27) obtained for the above quantum actions are presented in Figs. 7.12, 7.13, 7.14, and 7.15. Considering these figures, we mean that roots 1 and 2 (segments 1 and 2) describe the evolution of the low vacuum and the true vacuum. The change in the false vacuum is shown by the points.

1. First, it is assumed that $\lambda^{-1}f(\xi) = \widetilde{A}\text{argtan}[\exp(-\Omega\xi)]$ (Fig. 7.12, left), where $\widetilde{A} = A \cdot 10^{-30}$, $\Omega = 5 \cdot 10^{29}$, and $\overline{R} = -10^{-20}$ in (7.27). For this case, the solutions are shown in Fig. 7.12. It is seen that the scalar fields are independent if $A = 0.01$ or $A = 0.2$ in $\widetilde{A} = A \cdot 10^{-30}$. The segments (fields) 1 and 2 begin to interact when $A = 0.245$. The jumps of the scalar fields (segments 1 and 2) to the positive values take place when $A > 0.246$.

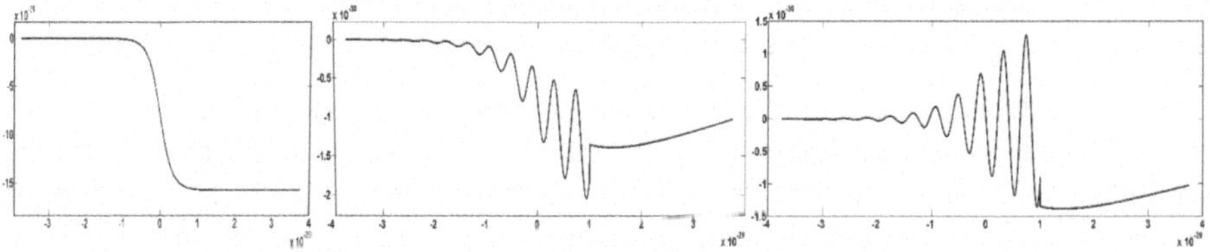

Fig. 7.11 The quantum action described by the expression $\lambda^{-1}f(\xi) = \widetilde{A}\text{argtan}[\exp(-\Omega\xi)]$ (left) and more complex curves describing the quantum action in (7.27) (center and right)

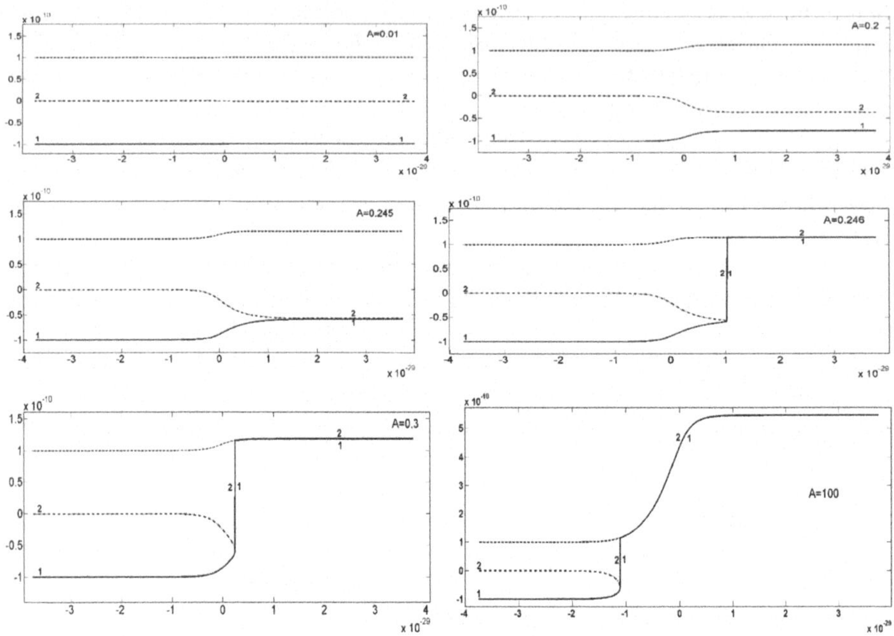

Fig. 7.12 Results of the quantum action $\lambda^{-1}f(\xi) = A \cdot 10^{-30}\text{argtan}[\exp(-\Omega\xi)]$ (Fig. 7.11, left) calculated for different coefficients A. The scalar fields 1 and 2 can jump up to the top level (the plateau) if $A \geq 0.246$

We have considered the case, when a long quantum action is not accompanied by quantum fluctuations of small amplitude. It is known [57, 59, 60] that their impact can be very important. These fluctuations determine the possibility of the destruction of the space-time of the pre-Universe. On the other hand, the small fluctuations can influence the eruption of the scalar field from the potential well. Below we present the results of the analysis of the impact of the small fluctuations on the nonlinear processes which are considered.

2. Let us consider the quantum fluctuation presented in Fig. 7.11 (center and right). The amplitude of the quantum fluctuation $\lambda^{-1}f(\xi)$ equals 10^{-30} and $\overline{R} = -10^{-20}$ in (7.27). Some results of the calculations are presented in Fig. 7.13. Curves A were calculated for the action presented qualitatively in Fig. 7.11 (center). Curves B were calculated for the action presented qualitatively in Fig. 7.11 (right).

Figure 7.13 demonstrates the existence of three scalar independent fields when the horizontal coordinate is smaller than -10^{-29}. These fields interact very strongly within an interval from -10^{-29} to 0 (A) and from 0 to 10^{-29} (B). The single composite field is formed after the interaction. Perhaps, this field corresponds to the origin of the Universe. The formation of the mushroom-like (elastica-like)

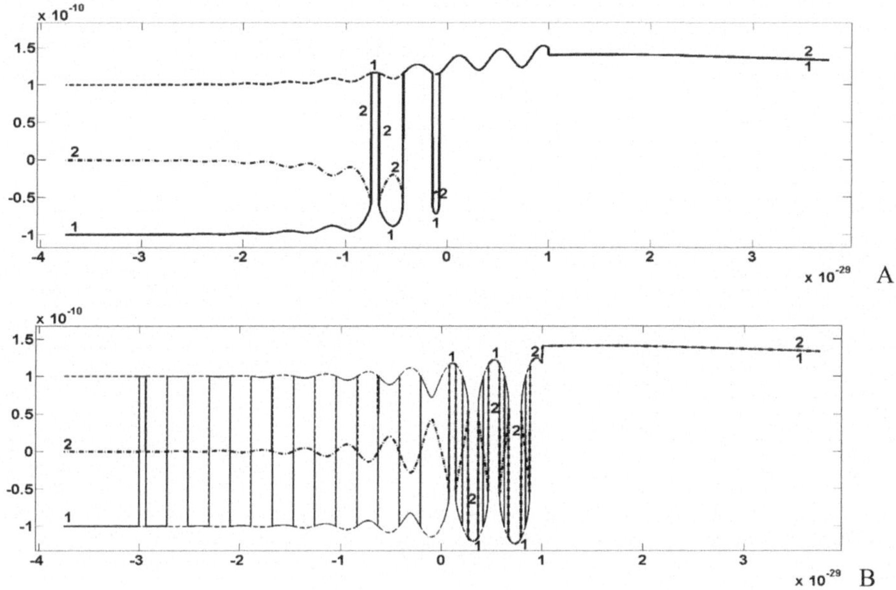

Fig. 7.13 Results of the quantum actions and effects of oscillations of quantum action profiles. The actions are accompanied by quantum oscillations of a small amplitude (see Fig. 7.11 (center and right))

structure [57, 59, 60] is shown in the picture b. This structure is a result of the strongly nonlinear interaction of the initially independent scalar fields.

3. We have considered cases when the coefficient $\overline{R}$ in (7.27) was very small. Now we study additionally cases when $\overline{R} = -10$ or $\overline{R} = -10^{20}$.

   3.1. Let us consider the case of the quantum action which is similar to the case presented in Fig. 7.11 (center). The calculation results are shown in Fig. 7.14.

   Here the curves A are calculated for $\overline{R} = -10$ and the amplitude of the quantum fluctuation $\lambda^{-1}f(\xi)$ equals $10^{1.1}$. Curves B are calculated for $\overline{R} = -10^{20}$ and the amplitude of the quantum fluctuation $\lambda^{-1}f(\xi)$ equals 10^{31}.

   3.2. Let us consider the case of the quantum action which is similar to the case presented in Fig. 7.11 (right). The calculation results are shown in Fig. 7.15.

   Here the curves A are calculated for $\overline{R} = -10$. Curves B are calculated for $\overline{R} = -10^{20}$.

Conclusion We described the resonant situations at which the three initially independent scalar fields begin to interact and form a new scalar field which can jump from the pre-Universe.

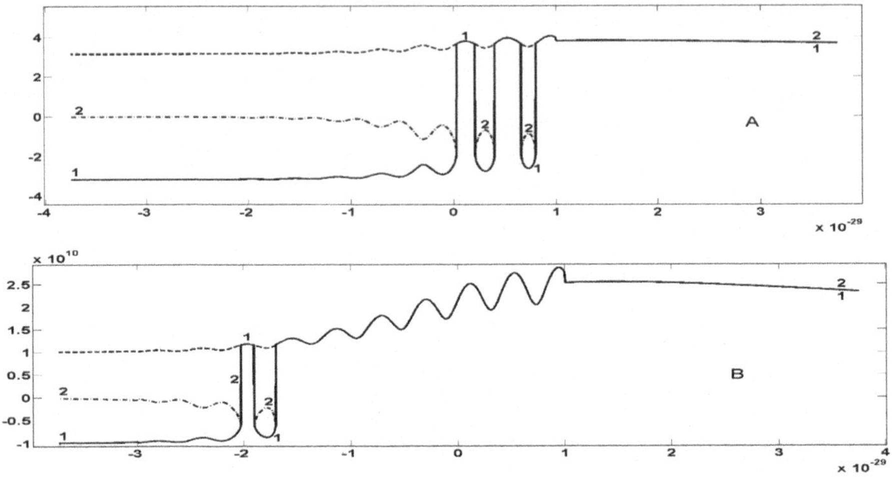

Fig. 7.14 Results of the quantum actions accompanied by quantum oscillations of a small amplitude. The curves A and the curves B are calculated for the case presented in Fig. 7.11 (centre)

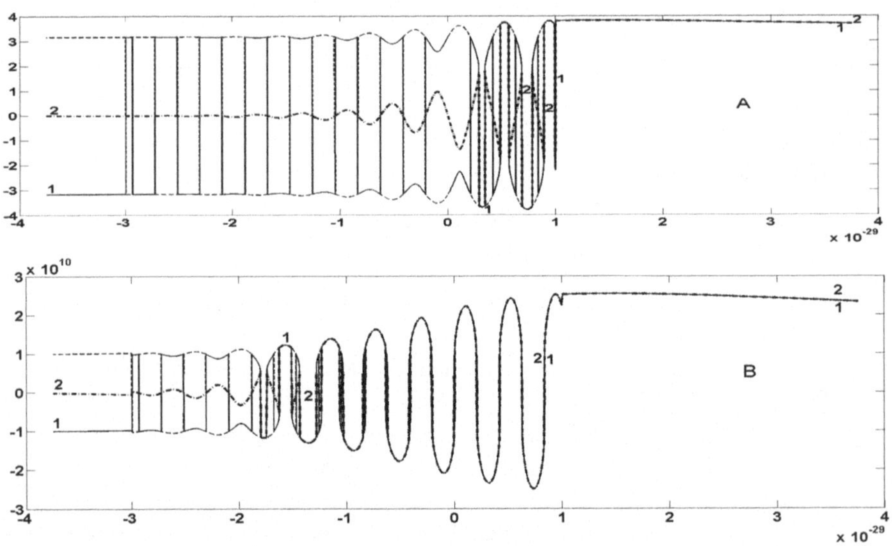

Fig. 7.15 Results of the quantum actions accompanied by quantum oscillations of a small amplitude. The amplitude of the quantum fluctuation $\lambda^{-1}f(\xi)$ equals -10^{-30}. The formation of the mushroom-like (elastica-like) structure [57, 59, 60] is shown. The curves A and the curves B are calculated for the case presented in Fig. 7.11 (right)

7.4 From a Multidimensional Bubble to a System of One-Dimensional Objects

We considered the structures of the landscapes of the scalar field and scalar potential. Generally speaking, the amplitudes of these reliefs may be quite different. This fact is very important for our research, since the origin of the Universe may be imagined as an eruption of the scalar field from some potential well. But how deep may these wells be and how intensive may the fields be? The replies are determined by the constants m^2 and λ.

7.4.1 Illustrations for the Jumps from Pre-Universes

We emphasize that we do not know the values of and for the scalar field that gave birth to our Universe. One might think that these values differ radically from those that can be expected for terrestrial conditions.

First of all, we can assume that the scalar field is distinguished by a very high density and a very high nonlinearity.

We carried out calculations showing the possibility of a very strong increase in the energy of a bubble (sphere and clot) located in one of the potential wells of the scalar field of the pre-Universe. After the jump, the value of the scalar field can greatly exceed the depth of the potential well. If this takes place, the field can erupt out of the pre-Universe. Then it can form the Universe.

Our next task is to model the future fate of this scalar field energy bubble. However, first we will try to make the leap process itself more visible.

Picture a shows the element of the landscape of the static potential. Pictures b and c show the sum of the static potential and the dynamic part of the scalar field at same moment of time. The scalar field is located in the valley of the potential. Picture b is the bird's-eye view, and picture c is the side view. Picture d is the bird's-eye view on the landscape after the jump of the scalar field. We think that the local jump of the landscape (Fig. 7.16) illustrates well the main results of Sect. 7.3. We stress that the local variation of the landscape which is presented in Fig. 7.16 is very instant and may be extremely large. We illustrate the main result of Fig. 7.16 through Figs. 2.14, 7.17, and 7.18.

Figure 7.17 shows qualitative results of quantum perturbations, bifurcations, and resonances [57, 58]. On the other hand, it corresponds qualitatively to the jump of a bubble of the scalar field to a new, very high energy level. It is interesting that the noted schemes resemble the painting presented in Fig. 7.18.

Of course, pictures like Figs. 7.16, 7.17, and 7.18 may be considered as some wild speculations connected with unsolved problems of science. Indeed, some cosmological theories may be considered as crazy. But society should get accustomed to them. The pictures like Figs. 7.16, 7.17, and 7.18 should help us understand the crazy ideas better. On the other hand, the scientific art which were showed in the

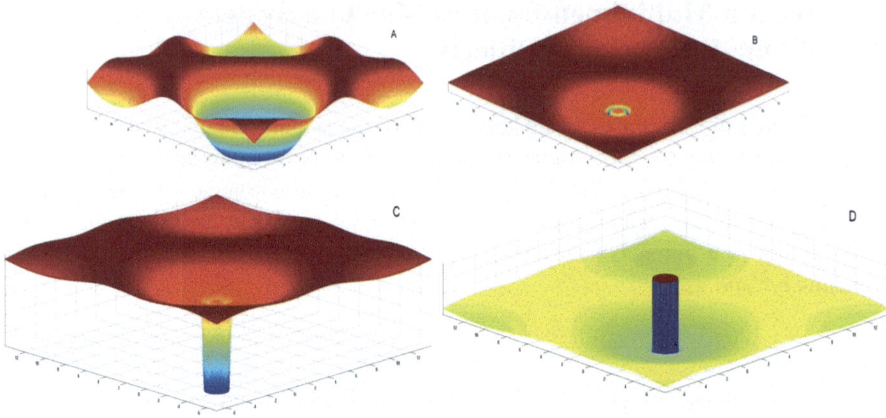

Fig. 7.16 Rough scheme of the jump. A sequence of pictures of the scalar potential (**a**) and the scalar potential plus the scalar field (**b**, **c**, **d**). The pictures demonstrate the landscapes before the jump (**a**, **b**, **c**) and the results of the jump (**d**)

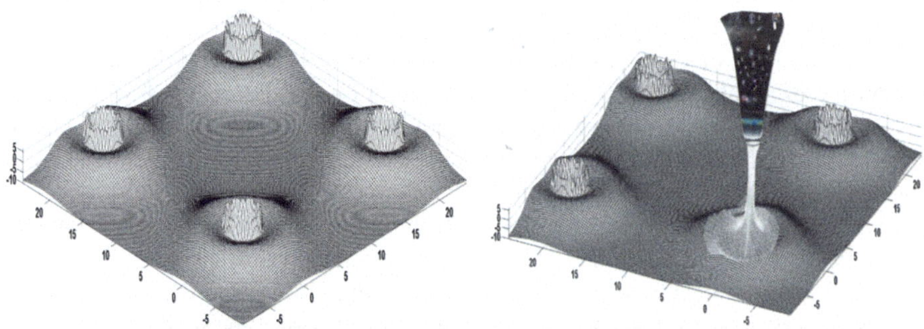

Fig. 7.17 The rough scheme of the development of the scalar field into the Universe. The two-dimensional landscape of the pre-Universe (left) and the scheme of the origin of the Universe (right) [57]

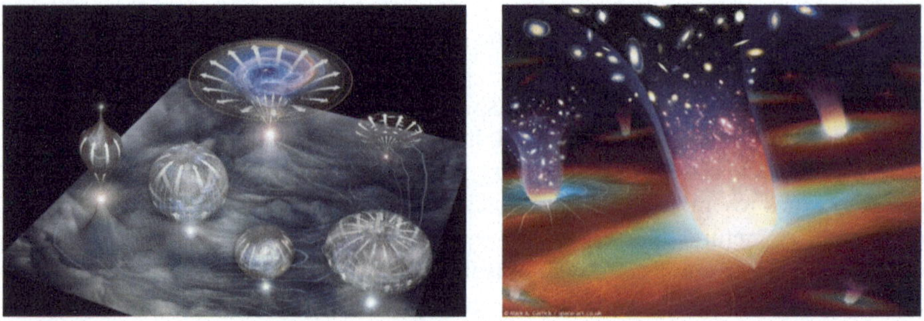

Fig. 7.18 Rough schemes of the eruption of Universe from the potential wells of a pre-Universe. (Panel (**a**) is taken from http://news.nationalgeographic.com/news/2014/03/140318-multiverse-inflation-big-bang-science-space/. Panel (**b**) was presented in a few sources, for example, in [57, 58])

last figures could popularize scientific ideas and results which are not supported up to this moment by experiments and observations [13, 44, 51, 57, 97].

Remark We recall that basic laws of quantum mechanics are applicable for our analysis. In particular, it is well known that the energy density and the energy pressure of the scalar field depend on derivatives Φ_i^2, Φ_t^2 and the scalar potential $V(\Phi)$ (see (1.13), (1.14), and (1.15)) [44, 83, 98, 99]. Thus, these density and pressure can increase infinitely in the points of discontinuity. Due to above circumstances, our Universe can emerge having practically infinite energy and mass.

7.4.2 To a System of One-Dimensional Objects

Suppose that the enhancement turned out to be so strong that the bubble appeared to be at the top of some hill of some scalar field (Fig. 7.19) [23, 100, 101].

According to many publications, the Universe arises in the process of rolling of the clot and its oscillations at the bottom of the landscape determined by the value of the true vacuum (a) [85–87].

This simple diagram is very useful in many ways. But here the diagram is not applicable.

The fact is that the bubble consists of bosons, which have no mass. Therefore, there is no reason for the bubble to roll down or rise up. However, as soon as the bubble left the space of the pre-Universe, which is stable, it began to crumble into smaller dimensions. In the theory under consideration, it is assumed that the bubble disintegrates into an almost infinite number of one-dimensional elements independent of each other (Fig. 7.19b).

Generally speaking, the initial energy field could have many dimensions. The string theory proclaims that the number of the dimensions may be different, for example, 5, or 11 or 26. As the number of dimensions reduces, the volume of the initial bubble (clot, sphere, and drop) can increase after the fragmentation very, very strongly.

The process could be visualized by imagining a three-dimensional drop of oil impacting the surface of water. As a result of the impact, the oil drop is separated into many elements (particles) which spread over the two-dimensional surface of water. These elements of oil occupy in the two-dimensional space much bigger volume

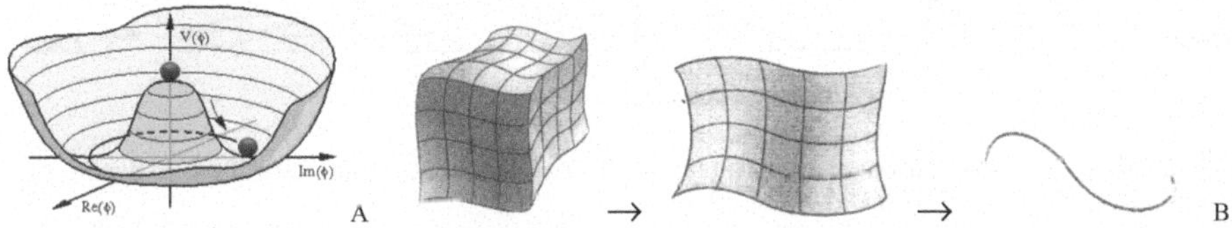

Fig. 7.19 As a result of the jump, the clot of energy appears on the energy hill (**a**). Elements of the space-time according to the string theory (**b**)

than they had in the initial moment. It is important that the elements became more isolated from each other compared to when they were inside the drop.

Qualitatively, the fragmentation of the space-time fabric of the bubble could remind us of the atomization of the water drop. Another model of multidimensional space scattering is shown in Fig. 7.19b.

It was assumed [57] that the elements became absolutely isolated without any connections. Generally speaking, it agrees with the "vanishing dimensions" theory [102, 103]. According to this theory, systems with higher energy have a smaller number of dimensions. The higher the energy, the smaller space-time dimensions. Thus, our theory implies that the number of dimensions in the Universe reduced during the eruption.

Using the terminology of the string theory, these elements may be defined as strings. They begin to vibrate (Fig. 7.20).

The space-time fabric was transformed and fragmented. The field energy has been increased strongly. It is assumed that the most stable state is one dimensional. Therefore, the multidimensional bubble crumbles into an unimaginably dense unity of independent one-dimensional structures. It is proposed to represent this unity as a conglomerate of strings or pipes in which one-dimensional gas vibrations occur.

But how do one-dimensional continuous structures transform into the particles that make up our Universe?

Let us recall that we have a huge number of calculations presented in Chaps. 2 and 4 showing the possibility of the emergence of isolated particles during oscillations of multivalued waves in one-dimensional resonators.

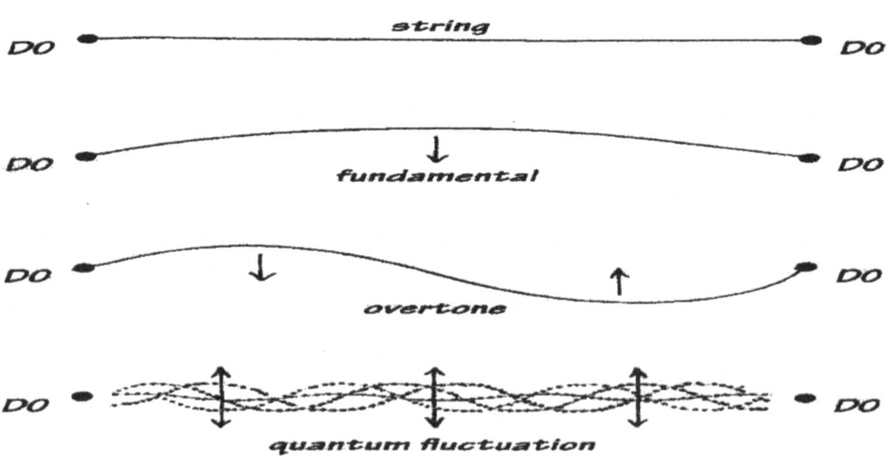

Fig. 7.20 Linear oscillations and a quantum fluctuation of a string stretched between two D0-branes [57, 104]

7.5 The Origin of the Particles of Energy and Matter as a Strongly Nonlinear Resonant Phenomenon

Calculations illustrating the appearance of particle-waves in resonators are presented especially fully in the first sections of Chap. 4. It is important that NKGE was used there. Therefore, here we will not repeat the theory of calculations but will present only some calculation results that illustrate the following provisions of the theory.

As a result of the instantaneous eruption from the pre-Universe, a very small region of one-dimensional space arose filled with an almost infinite number of one-dimensional elements of the scalar field. It is, by definition, a baryon field. Therefore, there should have been no special reasons for the elements to be dissatisfied with the cramped conditions. Therefore, there were no reasons for the indicated elements to begin a rapid expansion of the emerging, initial volume. However, having arisen, these elements began to vibrate according to their own frequencies. The nature of the resulting oscillations and particles could correspond to that shown in Fig. 7.21.

Particle-waves shown in Fig. 7.21 are formed by some scalar fields. These fields can be represented as some strings or as one-dimensional waves in the gas filling the pipe. Let us look at the latest version.

Let us consider the case when a pipe can be locally destroyed so much that a gas particle can be ejected from the pipe into a non-one-dimensional space. At the beginning of the process, there is no destruction. The pipes are very durable. But the energy of the waves in the pipe is also extremely high. Therefore, destruction of the pipe is very rare and local, but the energy of the ejected particles is also very high. But this is only energy. There is no gravity yet. Therefore, no "black holes" could form. Further, the vibration energy due to emissions decreases, but the amount of destruction in the pipe increases. Therefore, the number of particles "emitted" from the pipe is growing all the time, but their energy decreases accordingly.

At the end of some period of time, we would not see a pipe, but rather a formless cloud of particle-wave energy. And there are almost infinite sets of such formations in the area under consideration. Bosons emitted into non-one-dimensional space inevitably begin to interact. As a result, a new stable time-space may emerge. It is gradually filled with scalar field particles (bosons).

In the case of bosons, nothing prevents the particles from gathering together and they happily do so. We can say that an effective attraction occurs. Bosons can gather into a perfect sphere.

7.6 The Births of the Spherical Universe, Galaxies, and Wave Modes

As a result, the bosons gather into an ideal sphere. These particles have enormous energy. There are more and more bosons in a spherical volume. Their collisions will inevitably lead to the birth of massless fermions.

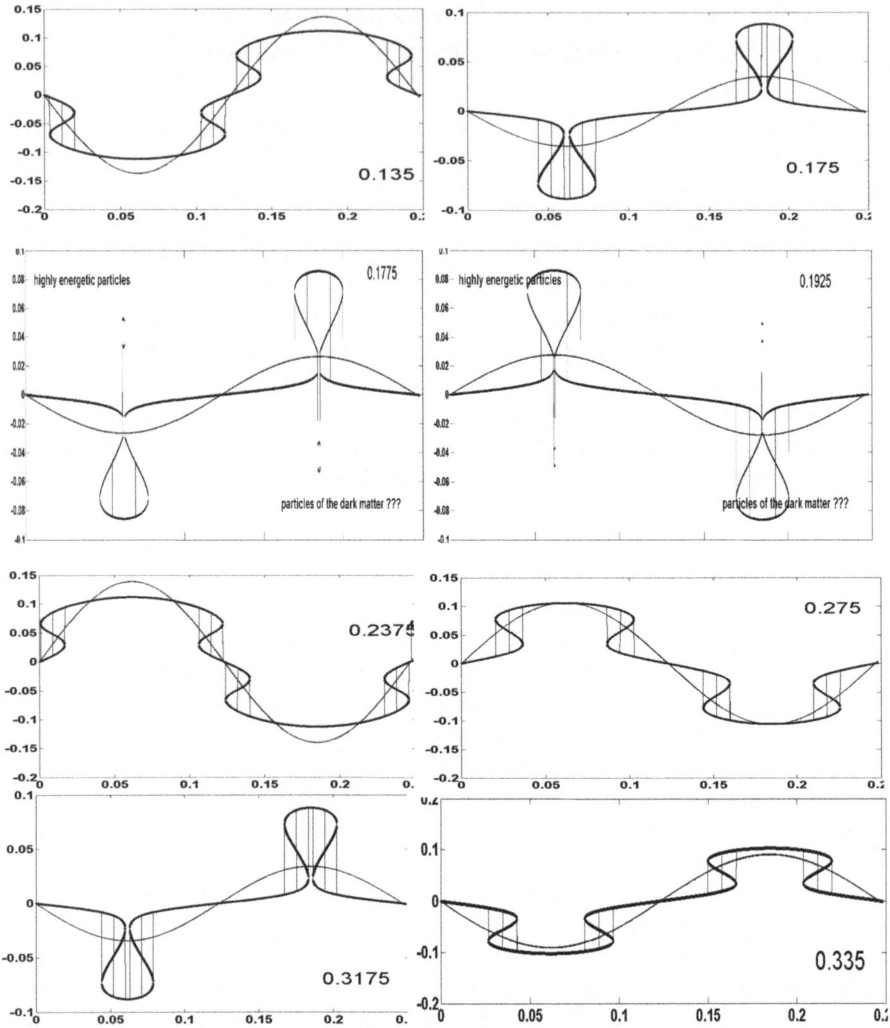

Fig. 7.21 Resonant continuous oscillations of the scalar field elements according to the second resonant form. Periodic formation of the particles of matter. The thin smooth lines correspond to the linear oscillations [57, 59, 60]

Apparently, this requires some level of kinetic energy during collisions. In 2004, physicists showed [105] that if bosons are forced to move in only one line, they acquire the properties of fermions. This effect is also true for bosons. By forcing a gas of bosons to expand in one direction, scientists observed a change in the momentum distribution from bosonic to fermionic [106]. Due to the enormous concentration density of bosons, the vast majority of bosons did not possess this energy. However, we believe that it was enough for just a few transitions of bosons to fermions to occur, and due to the incompatibility of fermions (two fermions cannot simultaneously be in the same quantum state (Pauli principle)), an increasing number of bosons acquired a sufficient amount of kinetic energy so that during collisions, they begin to turn into fermions. Although at first there were few fermions, there was still not enough space for them. Therefore, the initial sphere—let us make this assumption—begins to slowly expand. But more and more fermions were created. As a result, the original sphere expanded more and more rapidly. A chain, avalanche-like process of the emergence of matter (transition of energy into matter) began. This process is quite similar to the Breit-Wheeler process first formulated in 1934. According to it, a strong collision of two photons can produce matter [106].

This was the beginning of the start of the emergence of our Universe. Already in the very first moments of the expansion, the process of fermion formation began to weaken. Soon a certain equilibrium in the number of fermions and bosons arose, and conditions arose for the formation of the very first elementary particles.

Thus, the expansion does not start from a singularity. Due to the increasing number of fermions, the expansion rate becomes very high, and it continues to grow for some time. However, as the size of the sphere increases and the field density begins to decrease, the expansion rate begins to decrease.

Above, we built a theory trying to answer the following fundamental questions:

1. Where was the energy found for the origin of the Universe and how was it transformed?
2. Where did the so-called fine-tuning of the Universe come from?

   The issues considered are so important that we need to stop and summarize some of the research.

We assume that the evolution scenario of the scalar field subjected to quantum fluctuations presented above can be viewed as describing the origin of the Universe. In particular, we found that a scalar field can give birth to the billions of "seeds" of rapidly evolving Universes. Most of them are Planck-sized "flicker" Universes, which blink in and out of existence. But then, there are some Universes that evolve to a large size. One of them accidentally formed our Universe with the fundamental parameters which we have.

We think that the space-time of the bubble might undergo transformations in which its fabric was fragmented into the string-like vibrating elements. Then the elements began to emit different kinds of highly energetic particles. As a result after the tunneling the fabric repaired itself as new spatial dimensions.

Thus, the dynamic part of the scalar field—the bubble (clot) of energy (Fig. 7.18)—may be considered as the "seed" of the Universe. The Universe appeared after the tunneling of the "seed" from certain potential well when, as the result of the quantum action, the bubble obtained huge energy. During this process, the initial field was being fragmented into almost infinite quantity of highly energetic one-dimensional elements which oscillated at resonant frequencies. The waveforms of these resonant oscillations are highly nonlinear. These waveforms emit particles of energy as bosons. Their interaction yields fermions. The three-dimensional space and time emerged only after the appearance of enough fermions forming this space-time. However, the one-dimensional vibrations of the elements continued to generate particles of energy and matter. As a result, the total spatial dimensions of the bubble were increasing. Thus, the origin of the Universe was not instantaneous. The Universe had enough time to become homogeneous and isotropic and obtained the spherical form to the beginning of the very fast expansion.

7.6.1 Modeling of the Birth of Wave Modes

Let us repeat what has been said even more briefly. At the very beginning, the sphere of bosons could be quite compact. Indeed, bosons can literally be stacked on top of each other. But at a certain moment, energy inevitably begins to transform into matter, that is, into fermions. Fermions repel each other. As a result, a colossal repulsion appears inside the initial sphere and the Big Bang begins.

Everything said happens almost instantly. The situation can be compared, qualitatively, with a very short-term strike on a bell. At the same time, a colossal number of wave forms are excited in it. Something similar happened in the area in question. However, not harmonic, but highly nonlinear and even multivalued waveforms arose in it.

We do not have the opportunity to describe mathematically the very problematic, even incredible, scheme of the transition of bosons to fermions. It is no less difficult to describe mathematically the subsequent evolution of the resulting mixture of particles. However, we have the opportunity to assume that this mixture is described by the simplest nonlinear Klein-Gordon equation and discuss the dynamics of the mixture based on the results obtained in the previous sections.

First, we will give an example of calculating the natural nonlinear oscillations of a mixture shaped like an ideal sphere. Following Sect. 5.1.4, we can write down all the mathematical expressions necessary for this. In particular, we present here the expression (5.15):

$$4\pi r^2(\varphi_{1r}+\varphi_{2r}) = 4\pi r^2\Big[-r^{-2}(J+j)-r^{-1}J_\xi+r^{-1}j_\eta+\frac{1}{2}m^2c^{-2}(J+j)$$
$$+\frac{1}{2}m_2r^{-1}c^{-2}\left(J^2+j^2\right)-\frac{1}{2}\lambda r^{-2}c^{-2}\left(J^3+j^3\right)+\frac{1}{4}c^{-2}J^*(\xi)$$
$$+\frac{1}{4}c^{-2}j^*(\eta)\Big]. \tag{7.28}$$

Here $J^*(\xi)$ and $J^*(\eta)$ are arbitrary functions. We choose them accordingly to Sect. 5.1.4. Then we obtain an equation of the form (5.49):

$$J^3(\eta)-m^2\lambda^{-1}R^2J(\eta)=\lambda^{-1}R^2c^2\left[-l\cos\omega c^{-1}(\eta-R)+C\right]. \tag{7.29}$$

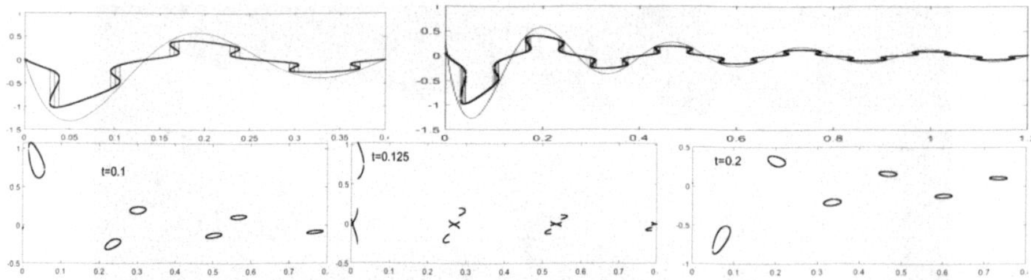

Fig. 7.22 Examples of wave modes and their dynamics calculated using Eq. (7.29). They complement the fluctuations presented in Figs. 5.6 and 5.7

The equation for another traveling wave is derived in a similar way. The sum of these waves determines the change in the scalar field along the radius of the sphere. Naturally, the calculations carried out resemble those presented in Figs. 5.6 and 5.7. They are shown in Fig. 7.22.

The top row of paintings shows two modes. They complement the paintings presented for the second fashion in Fig. 5.6. They are calculated for $C = 0$ in (7.29). The bottom row of pictures is calculated for the case $C = 1$ in (7.29). It also further illustrates the results of Figs. 5.6 and 5.7. It can be seen that in this case, continuous, although multivalued, curves break up into closed, ellipsoidal curves and single-valued, harmonic curves (Figs. 5.6 and 5.7). Unlike Fig. 5.7, we have given here a chain of ellipsoidal curves calculated for three moments in time.

It is not clear to us how the latest results can be interpreted for many reasons. In particular, the fact is that our original model assumes purely radial vibrations of the field. Consequently, according to calculations, on some segments of the radius, there are certain multivalued, ellipsoidal structures. Apparently, here we need to remember that in the calculation, we have curves of a scalar field, not space. Therefore, apparently, we are dealing with a situation similar to that existing for surface waves, when waves, drops, and underwater bubbles appear over a certain area of the water surface. Of course, in connection with the latest calculations, many other questions arise. We will return to them at the end of the section.

We have shown in Fig. 7.22 spherical nonlinear modes that could be excited in the sphere under consideration at the very beginning of its appearance, namely, when a certain equilibrium in the number of bosons and fermions began to be established. We assume that in the future the sphere continues to expand, but the number of modes there does not change. The shape of the modes, of course, changes as time-space itself expands inside the sphere. Thus, closely spaced points of space diverge further and further (this reminds, a little, the movement of a grid of Lagrangian coordinates relative to fixed Euler coordinates in problems of continuum mechanics). At the same time, some areas, in particular ellipsoidal structures, expanding, can oscillate in space, colliding with each other with some regularity (see bottom pictures Fig. 7.22). Note that due to spherical symmetry, ellipsoidal structures in space can look like some tori.

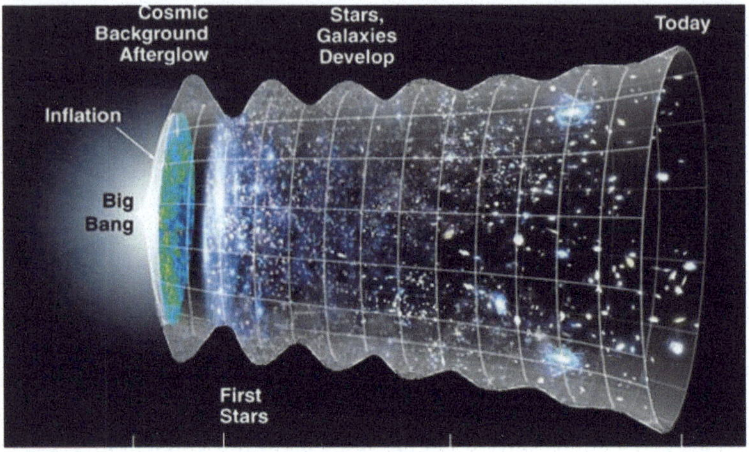

Fig. 7.23 The Universe itself may be oscillating through billions of years of cosmic time [58]

Unexpected consequences of the dynamics of transitions from bosons to fermions and vice versa are possible. We have already said that we associate the explosive nature of the expansion of the sphere with the transition of bosons to fermions. When the number of fermions reaches a critical value, the reverse process apparently begins. Some fermions begin to transform back into bosons. It seems to us that this to some extent determines the oscillatory nature of the expansion of the Universe, which is shown in Fig. 7.23.

Of course, these are all very rough assumptions that can hardly even be called hypotheses. However, a qualitatively similar assumption can be described mathematically following Sect. 5.3.

Let us use Eqs. (5.51) and (5.57). Let in (5.57).

$$\varphi = \varphi_1(r) + J(r). \tag{7.30}$$

Next, we write for $\varphi_1(r)$ an equation identical to (5.57). The equation for $J(r)$ can then be written as

$$C_1 J_{rr} + C_2 J_r + m^2 J - \lambda(\varphi_1 + J)^3 = 0. \tag{7.31}$$

We take the expression for $\varphi_1(r)$ in the form (5.58). Then we write the equation for $J(r)$ in the following:

$$J^3 - m^2 \lambda^{-1} J = \widetilde{A}\ \exp(3\Omega) \sin(\omega r). \tag{7.32}$$

Let us emphasize that the values $\widetilde{A}$, Ω, ω on the right side of Eq. (7.32) depend on the quantities C_1 and C_2 (5.56). That is, these quantities change when the radius of the sphere changes. It is very important. The fact is that it can be accepted as a model that the baryon medium can be attributed to a very high viscosity, which prevents its expansion. However, as the sphere expands, this viscosity rapidly decreases and the dynamic properties of the sphere are increasingly determined by dynamics of the content of baryon and fermions in the mixture (Fig. 7.24).

A qualitative example of changes in the magnitude of the scalar field calculated according to (7.32) is shown in Fig. 7.24 as a thick line. The thin line determines the amount of change on the right side of Eq. (7.32). We believe that the figure shows very roughly a change in the balance of the content of bosons and fermions in the first moments of the emergence of the Universe. First, there is a rapid increase in the number of fermions in the mixture. Then the reverse process occurs and damped oscillations are established (damped pendule). All this may have determined the evolution of the Universe. Possibly, echoes of this process are taking place even now, which is manifested in fluctuations in the dynamics of the expansion of the Universe (Fig. 7.23).

The increase in the amount of matter and the simultaneous increase in the radius of the sphere led to a complex law of expansion of the early Universe. At the beginning of the expansion, it expanded very quickly, perhaps exponentially. Then the expansion slowed down greatly, and perhaps became oscillatory (Fig. 7.23). There is a delicate balance between the matter produced and the rate of expansion of the Universe. It possibly leads to oscillations in the evolution of the Universe.

We proceed from the fact that an ideal sphere (the initial starting configuration of the Universe) is not a stable configuration, since it has a spectrum of natural oscillations. No matter how strong the connections between the particles of the medium are, the sphere can lose stability in one of the modes. The issue under consideration has naturally been condemned earlier. In Lamb's book [Hydrodynamics, pages 473–475], you can find a consideration of the natural vibrations of a spherical particle of a liquid surrounded by another liquid. The case of very strong vibrations of a drop leading to its atomization is shown in Fig. 2.14. It can be assumed that our Universe, during the process of its purely radial expansion, was fragmented on the expansion surface into components, which eventually filled a

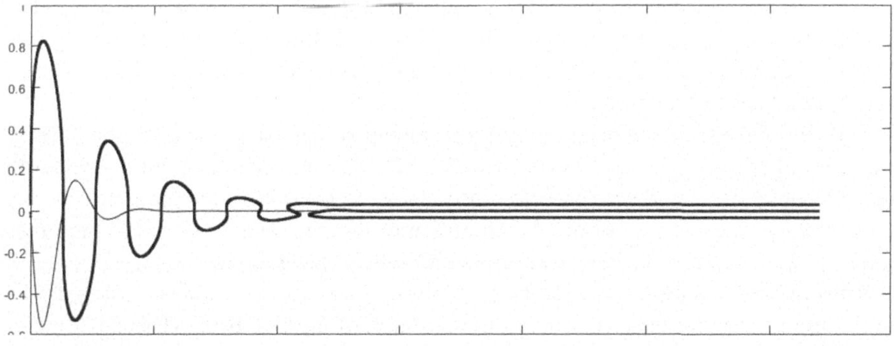

Fig. 7.24 The change in the magnitude of the scalar field calculated according to (7.32)

certain spherical region. This area continued to expand. This region then evolved into the modern Universe.

7.6.2 Multivalued Waves and the Richtmyer-Meshkov Instability as the Causes of the Formation of Galaxies

For thousands of years, people have observed the starry sky and wondered about the Universe and its origins. The ancient Greeks began to build speculative hypotheses explaining the laws of planetary motion and the immobility of stars. However, it was only after Galileo invented the telescope that what is now called observational astronomy arose. More and more advanced telescopes began to be created and observatories were built for them. All this made it possible to explore space more and more deeply. As a result, by the end of the 20th century, a fairly general scientific view of the origin and evolution of the Universe had formed.

A Brief Overview of Modern Cosmology and Some Results from the James Webb Space Telescope (JWST) The main provisions of modern cosmology are as follows: The Universe arose from a certain singularity, very small in size and shaped like an ideal sphere. At some point in time, for unknown reasons, the sphere began to rapidly expand, and a certain substance appeared in it. When the expansion slowed down somewhat, gravity emerged. Thanks to it, matter began to slowly gather around the disturbances in the density of matter that arose at the beginning of the expansion of the sphere.

These disturbances were very weak and extended in space, so astrophysicists believe that the process of grouping material into stars took a time of the order of hundreds of millions of years. Then the emerging stars, also under the influence of gravity, began to group into galaxies with an increasingly symmetrical appearance. The formation of the first galaxies took at least a billion years. Finally, galaxies began to form into gigantic chains and clusters of galaxies, between which there are giant voids where there is almost no matter.

All the processes described took, of course, many billions of years. In particular, few doubt that the formation of the first galaxies required at least a billion years. After all, their formation requires millions and millions of stars, the formation of which also takes a lot of time.

But the scheme of the origin and development of the Universe described above may be collapsing. The fact is that in 2022, JWST was launched into near-Earth space to observe very deep space [54–56]. Thus, to observe stars and galaxies that arose at the earliest stages of the development of the Universe, right up to the times when, as was supposed, there were no stars, much less galaxies! And suddenly, it was discovered that both stars and galaxies existed already at a time very close to the beginning of the Universe (that is, to the moment of the Big Bang (BB)).

In particular, two galaxies were discovered that are about 300 million years old if the years are counted from the supposed BB (the age of the Universe is estimated at 13.797 billion years). It is noted that these galaxies contain a quarter of the elements heavier than hydrogen and helium. Thus, the star systems contained there went through a long path of evolution, which led to the emergence of some heavy elements.

The so-called black holes were discovered that existed in the distant past. The earliest black hole formed just 400 million years after BB [56].

Thanks to this observation, it became possible to revise the picture of the origin of the Universe described above. This picture was created by generations of astronomers and astrophysicists. It is based on numerous careful observations, solutions to very complex equations, and a comparison of calculations and observations. Thousands and thousands of articles, several internationally recognized journals, and many books are devoted to the problem. Many scientists have received the Nobel Prize for outstanding services in this field.

However, noted observations of the Webb telescope are in no way consistent with the forecast of modern cosmology. The thought arises that we are standing on the threshold of a new revolution in cosmogony comparable to the revolutions of Copernicus and Galileo, hypotheses about the non-stationary Universe, and the Big Bang!

Of course, the possibility of such a revolution was in the air. Some theories that have emerged recently looked very vulnerable to criticism. For example, the theory of inflation. But the problem looked unstable also when viewed from other perspectives. Many astrophysicists began to look at the modern theory of cosmology as an almost completed picture of the Universe, in which only the interior decoration, such as dark matter or black holes, remained to be corrected. It was assumed that the frame of the building was built and it was unshakable. From a philosophical point of view, such an opinion is, of course, not justified. Indeed, we cannot say anything definite about the finitude or infinity of the Universe, but almost all philosophers agree that the process of cognition of it is endless. It will be truly surprising if, as a result of 600 years of development of modern science, humanity answers the question about the emergence of the Universe. According to many, the origin of the Universe will be the problem for millennia. Perhaps it lies beyond the possibility of solution by science and humanity!

At the same time, this does not mean that there will be no future attempts to improve cosmological models. This applies primarily to times when observations provide information that requires explanation. The development of science cannot stop. A number of scientists believe that noted observations discoveries a new cosmology. The whole picture of early galaxy formation is into question.

Galiev and Galiyev Model and the Birth of Galaxies The above applies to the generally accepted (standard) model of the development of the Universe. Along with this model, other models well known to specialists are developed. They try to explain the Webb observations. In recent years, models have appeared related to the possibility of an “eruption” of our Universe from the pre-Universe. One of them is developed by Galiev and Galiyev (MGG) [57–60].

All of the above provisions of the described theory are, of course, speculative. This makes them not much worse than all other theories of the origin of the Universe.

The MGG, which is extended to the case of galaxy formation, is proposed here. It is assumed that at the moment the expansion begins, the Universe is filled with a scalar field described by the nonlinear Klein-Gordon equation. The field is very dense. During the expansion process, the density rapidly decreases. Conditions are created for the growth of initial disturbances and the emergence of turbulence. In particular, at the front of the expanding sphere, due to the Richtmyer-Meshkov instability, multivalued waves and vortices arise. Qualitatively, this process could correspond to that presented in Fig. 7.26.

7.6.3 Modeling of the Birth of Galaxies

Next we will consider some questions regarding the subsequent evolution of the Universe.

We will further examine this phenomenon as the final element of our theory.

We examined in Sect. 5.1 the natural nonlinear waves arising in a spherical resonator. As in the case of linear waves, we assumed that standing waves are formed by the sum of oppositely traveling waves. As a special case, unidirectional waves follow from this approach. Let us consider the wave $J(\xi)$ propagating from the center of the sphere. The equation for this wave follows from (5.14):

$$-4c^2r^{-1}\varphi_{2\xi} - 2m^2r^{-1}J - 2m_2r^{-2}J^2 + 2\lambda r^{-3}J^3 = r^{-1}J^*(\xi). \tag{7.33}$$

Next, remember that $\varphi_1 \gg \varphi_2$. We take the latter into account and obtain the following model Eq. (7.34):

$$\varphi^3 - m_2\lambda^{-1}\varphi^2 - m^2\lambda^{-1}\varphi^3 = \frac{1}{2}\lambda^{-1}r^{-1}J^*(\xi). \tag{7.34}$$

Here we took into account that $\varphi_1 \gg \varphi_2$ and also introduced the notation $\varphi = r^{-1}J$. Next, we will consider the quantity $r^{-1}J^*(\xi)$ in Eq. (7.34) as a seed perturbation. Let us accept the seed perturbation in the form (5.58), $\frac{1}{2}\lambda^{-1}r^{-1}J^*(\xi) = -r^{-1}A\exp(\Omega\xi)\cos\omega\xi$, where A, Ω, and $\overline{\Omega}$ are constants. In the last expression r^{-1} changes much more slowly than $\sec h\Omega\xi$ and $\cos\omega\xi$, so for simplicity we assume $r^{-1} = 1$. As a result, (7.34) gives

$$\varphi^3 + M\varphi^2 + \overline{R}\varphi^3 + A\exp(\Omega\xi)\cos\omega\xi = 0. \tag{7.35}$$

We emphasize that the seed term is a solution to the linearized Eq. (5.57), which significantly depends on the viscosity of the scalar field. This viscosity depends on the initial viscosity. For example, the concept of vacuum viscosity is well known [111, 112]. We assume that as the volume of the sphere increases, the value of the viscosity coefficient decreases. The goal of the calculations was to demonstrate the emergence and development of instability of the front of a spherical wave as it propagates from a certain source.

Figure 7.25, constructed for a spherical wave $\varphi = r^{-1}J$, shows a traveling wave, as well as its front. You can see how, when approaching the front, the wave profile is distorted into Euler figures and can be transformed into ellipsoidal figures. As we have shown earlier, these figures can correspond to vortices upon more detailed study.

Calculations using Eq. (7.35) were carried out with the following parameters: $\omega = 0.75{*}10^{41}, \overline{R} = -0.5{*}10^{-14}, A = 10^{-21}, M = 0$ (left), or $M = 10^{-7}$ (right). In the calculations, it was assumed that the value Ω takes into account the influence of viscosity. Viscosity changes proportionally to time: $\Omega = 10/t$. The influence of quadratic nonlinearity on the wave structure was studied. The left curves in the

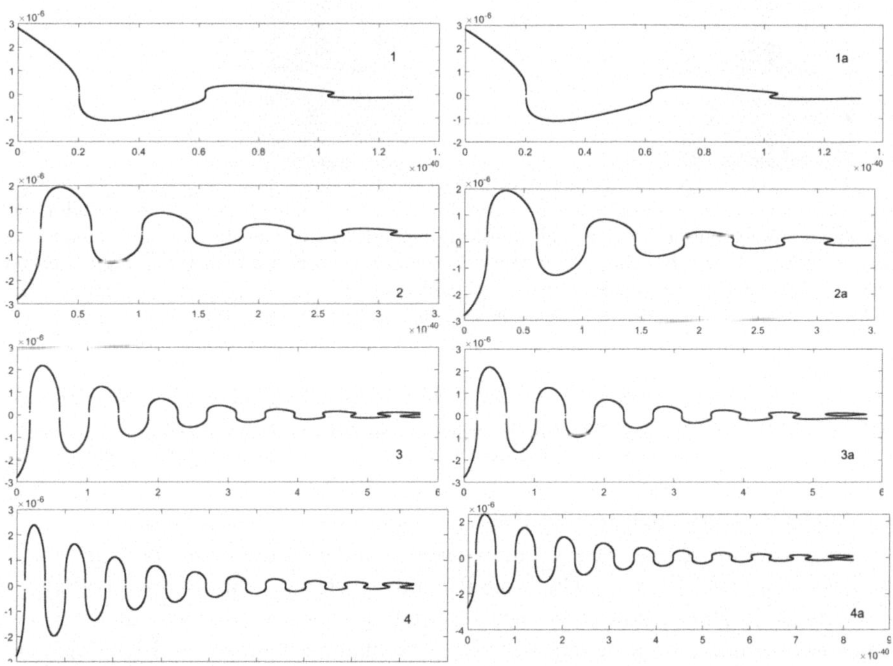

Fig. 7.25 Nonlinear change in the profile of a spherical wave calculated for various moments of dimensionless time: $t - \overline{R}/8$, $\overline{R}/3$, $\overline{R}/2$, and $0.75\overline{R}$. The upper curves correspond to $t = \overline{R}/8$. At this point, the wave has traveled about 0.125 of the radius of the sphere in question. The next two rows of curves correspond to the moments when the wave passed near $\overline{R}/3$ and $\overline{R}/2$. The lower curves correspond to $0.75\overline{R}$

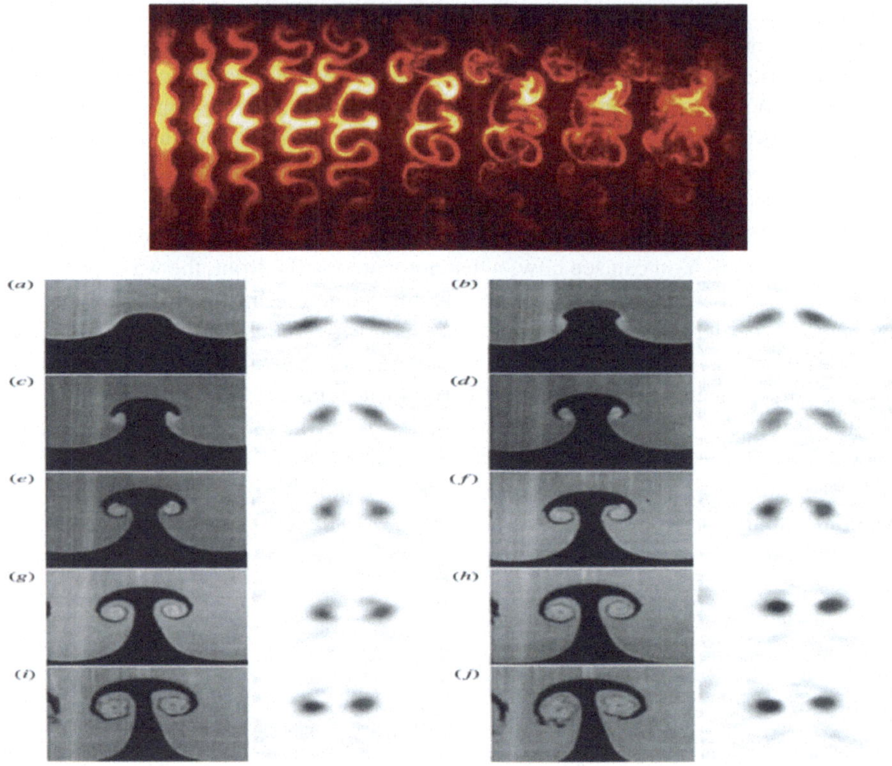

Fig. 7.26 This image was originally captioned "Coexistence of three patterns" and was additionally noted: "This poster was created for the 1996 APS-DFD Gallery of Fluid Motion by Paul Rightley, Robert Benjamin, and Peter Vorobieff. It won the Gallery of Fluid Motion Award." This image was obtained from http://cnls.lanl.gov/~azathoth/cover.html (upper). The Richtmyer-Meshkov instability in shock tube experiments [90, 91] (bottom)

figure are calculated for $M = 0$, and the right curves are calculated for $M = 10^{-7}$. It can be seen that, just as was shown earlier, quadratic nonlinearity significantly complicates the structure of the wave front. However, at first, the influence of quadratic nonlinearity does not appear. Only in the last two rows of curves at the wave front do ellipsoidal structures appear, additional to the Euler figures.

Along with the above, we present experimental data showing the propagation of the front of a plane wave on which the Richtmyer-Meshkov instability develops.

Figure 7.26 shows how structures resembling Euler figures arise at the front, which evolve into closed figures and vortex formations. Comparing Figs. 7.25 and 7.26, one can see that the calculations qualitatively correspond to Fig. 7.22 and Figs. 3.30 and 3.34.

Thus, our analysis is consistent with experimental data and perhaps provides new insight into the beginning of the formation of some structures of future galaxies [54–56, 109, 110, 113]. In our opinion, these structures could have arisen in the very first moments of the emergence and development of the Universe. In particular, if we take into account the fact that the formation of these structures occurred on a spherical surface, then, it seems to us, the formation of galaxies could well have occurred according to the scenario shown in Fig. 7.26.

It can be assumed that when the Universe emerged, it was in a state of dynamic equilibrium for a very long time. However, in some situation, the initial disturbance exceeded the critical disturbance, and the system went out of balance. This initial disturbance can be associated with the disturbances presented in Figs. 3.30 and 3.34, which led to the loss of equilibrium of the interface and the growth of disturbances shown there. In Figs. 3.30, 3.34, and 7.26, one can clearly see that during the process of instability, not only particles appear, which are so clearly visible in Fig. 2.14, but also vortices appear. Thus, from the described model, it follows that the seed waves and vortices around which clusters of material and galaxies can subsequently be organized arose at the very first moments of time. At moments when, perhaps, there were not stars, but even elementary particles.

Remark We emphasize that folds of spherical waves were discussed earlier in Chap. 5 (see Fig. 5.20). In particular, it was noted that the folds have been observed recently in waves of solar wind.

7.7 A Few Notes About the Theory

There is the most fundamental, in my opinion, question related to the origin of the Universe. How did it happen that the constants characterizing the fundamental physics of Nature are so successfully and carefully adjusted to each other?

This fine-tuning can be explained in different ways. Below I will give two versions:

1. Our Universe is part of some boundless Universe, in which daughter Universes randomly appear and disappear. The parameters of one of them accidentally coincided with the parameters of our Universe [18, 20].
2. There is a pre-Universe, which constantly radiates the "seeds" of Universes.

We develop this option in this chapter. We assumed that somewhere there is a "kitchen" for producing these "seeds." Outside the kitchen, these seeds die, but the parameters and laws of one "seed" turned out to be so successful that it grew into our Universe.

In conclusion, we recall that results of this chapter are based on an approximate analysis of Eqs. (7.1), (7.20), (7.25), and (7.30), the applicability of which to the problems considered is far from obvious. Moreover, the solutions to this equation were obtained with very strong assumptions and rough assumptions. The author understands this. His analysis pursues the main goal of this book, which is intended to show the possibility of describing, albeit very approximately, a very wide range of problems of the modern science of "physics" based on the simplest equation describing a scalar field. This approach is far from new. Perhaps the most consistent proponent of this approach in the twentieth century was Albert Einstein.

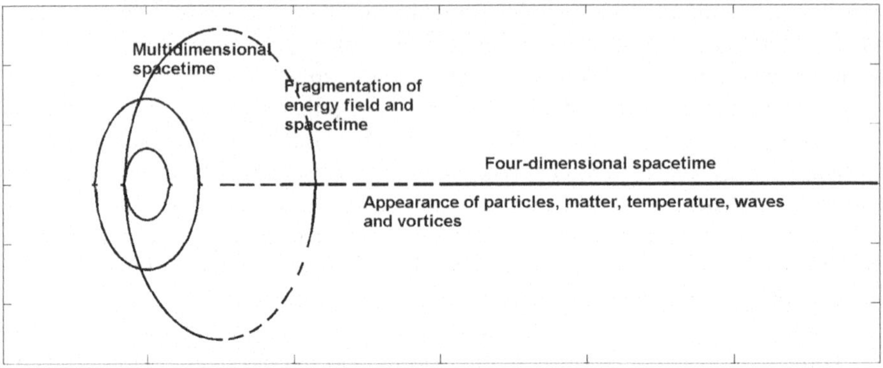

Fig. 7.27 A rough scheme of the results presented in the chapter. The moments from the emergence of the "seed" of the Universe to the vortices in the nascent galaxies are highlighted [60]. http://hdl.handle.net/2292/23783

Thus, in this chapter we did not try to describe the existing Universe. We were more interested in the following question. Which Universe's birth is described by the simplest scalar field equation used in the book?

It turned out that approximate solutions of this equation can be constructed in such a way that they describe a certain sequence of events that determine a certain picture of the emergence of the Universe. It includes the existence of an oscillating proto-Universe. As a result of quantum flocculations, the oscillations emit high-energy particles into the surrounding space. One of these particles had sufficient energy and such initial parameters that allowed it to evolve into the modern Universe. We tried to present the main stages of the described process in Fig. 7.27.

In the figure, the circles on the left symbolize the oscillating multidimensional pre-Universe filled with a nonlinear scalar field (Sect. 7.2). Under the action of quantum fluctuation, the field particle tunnels beyond its space-time dimensions (Sect. 7.3). There it loses stability and disintegrates into a system of one-dimensional oscillating string-like elements (the center of Fig. 7.27 and Sect. 7.4). The latter emit high-energy elementary particles that form a certain spherical volume of bosons (Sect. 7.5). At some point, the bosons begin to transform into fermions. A rapid expansion of the spherical volume begins (Sect. 7.6). This expansion can be compared with the Big Bang. During the expansion, events corresponding to those described by the standard cosmological model could have occurred.

References

1. Galiev SU (2009) Modelling of Charles Darwin's earthquake reports as catastrophic wave phenomena. The University of Auckland
2. Galiev SU (2010) Charles Darwin's geophysical reports as models of the theory of catastrophic waves. The University of Auckland, p 655

3. Hoyle F, Burbidge G, Narlikar JV (2000) A different approach to cosmology: from a static universe through the big bang towards reality. Cambridge University Press. Wetterich C (2014) Hot big bang or slow freeze? Phys Lett B 736:506–514; Ali F, Das S (2015) Cosmology from quantum potential. Phys Lett B 741:276
4. Friedmann A (1922) Uber die krümmung des raumes (on space curvature). Z Phys 10:377. Friedmann A (1924) Über die möglichkeit einer welt mit konstanter negativer krümmung des raumes. Zeitschrift fur Physik 21:326
5. Lemaître G (1931) Lemaître the beginning of the world from the point of view of quantum theory. Nature 127:706
6. Gamow G (1946) Expanding universe and the origin of elements. Phys Rev 70:572–573. Gamow G (1946) Rotating universe? Nature 158:549; Gamow G (1948) The evolution of the Universe. Nature 162:680–682; Alpher R, Bethe H, Gamow G (1948) The origin of chemical elements. Phys Rev 73(7):803–804; Gamow G (1948) The origin of elements and the separation of galaxies. Phys Rev 74:505–506; Gamow G (1952) The creation of the universe. Viking Press, New York
7. Sakharov AD (1965) The Initial stage of the expansion of the Universe and the occurrence of inhomogeneity of the distribution of matter. Zh Exp Theor Phys 49 (1): 345–357 (Сахаров АД. Начальная стадия расширения Вселенной и возникновение неоднородности распределения вещества // Журн. эксперим. и теорет. Физики 49 (1): 345–357 (1965))
8. Zeldovich YB, Sunyaev RA (1969) The interaction of matter and radiation in a hot-model universe. Astrophys Space Sci 4:301–316
9. Starobinsky AA (1980) A new type of isotropic cosmological models without singularity. Phys Lett B 91:99
10. Mukhanov VF, Chibisov GV (1981) Quantum fluctuation and nonsingular Universe. JETP Lett 33:532. [Pisma Zh. Eksp. Teor. Fiz. 33:549 (1981) (in Russian)]
11. Cai YF, Easson DA, Brandenberger R (2012) Towards a nonsingular bouncing cosmology. J Cosmol Astropart Phys 2012(08):020
12. Tryon EP (1973) Is the universe a vacuum fluctuation? Nature 246:396
13. Starobinsky AA (1979) Spectrum of relict gravitational radiation and the early state of the universe. JETP Lett 30:682. [Pisma Zh. Eksp. Teor. Fiz. 30:719 (1979) (in Russian)]
14. Guth AH (1981) The inationary universe: a possible solution to the horizon and flatness problems. Phys Rev D 23:347
15. Linde AD (1982) A new inflationary universe scenario: a possible solution of the horizon, flatness, homogeneity, isotropy and primordial monopole problems. Phys Lett B 108:389
16. Albrecht AJ, Steinhardt PJ (1982) Cosmology for grand unified theories with radiatively induced symmetry breaking. Phys Rev Lett 48:1220
17. Atkatz D, Pagels H (1982) Origin of the universe as a quantum tunneling event. Phys Rev D 25:2065
18. Vilenkin A (1982) Creation of universe from nothing. Phys Lett B 117:25
19. Krauss LM (2012) A universe from nothing. Atria Paperback, New York
20. Vilenkin A (2006) Many worlds in one. Hill and Wang, New York
21. Vilenkin A (1983) The birth of the inflationary universes. Phys Rev D 27:2848
22. Linde AD (1986) Eternally existing self-reproducing chaotic inflationary universe. Phys Lett B 175:395
23. Linde A (1990) Particle physics and inflationary cosmology. CRC Press
24. Hawking SW, Moss IG, Stewart JM (1982) Bubble collisions in the very early universe. Phys Rev D 26:2681
25. Guth AH (2013) Quantum fluctuations in cosmology and how they lead to a multiverse, arXiv:1312.7340
26. Linde A (2017) A brief history of the multiverse. Rep Prog Phys 80(2):022001
27. Salem MP (2012) Bubble collisions and measures of the multiverse. JCAP 1201:021
28. Garriga J, Vilenkin A (1998) Recycling universe. Phys Rev D 57:2230

29. Bojowald M (2001) Absence of a singularity in loop quantum cosmology. Phys Rev Lett 86:5227
30. Steinhardt PJ, Turok N (2002) A cyclic model of the universe. Science 296:1436
31. Bojowald M (2007) What happened before the big bang? Nat Phys 3:523
32. Steinhardt PJ, Turok N (2007) Endless universe. Doubleday, New York
33. Ashtekar A, Corichi A, Singh P (2008) Robustness of key features of loop quantum cosmology. Phys Rev D 77:024046
34. Penrose R (2010) Cycles of time: an extraordinary new view of the universe. Bodley Head, London
35. Brandenberger R (2012) The matter bounce alternative to inflationary cosmology, arXiv:1206.4196v1 [astro-ph.CO]
36. Barrau A, Linsefors L (2014) Our universe from the cosmological constant. J Cosmol Astropart Phys 2014(12):037
37. Gielen S, Turok N (2016) Perfect quantum cosmological bounce. Phys Rev Lett 117:021301
38. Smolin L (1992) Did the universe evolve? Class Q Grav 9:173
39. Smolin L (1994) The fate of black hole singularities and the parameters of the standard models of particle physics and cosmology, arXiv:gr-qc/9404011v1
40. Gasperini M, Veneziano G (2007) String theory and pre-big bang cosmology. arXiv:hep-th/0703055v3; Veneziano G (1998) A simple/short introduction to pre-Big-Bang physics/cosmology, arXiv:hep-th/9802057v2; Klebanov IR, Maldacena JM *(2009) Solving quantum field theories via curved spacetimes. Phys Today 62(1):28–33;* Veneziano G *(1991) Scale factor duality for classical and quantum strings.* Phys Lett B *265(3–4):287–294*
41. Gonzalez-Mestres L (2015) BICEP2, Planck, spinorial space-time, pre-Big Bang. In: EPJ web of conferences, vol 95. EDP Sciences. https://indico.cern.ch/event/277650/contributions/629783/attachments/505847/698394/ICNFP2014talknewter.pdf
42. Garriga J, Vilenkin A, Zhang J (2016) Black holes and the multiverse. J Cosmol Astropart Phys 2016(02):064
43. Susskind L (1995) The world as a hologram. J Math Phys 36:6377
44. Ijjas A, Steinhardt PJ, Loeb A (2013) Inflationary paradigm in trouble after Planck2013. Phys Lett B 723:261–266
45. Ijjas A, Steinhardt PJ (2016) Implications of Planck2015 for inflationary, ekpyrotic and anamorphic bouncing cosmologies. Class Q Grav 33:044001
46. Afshordi N, Magueijo J (2016) The critical geometry of a thermal big bang. Phys Rev D 94: 101301. Carroll SM (2014) In what sense is the early universe fine-tuned? arXiv:1406.3057v1 [astro-ph.CO]
47. Ade PA, Aghanim N, Alina D, Alves MI, Armitage-Caplan C, Arnaud M, Arzoumanian D, Ashdown M, Atrio-Barandela F, Aumont J, Baccigalupi C, Planck intermediate results. XIX. (2015) An overview of the polarized thermal emission from galactic dust. Astron Astrophys 576:A104
48. Gott JR, Statler TS (1984) Constraints on the formation of bubble universes. Phys Lett B 136: 157
49. Aguirre A, Johnson MC (2008) Towards observable signatures of other bubble universes II: exact solutions for thin-wall bubble collisions. Phys Rev D 77:123536
50. Johnson MC, Wainwright CL, Aguirre A, Peiris HV (2016) Simulating the universe (s) III: observables for the full bubble collision spacetime. J Cosmol Astropart Phys 2016(07):020
51. Chary R (2016) Spectral variations of the sky: constraints on alternate universe. Astrophys J 817:33
52. Cho A (2016) Controversial test finds no sign of a holographic universe. Science 350:1303
53. Chiao M (2019) Waves from the early universe. Nat Astron 3:960. Kusenko A et al (2020) Fundamental forces and scalar field dynamics in the early universe. Phys Lett B 804:135369; Raushan R et al (2020) Universe with quadratic equation of state: a dynamical systems perspective. Int J Geom Methods Mod Phys 17(5):2050064

54. Menzel M et al (2023) The design, verification, and performance of the James Webb space telescope. PASP 135:058002DOI. https://doi.org/10.1088/1538-3873/acbb9f
55. Costantin L et al (2023) A milky way-like barred spiral galaxy at a redshift of 3. Nature. https://doi.org/10.1038/s41586-023-06636-x
56. Silk J, Begelman MC, Norman C, Nusser A, Wyse RF (2024) Which came first: supermassive black holes or galaxies? Insights from JWST. Astrophys J Lett 961(2):L39
57. Galiev SU (2015) Darwin, geodynamics and extreme waves. Springer
58. Galiev SH (2020) Evolution of extreme waves and resonances. CRC Press, p 508
59. Galiev SU, Galiyev TS (2014) Nonlinear scalar field as a model describing the birth of the Universe. Известия Уфимского Научного Центра Российской Академии Наук (Herald of Ufa Scientific Center, Russian Academy of Sciences (RAS)) 2:7; Galiev SU, Galiyev TS (2016) Simple model of the origin of the Universe. Известия Уфимского Научного Центра Российской Академии Наук (Herald of Ufa Scientific Center, Russian Academy of Sciences (RAS)) 4:27; Galiev SU, Galiyev TS (2021) Modeling of an appearance of particle-waves and the Universe as the extreme wave resonant events. Известия Уфимского Научного Центра РАН ((Herald of Ufa Scientific Center, Russian Academy of Sciences (RAS)) 2:5–23. http://journal.ufaras.ru/wp-content/uploads/2021/10/5-23.pdf
60. Galiev SU, Galiyev TS (2013) Scalar field as a model describing the birth of the Universe. InterNet. Galiev – Galiyev 18-12–2013. https://researchspace.auckland.ac.nz/docs/uoa-docs/rights.htm; Galiev SU, Galiyev TS (2014) Coherent Model of the Births of the UNIVERSE, the Dark Matter and the Dark Energy. http://hdl.handle.net/2292/23783; Galiev SU, Galiyev TS (2016) Eruption of the Universe Out of a Pre-universe. https://researchspace.auckland.ac.nz/handle/2292/30899; Galiev SU, Galiyev TS (2017) Eruption of Matter, Dark Energy and the Universe from a Pre-universe. https://researchspace.auckland.ac.nz/handle/2292/31792; Galiev SU, Galiyev TS (2017) Nonlinear Quantum Waves in the Light of Recent Slit Experiments. https://researchspace.auckland.ac.nz/handle/2292/36751; Galiev SU, Galiyev TS (2017). 'Elastica'-like waves and particles: from Bernoulli, Euler, Laplace and Faraday to the eruption of the Universe. http://hdl.handle.net/2292/31785
61. Bellini M (2011) Pre-big bang collapsing universe from modern Kaluza-Klein theory of gravity. Phys Lett B 705:283
62. Hawking SW (1982) The development of irregularities in a single bubble inflationary universe. Phys Lett B 115:295
63. Ball P (2001) Tabletop astrophysics. Nature 411:628. Ohlsson T (2011) Relativistic quantum physics. Cambridge University Press
64. Gasiorowicz S (1996) Quantum physics. Wiley, New York. Ohlsson T (2011) Relativistic quantum physics. Cambridge University Press, Cambridge
65. Mureika J, Stojkovic D (2011) Detecting vanishing dimensions via primordial gravitational wave astronomy. Phys Rev Lett 106:101101
66. Gleiser M (1994) Pseudostable bubbles. Phys Rev D 49:2978
67. Umbanhowar PB, Melo F, Swinney HL (1996) Localized excitations in a vertically vibrated granular layer. Nature 382:793–796
68. Belova TI, Kudryavtsev AE (1997) Solitons and their interactions in classical field theory. Phys Usp 40:359
69. Cerda E, Melo F, Rica S (1997) Model for subharmonic waves in granular materials. Phys Rev Lett 79(23):4570. https://doi.org/10.1103/PhysRevLett.79.4570
70. Clement E, Labous L (2000) Pattern formation in a vibrated granular layer: the pattern selection issue. Phys Rev E 62(6):8314–8323. Clement E, Vanel L, Rajchenbach J, Duran J (1996) Pattern formation in a vibrated granular layer. Phys Rev E 53:2972–2975
71. Galiev SU (1999) Topographic effect in a Faraday experiment. J Phys A Math Gen 32:6963–7000
72. Graham N, Stamatopoulos N (2006) Unnatural oscillons lifetimes in an expanding background. Phys Lett B639:541–545. https://doi.org/10.1016/j.physletb.2006.06.070

73. Stock S, Bretin V, Chevy F, Dalibard J (2004) Shape oscillation of a rotating Bose-Einstein condensate. Europhys Lett 65:594–600
74. Laposky BF (1969) Oscillons: electronic abstractions. Leonardo 2(4):345–354
75. Alexeeva NV, Barashenkov IV, Bogolubskaya AA, Zemlyanaya EV (2023) Understanding oscillons: standing waves in a ball. Phys Rev D 107(7):076023
76. Galiev SU (1998) Resonant oscillations of gas sphere. In: Proceedings of the 13th Australasian fluid mechanics conference (Melbourne, 13–18 December 1998). Monash University, p 1027–30; Galiev SU (1998) Is there some new kind of nonlinear surface waves? Rep. 584, School of Engineering, The University of Auckland; Galiev SU (1999), Unfamiliar waves excited due to parametric and resonant effects. In: Proceedings of the 2nd international conference on unsolved problems of noise and fluctuations (UPoN'99). Adelaide: University of Adelaide, p 321–6
77. Galiev SU (2000) Unfamiliar waves excited due to parametric and resonant effects. Proc Am Ins Phys 511:361
78. Graham NN (2007) An electroweak oscillon. Phys Rev Lett 98:101801
79. Farhi E, Graham N, Guth AH, Iqbal N, Rosales RR, Stamatopoulos N (2008) Emergence of oscillons in an expanding background. Phys Rev D 77:085019
80. Amin MA, Easther R, Finkel H, Flauger R, Hertzberg MP (2012) Oscillons after inflation. Phys Rev Lett 108:241302
81. Hau LV, Harris SE, Dutton Z, Behroozi CH (1999) Light speed reduction to 17 metres per second. **Nature** 397:594
82. Kobzarev IY, Okun LB, Voloshin MB (1974) Bubbles in metastable vacuum. Yad Fiz. 20: 1229 (in Russian) (Sov J Nucl Phys 0:644 (1975)); Coleman S (1977) Fate of the false vacuum: semiclassical theory. Phys Rev D 15(10):2929–2936
83. Galiev SU, Galiyev TS (2001) Nonlinear trans-resonant waves, vortices and patterns: from microresonators to the early universe. Chaos: an interdisciplinary. Journal of Nonlinear Science 11(3):686–704
84. Linde AD (1983) Decay of the False Vacuum at Finite Temperature. Nucl. Phys. B 216:421. [Erratum: Nucl.Phys.B 223, 544 (1983)]; Coleman SR (1977) The fate of the false vacuum. 1. Semiclassical theory. Phys Rev D 15:2929; Turner MS (1982) Wilczek F. Is our vacuum metastable? Nature 298:633; Sato K, Kodama H, Sasaki M, Maeda K-I (1982) Multi-production of universes by first-order phase transition of a vacuum. Phys Lett B 108:103; Riek C et al (2015) Direct sampling of electric-field vacuum fluctuations. Science 350:420
85. Wilczek F (n.d.) The materiality of vacuum. https://pa.as.uky.edu/video/frank-wilczekmateriality-vacuum; Bass SD (2015) Vacuum energy and the cosmological constant. Mod Phys Lett A 30(22):1540033
86. Lagnese G, Surace FM, Kormos M, Calabrese P (2021) False vacuum decay in quantum spin chains. Phys Rev B 104:L201106
87. Milsted A, Liu J, Preskill J, Vidal G (2022) Collisions of false-vacuum bubble walls in a quantum spin chain. PRX Q 3:020316
88. Hogan C (1986) Gravitational radiation from cosmological phase transitions. Mon Not R Astron Soc 218:629
89. Shaposhnikov ME (1987) Baryon asymmetry of the universe in standard electroweak theory. Nucl Phys B 287:757
90. Feeney SM, Johnson MC, Mortlock DJ, Peiris HV (2011) First observational tests of eternal inflation: analysis methods and wmap 7-year results. Phys Rev D 84:043507
91. Zenesini A et al (2024) False vacuum decay via bubble formation in ferromagnetic superfluids. Nat Phys. https://doi.org/10.1038/s41567-023-02345-4. Mazumdar A, White G (2019) Review of cosmic phase transitions: their significance and experimental signatures. Rept Prog Phys 82:076901; Hindmarsh M, Lüben M, Lumma J, Pauly M (2021) Phase transitions in the early universe. Sci Post Phys Lect Notes 24
92. Minev ZK et al (2019) To catch and reverse a quantum jump mid-flight. Nature 570:200–204

93. Ramos R et al (2020) Measurement of the time spent by a tunnelling atom within the barrier region. Nature 583:529–532
94. Kocsis S et al (2011) Observing the average trajectories of single photons in a two-slit interferometer. Science 332:1170–1173
95. Menzel R, Puhlmann D, Heuer A, Schleich WP (2012) Wave-particle dualism and complementarity unraveled by a different mode. PNAS 109(24):9314–9319
96. Mahler DH et al (2016) Experimental nonlocal and surreal Bohmian trajectories. Sci Adv 2(2): e1501466
97. Smolin L (2006) The trouble with physics: the rise of string theory, the fall of a science, and what comes next. Houghton Mifflin Harcourt
98. Jacques V et al (2007) Experimental realization of Wheeler's delayed-choice gedanken experiment. Science 315:966–968
99. Manning AG et al (2015) Wheeler's delayed-choice gedanken experiment with a single atom. Nat Phys. https://doi.org/10.1038/nphys3343
100. Lyth DH, Liddle AR (2009) The primordial density perturbation. Cosmology, inflation and origin of structure. Cambridge University Press
101. Easther R (1993) The evolution of scalar fields and inflationary cosmology. A thesis submitted in partial fulfilment of the requirements for the Degree of Doctor of Philosophy in Physics in the University of Canterbury by Richard John Maddock Easther, University of Canterbury; Easther R, Parry M (2000) Gravity, parametric resonance, and chaotic inflation. Phys Rev D 62:103503
102. Stojkovic D (2013) Vanishing dimensions: theory and phenomenology, arXiv:1304.6444 (hep-th)
103. Stojkovic D (2013) Vanishing dimensions: a review. Mod Phys Lett A 28(37):1330034
104. Gubser SS (2010) The little book of string theory. Princeton University Press, Princeton
105. Kinoshita T, Wenger T, Weiss DS (2004) Observation of a one-dimensional Tonks-Girardeau gas. Science 305(5687):1125–1128
106. Wilson JM, Malvania N, Le Y, Zhang Y, Rigol M, Weiss DS (2020) Observation of dynamical fermionization. Science 367(6485):1461–1464. Adam J, Adamczyk L, Adams JR, Adkins JK, Agakishiev G, Aggarwal MM, Ahammed Z, Alekseev I, Anderson DM, Aparin A, Aschenauer EC (2021) Measurement of e+ e-momentum and angular distributions from linearly polarized photon collisions. Phys Rev Lett 127(5):052302
107. Lamb H (1932) Hydrodynamics, 6th edn. Dover Publications, New York
108. Galiev SU, Galiyev TS (2017) 'Elastica'-like waves and particles: from Bernoulli, Euler, Laplace and Faraday to the eruption of the Universe. http://hdl.handle.net/2292/31785
109. Morgan RV et al (2012) On the late-time of the two-dimensional Richtmyer-Meshkov instability in shock tube experiments. JFM 712:354–383
110. Galiev SU (2024) Multivalued waves and the Richtmyer-Meshkov instability as the causes of the formation of galaxies, vol 1. Herald of Ufa Scientific Center Russian Academy of Sciences, p 5–17. https://doi.org/10.31040/2222-8349-2024-0-1-5-17
111. Sonnleitner M, Trautmann N, Barnett SM (2017) Will a decaying atom feel a friction force? Phys Rev Lett 118(5):053601
112. Barnett SM, Sonnleitner M (2018) The vacuum friction paradox and related puzzles. Contemp Phys 59(2):145–154. https://doi.org/10.1080/00107514.2018.1439333
113. Eckel S, Kumar A, Jacobson T, Spielman I, Campbell G (2018) A rapidly expanding Bose-Einstein condensate: an expanding universe in the lab. Phys Rev X 8:021021

Chapter 8
Everything Is Determined by Everything

Many years ago, I began researching multivalued waves, and as it seemed to me then, promising results were immediately obtained. But at the same time, great doubts arose both in the results obtained and in the mathematical models used. These doubts were reflected in the publications of those years and did not allow me to collect the results obtained for multivalued waves in one book. I was given additional courage by the publication of a remarkable article [1], which greatly strengthened my confidence in the validity of the results obtained and in the name "multivalued" wave.

Another circumstance that previously greatly limited me was related to the area of fairness and application of the original results obtained. The area of multivalued solutions and resonances that I considered was until recently outside the area of interest of science and its applications. Indeed, the area of traditional areas of human experience and interests is the area of linear processes. Our whole life proceeds in an environment of repeating processes, where their possible ambiguity rarely appears. We also note that wave processes in resonance regions have been practically not studied. Resonances were seen as areas of singularities to be ignored or exploited, rather than as areas where extreme waves occurred.

Linear thinking leads to the fact that when we encounter something new, we use familiar tools and concepts to describe it, if only because there are no generally accepted concepts for this new thing! Even after introducing and widely using the concept of "quantum," physicists continued to use differential equations to describe quantum effects and some effects of the general theory of relativity (GTR), which, by definition, are valid only for sufficiently smooth processes.

I will give examples to illustrate what has been said:

1. When Rutherford received the data from the famous experiments (Fig. 6.15), he came up with the idea of atomic structure similar to the structure of the solar system. The validity of this model is still debated.
2. Niels Bohr, perhaps, when creating his model of the electronic orbits of the atom, intuitively drew parallels with the theory of the ancient Greeks about the

S. U. Galiev, *Extreme Multivalued Waves as a Model of Everything*,
https://doi.org/10.1007/978-3-031-67027-5_8

existence of a family of concentric spheres around the Earth and about their joint sound!

3. The creators of wave quantum mechanics, apparently, when building their linear theory, proceeded from the foundations of Fourier theory. Therefore, the concepts of frequency and wave spectrum naturally entered into the theory of them. That is, they used an idea dating back to Pythagoras, that oscillations and waves can be characterized by a well-defined sequence of integers. If we keep this in mind, then even Planck's introduction of the concept of an energy quantum does not seem anything too unexpected. Indeed, if a string can only vibrate with certain frequencies and in certain shapes, then why can't something similar be extended for energy and space.

8.1 The Great Tragedy of Science and a New Understanding of the General Theory of Relativity (GTR)

Human thought has always sought to explain and unite everything as simply as possible. Apparently, the greatest success in this direction has been achieved by religious thinkers. Studying their theories, you understand how well they understood the desires and needs of people and how fully their constructions met people's aspirations.

When constructing a scientific theory, the situation is much more complicated, since in this case, the researcher does not strive to satisfy the aspirations of people, but to describe a reality outside of them that does not depend on them and which is in no way interested the most of people.

1. **Birthmarks of quantum mechanics and GTR.** It is very important that, unlike many religious thinkers, a scientific researcher is always limited in his research. In particular, the experimenter is limited by the level of possibility of creating experimental equipment. The theorist is almost always limited by the level of development of mathematics and the level of knowledge of contemporary science.

Literally, the rarest researchers have managed to overcome these limitations by combining in themselves both an experimenter and a theorist who creates mathematics corresponding to these experiments. The two most famous of these highest level researchers are Archimedes and Newton.

An ideal researcher who strives to cover the maximum number of phenomena with his theory must combine in himself, as it were, two principles—an experimenter and a theorist. Usually, the theorist is limited by the level of mathematical knowledge known to him. In particular, until today, there is no mathematical tool that allows one to easily study nonlinear, heterogeneous, multivalued, and discontinuous wave processes. The need to study such processes arose, in my opinion, at the beginning of the last century. A. Einstein and the creators of quantum mechanics encountered this problem almost simultaneously. And for one mystery of their problems, they made the same solution. Thus, when describing problems mathematically, they decided to ignore the effect of heterogeneity and possible discontinuity of processes and environments. Regarding the influence of nonlinearity, they took different paths. Einstein, following Richard Riemann, took into account the nonlinearity of changes in the curvature of space. In contrast, Schrödinger derived the linear wave equation, which is still widely used in calculations.

In particular, this is why we now have two theories, which, each in its own area, are successful in different ways. It seems to me that GTR is more successful, although it is immeasurably less used in applications. She is like a successful child with whom her parents have no problems of growing up. In contrast, the parents of quantum theory and its users, as is known, had and still have a lot of misunderstandings. Some of them are mentioned in the book.

However, no matter how successful the child is, the mistakes of the parents in his upbringing or heredity that came from the depths of time can unexpectedly affect the child. The same can be said about GTR. Now it is becoming more and more clear that the assumptions inherent in GTR about the homogeneity, smoothness, and continuity of all quantities included in it are beginning to contradict observations to one degree or another. That is, those facts that were completely ignored when deriving the Schrödinger equation and which led to difficulties in the creation and development of quantum mechanics began to assert themselves, in the light of new cosmological observations, and in general relativity. It is possible that up to this moment, both of these theories develop (move in their development) along the "slippery" slope of using assumptions about the homogeneity, smoothness, and continuity of the functions used in the theories.

2. **Unexpected cosmology and the Sloan Digital Sky Survey.** Modern cosmology is built on the assumption that the space-time of the Universe is smooth, homogeneous, and isotropic. The validity of these assumptions is not obvious. In particular, the Sloan Digital Sky Survey (SDSS) recently discovered an arc of galaxies about 330 million light-years wide and 3.3 billion light-years long [2–7]. The Hercules-Corona Borealis Great Wall (HCBGW) spans about 10 billion light-years. Its width is 7.2 billion light-years and its thickness is about 1 billion light-years.

Hercules-Corona Borealis Great Wall. (https://kulturologia.ru/blogs/190922/54269/ (left). The map which shows more than 6 million galaxies (right)

Today, it is discovered that galaxies exist in the form of clusters and filaments. According to the latest star maps in the Universe, you can see a structure covering an area much larger than the HCBGW. It has a round shape. Most likely, this is a supercluster of galaxies, not a wall. It crosses more than 20 constellations and covers 125 degrees of the sky (a third of the total). Certainly, his size categorically does not correspond to the so-called cosmological principle, which states that the matter of the Universe is, on average, distributed throughout it uniformly.

The existence of these structures means that humanity will have to reconsider the foundations of modern cosmology and the general theory of relativity.

The standard model of the Universe is based on Einstein's equations of general relativity which were derived using two big assumptions. First, that the Universe is homogeneous, looking roughly the same in all locations. Second, it is isotropic, looking roughly the same in all directions from any standpoint. These assumptions led to the big bang model and the cosmic inflation theory. Einstein's equations become very complex and the Big Bang and the inflation become slippery concepts if these assumptions are not valid. "The great tragedy of science," as Thomas Huxley observed, is "the slaying of a beautiful hypothesis by an ugly fact." He was talking about the origins of life, but it may be true for the starting point of the Big Bang model and the cosmic inflation theory.

Even 100 years ago, knowledge about the structure of the Universe was quite vague and foggy. Assumptions about the smoothness and homogeneity of processes seemed no worse than other assumptions and were accepted not in accordance with deep knowledge, but rather from considerations of constructing the simplest and most integrable model of a particular process or observation. Then sometimes the accepted assumptions and models became so familiar that they were considered obvious.

If modern cosmology changes significantly, it will not be the only example of a revolution in science. At the very beginning of their development, ideas about geocentric and heliocentric pictures of the world were equally promising. But then, for almost two millennia, the heliocentric system of the world was almost not considered. The entire experience of humanity, which knew only one Universe—our planet Earth, convinced him of the geocentricity of the world. By accepting this point of view, scientists created theories of amazing beauty and predictability, which did not become any more correct. It was only with the expansion of human experience beyond the Earth, after the invention of the telescope, that the theory of the heliocentric world began to develop, although the merits of this theory were obvious 2000 years before the birth of Galileo Galilei.

The same experience exists with theories about the structure of matter. For more than 2,000 years, the generally accepted explanations went back to Aristotle. Atomic theory completely triumphed only 100 years ago.

3. **Dialectics of Nature, "emptiness" and the emergence of the Universe.** The ancient Greeks believed that nature does not like emptiness (disturbances). If a void has arisen, then the surrounding bodies strive to fill this void, that is, to destroy the resulting disturbance. I don't think this old idea is so bad when you

consider the movement. Taking it into account allows us to qualitatively understand the possible trajectory of a body falling into the void. As a matter of fact, this idea underlies the GTR modification proposed in Chap. 5.

Everything is determined by movement. According to the quantum space-time model, each moving particle seems to be followed by a "trace" or "emptiness." Therefore, the more moving particles there are in a body, the more "empty" it is. Mass and "emptiness" are like different sides of the same coin. It is following from the basics of dialectics, which affirms the "unity and struggle of opposites." The more massive a body contains moving particles (that is, it contains more matter), the "emptier" it is and the more strongly it has attraction, that is, gravity. At the moment of the emergence of the Universe, according to the model of Galiev and Galiyev (MGG), all particles (bosons) were packed so tightly that they did not move and there were no "emptinesses." But such symmetry had to be broken sooner or later. A fairly significant "emptiness" could arise according to the Heisenberg principle. As a result, one boson could gain enough motion to transform into a fermion. Thus, a fermion appears surrounded by bosons, in particular photons that arose simultaneously with the "emptiness." Let us repeat what has been said, adding a little to it. At $t = 0$, the packing of bosons and one-dimensional strings was so dense that there were no space-time quanta (QST). However, situations arose when a boson went into a fermion and back due to quantum perturbations. In some cases, the emergence of a fermion was accompanied by the appearance of space-time quanta (QST) and the fermions existed for several QST. At the same time, several fermions could arise simultaneously at a very close distance, sufficient for their interaction to begin. They began to repel each other through photons. As a result, more and more "emptinesses" and fermions appeared. The process became more and more avalanche-like. At the same time, more and more QSTs emerged. Finally, these quanta created a certain structure that could be perceived by us as some kind of continuous space-time. That is, with some accumulation of QST, some, apparently spherical, structure could arise. It began to expand more and more quickly. At its initial moment, this structure can be defined as the embryo of the Universe. Of course, we do not claim in any way that the described process corresponds to any reality. We introduced it into consideration in order to introduce particle-wave tracks into the MGG and into the above understanding of the results of general relativity.

In my modification of the GTR, I am trying to return, to some extent, to the ancient Greeks. The essence of the new understanding of GTR is that all bodies emitting energy or consisting of moving particles create peculiar "voids" (areas of loosening or "holes") in quantum space-time. All bodies interact through these "voids" and roll to one degree or another into these voids. If the speed of the body is low, then they fall and connect with the "emptiness." That is, they increase looseness, that is, they reduce density, "voids." If the initial speed of the body is high enough, then it begins to rotate around the "emptiness." For example, this is the case of planets in the solar system or electrons in an atom. Finally, the speed of a body can be so high that in the vicinity of the "emptiness," the body only slightly changes the trajectory of its movement. This occurs, apparently, for photons moving near stars or galaxies.

We emphasize that everything said above rests on the concept of extreme multivalued waves developed in the book. They can exist not only in the macrocosm, but the author believes that they are excited and exist, first of all, in space and the microcosm.

Humanity has vast experience in using wave processes in the macrocosm. This experience is primarily associated with linear and unambiguous waves. However, this experience may not be enough in the future, since we do not know how widespread multivalued waves are in the Universe. We also do not know what their contribution to the physical processes in the Universe is. This book represents the first attempt to address these contributions.

8.2 Difficulties and Features of Studying Multivalued Waves

Maybe multivalued structures and waves are the most capacious ways of packaging information? Maybe it was them that Nature used at the moment of the birth of the Universe so that this information would then organize all the force and material fields of the world around us? Maybe this information served as a code for the further development of the Universe, similar to how the genes in the double helix largely determine the evolution of living beings?

We do not know all the possibilities of Nature; we cannot imagine all cases of resonant amplification or suppression of perturbations in it and in biological communities. The advantage of multivalued, resonant processes over others is that they can exist while consuming a minimum of energy. Probably, Nature did not lose sight of the indicated advantages of resonant processes. Perhaps Nature used resonances, both in creating new and diverse forms and in creating systems that minimize the energy required for their functioning. However, the possibilities of studying them are often beyond our capabilities. In our attempts at analysis, we resort to nonlinear equations that have a large number of solutions in singular and resonant regions.

Singularities and resonances are one of the central problems of modern science [8]. To some extent, they remain mysterious. How to understand areas in which small can become big and the past can become future? Of course, it is difficult to expect that in these areas and their surroundings the homogeneity and smoothness of the functions describing them are fulfilled. There may be zigzags, creases, and bends of space-time. On the other hand, even over long distances, the Universe is not homogeneous [2–7]. The energy of the very, very early Universe was many tens of orders of magnitude higher than that which humanity has now mastered. How did the initial energy evolve without the use of multivalued wave processes? Let us assume that the initial ball of energy is described by a simple scalar field in such a way that all forces and types of matter are captured. Then how to write this information without the help of multivalued waves? It's still amazing to me that Nature managed to write down hereditary information (double helix) without using multivalued waves!

It seemed to me from the very beginning of the research, that it is in singular, resonant situations that multivalued waves can arise and exist [9].

Let us clarify this idea once more. A huge number of processes and phenomena occurring around us can be approximately interpreted as linear. They do not change qualitatively over short periods of time and space. They are often repeated. We use them often. These processes are usually well described by linear equations, in particular the linear algebraic equations that we remember from school. The solution to every linear equation is unique, smooth, and continuous. Very often, it describes the process being studied well. However, the applicability of the solution is usually limited. Figuratively speaking, the solution can only describe one cat of a certain color. The solution does not describe the color change. The cat can change its size, but not very much.

For example, a linear equation cannot describe the transformation of a cat into a tiger or the evolution of a smooth, imperceptible wave into a relentless, terrifying tsunami.

If we want to describe reality more accurately, we must use nonlinear equations. Unlike linear equations, they can have many solutions. For example, the quadratic equation known from the school curriculum has two solutions. Returning to the imaginary cat, we can interpret these decisions as describing a black and also a white cat, or even a black and white cat. But we have no answer as to which color corresponds to the cat being studied. Thus, nonlinear equations can have many solutions, and the researcher is faced with the task of choosing the solution that corresponds to the real process. If the process is far from multivalued (extreme, catastrophic), then the solution can be close to linear, which allows us to settle on the so-called weakly nonlinear solutions. In general, the researcher has to use observational and experimental data or rely on intuition and a sense of beauty when choosing from a variety of solutions a solution that corresponds to the process being studied.

The above determines the main difficulty in studying nonlinear wave processes. Another difficulty is determined by the fact that our mathematical models always only approximately describe reality. There is always the possibility that we have missed some important effects in our analysis. You can use the well-known analogy with a butterfly. There is always the danger that we have not taken into account the decisive influence of its wing flapping when modeling.

We examined two difficulties in calculating and predicting multivalued waves, which are largely related to the research methodology.

Finally, let us remember that we live in conditions of incomplete knowledge. The conditions under which a particular process arises and develops are always only approximately known. For example, the paramcters of the movement of masses of the Earth and ocean, leading to catastrophic tsunamis, are almost always unknown. These reasons do not allow us to give accurate forecasts for many natural wave catastrophic phenomena. The forecast is usually associated with the solution of certain equations.

It is known that there are two main methods for solving nonlinear problems of mathematical physics: the numerical method and the perturbation method. The book widely uses a certain modification of the perturbation method, which is valid in resonance areas. There, the use of the traditional perturbation method leads to the appearance of small divisors (or resonances) in the solutions (see Sect. 2.4). To avoid this, we accepted that near exact resonances, the influence of nonlinearities increases, and the influence of the d'Alembert operator decreases. This greatly simplifies the research, up to reducing the research to solving an algebraic equation. In particular, this occurs in finite resonators, when an integer number of waves fits along the length of the resonator. Then the influence of linear terms in the corresponding wave equation can disappear near the boundaries.

Similar possibilities exist for all media and waves. However, the theory is illustrated by surface waves, compression waves, and waves described by the Klein-Gordon equation.

It is easy to see that the wave equations used in the book are similar. At the same time, they are representatives of a wide class of wave equations that are valid in various situations and environments. This, as well as the generality of the developed method for analyzing multivalued waves, made it possible to describe a wide range of observations and experiments.

Note In the broadest sense, the waves discussed in the book can be called transresonant. Indeed, waves excited far from resonances, when the d'Alembert operator is far from zero throughout the entire research area, were practically not considered. At the same time, the book devotes a lot of space to resonant waves, which are generated when the corresponding resonant condition is met. In particular, finite resonators are considered when the forced frequency is close to a certain resonant frequency of the resonator. In this case, frequency is a transresonant parameter. Transresonant multivalued waves arise in finite resonators if the forced frequency is not constant, but varies within the resonant band. In other cases, the transresonant parameter is the speed of propagation of disturbances or other parameters of the scalar field.

8.3 Multivalued Waves and Discreteness

Since the beginning of the twentieth century, the progress of science began to be increasingly associated not only with the development of the technical and experimental capabilities of mankind but also with the development of mathematical methods for studying nature. Humanity has begun to study closely the question of the structure and origin of the Universe, as well as the structure of matter. That is, science has entered into areas of enormous energies, which were impossible to study on the basis of experiment alone. Therefore, the creators of new fields of physics were primarily theorists with a good command of the mathematics of the early twentieth century. These were primarily linear differential equations. When solving them, Fourier methods were usually used. Therefore, quantum mechanics immediately began to master and use the concepts of eigenfunctions and numbers, frequency, and wave spectrum of those methods. The concepts were used to describe the discreteness of matter, space, and energy. At the same time, the linear mathematical apparatus did not correspond well to the description of processes with a huge energy density that characterize atoms, electron clouds, atomic nuclei, and even more so the processes characterizing the emergence of the Universe. Accordingly, the results of the theory did not correspond well to the experimental data, so a new understanding of them was introduced, new, nontraditional methods for processing calculation data, and the like. Finally, in many cases, unsuccessful calculation results and their limitations were not advertised, while successful results were emphasized.

In my opinion, the problem each time came down to the fact that linear (unambiguous) solutions had to be given nonlinear, multivalued interpretations, hence the uncertainty in particle trajectories, the introduction of the principle of quantum superposition, and the like.

The difficulties encountered may be related to two circumstances:

1. Physicists did not understand that they had begun to study nonlinear processes and effects.
2. While claiming that they were studying corpuscular, discontinuous effects, they used differential relations that are not applicable to the description of such phenomena. Therefore, further, the conditions and criteria are introduced that make solving differential equations acceptable in practice.

In particular, algebraic relations were successfully introduced, for example, in the collision of particles. The success of using such corrections of the general theory was, in some cases, passed off as the success of the linear theory.

Consider the Schrödinger equation, which allows us to calculate the "wave function" of an electron. The wave function in turn gives the "probability amplitude," which squared gives the probability that you will find an electron at a certain location. When you learn about the described, not simple, procedure and hear admiration for the success of the equation in describing reality, you think that all the problems in wave mechanics have been solved. But that's not true! Although the Schrödinger equation does an excellent job of modeling the hydrogen atom, it already has significant difficulty describing the helium atom, which has only two electrons! Physicists extol the beauty and elegance of the Schrödinger equation. But it corresponds to experimental data only with the help of complex patches and approximations.

To me, the difficulties of using the Schrödinger equation to describe more and more complex atoms were reminiscent of the difficulties encountered by Ptolemy's theory in describing more and more planets. His geocentric model of the solar system with its circles within circles, in its simplest form, perfectly described the movements of the Sun around the Earth. However, in general case, epicycles, deferents, eccentrics and equants introduced in the theory for more and more exact description of the planets' motions around the Earth very strongly complicated the calculations. As a result, the theory became scandalously complex. "Let no one, seeing the difficulties of our devices, find trouble some such hypothesis," Ptolemy wrote. But few people agreed with him! For example, Alfonso the Wise, the thirteenth-century king of Castile (1221–1284), was rather skeptical about the theory: "If the Lord Almighty had consulted me before embarking upon creation, I should have recommended something simpler."

In universities, students are taught to use mathematical knowledge so well that they can apply it mindlessly, like an automaton. Within the framework of quantum physics, this approach is defined as the order "Shut up and count." That is, don't worry about understanding the processes being calculated, just follow the standard rules for solving the Schrödinger equation, which in general can be very complex and time-consuming to solve.

In this book, I also start from the wave differential equation. Accordingly, it, like the Schrödinger equation, is not acceptable for describing quantum processes. However, a closer look reveals that it is used mainly in resonance zones. There it is reduced to algebraic relations reminiscent of those used to study collisions of hard balls. Thus, a version of the nonlinear Klein-Gordon equation is obtained that is valid for nonsmooth, discontinuous, discrete, and multivalued processes.

Our goal in the book is to achieve, albeit a very approximate, but understandable description of some physical processes based on the introduction of the concept of a resonant scalar field, that is, described by a nonlinear algebraic equation.

The well-known fact is that physicists have created the mathematical apparatus of quantum mechanics; however, the reasons for its success are not entirely clear even to the creators of the theory themselves. This opinion was perhaps expressed most clearly by Feynman (see Sects. 4.7 and 6.1). But de Broglie already understood this fact well. He proposed a hidden parameter model to describe quantum processes. It was supported by many prominent physicists. But it was not clear where these parameters were hidden and where to look. Perhaps they are hidden in the nonlinearities and solutions of the corresponding solutions. In the case of multivalued waves under consideration, many solutions appear that have their own parameters. These parameters determine reality, which is one, but has many possibilities that represent individual solutions. Moreover, such combinations of solutions in space and time are possible that the resulting picture describes a single continuous, but very difficult for human perception, reality.

Thus, perhaps there is a reality of multivalued solutions to nonlinear scalar field equations, which can be understood as a reality hidden behind hidden parameters. Classical quantum mechanics tries to describe (or, one might say, hide) this reality behind linear solutions wrapped in a probabilistic interpretation.

We showed in the book that scalar fields can be discrete and contain multidimensional, multivalued spheres, rings, and loops. All this is in motion. All these dynamics are described as multivalued waves (see Sects. 2.5, 2.6, 3.4, 3.5, and 3.6, Chap. 4, and Sect. 5.4).

Our study does not attribute these properties to space-time itself. The latter is usually used to describe the characteristics of traveling waves. However, various wave figures determine the variety of possible landscapes of scalar nonlinear fields. In particular, a discrete particle-wave (see, e.g., Figs. 5.15 and 5.16) can define a spherical local hill or hole in the spatial landscape of the scalar field.

There is a possibility that the path of development (evolution) of the Universe is not unambiguous. If it is nonlinear, then at every moment there are many possibilities, and literally the slightest, virtual, possibilities determine the further path in some moments of development.

8.4 Shortly About the Book

The book consists of two parts: in the first part, the theory of constructing multivalued waves is developed, and its accuracy is tested using examples of surface waves and waves in resonators. In the second part, the developed approach is applied to the description of some fundamental physical problems and observations. Emphasizing that perhaps the appearance of these problems is determined by the nonlinear and resonant essence of the world around us, which exists in a non-smooth and quantum heterogeneous space-time. These entities are as follows:

1. Condensations of fields interpreted by us as quantum particle-waves (QPWs) arise and exist only when certain resonance conditions are met. Accordingly, all quantum-mechanical processes in one form or another, explicitly or implicitly, are associated with resonances and transresonance processes. In particular, all QPWs correspond to certain natural frequencies, and this may determine the nature of their interaction with other QPWs. Apparently, agreement in the values of natural frequencies is an important condition for the formation of various structures from QPWs. Thus, the resonant emergence of particle-waves in resonant scalar fields, which is described in the book, opens the way for the emergence of complex structures. For example, these can be atoms and structures consisting of atoms. Therefore, the resonant conditions of particle emergence should underlie the understanding of the origin and functioning of the Universe. Moreover, the behavior of quantum waves, many bodies, and even living beings should be studied from the standpoint of resonator synchronization and resonance interaction. Perhaps only optimally synchronized resonant systems survived. As a result, structures arose that did not require energy or required a minimum of energy. Systems less capable of such synchronization disappeared in the process of evolution [10–12].
2. Until the end of the twentieth century, ambiguous waves were not of much interest to researchers. Even the great interest in extreme waves that arose at that time did not change the situation much. They were studied as single-valued waves of very large amplitude. As an exception to this approach, studies can be cited [13–17]. In particular, Sibgatullin [14, 15] connected the appearance of multivalued waves with the resonant amplification of waves. The situation slowly began to change in 2001 when a large article appeared [9]. In it, the authors

presented an overview of nonlinear transresonant waves, vortices, and patterns and also presented a long-term program for studying multivalued waves. As a matter of fact, this book is some kind of completion of the program presented more than 20 years ago. Therefore, the article [9], as the beginning of a systematic study of multivalued waves, has special significance for us. Therefore, I will provide an abstract from this article so that the reader can see what ideas motivated the author of this book over the next 20 plus years. "Perturbed wave equations are considered. Approximate general solutions of these equations are constructed, which describe wave phenomena in different physical and chemical systems. Analogies between surface waves, nonlinear and atom optics, field theories, and acoustics of the early Universe can be seen in the similarities between the general solutions that govern each system. With the help of the general solutions and boundary conditions and/or resonant conditions, we have derived the basic highly nonlinear ordinary differential equation or the basic algebraic equation for traveling waves. Then, approximate analytic resonant solutions are constructed, which describe the transresonant transformation of harmonic waves into traveling shock-, jet-, or mushroom-like waves. The mushroom-like waves can evolve into cloud-like and vortex-like structures. The motion and oscillations of these waves and structures can be very complex. Under parametric excitation, these waves can vary their velocity, stop, and change the direction of their motion. Different dynamic patterns are yielded by these resonant traveling waves in the x-t and x-y planes. They simulate many patterns observed in liquid layers, optical systems, superconductors, Bose-Einstein condensates, micro-, and electron resonators. The harmonic excitation may be compressed and transformed inside the resonant band into traveling or standing particle-like waves. The area of application of these solutions and results may possibly vary from the generation of nuclear particles, acoustical turbulence, and catastrophic seismic waves to the formation of galaxies and the Universe. In particular, the formation of galaxies and galaxy clusters may be connected with nonlinear and resonant phenomena in the early Universe." In the book, the scientific area defined in the last sentences of the above abstract is considered from the standpoint of extreme multivalued resonant waves. In addition to the above list, the book examines the origin and existence of gravity. According to the presented results, gravity is not a fundamental phenomenon. Gravity is a consequence of the existence of a fundamental background in which there are rarefaction zones created by rapidly moving quantum elementary particle-waves. The mutual attraction of two bodies is stronger, and the speed of the particles that comprise them is greater. By increasing the speed of these particles, the gravity that binds them can be strengthened.

3. According to the results of the book, only in resonant situations are created of QPWs and it becomes possible to talk about quanta, quantum mechanics, structures, and origin of matter. The author, putting forward these assumptions for discussion, of course, does not hope for their verification in the near future in experiments. Science has become a very expensive pleasure in recent years, in terms of experimental testing of hypotheses. There are more and more brilliant

theoretical studies, the verification of which lies far beyond the capabilities of the experiments. Therefore, to a certain extent, this book might have been considered as a collection of ideas, and results not confirmed by experiment. This explains the fact that my long-term coauthor (Timur Galiyev) and I, having obtained the main results describing multivalued, resonant waves already in 2001 [9], did not seek to advertise them. Experimental data were insufficient for this. Experimental data published in 2003 [18] and, especially, in 2022 [19] gave the author the courage to begin work on this book.

Conclusion In the preface, we emphasized the well-known statement that everything around consists of clots of scalar fields that oscillate, generating the sounds of the boundless Universe. This statement goes back to Einstein [20]. But almost 100 years ago it was more of a brilliant declaration than a substantiated assertion. Einstein assumed that "Our experience hitherto justifies us in believing that nature is the realization of the simplest conceivable mathematical ideas." In this book, we have tried to bring some semblance of a theory under this assertion. We have reduced the study of nonlinear waves propagating in space-time to the study of traveling waves propagating in resonant fields. We have shown, of course, very sketchily, the connection of everything with resonant, multivalued processes. We very much hope that this connection is not only the result of mathematical manipulation, but reflects to some extent the nonlinear, quantum, wave essence of Nature and the Universe. Thus, multivalued waves, particle-waves and resonator ensembles are, on the one hand, instruments, and on the other hand, sounds filling the Universe.

Multivalued Universe Nature is a constantly performing, endless-sounding symphony of various waves. Multivalued waves sound very strongly in it, although their meaning has not yet been sufficiently assessed by modern science.

References

1. Liu D, Lin P (2022) Two-layer liquids interface instabilities in Faraday waves of two-layer liquids with free surface. J Fluid Mech 941:A33. https://doi.org/10.1017/jfm.2022.259
2. Galiev SU (1981) Dynamics of hydroelastoplastic systems. Naukova Dumka, Kyev (1981) (in Russian); Galiev SU (1988) Nonlinear waves in bounded continua. Naukova Dumka, Kyiv (in Russian); Galiev SU (1997) The influence of cavitation upon anomalous behavior of a plate/liquid/underwater explosion system. Int J Impact Engng 19:345–359; Galiev SU, Flay RGJ (2014) Interaction of breaking waves with plates: the effect of hull cavitation. Ocean Eng 88: 27–33; Galiev SU (2000) Extreme waves and shock-excited processes in structures and space objects. CRC Press
3. Li Q et al (2018) Measurements of the gravitational constant using two independent methods. Nature 560:582–588. https://doi.org/10.1038/s41586-018-0431 5
4. Švančara P et al (2024) Rotating curved spacetime signatures from a giant quantum vortex. Nature 628:66–70
5. Burulo C, Liberati S, Visser V (2022) Analogue gravity. Living Rev Relativ 14:3

6. Glazebrook K, Nanayakkara T, Schreiber C et al (2024) A massive galaxy that formed its stars at z ≈ 11. Nature 628:277–281. https://doi.org/10.1038/s41586-024-07191-9
7. Rao R (2024) JWST is most in-demand telescope ever – leaving many astronomers in the cold. Nature 626:464–465. https://doi.org/10.1038/d41586-024-00315-1
8. Castelvecchi D (2024) Could JWST solve cosmology's big mystery? Physicists debate universe-expansion data. Nature 628:697. https://doi.org/10.1038/d41586-024-01115-3
9. Galiev SU, Galiyev TS (2001) Nonlinear trans-resonant waves, vortices and patterns: from microresonators to the early universe. Chaos: an interdisciplinary. Journal of Nonlinear Science 11(3):686–704
10. Galiev SU (2009) Modelling of Charles Darwin's earthquake reports as catastrophic wave phenomena. http://researchspace.auckland.ac.nz/handle/2292/4474
11. Galiev ShU Charles Darwin's geophysical reports as models of the theory of catastrophic waves. Moscow. 655 pp (in Russian)
12. Galiev SU (2015) Darwin, geodynamics and extreme waves. Springer
13. Galiev SU (2020) Evolution of extreme waves and resonances. CRC Press. 508 pp
14. Сибгатуллин НР (1972) О нелинейных поперечных колебаниях при резоансе ы упругом слое и слое идеальной проводящцкй жмдкостю ПММ, том 36. Стр. 79
15. Сибгатуллин НР (1984) Колебания и волны в сильных гравитационных и электромагнитных полях. Наука, Москва. (Sibgatullin N.R. Oscillations and Waves in Strong Gravitational and Electromagnetic Fields. Springer. 1991)
16. Филиппов АТ, Многоликий солитон М, Наука, 1986 (in Russian) (A.T. Filippov, The Versatile Soliton, Birkh user, Boston (2010))
17. Chester W, Moser A (1982) On an equation which describes the resonant oscillations of a gas in an open-ended tube. Proc R Soc Lond A 380:409–415
18. Niederhaus CE, Jacobs JW (2003) Experimental study of the Richtmyer-Meshkov instability of incompressible fluids. J Fluid Mech 485:243
19. Liu D, Lin P (2022) Two-layer liquids Interface instabilities in Faraday waves of two-layer liquids with free surface. J Fluid Mech 941:A33. https://doi.org/10.1017/jfm.2022.259
20. Bartusiak M (2000) Einstein's unfinished symphony: listening to the sounds of space-time. National Academies Press, Washington DC

The manufacturer's authorised representative in the EU is Springer Nature Customer Service Centre GmbH, Europaplatz 3, 69115 Heidelberg, Germany. If you have any concerns regarding our products, please contact ProductSafety@springernature.com

Printed and bound by CPI Group (UK) Ltd, Croydon, CR0 4YY
30/09/2026
02225600-0017